# Physicochemical Aspects of Polymer Surfaces
Volume 1

# Physicochemical Aspects of Polymer Surfaces
## Volume 1

Edited by

### K. L. Mittal
IBM Corporation
Hopewell Junction, New York

Plenum Press • New York and London

Library of Congress Cataloging in Publication Data

International Symposium on Physicochemical Aspects of Polymer Surfaces (1981: New York, N.Y.)
  Physicochemical aspects of polymer surfaces.

  "Proceedings of the International Symposium on Physicochemical Aspects of Polymer Surfaces held as part of the American Chemical Society meeting, August 23–28, 1981, in New York City, New York"— T.p. verso.
  Includes bibliographical references and indexes.
  1. Polymers and polymerization—Surfaces—Congresses. I. Mittal, K. L., 1945–      . II. American Chemical Society. III. Title.
  QD380.I589  1981            547.8'404545            82-18960
  ISBN 0-306-41189-X (v. 1)
  ISBN 0-306-41190-3 (v. 2)

Proceedings of the International Symposium on Physicochemical Aspects of Polymer Surfaces held as part of the American Chemical Society meeting, August 23–28, 1981, in New York City, New York

© 1983 Plenum Press, New York
A Division of Plenum Publishing Corporation
233 Spring Street, New York, N.Y. 10013

All rights reserved

No part of this book may be reproduced, stored in a retrieval system, or transmitted in any form or by any means, electronic, mechanical, photocopying, microfilming, recording, or otherwise, without written permission from the Publisher

Printed in the United States of America

PREFACE

This and its companion Volume 2 document the proceedings of the International Symposium on Physicochemical Aspects of Polymer Surfaces held under the auspices of the American Chemical Society in New York City, August 23-28, 1981. This event was sponsored by the Division of Colloid and Surface Chemistry and the Divisions of Organic Coatings and Plastics Chemistry, and Industrial and Engineering Chemistry were the cosponsors.

The study of polymer surfaces is important from both fundamental and applied points of view. The applications of polymers are legion and wheresoever polymers are used, their surface characteristics, inter alia, are of great concern and importance; and the areas where polymers find applications range from microelectronics to prosthetics. In the last decade or so, the availability of various sophisticated surface analytical techniques, particularly ESCA, has been a boon in enhancing our knowledge of polymer surfaces.

This Symposium was designed to bring together scientists and technologists interested in all aspects of polymer surfaces, to provide a forum for discussion of various ramifications of polymer surfaces, to discover the latest developments, to provide an opportunity for cross-pollination of ideas, and to highlight areas which are in a state of rapid development and those which need intensified efforts. If the comments from attendees is any barometer of the success of an event, then this Symposium was a grand success and the above objectives were amply fulfilled.

Although the topic of polymer surfaces had been a subject of discussion in various meetings, this Symposium was hailed as the most comprehensive event ever held on this topic; and it was a veritable international event. The program contained 84 papers from 21 countries and the authors represented many and varied backgrounds and interests. The program comprised both invited overviews and contributed original research papers. The invited speakers were selected so as to represent widely differing disciplines and interests and they hailed from academic, governmen-

v

tal and industrial research laboratories. Also the inter-, multi- and transdisciplinary nature of the topic was quite patent as the papers ranged from surface and interface analysis to tribology to bioadhesion to metallized plastics.

With regard to these proceedings volumes, it should be pointed out that, for a variety of reasons, 17 papers (out of 84) are not included. However, it should be noted that three papers are included which were not in the program, and two of these papers (by Dr. Vroman, and by Drs. Nyilas and Chiu) were specially commissioned by the Editor. So these volumes contain a net total of 70 papers by 142 authors from 20 countries. It should be emphasized that all papers were reviewed by qualified reviewers and concomitantly many papers were sent back to the authors for suitable revision, and some were not accepted at all.

In these volumes, the papers have been somewhat rearranged (from the order in the program) to fit them better and have been grouped in nine Parts: Spectroscopic Analysis; Contact Angle, Wettability and Surface Energetics; Reactions and Interactions at Polymer Surfaces; Tribology and Triboelectrification; Crazing, Fracture and Morphology; Adsorption and Adhesion; Modification of Polymer Surfaces; Biomedical Aspects and Bioadhesion; and Polymer-Metal Interfaces. Volume I contains Parts I-V and Parts VI-IX constitute Volume 2. The topics covered include: surface analysis of polymers using a variety of techniques, e.g., Auger, ESCA, surface enhanced Raman; wettability characteristics of polymers, surface free energy determination of polymers, surface thermodynamics of liquid polymers, and adhesion of ice to polymer surfaces; reactions and interactions at polymer surfaces; surface phenomena in latex autohesion and vulcanization of latex films; tribological and trielectrification aspects of polymers, electrical conduction mechanism in polymers; adsorption on polymers, role of acid-base and electrostatics in polymer adhesion, and welding of polymers; fracture and morphological aspects of polymers, variation of polymer morphology and structure through surface interactions; various ways (RF plasma, microwave, chemical, graft copolymerization and use of monolayers) to modify polymer surfaces; biomedical aspects of polymer surfaces and bioadhesion, role of surface energetics in biological adhesion, protein adsorption on polymers, and blood-polymer surface interaction; investigation of interfacial interactions between polymers and other materials (particularly metals), study of metal-polymer boundaries using ultrasonic interface waves, adhesive joining of metals to engineering plastics, and metallization of plastics.

It is clear from the above list of topics that there is a brisk activity taking place in the analysis, understanding and tailoring of polymer surfaces to suit a variety of needs and applications. This two-volume set (~1200 pages) should be a useful

# PREFACE

source of information to both the seasoned researcher as well as to the neophyte in this arena. It should be pointed out that the Editor had hoped to include discussion in these volumes, but, unfortunately, in spite of constant exhortation, the number of written questions received did not warrant undertaking such endeavor. However, it should be recorded that there were many brisk and enlightening (but not exothermic) discussions both formally (in the auditorium) and informally in the corridors and other suitable places.

Acknowledgments. First of all, I am thankful to the appropriate officials of the various Divisions for sponsoring or co-sponsoring this event, and to the appropriate management of IBM Corp. for permitting me to organize this Symposium and to edit these volumes. The generous support of the Petroleum Research Fund of the American Chemical Society in providing travel monies for certain invited overseas speakers is gratefully acknowledged. Also I must take this opportunity to express my sincere thanks to the unsung heroes (reviewers) for their time and valuable comments, as comments from the peers are important in maintaining the quality of scientific publications.

On a personal side, I would like to acknowledge the assistance (in many ways) and patience of my wife (Usha) and my kids (Anita, Rajesh, Nisha and Seema) for maintaining a low entropy state in the house so as to render it conducive to work. Special thanks are due to Mr. Jim Busis, Plenum Publishing Corp., for his continued interest in this project and for lending a helping hand whenever the need arose. Also I would like to express my appreciation to Barbara Mutino for her heroism at the typewriter and for meeting various typing deadlines. It would be a remiss if I do not acknowledge the cooperation, enthusiasm and patience of the contributors without which this set would not have seen the light of day.

K. L. Mittal

IBM Corporation
East Fishkill Facility
Hopewell Junction, New York 12533

CONTENTS OF VOLUME 1

PART I.  SPECTROSCOPIC ANALYSIS

The Modification, Degradation and Synthesis of
Polymer Surfaces Studied by Means of ESCA
    D. T. Clark........................................... 3

Spectroscopy of Polymer Surfaces using the Surface
Enhanced Raman Effect
    D. L. Allara, C. A. Murray and S. Bodoff............ 33

X-Ray Photoelectron Spectroscopic Study of the Electronic
Structure of Poly-p-Phenylene Sulfide
    J. Riga, J. P. Boutique, J. J. Pireaux and
    J. J. Verbist....................................... 45

XPS Analysis of Fluorocarbon Films Produced by Sputtering
of a PTFE Bulk Cathode
    J. J. Pireaux, J. P. Delrue, A. Hecq and
    J. P. Dauchot....................................... 53

Localized Auger States in Polyethylene
    J. A. Kelber, R. R. Rye, D. R. Jennison and
    J. C. Houston....................................... 83

The Influence of a Substrate on the Surface
Characteristics of Silane Layers
    H. Ishida, S. Naviroj and J. L. Koenig.............. 91

PART II.  CONTACT ANGLE, WETTABILITY AND
SURFACE ENERGETICS

Wettability of Polymer Surfaces
    M. A. Fortes........................................ 107

Novel Methods of Studying Polymer Surfaces by
Employing Contact Angle Goniometry
    F. J. Holly...................................... 141

The Solidification Front Technique: Its Scope and
Use to Determine Interfacial Tensions
    R. P. Smith, S. N. Omenyi and A. W. Neumann......... 155

Surface Thermodynamics of Liquid Polymers: Theory
    I. C. Sanchez and C. I. Poser....................... 173

Characterization of High Surface Area Polyester
Filaments by Means of Wetting Force Measurements
    B. Miller, H.-D. Weigmann and D. Simonetti.......... 183

Surface Free Energy of Plasma-Deposited Thin
Polymer Films
    A. M. Wrobel....................................... 197

Properties of n-Alkane Films in the System: Teflon/
n-Alkane-Water
    E. Chibowski, B. Janczuk and W. Wojcik.............. 217

Interfacial Free Energies of Cells and Polymers in
Aqueous Media
    D. F. Gerson....................................... 229

Adhesion of Ice to Polymers and Other Surfaces
    K. Itagaki......................................... 241

## PART III. REACTIONS AND INTERACTIONS AT POLYMER SURFACES

Surfaces (Interfaces) and Polymer Stability
    H. H. G. Jellinek.................................. 255

Surface Oxidation Reactions of Unsaturated Polymers
    B. Rånby, J. F. Rabek and J. Lucki................. 283

Photooxidative Degradation of Clear Ultraviolet
Absorbing Acrylic Copolymer Surfaces
    A. Gupta, R. H. Liang, O. Vogl, W. Pradellok,
    A. L. Huston and G. W. Scott....................... 293

# CONTENTS

Interactions Between Several Radiation Sources and Certain Polymer Surfaces; Reflectance - Transmittance Characteristics
    J. R. Hallman, C. M. Sliepcevich and
    J. R. Welker............................................. 305

Characterization of Dense Sulfonated Polysulfone Membranes
    M. A. Dinno, Y. Kang, D. R. Lloyd, J. E. McGrath and J. P. Wightman................................. 347

Surface Phenomena in Latex Films Vulcanization
    O. Shepelev and M. Shepelev........................... 367

Factors Influencing Latex Autohesion
    M. Gur-Aryeh and M. Shepelev......................... 377

## PART IV. TRIBOLOGY AND TRIBOELECTRIFICATION

Tribology of Polymers: State of an Art
    B. J. Briscoe............................................. 387

Tribological Properties of Rubberlike Materials
    D. F. Moore.............................................. 413

A Survey of the Adhesion, Friction and Lubrication of Polyethylene Terephthalate Monofilaments
    M. J. Adams, B. J. Briscoe and S. L. Kremnitzer..... 425

Surface Energetics and Tribological Properties of Miniature Polymer Elements
    Z. Rymuza................................................ 451

Charge States in Polymers: Application to Triboelectricity
    C. B. Duke............................................... 463

Triboelectrification of Polymers - A Chemist's Viewpoint
    D. A. Seanor............................................. 477

Electrical Contact Performances and Electrical Conduction Mechanisms of an Elastomeric Conductive Polymer
    T. Tamai................................................. 507

## PART V. CRAZING, FRACTURE AND MORPHOLOGY

Surface Free Energies and Fracture Surface Energies of
Glassy Polymers
    L. H. Lee............................................... 523

The Work of Adhesion and the Fracture Surface Energy
of Epoxy-Polycarbonate Adhesive Joints
    B. W. Cherry.......................................... 545

The Variation of Polymer Morphology and Structure
Through Surface Interactions
    J. B. Lando........................................... 559

About the Contributors..................................... 569

Subject Index.............................................. xvii

CONTENTS OF VOLUME 2

PART VI.  ADSORPTION AND ADHESION

Acid-Base Interactions in Polymer Adhesion
    F. M. Fowkes........................................ 583

Role of the Molecular and the Electrostatic Forces in the Adhesion of Polymers
    B. V. Derjaguin and Yu.P. Toporov................... 605

Adsorption and Contact Angle Studies. IV. Alcohols and Water on Polypropylene and Polycarbonate
    B. C. Nayar and A. W. Adamson...................... 613

Adsorption on Modified Silicone Surfaces
    F. J. Holly and M. J. Owen......................... 625

The Physicochemical Surface Properties of Fiber-Forming Polymers
    H.-J. Jacobasch.................................... 637

Surface Characterization of Polymers by Inverse Gas Chromatography. Poly(ethylene terephthalate) Film
    J. Anhang and D. G. Gray........................... 659

Elastic and Viscoelastic Adhesion
    G. A. D. Briggs.................................... 669

Autoadhesion of Etched Polyethylene Films
    M. Salkauskas and M.-G. Klimantavičiūtė............. 689

Development Trends with Regards to Welding Techniques for Plastics
    H. Potente......................................... 699

Welding Technology and Durability of Plastic Film
Welded Joints
    M. G. Dodin............................................. 717


PART VII. MODIFICATION OF POLYMER SURFACES

Acid-Base Considerations of Surface Interactions in
Polymer Systems; Control by Microwave Plasma Treatment
    H. P. Schreiber, C. Richard and M. R. Wertheimer.... 739

An ESCA Investigation of the Plasma Oxidation of
Poly (p-Xylylene) and its Chlorinated Derivatives
    A. Dilks and A. VanLaeken........................... 749

Mild Direct Fluorination of Polymers Studied by X-Ray
Photoelectron Spectroscopy
    M. Millard, J. Burns and H. Sachdev................. 773

Chemical Characterization of Surface-Activated Polymer
Films Using the ESCA Technique
    T. Ohmichi, H. Tamaki, H. Kawasaki and
    S. Tatsuta........................................... 793

Introduction of Reactive Groups onto Polymer Surfaces
    R. K. Samal, H. Iwata and Y. Ikada.................. 801

Oriented Monolayer Assemblies to Modify Fouling
Propensities of Membranes
    L. M. Speaker and K. R. Bynum....................... 817

Flexural Fatigue and Abrasion Resistance Effects of
Sunlight-Induced Modification of the Surface of Nylon 6
Fibers
    M. S. Ellison, Y. Fujiwara and S. H. Zeronian....... 843

Chemical Modification of Kevlar® Surfaces for Improved
Adhesion to Epoxy Resin Matrices: I. Surface
Characterization
    T. S. Keller, A. S. Hoffman, B. D. Ratner and
    B. J. McElroy........................................ 861

Graft Copolymerization of Vinyl Monomers onto Wool by Use
of (TBHP-FAS) System as Redox Initiator
    B. N. Misra and D. S. Sood.......................... 881

## PART VIII. BIOMEDICAL ASPECTS AND BIOADHESION

Surface Energetics and Biological Adhesion
    R. E. Baier and A. E. Meyer........................ 895

Polymer-Water Interface Dynamics
    J. D. Andrade, D. E. Gregonis and L. M. Smith....... 911

Polymer Surface Modification to Attain Blood Compatibility of Hydrophobic Polymer
    M. Suzuki, Y. Tamada, H. Iwata and Y. Ikada......... 923

Polymer Surface Interactions in the Biological Environment
    P. Y. Wang and M. J. Bazos.......................... 943

XPS Analysis of Segmented Polyether Polyurethane-Ureas: Assessment of Surface Activity Toward Blood Platelets
    N. A. Mahmud, S. Wan, V. Sa da Costa, V. Vitale, D. Brier-Russell, L. Kuchner, E. W. Salzman and E. W. Merrill........................................ 953

ESCA Studies of Extracted Polyurethanes and Polyurethane Extracts: Biomedical Implications
    B. D. Ratner........................................ 969

Interactions of Blood with Multiphase Polymers; Effect of Sulfonate-Ester-Containing Domains on Platelet Reactivity
    S. J. Whicher and J. L. Brash...................... 985

Specific Protein Interactions as a Possible Explanation for Unexpected Behavior of Blood at Interfaces
    L. Vroman.......................................... 1003

Fundamentals of Native Plasma Protein Adsorption on Polymer Surfaces
    E. Nyilas and T-H. Chiu............................ 1011

## PART IX. POLYMER-METAL INTERFACES

Interfacial Interactions Between Polymers and Other Materials and Their Effects on Bond Durability
    W. J. van Ooij..................................... 1035

Chemical Interactions at Polymer-Metal Interfaces: Studies with X-Ray Photoemission
    J. M. Burkstrand................................... 1093

Analysis of Metal-Polymer Boundaries using Ultrasonic
Interface Waves
  R. O. Claus and R. T. Rogers......................... 1101

Effect of Filler Size and Plasticizer on the Adhesion of
Electroless Copper to PVC
  C.-C. Wan............................................ 1115

Adhesion of Metal Deposits on Plastics Based on Acrylic
Polymers
  M. Kadreva........................................... 1125

Adhesive Joining Aluminum to Engineering Plastics: I.
Polyester Fiberglass Composite
  J. D. Minford........................................ 1139

Adhesive Joining Aluminum to Engineering Plastics: II.
Engineering Grade Styrene and Cross-Linked Styrene
  J. D. Minford........................................ 1161

ESCA Studies of Interfacial Degradation Between Ethylene-
Acrylic Acid Copolymers and Lead/Tin Alloys
  F. Yamamoto, S. Yamakawa and M. Wagatsuma........... 1181

Analysis of Vinyl Copolymer Surfaces by XPS and Surface
Reactions
  J. F. M. Pennings.................................... 1199

Electroless Metallization on Sensitized Polymer Powders
Fused onto a Polymer Surface
  L. G. Svendsen and G. Sørensen....................... 1213

About the Contributors....................................... 1225

Subject Index................................................ 1237

# Part I
# Spectroscopic Analysis

THE MODIFICATION, DEGRADATION AND SYNTHESIS OF POLYMER SURFACES

STUDIED BY MEANS OF ESCA

D.T. Clark

Department of Chemistry
University of Durham
South Road, Durham DH1 3LE, U.K.

The application of ESCA to the study of structure, bonding and reactivity is exemplified by recent areas of research in the Durham Laboratories encompassing:

(i)   the direct determination of electron mean free paths at high kinetic energy;

(ii)  the synthesis of plasma polymer films;

(iii) the photo-degradation of polycarbonate.

## INTRODUCTION

Organic based polymers represent extremes in terms of both complexity of structure and sensitivity to interrogation and the single most powerful tool which has emerged for surface investigations has proved to be ESCA.[1] Detailed studies over the past decade have revealed the range of information levels available from a single experiment and have delineated most of the areas of likely applicability of the technique. The progress which has been made in this period has been spectacular and the varied nature of the applications of ESCA reported in this proceedings volume attest to the versatility of the technique.

The information levels available from a single ESCA experiment endow the technique with wide ranging applicability in the study of the surface structure of polymers, of surface modification and reactivity in general and of interfacial synthesis. In the present article we consider two representative applications of the technique to research programmes of considerable current technological importance namely the plasma synthesis of ultrathin polymer films and the environmental modification of polymers studied from the surface aspect. Before considering these topics, however, we briefly consider some recent experiments pertinent to the quantification of ESCA data namely the measurement of electron mean free paths as a function of kinetic energy.[2]

## ELECTRON MEAN FREE PATHS AS A FUNCTION OF KINETIC ENERGY

We have previously described in detail the direct measurement by the substrate-overlayer technique of electron mean free paths in linear and cross linked polymers as well as in ordered Langmuir films of simple and polymeric fatty acids.[3] These studies have been confined to the kinetic energy range available to the commonly used $Mg_{K\alpha_{1,2}}$ and $Al_{K\alpha_{1,2}}$ the corresponding photon energies being 1253.7 eV and 1486.6 eV respectively. We have recently pointed out the virtues of complementary studies using harder X-ray sources such as $Ti_{K\alpha_{1,2}}$ ($h\nu$ 4510 eV) which with the concomitant increase in sampling depth provides a convenient essentially non-destructive means of depth profiling polymers in fibre film or powder form.[1] The quantification of ESCA data requires, however, a knowledge of mean free paths in this higher kinetic energy range and we report here a preliminary analysis of data for polyparaxylylene "parylene"† films studied under closely similar conditions to those previously described in an investigation employing the softer X-ray sources.

---

† Parylene is a trade mark of Union Carbide Corporation.

Figure 1 shows wide scan ESCA spectra for a gold substrate on which has been deposited a monolayer or so of parylene N. The $Mg_{K\alpha_{1,2}}$ excited spectra covering the kinetic energy range 0-1253.7 eV show the large cross section for photoemission from the $Au_{4f}$ levels and the distinctive signal arising from the polymer overlayer. The corresponding $Ti_{K\alpha}$ excited spectrum is also shown in the Figure and reveals the considerable change in cross section as a function of photon energy. This is most clearly evident for the 4d and 4f levels (cf. $Mg_{K\alpha_{1,2}}$ spectra) which correspond to photoemitted electrons of comparable kinetic energy.* The $C_{1s}$ levels are not readily apparent from the wide scan $Ti_{K\alpha}$ spectra, however, there are both direct photoionisation and Auger lines spanning the region from 1500-4500 eV which may be monitored for

---

*The $Au_{4d}$, $Au_{4f}$ and $C_{1s}$ levels at low KE indicated in the $Ti_{K\alpha}$ spectra arise from "spill over" from the $Mg_{K\alpha_{1,2}}$ source in the dual anode configurations. The spectra were recorded in the Fixed Retardation Ratio mode and low kinetic energy electrons are therefore discriminated against.

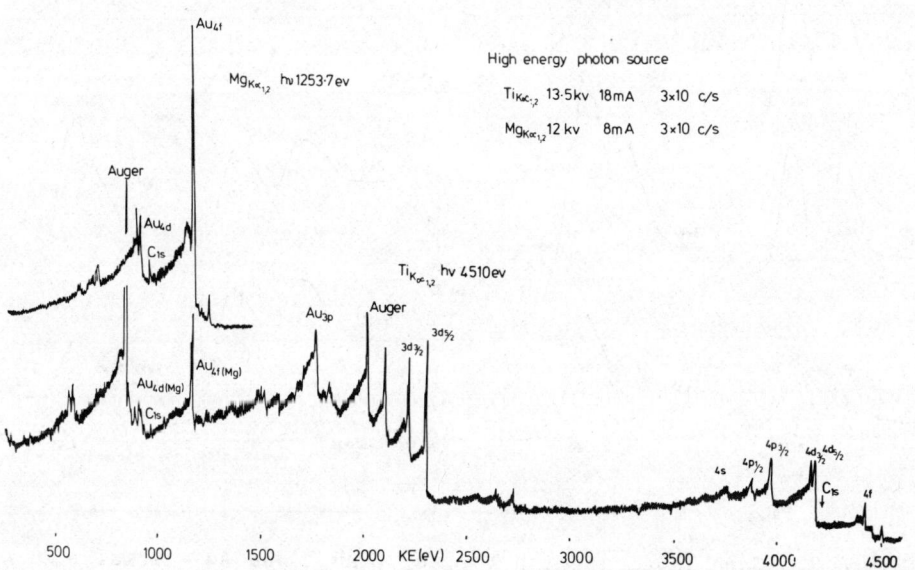

Figure 1. Wide survey scan ESCA spectra for a gold substrate with thin polyparaxylylene overlayer employing $Mg_{K\alpha_{1,2}}$ and $Ti_{K\alpha_{1,2}}$ photon sources.

substrate overlayer attenuation measurements. The fact that the absolute cross sections are lower for photoemission from the 4d and 4f levels compared with the $Mg_{K\alpha_{1,2}}$ photon source coupled with the inherent width of the $Ti_{K\alpha_{1,2}}$ photon source makes such measurements by no means straightforward. Figure 2 shows representative data for the $Au_{3d}$, $Au_{4d}$ and $Au_{4f}$ levels ($Ti_{K\alpha}$) and for comparison the $Au_{4f}$ levels excited by means of $Mg_{K\alpha_{1,2}}$.

Considering firstly the $Au_{4f}$ levels; for the $Mg_{K\alpha_{1,2}}$ photon source comparison of the initial substrate core level intensity with that corresponding to a deposited overlayer of $\sim 90$Å of parylene N shows an attenuation for photoemitted electrons of KE $\sim$ 1170 eV of a factor of $\sim 50$. For the 4f levels photoemitted by $Ti_{K\alpha}$ corresponding to a KE of $\sim 4430$ eV the corresponding attenuation is only by a factor of 5 illustrating the strong dependence of the escape depth on kinetic energy. Analysis of the data in Figure 2 provides the estimated mean free paths given in Table 1.

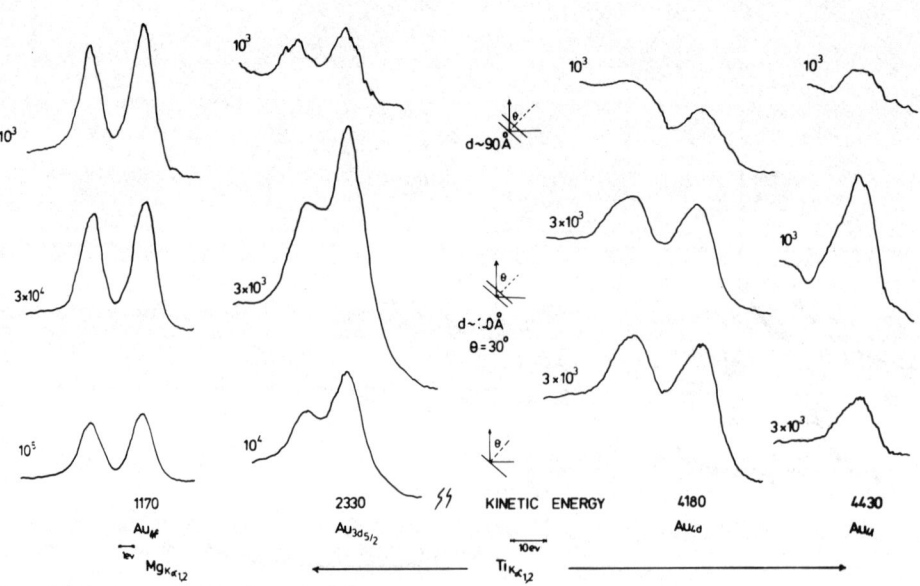

Figure 2. Core level spectra for $Au_{4f}$, $Au_{3d}$ and $Au_{4d}$ levels excited using $Mg_{K\alpha_{1,2}}$ and $Ti_{K\alpha_{1,2}}$ photon sources showing the distinctive change in signal attenuation.

Table I. Preliminary estimates of mean free paths from $Mg_{K\alpha_{1,2}}$ and $Ti_{K\alpha_{1,2}}$ studies of polyparaxylene overlayers on gold.

| Level | KE | $\lambda(\text{Å})$ | Photon source |
|---|---|---|---|
| $C_{1s}$ | 970 | 15 | |
| $Au_{4f}$ | 1170 | 20 | $Mg_{K\alpha_{1,2}}$ |
| $Au_{3d}$ | 2310 | 32 | |
| $Au_{4d}$ | 4180 | 40 | $Ti_{K\alpha_{1,2}}$ † |
| $Au_{4f}$ | 4410 | 46 | |

† Preliminary data

The mean free path as a function of KE for parylene evaluated from this and other data show that the typical sampling depth (taken as $\sim 3 \times \lambda$) more than doubles from $\sim 50$Å at $\sim 1000$ eV to $\sim 120$Å at $\sim 4000$ eV.

This illustrates the great importance of complementary studies with the differing photon sources as a means of depth profiling particularly for polymer surfaces of samples in powder or fibre form for which routine angular dependent studies are not available.

## PLASMA SYNTHESIS OF ULTRA THIN POLYMER FILMS

### Introduction

Although there are excellent well established routes from functional monomers to polymers[4] the range of materials which can be synthesized is greatly expanded by the addition of plasma techniques to the synthetic chemists armoury. The particular virtues of plasma techniques[5-8] with the capability under appropriate conditions of polymerizing any monomer and of producing pore free, uniform films of unusual physical, chemical, electrical and mechanical properties at first sight make it difficult to understand why the field has been so slow to develop.

It is now more than a century ago since Crookes[9] prophesied that the investigation of the 4th state of matter (subsequently denoted as the plasma[10] state) would be one of the most exciting growth points in chemistry and physics. Progress in the intervening period has been slow, however considerable advances have been made in the past few years and this can be traced directly to the possibility of direct interrogation of polymer structure by means of ESCA. The characterization of plasma polymer films which are generally of a highly crosslinked nature and are therefore insoluble presents particular difficulties. Since the interest is also in the deposition of ultra thin films < 1μ the development of the field has had to await the arrival of a technique capable of allowing direct investigation of such films in situ on the substrates on which the films are deposited. We outline here representative examples of how ESCA has allowed considerable progress to be made in this important field.

Background Information

Although plasmas may be excited under a variety of conditions; in studying the main features relating "monomer" structure to that of the plasma polymer the most convenient instrumentation involves low power inductively coupled plasmas excited in flow systems.[7] In such an experiment an important parameter which must be considered is the total power dissipated in the plasma per unit weight of material. Thus if W is the total power input to the plasma, F is the flow rate, and M is the molecular weight of the "monomer" in which the plasma is excited then W/FM is a convenient definable parameter[6] in the investigation of synthetic routes. The dynamic processes obtained in a plasma experiment are indicated schematically in Figure 3 where we propose the use of the symbol ⌇⌇⌇→ to denote a plasma reaction. Excitation of the plasma creates a pool of reactive species including ions, radicals and excited states and these can undergo gas phase reactions to produce gas phase products. Indeed under appropriate conditions gas phase homogeneous reactions can be the dominant process and there is considerable potential[11] for effecting cost effective cyclizations, isomerizations and eliminations in one step using electrical power

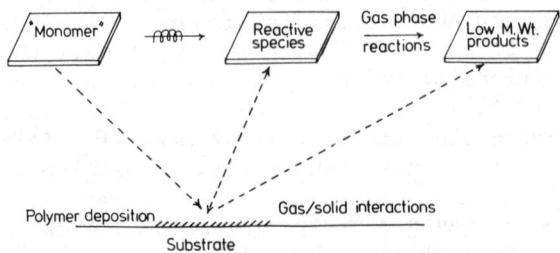

Figure 3. Schematic of reaction sequences in plasma polymerizations.

as opposed to multi-step reactions requiring skilled synthetic chemists. A competitive process of stepwise build up of polymer from reaction of "monomer" and reactive species in the plasma and of ablation of the polymer also by reactive species in the plasma implies that under certain conditions ablation will dominate so that no polymer is deposited whilst under other conditions polymerization may be extremely rapid.

A schematic of the typical behaviour of the deposition rate for polymer versus the W/FM parameter is shown in Figure 4. At low power inputs and high flow rates it may be difficult to sustain a plasma and the deposition rate may therefore be negligible. At high power inputs and low flow rates the power dissipated per unit mass of "monomer" may be sufficiently high to lead to ablation of the surface and the deposition rate therefore again drops to zero. There are therefore optimum conditions for polymer deposition. Under certain conditions nucleation can occur in the gas phase to give rise to powdery product whilst under other conditions lower molecular weight oily materials can be obtained. In the particular examples described here the deposition corresponds to the production of clear films in the region close to the maximum in the rate of deposition for a given monomer.

We may draw a distinction in basic mechanism of formation of polymers and this is indicated schematically in Figure 5. For functionalised monomers e.g. an alkene it is possible under certain conditions to produce linear polymers. These arise from conventional cationic or radical addition mechanism the plasma then acting as a source of reactive intermediates to initiate the plasma. This is termed plasma induced polymerizations (PIP).[12] It is invariably the case however that in the region of maximum glow where the density of reactive species is highest polymerization occurs in a stepwise manner by a competitive

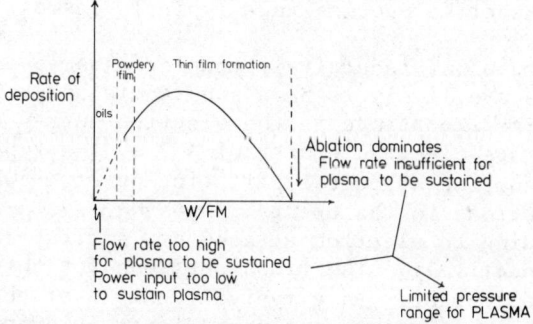

Figure 4. Schematic of dependence of rate of deposition on W/FM parameter.

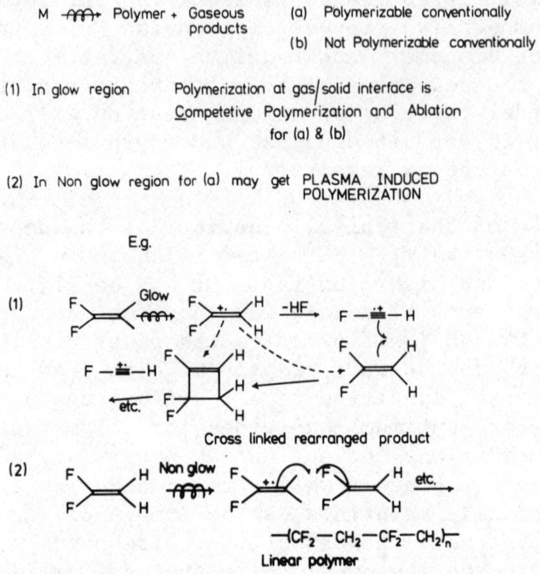

Figure 5. Competitive ablation and polymerization (CAP) and plasma induced polymerization (PIP) mechanisms.

process of polymerization and ablations (CAP process) even for functionalised monomers. By contrast for such systems PIP may well be of importance in other regions of the plasma.

In the work described below the emphasis has been on the synthesis of fluoropolymer films. In addition to having particularly interesting properties such films are amongst the most straightforward to characterize by ESCA consequent upon the distinctive nature of the core level spectra.

## Reactive Species in a Typical Plasma

For the low power inductively coupled plasmas of interest in this work there are likely to be a substantial number of reactive species. Of prime importance is the electron energy distribution and for an average electron energy of $\sim$ 2 eV the typical distribution function suggest a partitioning of $\sim$ 60%, $\sim$ 40% and $\sim$ 0.5% for electrons in the energy ranges 0-3 eV, 3-8 eV and > 8 eV corresponding to electron attachment, excitation and ionization respectively. The cross sections for these processes can be widely different. As a typical example of the species which can be present in such a plasma Figure 6 indicates the excited states which are available by electron impact in the typical energy range available.[13] This includes the lowest singlet

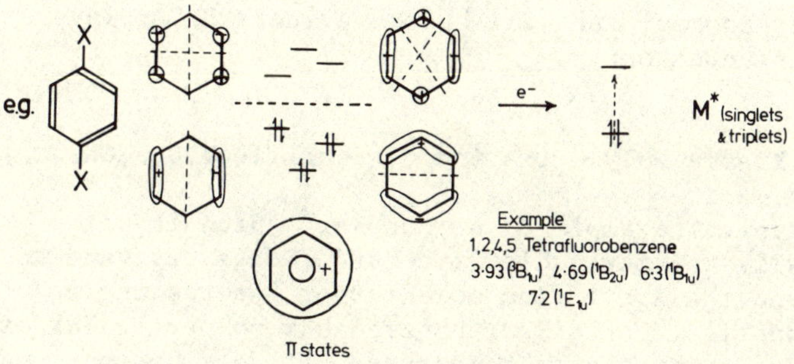

Figure 6. Typical excited state for benzenes available by electron impact.

and triplet states and a considerable body of data is available on the chemical transformations of such states.[14] The nature of the states in the ion manifold is indicated in Figure 7. For the benzenes, evidence has been presented[15] for the π radical anion nature of the 3 lowest energy states available from electron capture whilst recent studies of the emission spectra of radical cations makes it extremely likely that the three lowest π radical cation states are also populated. With such a range of excited species one might expect a priori that plasma polymerization would be a complex process. This does indeed appear to be the case but the important point to emerge from the investigations carried out to date is that polymerization does proceed by a well defined reproducible route and our goal is to understand the process as a

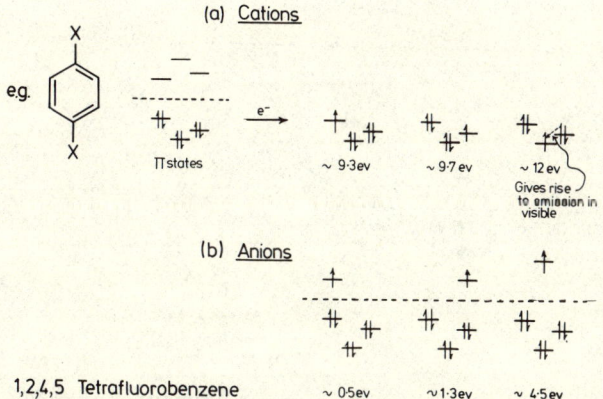

Figure 7. Typical cation and anion states for benzene derivatives in a plasma.

function of monomer. As will become evident ESCA plays a dominant role in this endeavour.

## Plasma Polymerization of Pentafluorobenzene

As a typical example of a "monomer" which is not conventionally polymerizable we consider data relevant to pentafluorobenzene.[16] Apart from some rather interesting reactions with nucleophiles, pentafluorobenzene has no particular excitement as far as its chemistry is concerned. It does however undergo plasma polymerization rather easily and the $C_{1s}$ and $F_{1s}$ core levels for a thin plasma polymer film are displayed in Figure 8.

It is evident from this that fluorine is retained in the polymer. Indeed the C:F stoichiometry determined independently from the integrated $C_{1s}$ and $F_{1s}$ intensity ratios and from the components of the $C_{1s}$ levels corresponds closely with that of the starting monomer. The $C_{1s}$ profile shows extensive fine structure extending to high binding energy corresponding to $\underline{C}F_3$, $\underline{C}F_2$, $\underline{C}F$ and $\underline{C}$ structural features. The component at lowest kinetic energy (highest binding energy) derives from π→π* shake up satellite of the $\underline{C}F$ component which dominates the spectrum at low binding energy. This reveals the conjugative unsaturation retained in the polymer. This illustrates the fact that polymerization is accompanied by substantial molecular rearrangement.

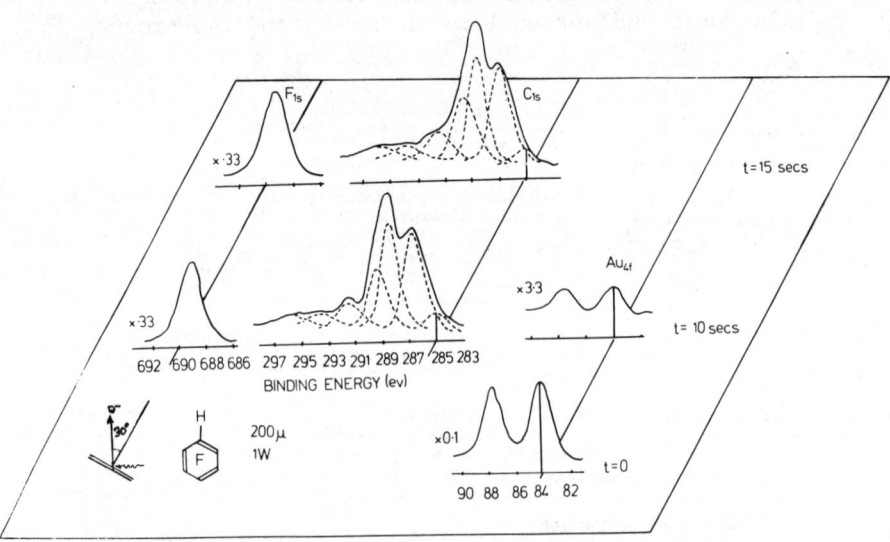

Figure 8. $F_{1s}$ and $C_{1s}$ core level spectra for pentafluorobenzene plasma polymer deposited on gold.

The gross structure of the polymer as delineated by the components of the $C_{1s}$ levels is essentially independent of the operating parameters in the pressure and power range of 100-200μ and 0-30 watts respectively as may be inferred from the data in Figure 9.

The capability of being able to look at extremely thin films (consequent upon the very short mean free paths for photoemitted electrons of interest in ESCA) endows the technique with great potential for the direct investigation of the initial rate of deposition of plasma polymer films. As an example Figure 10 pertains to rates of deposition of the plasma polymer from pentafluorobenzene as a function of power and pressure. One of the great advantages of plasma polymer films is evident from this in the very close degree of control which may be exercised over

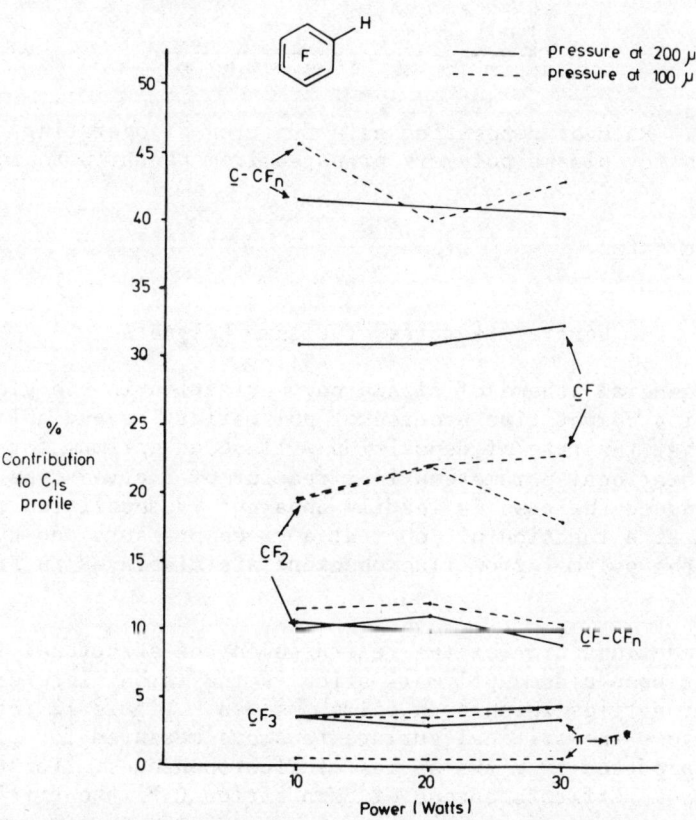

Figure 9.  Component analysis for plasma polymers from pentafluorobenzene as a function of power and pressure.

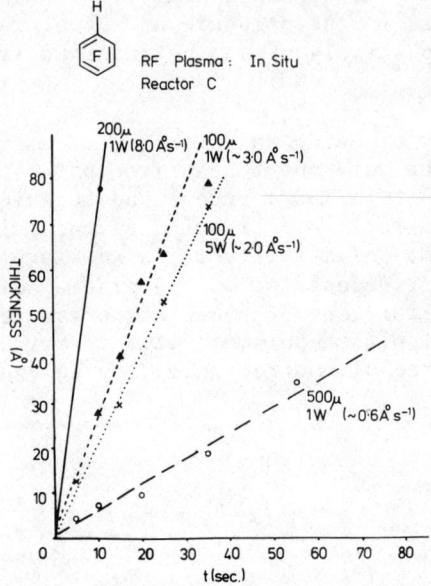

Figure 10. Rate of deposition as a function of operating parameters for plasma polymers prepared from pentafluorobenzene.

deposition rates.

## General Features of Plasma Polymers

The general scheme of plasma polymerization in the glow regions of a competitive process of polymerization and ablation implies that the rate of deposition will be an optimum for a given set of operational parameters in a reactor of a given type. That this is indeed the case is readily apparent if deposition rate is monitored as a function of power at a given pressure and typical data for the polymer from fluorobenzene are displayed in Figure 11.[17]

One manifestation of the rearrangement of structural features consequent upon plasma polymerization is the rather interesting surface properties which such films possess. Figure 12 for example shows the critical surface tensions measured for plasma polymers produced from the series of fluorobenzenes. For the polymer from perfluorobenzene of composition $C_1F_1$ the critical surface tension of 20 dynes may be compared with that for $PVF_2$ of the same C:F stoichiometry of 27 dynes. The critical surface tension increases as the fluorine content decreases.

# MODIFICATION, DEGRADATION AND SYNTHESIS

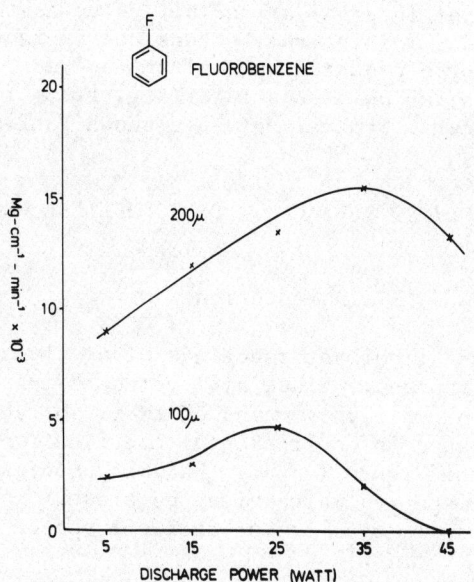

Figure 11. Rate of deposition as a function of power at different pressures for fluorobenzene monomer.

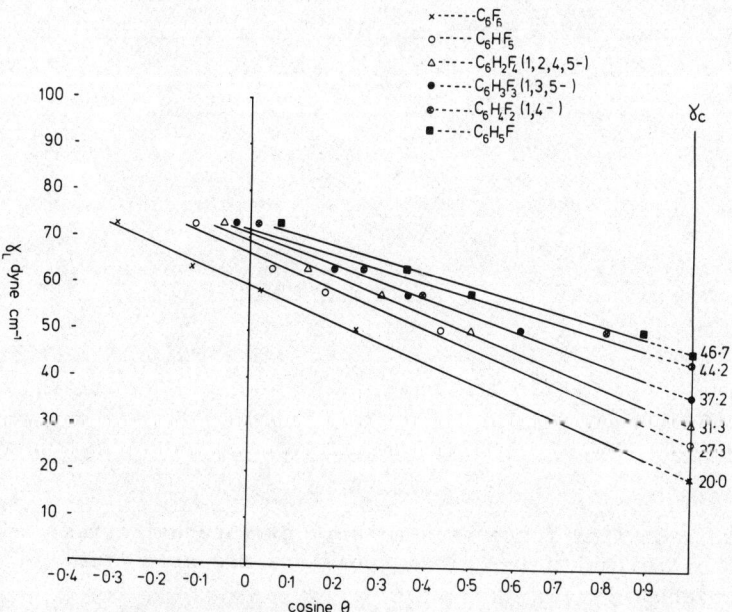

Figure 12. Critical surface tension for plasma polymers prepared from the isomeric fluorobenzenes.

With a cross linked structure, it is often characteristic of plasma polymers, that internal stress builds up as the film thickness increases. The internal stress can in fact be measured from the curling force produced when plasma polymer films of known thickness are deposited on a substrate (e.g. LDPE) for which the bulk modulus is known. Typical data are shown in Figure 13.

## REACTIONS AT POLYMER SURFACES; PHOTOCHEMICAL DEGRADATION

### Introduction

One of the most important reactions of a polymer system is that with its environment. Since all solids communicate with the rest of the world primarily by means of their surfaces, the surface is of prime importance in environmental modification. Thus it should be clear from Figure 14 that the partial pressure of oxygen and other reactive species which might be present in the atmosphere is highest at the surface as is the incident photon flux and any precipitation.

In view of this it would be somewhat surprising if the reactions at the surface were entirely representative of those in the bulk. It is conceivable however that with rapid surface reactions leading to low molecular weight molecules which can

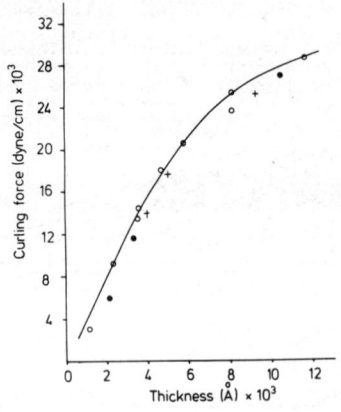

Figure 13. Curling force as a measure of internal stress as a function of thickness for plasma polymers prepared from fluorobenzenes.
200μ  o....perfluorobenzene
10W   +....1,2,4,5-tetrafluorobenzene
      •....1,4-difluorobenzene

# MODIFICATION, DEGRADATION AND SYNTHESIS

Photochemical Degradation of Polymers
(a) Model Studies
    Wavelength, photon flux, partial pressure of oxygen
(b) Natural Weathering

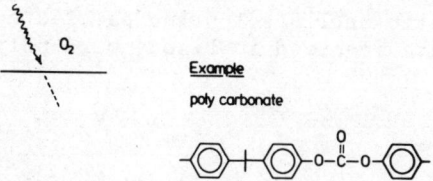

Figure 14. Schematic of the importance of surfaces in photodegradation, indicating greater photon flux, precipitation and partial pressure of oxygen in the surface regions.

readily desorb from the surface that the surface reactions could dominate the overall degradation of the sample particularly if this had a high surface area to volume ratio (e.g. a film).

In recent papers we have shown[18-20] how ESCA may be used to study the surface aspects of environmental modification of commercial high tonnage polymers. We describe here a particularly interesting example of the photochemical degradation of Bisphenol A based polycarbonate studied under controlled conditions in model experiments for environmental degradation of this important polymer.

## Photophysical Aspects

There is a considerable literature on the photochemical degradation of polycarbonates ranging from fundamental photophysics studies[21] to investigations of molecular weight changes during irradiation.[22] The majority of investigations have involved solution phase studies at wavelengths < 290 nm and are not thus strictly relevant to the solid state photodegradation of polycarbonate in sunlight. The main features of the main routes for photochemical degradation which have been described in the literature together with a summary of the available data on the nature of the excited state involved is set out in Figure 15.

Evidence has been presented[23] for the dominant role of photo-

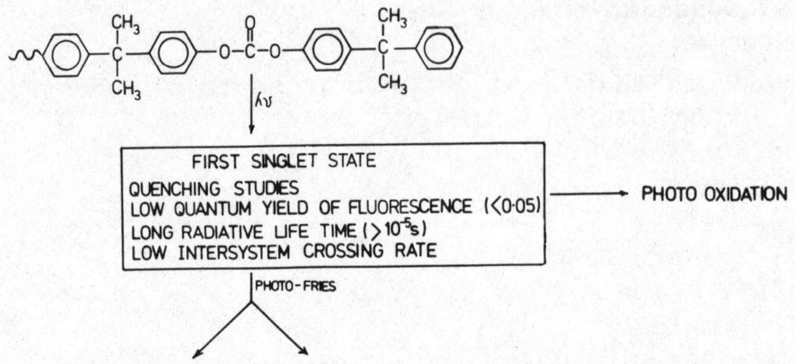

Figure 15. Schematic outline of principal features of the nature of the excited state involved in Bisphenol A polycarbonate photodegradation.


Fries rearrangements by both inter and intramolecular processes and the classic text by Ranby and Rabek also emphasizes the importance of photoFries rearrangements. However the detailed studies recently reported[24] by Factor and Chu suggest that photooxidative pathways involving the gem dimethyl and phenyl groups are of greater importance in photodegradation of polycarbonate in the solid state in oxygen rich atmospheres and at wavelengths corresponding somewhat more closely to sunlight than in previous studies.

Before considering the experimental data, we may briefly consider some recent theoretical computations[25] which shed new light on the nature of the excited states involved in photodegradation of polycarbonates.

The most complete studies relating to samples studied in the solid state per se arise from the work of Moacanin and coworkers.[21] On the basis of the low quantum yield of fluorescence, on the long radiative lifetime and the lack of triplet quenching it has been inferred that the excited state involved is probably $n \to \pi^*$ singlet in character. There have been no detailed studies of the conformational energy preferences for either the ground or excited states of Bisphenol A polycarbonate and most discussions have therefore centred on the early suggestions of Flory and Williams[25] which suggested an energetically preferred trans-trans configuration with a strong possibility of an orthogonal arrangement of the phenyl groups with respect to the carbonate group in the ground state. A knowledge of the electronic configurations of simple carbonyl, carboxyl and carbonate esters would suggest that a $n \to \pi^*$ designation for the excited state involved in polycarbonate photodegradation is unlikely. Thus if the ground state does indeed involve an orthogonal configuration then the electronic effects of the phenyl groups as far as the carbonyl groups (of the carbonate) are concerned are equivalent to alkyl groups. It is well known that the effect of replacing an alkyl group by an oxygen functionality in going across the series ketone, carboxy ester, carbonate ester is that whereas the n orbital remains roughly constant in energy the π orbital increases in energy.[26] Indeed UPS valence band studies reveal this trend in a graphic manner.[27] One manifestation of this is that the n→π* states move to progressively higher energy across the series and the lowest excited states of carbonate esters are therefore normally different from those for simple carbonyl compounds. UPS data also reveals that the n orbital of a simple carbonyl group is at a higher ionization potential than the ring π orbitals of a phenyl ring. This all adds up to considerable uncertainty in the assignment and nature of the excited states of Bisphenol A polycarbonate.

To shed some light on this we have recently completed[20] a detailed MNDO SCF MO study of aspects of the energy hypersurface for diphenyl carbonate as a model for the chromophore system of Bisphenol A polycarbonate. We report here the main outline of this work.

The relative energies of various orthogonal and planar configurations of diphenyl carbonate are shown in Figure 16. It is clear from this that the orthogonal trans-trans and cis-trans configurations are of similar energy, substantially below the corresponding planar configurations. There is a small energetic preference for the trans-trans configuration and the cross section (Figure 17) through the potential energy surface relating this to the cis-trans configuration shows that this is a low energy process. Inspection of the relevant eigenvalues reveals that the lowest excited state is π→π* in character with considerable phenoxy character. Whilst the ground state involves orthogonal arrangement of phenyl and carbonate groups the excited state probably involves an equilibrium geometry closer to a planar arrangement. This is outlined schematically in Figure 18. The identification of the π→π* singlet nature of the excited state and the change in potential energy surface is entirely consistent with the experimental data and also falls into a consistent picture with the extensive background information on substituted benzenes and simple carbonyl, carboxyl and carbonate chromophores.

Model Photodegradation Studies on Bisphenol A Polycarbonate

We outline here the main features of studies on the photodegradation of polycarbonate viewed from the surface point of view with particular reference to wavelengths > 290 nm.

To facilitate such studies a special reactor has been constructed which enables samples to be irradiated in a controlled atmosphere whilst mounted on an ESCA probe so that samples may be

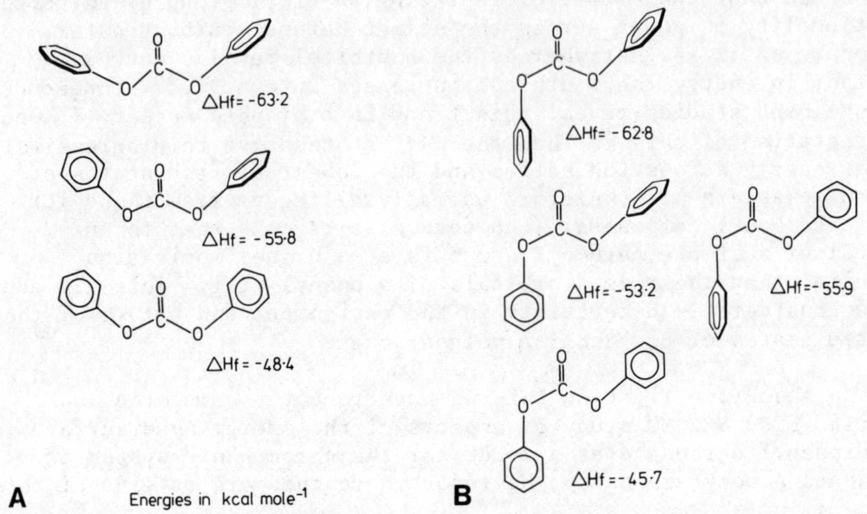

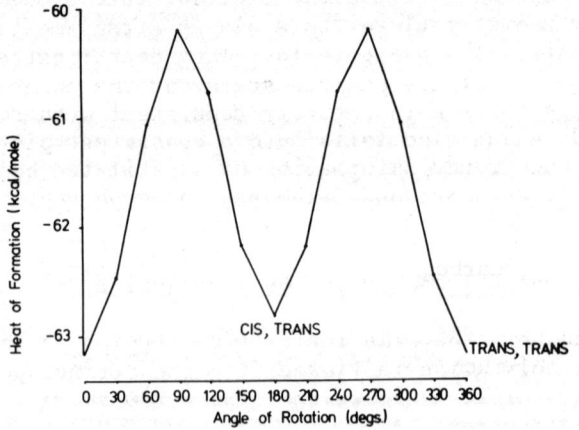

Figure 16. MNDO SCF MO computed relative heats of formation for various conformers of diphenyl carbonate as a model for the polycarbonate chromophore ((a) trans-trans (b) cis-trans).

Figure 17. Cross section through the potential surface relating the orthogonal configurations of the cis-trans and trans-trans conformers of diphenyl carbonate.

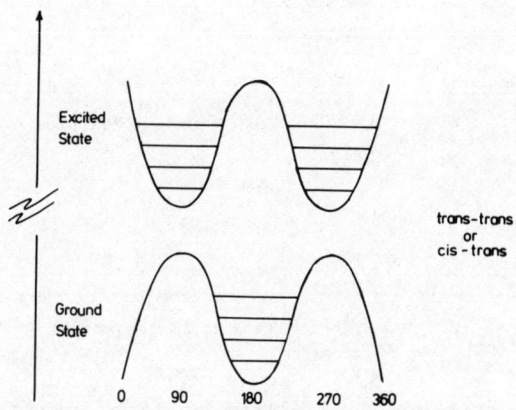

Figure 18. Possible nature of cross sections through potential energy surfaces for the ground and excited states of the diphenyl carbonate chromophore showing the energetic preference for orthogonal configurations for the former and planar for the latter state.

directly studied after irradiation (Figure 19). The dosage under a given set of conditions may readily be monitored by following the absorption changes at 370 nm in polysulphone films exposed under identical conditions as described by Davis.[28] For comparison purposes, the dosages using a 500 watt medium pressure lamp at distances of 18 cm or 50 cm from the sample position produce the dosages shown in Figure 20. Fifteen minutes irradiation at 50 cm corresponds roughly to 1½ hrs. exposure at midday on a sunny day in

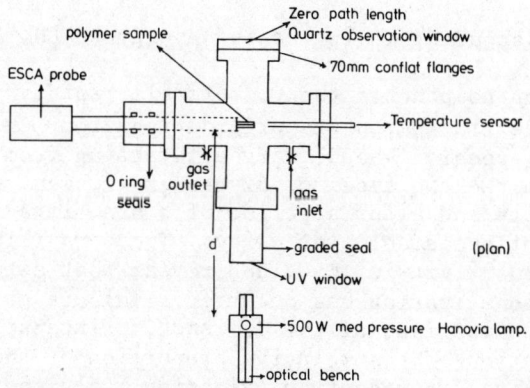

Figure 19. Schematic of reaction chamber used for model photo-oxidation studies of polycarbonate.

Lamp intensities > 290 nm
(monitored at 330 nm by polysulphone)

| | | whm$^{-2}$h$^{-1}$ | Equivalent times |
|---|---|---|---|
| Hanovia | 50cms | 5.7 | 15mins ≡ 90mins PERME, 54mins JTTRE |
| 500 Med pressure Lamp | 18cms | 52.5 | |
| | | whm$^{-2}$ d$^{-1}$ | |
| PERME (June) | | 9 | |
| | | | 15mins at 18cms gave same charge at |
| JTTRE (Australia) (October) | | 15 | 330nm as 2days in Dhahram, Saudi Arabia (September) |

Figure 20. Photon flux data for lamp configurations employed in the model photooxidation studies. For comparison purposes, photon fluxes for natural weathering at different sites are also included.

Southern England. Changes produced by such a short exposure are not apparent from conventional multiple attenuated total internal reflectance i.r. studies but are readily detected by ESCA as will become apparent.

As a starting point for the investigation of changes in surface chemistry we consider the irradiation of polycarbonate films in an oxygen atmosphere for varying periods at a dosage of 5.7 whm$^{-2}$h$^{-1}$. The ESCA data shown in Figure 21 are distinctive and reveal the extensive changes in surface chemistry under these experimental conditions. Starting from a $C_{1s}$ profile showing four components corresponding in increasing energy to $\underline{C}H$, $\underline{C}$-O, $O\diagdown_{\underline{C}}\diagup^O_O$

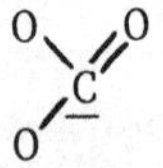

and π→π* shake up components the $C_{1s}$ profile rapidly increases in complexity with extra components originating from $\diagup C=O$ and $O-C=O$ functionalities appear. The $O_{1s}$ signal starting from a 2:1 doublet corresponding to the two types of oxygen environment increases in relative intensity and after a period of 6 hrs. irradiation the two components are of comparable intensity. The data illustrate the great surface sensitivity of ESCA and reveal that extensive oxidative functionalisation has occurred. This is more readily apparent from a comparison of the $O_{1s}$ and $C_{1s}$ intensity ratios and from a consideration of the relative proportion of the $C_{1s}$ signal arising from carbons representing oxidation functionalities and this is shown in Figure 22. It is clear from this that the photoFries rearrangement cannot represent the main contribution to the photo-degradative process. At an electron take off angle of 30° 95% of

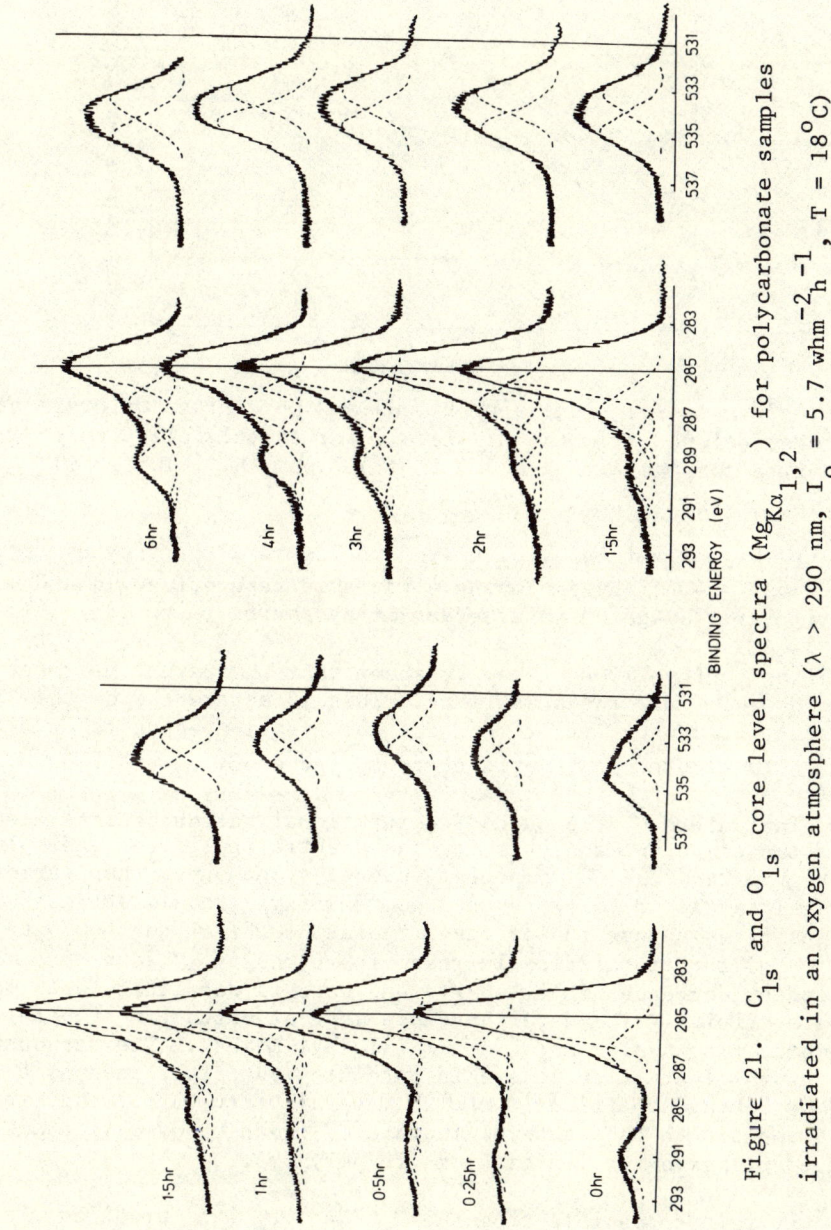

Figure 21. $C_{1s}$ and $O_{1s}$ core level spectra ($Mg_{K\alpha_{1,2}}$) for polycarbonate samples irradiated in an oxygen atmosphere ($\lambda > 290$ nm, $I_o = 5.7$ whm$^{-2}$h$^{-1}$, T = 18°C)

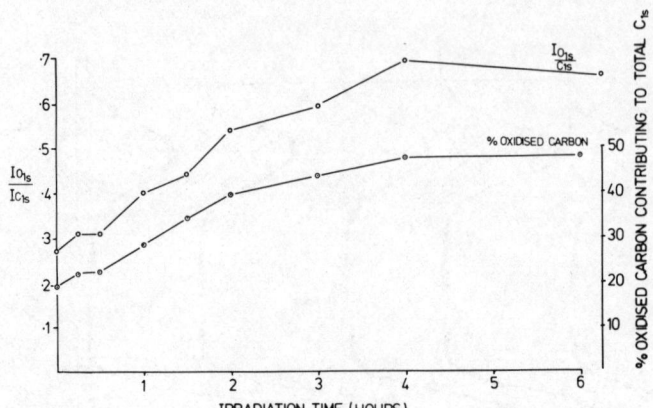

Figure 22. $C_{1s}/O_{1s}$ intensity ratios and percentage of oxygenated features defined from the $C_{1s}$ levels for photooxidized polycarbonate samples ($\lambda > 290$ nm, $I_o = 5.7$ whm$^{-2}$h$^{-1}$, T = 18°C).

the $C_{1s}$ signal intensity displayed in Figure 21 originates from the outermost $\sim 40$Å. By employing a variable take off angle and more particularly by going to a harder X-ray source (e.g. $Ti_{K\alpha_{1,2}}$) sampling depth $\sim 100$Å it may be shown that the oxidative reaction extends well into the subsurface. This is apparent from the substantial changes in $O_{1s}$ and $C_{1s}$ core level spectra taken with a $Ti_{K\alpha}$ X-ray source as is evident from Figure 23.

The nature of the oxidative functionalisation becomes clear from the component analysis for the spectra in Figure 21 displayed in Figure 24. The CH component (phenyl group other than carbon directly attached to oxygen and gem dimethyl group) decreases in intensity as a function of the irradiation whilst the C-O, >C=O and O-C=O functionalities increase in intensity. The π→π* component decreases in intensity and this is consistent with loss of aromaticity. The data therefore suggest oxidation of both the aromatic ring system and the gem dimethyl moiety. The carbonate structural feature decreases in intensity but stabilizes at a low percentage contribution to the overall structure. This in fact appears to be a general feature of the photooxidation and indeed plasma oxidation of aromatic polymers.

It is not possible from the $C_{1s}$ and $O_{1s}$ line profiles alone to delineate the importance of hydroperoxide (-O-O-H) structural features and we have accomplished this by allowing the irradiated films to be exposed to a stream of $SO_2$ to convert the hydroperoxide to $-SO_4H$ functionalities (Figure 25).

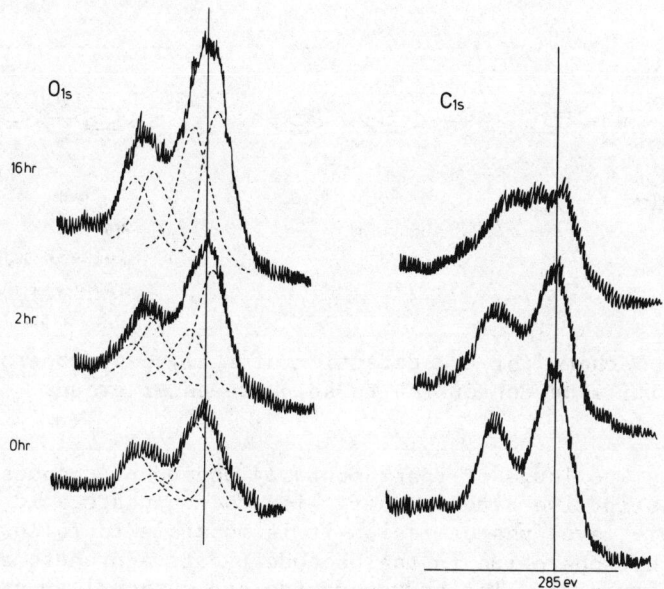

Figure 23. $C_{1s}$ and $O_{1s}$ core level spectra for photooxidized polycarbonate obtained with $Ti_{K\alpha_{1,2}}$ photon source ($I_o$ = 5.7 whm$^{-2}$h$^{-1}$, T = 18°C).

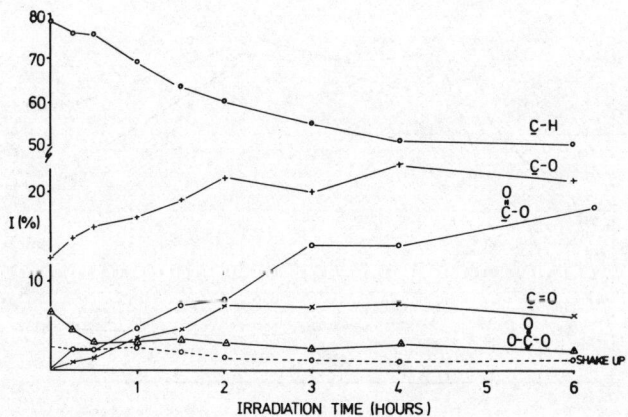

Figure 24. Component contributions to $C_{1s}$ line profile for the core level spectra displayed in Figure 21 for photooxidized polycarbonate ($\lambda$ > 290 nm, $I_o$ = 5.7 whm$^{-2}$h$^{-1}$, T = 18°C).

Figure 25. Scheme for the detection of surface hydroperoxide functionalities by conversion to sulphate ester groups.

With a knowledge of instrumentally dependent response factors and the distinctive binding energy for the sulphate acid group in the $S_{2p}$ core level photoemission it is possible to follow the role played by hydroperoxide in the photodegradative process and this is shown in Figure 26. The hydroperoxide and carbonyl functionalities build up to a maximum and then decrease slowly as is apparent from the component analysis in Figure 26. The induction period is not apparent from studies at higher lamp intensities since the hydroperoxide builds up much more rapidly initially and this is well illustrated by comparison of the data in Figure 26 and Figure 27. The increased rate of oxidative functionalisation with increased UV dosage is illustrated by the data in Figure 28. At higher dosages there is a tendency for the sample temperature to increase

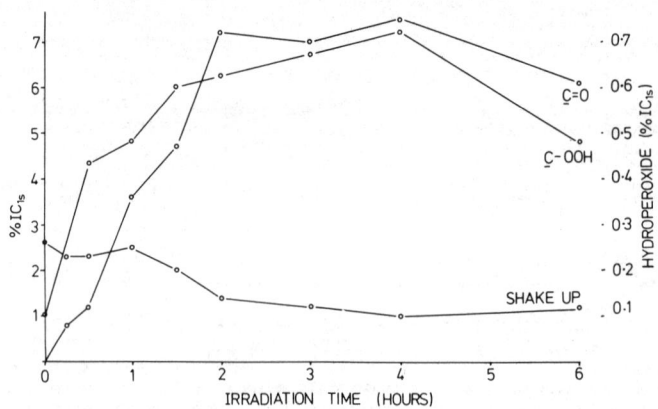

Figure 26. $>C=O$, $-COOH$ and $\pi \rightarrow \pi^*$ shake up components for photo-oxidized polycarbonate films ($\lambda > 290$ nm, $I_o = 5.7$ whm$^{-2}$h$^{-1}$, T = 18°C).

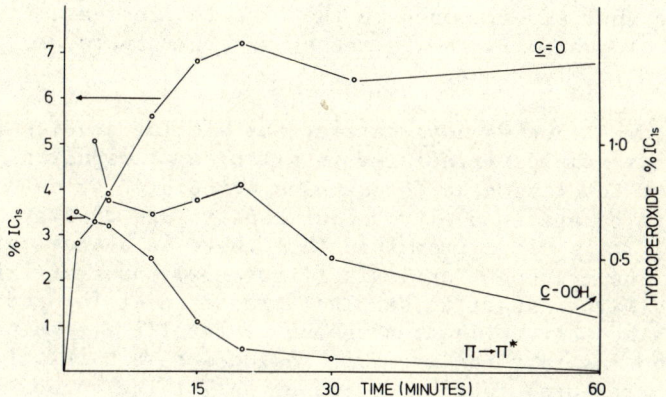

Figure 27. Components as for Figure 26 with higher photon flux ($I_o$ = 52.5 whm$^{-2}$h$^{-1}$, θ = 30°).

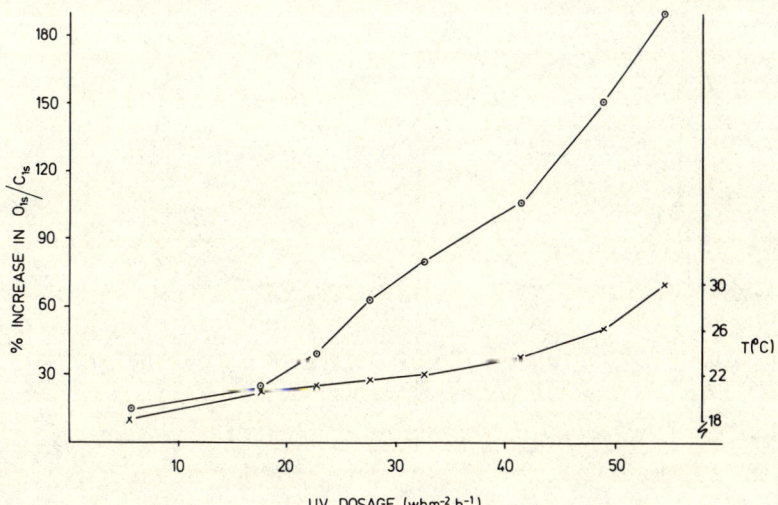

Figure 28. % increase in $O_{1s}/C_{1s}$ intensity ratios for polycarbonate samples irradiated in oxygen as a function of UV dosage. (Also shown is the sample temperature as a function of UV dosage.)

somewhat so that superimposed on the expected increase in rate as a function of dosage is the increased sample temperature and this is indicated in Figure 28.

As a comparison we have carried out similar studies for polycarbonate samples irradiated in nitrogen atmospheres. The $C_{1s}$ and $O_{1s}$ core level spectra for samples irradiated at an intermediate dosage in a nitrogen atmosphere are displayed in Figure 29. It is clear from this that there is a small oxygen uptake and the component analysis (Figure 30) shows striking differences with respect to samples irradiated at lower dosage in an oxygen atmosphere. The $\pi \to \pi^*$ shake up satellite remains roughly constant showing that the aromatic residues remain intact. The carbonate structural feature decreases whilst the carboxylate intensity increases and this is most readily accommodated by a photoFries rearrangement mechanism. There is a small contribution from $\rangle C=O$ structural features suggesting a low level of oxidation of gem dimethyl groups presumably arising from the low level of dissolved oxygen in the polymer film. This provides a solid state analogue of the solution phase photochemistry studies which have received so much attention in the literature and which also involves

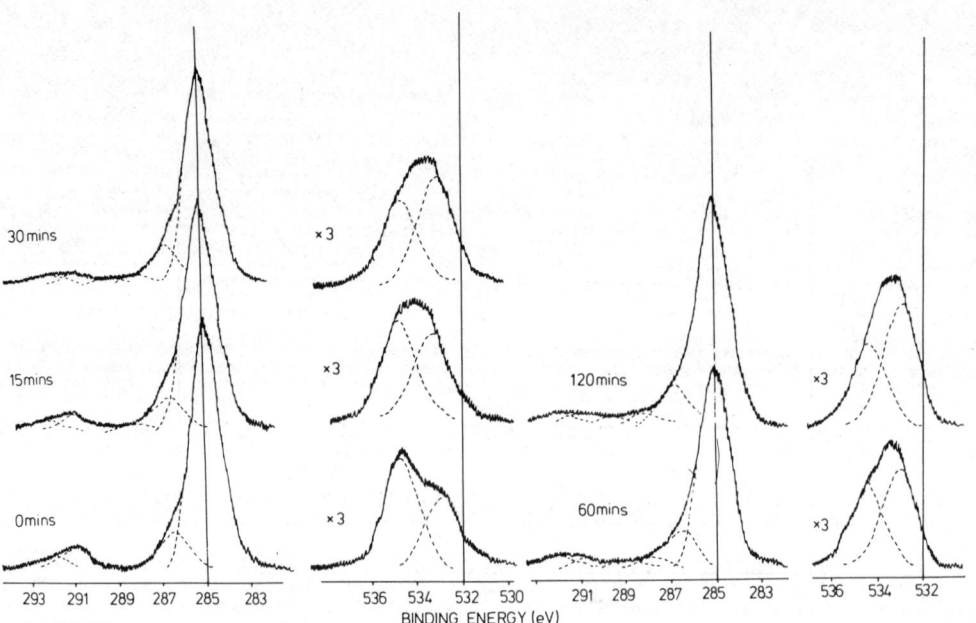

Figure 29. $C_{1s}$ and $O_{1s}$ levels ($Mg_{K\alpha_{1,2}}$) as a function of irradiation time for polycarbonate samples irradiated in a nitrogen atmosphere ($\lambda > 290$ nm, $I_o \sim 30$ whm$^{-2}$h$^{-1}$).

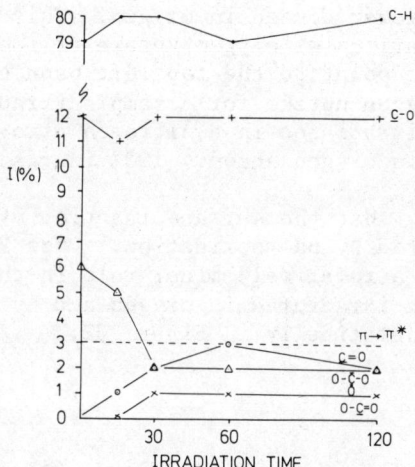

Figure 30. Components of the $C_{1s}$ levels for polycarbonate irradiated in a nitrogen atmosphere ($\lambda < 290$ nm, $I_o \sim 30$ whm$^{-2}$h$^{-1}$).

a photoFries rearrangement since the level of dissolved oxygen is also low. The low degree of oxygen uptake for the samples irradiated in nitrogen is also shown by comparison of the relative intensities of the $O_{1s}$ and $C_{1s}$ levels and this is shown in Figure 31. The level of oxygen incorporation for the samples irradiated in a nitrogen atmosphere is somewhat below that for samples

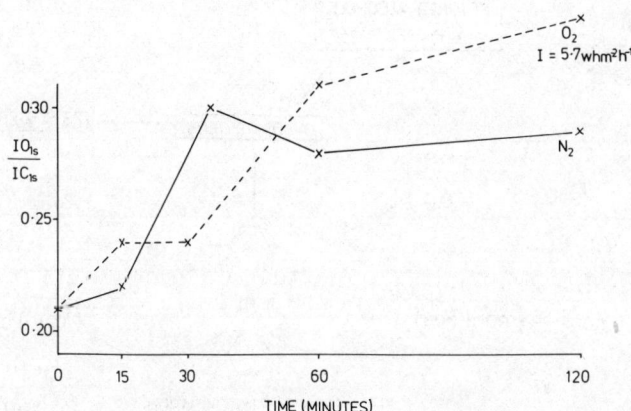

Figure 31. $O_{1s}/C_{1s}$ intensity ratio as a function of irradiation time for polycarbonate in a nitrogen atmosphere. Also shown are corresponding ratios for irradiation in oxygen ($\lambda > 290$ nm, $I_o \sim 30$ whm$^{-2}$h$^{-1}$).

irradiated at much lower dosage in oxygen. This difference is highlighted by comparison of oxygen uptake at identical UV dosages for 15 minutes. The point to the top left hand corner of the figure shows the oxygen uptake for a sample irradiated in oxygen. Whereas the sample irradiated in a nitrogen atmosphere shows little oxygen uptake that in oxygen shows ~ 100% increase.

These data show that the surface reaction of polycarbonate is rapid and is dominated by photooxidation. PhotoFries rearrangements form a relatively minor role in these surface reactions of samples irradiated in oxygen and by inference air and this is outlined schematically in Figure 32.

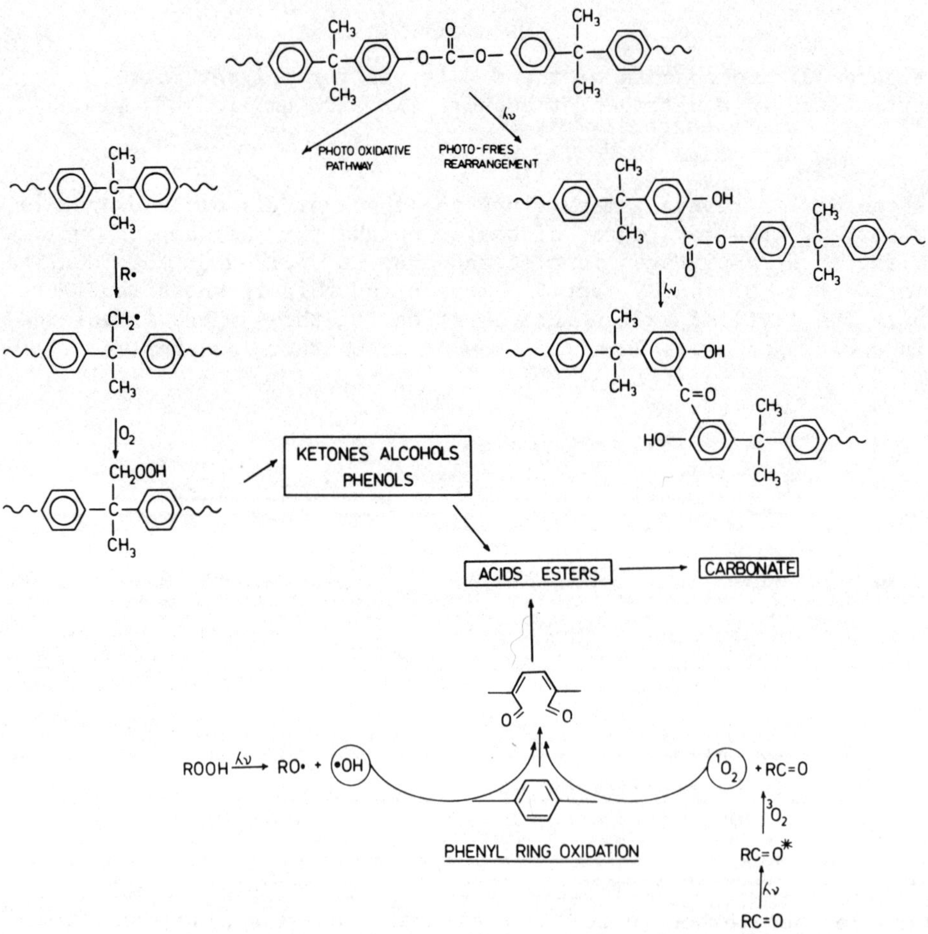

Figure 32. Summary of main photodegradative pathways for polycarbonate.

The particular virtue of ESCA is that these reactions can be monitored at a very early stage. These studies provide strong confirmatory evidence for the scheme outlined by Factor[24] who by contrast studied the bulk phase after extended irradiation periods at elevated temperatures. The results presented here show that such reactions occur in the surface regions at ambient temperatures and at relatively low UV dosages.

Information on the characterization of polymer surfaces exposed to environmental degradation has not previously been amenable prior to the development of ESCA as a powerful spectroscopic tool for such surface studies.

## ACKNOWLEDGEMENTS

Thanks are due to the Science Research Council U.K. for provision of equipment without which these programmes would not have been possible. Thanks are also due to a succession of exceptional research students in the Durham group who have enabled the ESCA programme to be developed in the past decade or so. Particular thanks are due to Mohammed AboShebak, Bill Brennan, Hugh Munro and Zaki AbRahman who are primarily responsible for the work described in this article.

## REFERENCES

1. (a) D.T. Clark, in "Advances in Polymer Science", H.J. Cantow, Editor, Vol. 24, p. 125, Springer-Verlag, Berlin, 1977.
   (b) D.T. Clark, in "Photon, Electron and Ion Probes of Polymer Structure and Properties", D.W. Dwight, T.J. Fabish and H.R. Thomas, Editors, A.C.S. Symposium Series No. 162, Chap. 17, American Chemical Society, Washington, D.C., 1981.
2. D.T. Clark, M. AboShebak and W. Brennan, (1982), unpublished data.
3. (a) D.T. Clark and H.R. Thomas, J. Polym. Sci., Polym. Chem. Ed., $\underline{15}$, 2843 (1977).
   (b) D.T. Clark and D. Shuttleworth, J. Polym. Sci., Polym. Chem. Ed., $\underline{16}$, 1093 (1978).
   (c) D.T. Clark, Y.C.T. Fok and G.G. Roberts, J. Elec. Spec., $\underline{22}$, 173 (1981).
4. H.G. Elias, "Macromolecules", J. Wiley and Sons Ltd., London, 1977.
5. J.R. Hollahan and A.T. Bell, Editors, "Techniques and Applications of Plasma Chemistry", J. Wiley and Sons Ltd., New York, 1974.
6. H. Yasuda and J. Hirotsu, J. Polym. Sci., Polym. Chem. Ed., $\underline{16}$, 743 (1978).

7. D.T. Clark, A. Dilks and D. Shuttleworth, in "Polymer Surfaces", D.T. Clark and W.J. Feast, Editors, J. Wiley and Sons Ltd., London, 1978.
8. M. Shen and A.T. Bell, Editors, "Plasma Polymerization", A.C.S. Symposium Series No. 108, American Chemical Society, Washington, D.C., 1979.
9. W. Crookes, Phil. Trans., Part I, 152 (1879).
10. K.T. Crompton and I. Langmuir, Electrical Discharges in Gases, I, Rev. Modern Physics, $\underline{2}$, 123 (1930).
11. H. Suhr, in "Techniques and Applications of Plasma Chemistry", J.R. Hollahan and A.T. Bell, Editors, Chap. 2, Wiley, New York, 1974.
12. Y. Osada, A.T. Bell and M. Shen, J. Polym. Sci., Polym. Chem. Ed., $\underline{16}$, 309 (1978).
13. R.P. Frueholz, W.M. Flicker, O.A. Mosher and A. Kuppermann, J. Chem. Phys., $\underline{70}$, 3057 (1979).
14. J.A. Baltrop and J.D. Coyle, Editors, "Excited States in Organic Chemistry", Chap. 9, Wiley, London, 1975.
15. J.R. Frazier, L.G. Christophorou, J.G. Carter and H.C. Schweinter, J. Chem. Phys., $\underline{69}$, 3807 (1978).
16. D.T. Clark and M.Z. AbRahman, J. Polym. Sci., Polym. Chem. Ed., $\underline{19}$, 2129 (1981).
17. D.T. Clark and M.Z. AbRahman, (1981), J. Polym. Sci., Polym. Chem. Ed., submitted for publication.
18. J. Peeling and D.T. Clark, Polym. Degrad. and Stab., $\underline{3}$, 97 (1981); Polym. Degrad. and Stab., $\underline{3}$(3), 177 (1981).
19. A. Dilks and D.T. Clark, (1981), J. Polym. Sci., Polym. Chem. Ed., submitted for publication.
20. D.T. Clark and H.S. Munro, (1981), Polym. Degrad. and Stab., submitted for publication.
21. (a) A. Gupta, A. Rembaum and J. Moacanin, Macromolecules, 11, 1285 (1978).
    (b) A. Gupta, R. Liang, J. Moacanin, R. Goldbeck and D. Kliger, Macromolecules, $\underline{13}$, 262 (1980).
22. E. Ong and H.E. Bair, Polymer Preprints, $\underline{20}$, 945 (1979).
23. B. Ranby and J.F. Rabek, "Photodegradation, Photooxidation and Photostabilization of Polymers", J. Wiley and Sons Ltd., London, 1975.
24. A. Factor and M.L. Chu, Polym. Degrad. and Stab., $\underline{2}$(3), 203 (1980).
25. A.D. Williams and P.J. Flory, J. Polym. Sci., A-2, $\underline{6}$, 1945 (1968).
26. (a) J.N. Murrel, "The Theory of Electronic Spectra of Organic Molecules", Methuen and Co. Ltd., London, 1963.
    (b) M.M. Jaffe and H. Orchin, "Theory and Application of Ultraviolet Spectroscopy", Wiley, New York, 1962.
27. (a) J.F. Meeks, J.F. Arnett, D. Larson and S.P. McGlynn, Chem. Phys. Lett., $\underline{30}$, 190 (1975).
    (b) D.W. Turner, "Molecular Photoelectron Spectroscopy", Wiley Interscience, Chichester, England, 1970.

SPECTROSCOPY OF POLYMER SURFACES USING THE SURFACE ENHANCED RAMAN EFFECT

D. L. Allara, C. A. Murray and S. Bodoff

Bell Laboratories

Murray Hill, New Jersey 07974

The conventional methods for obtaining vibrational spectra of polymer surfaces depend upon the internal reflection technique. The physical nature of this technique is discussed and in particular the reason for the inability to see exclusively the near surface regions (10-100Å). The recently discovered phenomenon of surface enhanced Raman spectroscopy induced by rough surfaces or particles of certain metals, primarily silver, offers interesting new possibilities for polymer surface analysis. The necessary roughness features fall in the range of below 20Å to nearly 1000Å and appear to be correlated with the dependence of the scattering intensity of the molecule on the surface-molecule distance. Results with extremely rough and nearly continuous silver films (up to ~ 1000Å features) show that enhanced Raman scattering can come from as far as ~ 200Å into a polymer film. However, with flattened ellipsoidal shaped islands of silver, ~100-300Å diameter, on glass substrates, enhanced scattering only appears to come from the first 20Å or less of the polymer surface at the silver interface. These results suggest that surface enhanced Raman spectroscopy can be a useful tool in examining the surface regions of polymer films.

# INTRODUCTION
## PROBLEMS OF CONVENTIONAL INFRARED AND RAMAN TECHNIQUES

The typical approach to characterization of polymer surfaces is to use ion and electron spectroscopies which probe to depths of ~10-100Å depending upon the type of technique. These techniques, in particular x-ray photoelectron emission spectroscopy have yielded much useful information about polymer surfaces but there have been some associated problems. In many cases subtle details of organic bonding cannot be determined. In addition, the requirement of medium to high vacuum conditions for analysis and the possibility of sampling beam damage restrict the ability to do certain types of <u>in situ</u> exposure studies and cause obvious problems with the interpretation of the chemical bonding structures, respectively.

Over the years, infrared and Raman spectroscopies have been popular photon probes for the vibrational structures of organic polymers. However, their application to surface analysis of bulk materials has been limited almost exclusively to internal reflection infrared spectroscopy (IRIRS) which has been used in a number of interesting studies. IRIRS analysis as well as other photon in-photon out probes in contrast to the ion and electron spectroscopies, need not be carried out under the often inconvenient conditions of vacuum. Furthermore the probing radiation is non-destructive to the sample. However, a drawback is the large probing depth, generally from one to several micrometers, of the measurement into the bulk of the sample. Because of this factor it follows that vibrational features of the outermost surface regions, Angstroms away from the ambient interface, cannot be easily separated from features arising from up to micrometers into the material. Some measurement of concentration-depth profiles is possible[1] but the results almost always are very qualitative.

The physical reason for the lack of "surface sensitivity" of IRIRS is well known[1,2] and will be briefly discussed here. A description of the experiment is shown in Figure 1. Using multiple internal reflections a beam of infrared radiation is propagated along the long axis of an IRIRS element of real refractive index $n_2$. We assume an infinitely thick sample rests on the element. At any particular reflection point with reflection angle $\theta$ an evanescent wave exists in the sample medium, of real refractive index $n_1$, and the electric field intensity of this wave decays with increasing distance from the interface according to Equation 1.[2]

$$E^2 \alpha \; e^{-2Z/d_p} \tag{1}$$

$$d_p = \frac{\lambda_2}{2\pi(\sin^2\theta - n_1^2/n_2^2)^{1/2}} \tag{2}$$

The quantity $d_p$ is the depth of penetration and is defined in Equation (2) above where $\lambda_2$ is the wavelength of the light propagating in the element. Using extremes of the realistic ranges of values for the parameters in Equation 2 one can calculate that it is virtually impossible to reduce $d_p$ to values much below a few tenths of a micrometer for mid-IR (400-4000 cm$^{-1}$) experiments

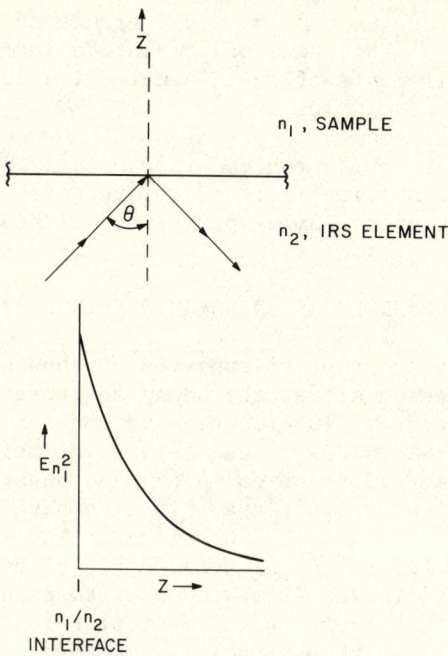

Figure 1. Schematic drawing of an internal reflection experiment and a plot of the decaying electric field as a function of distance from the element surface.

Recent studies have shown that it is also possible to obtain vibrational spectra of polymers with internal reflection Raman spectroscopy[3,4] (IRRS). The exciting radiation beam, normally visible laser light, is propagated along the internal reflection element, usually a transparent metal oxide. An evanescent wave arises beyond the element boundary, as with the infrared experiment, and couples with vibrational modes of material on the element face to generate Raman shifted radiation which is observed away from the element face. The physical description of the evanescent field is essentially that given in Figure 1 for the IR. However, given the shorter wavelength of the light in the Raman experiment $d_p$ in Equation 2 can be as small as ~1000Å for excitation by blue-green wavelengths. A closely related technique is waveguide Raman spectroscopy[5] in which the excitation beam is propagated via hundreds of internal reflections along the substrate (e.g., glass)-film interface. The physical principles and probing depths are similar to the IRRS technique. Although it follows that a large improvement can be made on the surface sensitivity, viz, reduced $d_p$, of vibrational spectra using internal reflection and waveguide Raman rather than IR techniques an additional factor of significance is that the quality of spectra obtainable from polymer films of about 100Å down to monolayer thickness is considerably poorer for Raman than for IR due to the large difference in cross sections for the techniques. Thus none of the internal reflection techniques fit the dual requirements of small penetration depth and good signal-noise characteristics.

Our approach to this problem of surface sensitivity has been to investigate the applicability of the recently discovered phenomenon of surface enhanced Raman spectroscopy (SERS).

## SURFACE ENHANCED RAMAN SPECTROSCOPY

Several years ago reports appeared[6,7] showing that for pyridine monolayers adsorbed on silver the Raman scattering intensity is enhanced by a factor of $\sim 10^6$ relative to the normal cross section of free pyridine molecules. These reports stimulated a number of interesting studies and theories which have been reviewed elsewhere[8,9,10]. The majority of these studies have been concerned with organic molecules adsorbed on silver. The most general conclusions are that for the metal some degree of surface roughness or particulate nature must exist in order to produce the giant enhancements and that there is no particular restriction on the type of molecule other than it be Raman active. Enhancements have also been observed on Au and Cu[9,10], Ag-rich Ag-Pd alloys[11], alkali metals[12], Pt[13], Ni[13,14] and liquid Hg[15,16].

Mechanisms have been proposed[7] to explain these various results and generally fall into two classes: direct, specific molecule-surface interactions and classical electromagnetic field enhancements. Whereas for cases such as liquid Hg, Pt and Ni the former mechanisms seem applicable, the latter appear adequate for the large body of results based on studies using silver exhibiting characterized roughness features or particle sizes of roughly between 20 to 1000Å. The electromagnetic mechanism provides some interesting and useful features with regards to the applicability of surface enhanced Raman to analyzing polymer surfaces. Among the basic characteristics of this type of mechanism are that Raman scattering from any Raman active molecule should be enhanced in intensity, that no specific surface-adsorbate chemical interactions are needed between the molecules and the silver or other enhancing metal surface, that the enhancement should occur even for molecules located some distance away from the metal surface and finally that the Raman intensity should fall off sharply with the molecule-surface distance over a distance range of 10-1000Å depending upon the particular roughness nature of the silver. These characteristics suggested to us that if one could prepare appropriately roughened silver at the surface of an organic film it should be possible to obtain quite intense Raman spectra from the near-surface regions of the film, hopefully with probing distances similar to typical ion and electron spectroscopies.

The physical basis of the electromagnetic mechanism for SERS from rough surfaces has been developed by several groups of workers.[10] The expressions are more complicated than for internal reflection IR and Raman which involve a rather straightforward evanescent electromagnetic field at a flat surface. In Figure 2 is shown a classical point dipole molecule located at a distance r from the center of an ellipsoidal metal protrusion tip of local radius of curvature a. When the metal roughness features, modeled here by the protrusion, are much smaller than the wavelength of the incident light the problem is reduced to the classical electrostatic system of an ellipsoid in a uniform applied field. At certain frequencies, depending upon the dielectric function of the metal, the incident light is resonant with a dipole plasma oscillation in the metal and the local electric field at the protrusion tip is enhanced. Further, the re-radiation of the Raman shifted light is now much more efficient because of this resonance (an "antenna" effect). Both of these above factors combine to give dramatic enhancements in the Raman scattering intensity. This intensity is proportional to, among other quantities, $(a/r)^{12}$, where r is the molecule-protrusion center distance, and thus the distance dependence or "penetration depth" should fall off steeply with increasing molecule-surface separation (r-a). The intensity from a uniform film surrounding the tip will fall off as $\sim(a/r)^{10}$

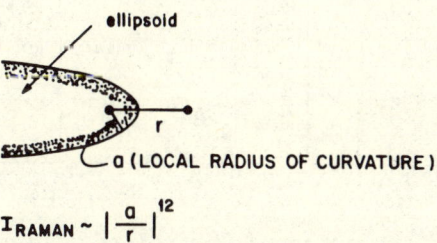

Figure 2. Schematic drawing of a roughness feature inducing an enhanced Raman effect.

because of the $r^2$ dependence on the exposed surface area of the film as the film is moved away. For example, for a protrusion or roughness feature with a 200Å radius the Raman intensity would be reduced to 0.1 of its surface value at a surface-film separation of 52Å. Thus SERS has potential to be a short range surface probe in cases where the electromagnetic mechanism predominates. In order to test the validity of the electromagnetic mechanism and the possible application to polymer surface analysis we performed a number of distance dependence experiments using multilayered "sandwich" samples. The details of this work have been published elsewhere[17,18] and the relevant results are reviewed here.

Two types of experiments were performed on samples with two types of multilayer structure as shown in the insets of Figures 3 and 4. Raman intensity enhancements for these samples fell in the range of $10^4$ to $10^7$ relative to the bulk materials. In the experiments in Figure 3 we first deposited $CaF_2$ films, usually 800Å thick, onto silicon substrates. These $CaF_2$ films exhibit roughness features averaging several hundred Angstroms. Then ~2000Å of aluminum was thermally deposited and the native oxide layer (~20-25Å) allowed to form at ambient conditions. Then either a solution formed monolayer of p-nitrobenzoic acid (PNBA) or a spin-cast film of poly(p-nitrostyrene) (PPNS) was deposited. In the

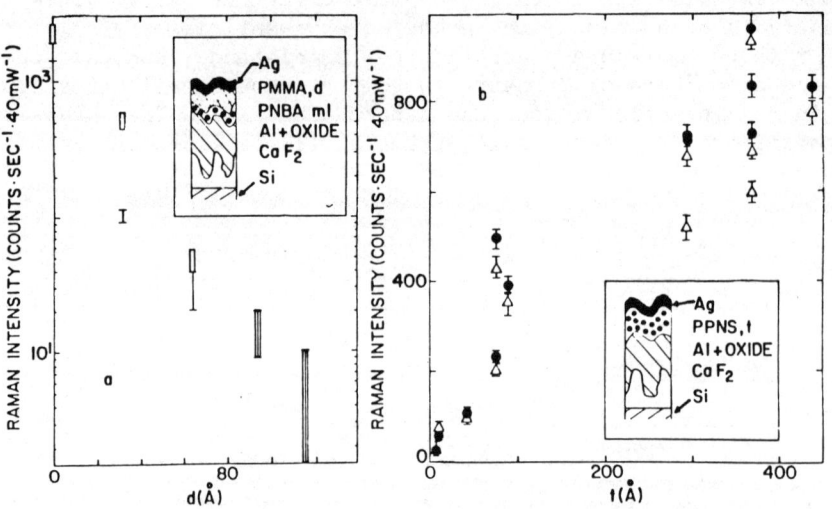

Figure 3. Plots of the enhanced Raman intensities as a function of PMMA spacer (d) or PPNS polymer (t) thickness and sample structure schematics. See text for details. Taken from reference 18.

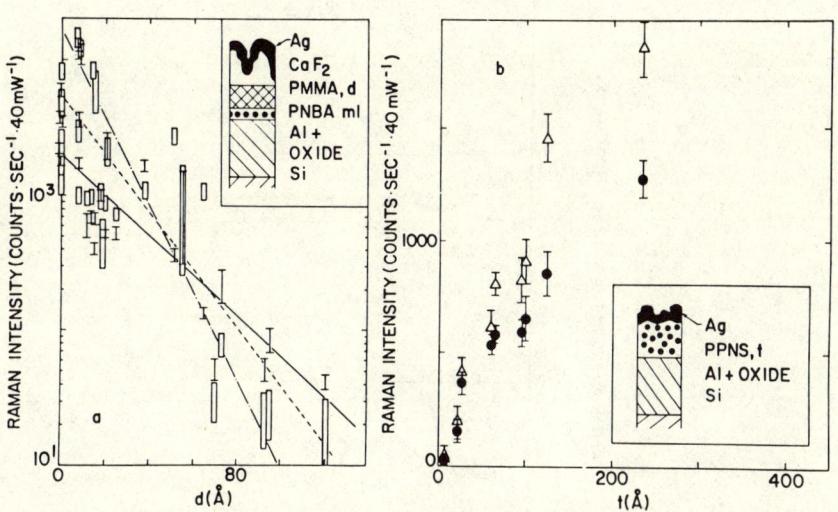

Figure 4. Plots of the enhanced Raman intensities as a function of PMMA spacer (d) or PPNS polymer (t) thickness and sample structure schematics. See text for details. Taken from reference 18.

case of the p-nitrobenzoic acid (adsorbed as the p-nitrobenzoate ion) monolayer various thickness poly(methyl methacrylate) (PMMA) layers were spin-cast over the monolayer. Finally for both types of samples a 200Å layer of replicating silver was thermally deposited onto the substrate cooled to ~77K. The Raman scattering intensity of the 1595 $cm^{-1}$ and 1425 $cm^{-1}$ peaks of the adsorbed PNBA were measured as a function of the PMMA spacer thickness and the 1596 $cm^{-1}$ and 1346 $cm^{-1}$ peaks of PPNS were measured as a function of PPNS thickness. It is clear from the distance dependence plots in Figure 3 that the Raman effect persists to distances of ~50-200Å before effectively decaying away.

In the experiments in Figure 4 ~2000Å of aluminum was thermally deposited onto silicon substrates to give a "smooth" substrate, the native oxide again allowed to form and as before PNBA with PMMA spacer layers or PPNS appropriately deposited. The PNBA samples were then covered with ~400Å of "rough" $CaF_2$ and finally with ~200Å of a replicating layer of silver onto the 77K substrate. The PPNS samples had some roughness features of their own and in addition the silver was deposited onto the films warmed up somewhat from 77K to produce a rough silver layer without using

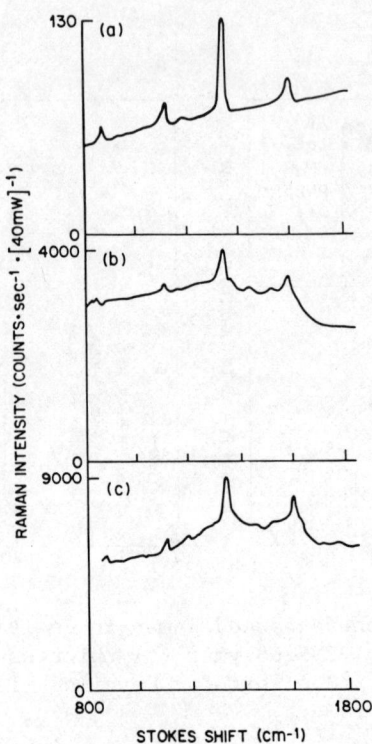

Figure 5. Representative Raman Spectra of PPNS (a) ordinary Raman of an 850Å film on aluminum; (b,c) spectra from Figure 3b for 29 and 127Å films respectively. Note the much higher count rates for the enhanced films. Taken from reference 18.

$CaF_2$. The distance dependence plots results are similar to those in Figure 3. Typical spectra of PPNS samples are shown in Figure 5.

From transmission and scanning electron microscopy[19] we have characterized the mean roughness features of the silver films above to be of the order of several hundred angstroms. These films were modeled as highly interconnected islands of randomly shaped ellipsoidal silver particles. On this basis the $(a/r)^{10}$ dependence should hold as discussed above. In Figure 4a lines corresponding to different values of $(a/r)^{10}$ are drawn through the points where the spacer layer distances is taken as r-a and values of 130, 180 and 260Å are taken for a. From a comparison with data the roughness features of the silver films responsible for SERS have radii of about 150-250Å in agreement with the existing features shown by electron microscopy.[19]

It is of interest to see if a silver film of a different morphology will lead to a different distance dependence. We recently investigated this point in a series of experiments[20] in which polystyrene films of various thicknesses were spin cast on top of Ag island structures prepared by thermal deposition on glass substrates. The silver island films ranged from 50-200Å in average or mass thickness. Transmission electron microscopy for the 100Å film showed the structure to consist of crowded oblate ellipsoids about 300Å in diameter by 100Å thick with their centers separated by about 300Å. In Figure 6 is a plot of the Raman intensity of the 1000 cm$^{-1}$ ring breathing mode of polystyrene as a function of the polymer thickness for a silver island structure of 100Å average thickness. The results show a saturation of intensity below 20Å polystyrene thickness and thus the electromagnetic field range from these structures is considerably shorter than for the $CaF_2$ and rough PPNS-Ag structures discussed earlier. These island films differ in structure from the latter in that the silver shapes are not completely interconnected and as the islands are flat, the radii of curvature can be as small as ~10Å at the edges of the ellipsoids. These sized features would give very steep fall-off behavior according to the $(a/r)^{10}$ relation. In contrast, the overlayer silver films for the earlier experiments show large round or egg-shaped silver blobs exhibiting much larger radii of curvature than the island film ellipsoids.

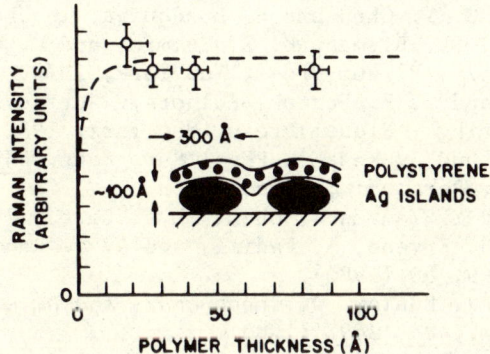

Figure 6. Plot of Raman intensity of polystyrene films on silver islands on glass substrates as a function of polystyrene film thickness. The silver island film has an average (mass) thickness of 100Å.

It is clear from these experiments that good quality vibrational spectra of very thin polymer films can be obtained with high sensitivity using the SERS effect. By varying the morphology of the silver, and presumably other appropriate metals, one can change the effective penetration depth of the Raman probe. Further, for the silver films we have investigated the penetration depth ranges from less than 20Å to somewhere in the range of ~50 to 200Å. These values are considerably shorter than any obtained by conventional internal reflection IR and Raman techniques. The enhancements are on the order of $10^4$ to $10^7$ compared to bulk materials so that the spectrum from the near surface region of a bulk polymer would be magnified greatly compared to the same region in the bulk. The SERS phenomenon thus should be a useful tool for polymer surface analysis and experiments are in progress toward this end.

## REFERENCES

1. H. G. Tompkins, Appl. Spectrosc., **28**, 335 (1974).
2. N. J. Harrick, "Internal Reflection Spectroscopy", John Wiley, New York, 1967.
3. R. I. Iwamoto, M. Miya, K. Ohta and S. Mima, J. Chem. Phys., **74**, 4780 (1981).
4. M. Delhaye, M. Dupeyrat, R. Dupeyrat and Y. Levy, J. Raman Spectrosc., **8**, 351 (1979).
5. J. F. Rabolt, R. Santo and J. D. Swalen, Appl. Spectrosc., **34**, 517 (1980).
6. M. G. Albrecht and J. A. Creighton, J. Am. Chem. Soc., **99**, 5215 (1977).
7. D. L. Jeanmaire and R. P. Van Duyne, J. Electroanal. Chem., **84**, 1 (1977).
8. T. E. Furtak and J. Reyes, Surface Science, **93**, 351 (1980).
9. E. Burstein, C. Y. Chen and S. Lundquist, in "Light Scattering in Solids", J. L. Birman, H. Z. Cummins and K. K. Rebane, Editors, p. 479, Plenum Press, New York, 1979.
10. R. K. Chang and T. E. Furtek, Editors, "Surface Enhanced Raman Scattering", Plenum Press, New York, 1982.
11. T. E. Furtak and J. Kester, Phys. Rev. Lett., **45**, 1652 (1980).
12. M. Moskovits, Personal Communication.
13. H. Yamada and Y. Yamamato, Chem. Phys. Lett., **77**, 520 (1980).
14. W. Krasser, H. Ervens, A. Fadimi, and A. J. Renouprex, J. Raman Spec., **9**, 80 (1980).
15. R. Naman, S. J. Buelow, O. Cheshnovsky and D. R. Hershbach, J. Phys. Chem., **84**, 2692 (1980).
16. L. A. Sanchez, R. L. Birke and J. R. Lombardi, Chem. Phys. Lett., **79**, 219 (1981).
17. C. A. Murray, D. L. Allara and M. Rhinewine, Phys. Rev. Lett., **46**, 57 (1981).
18. C. A. Murray and D. L. Allara, J. Chem. Phys., **76**, 1290 (1982).
19. C. A. Murray, D. L. Allara, A. F. Hebard and F. J. Padden, submitted to Surf. Science.

20. D. L. Allara, C. A. Murray and S. Bodoff, to be published. A preliminary account of this work also has been reviewed by C. A. Murray in "Surface Enhanced Raman Scattering", R. K. Chang and T. E. Furtak, Editors, Plenum Press, New York, 1982.

# X-RAY PHOTOELECTRON SPECTROSCOPIC STUDY OF THE ELECTRONIC STRUCTURE OF POLY-P-PHENYLENE SULFIDE

J. Riga, J.P. Boutique[*], J.J. Pireaux[**], J.J. Verbist

Laboratoire de Spectroscopie Electronique
Facultés Universitaires Notre-Dame de la Paix
B-5000 Namur, Belgium

Core and valence levels of poly-p-phenylene sulfide have been examined by X-ray photoelectron spectroscopy. Atomic charges distribution and characteristics of the C-S bond are obtained by comparison with related molecules such as (di)phenyl sulfide and (di)benzyl sulfide. The degree of the π electron delocalization in the polymeric chain and the amplitude of the energy gap are compared with previous results on poly-p-phenylene.

---

[*] Holder of a fellowship of I.R.S.I.A. (Belgium)
[**] Chercheur qualifié of the "Fonds National de la Recherche Scientifique".

## INTRODUCTION

Since the work of Walatka et al [1] on polymeric $(SN)_x$ crystals, the 1970's have seen a resurgence of interest in conducting polymers [2]. Generally, pristine polymers exhibit low electrical conductivities related to the reduced dimensionality of the electronic motion (molecular character) and to the lack of perfect periodic order (occurrence of localized states) [3]. Therefore numerous attempts have been made to dope the insulating polymeric materials such as polyacetylene [4] or poly-p-phenylene (PPP) [5]. By oxydation of cis-$(CH)_x$, the conductivity increases by as much as 11 or 12 orders of magnitude to $\sigma \simeq 10^2 - 10^3 \, \Omega^{-1} \, cm^{-1}$ at 298 K, a value which is only three orders of magnitude less than copper.

A new promising family of conductors is formed by phenyl chalcogenides. Presently the most interesting derivative is the poly-p-phenylene sulfide (PPS), the first melt-and solution-processable polymer, which can be doped by $AsF_5$ to conductivities of about $3 \, \Omega^{-1} \, cm^{-1}$ [6,7]. The crystal structure of PPS is known [8], but after doping the material becomes very poorly crystalline. As in the case of other polymers, doped PPS appears as brittle and very sensitive to air [7]. A remarkable point is that partial oxidation of a material in which the arrangement of the phenyl rings is orthogonal is sufficient to give rise to electronic conductivity. In a first step towards a knowledge of the electronic structure of doped conducting polymers, we report in this paper the X-ray photoelectron spectroscopic (XPS) study of pristine PPS. The characteristics of the electronic structure are deduced by comparison with two related molecules, (di)phenyl sulfide (PS), $C_6H_5-S-C_6H_5$ and (di)benzyl sulfide (BS), $C_6H_5-CH_2-S-CH_2-C_6-H_5$. The degree of the $\pi$ electron delocalization in the polymeric chain is compared to PPP results [9].

## EXPERIMENTAL PART

X-ray photoelectron spectra have been recorded on a Hewlett-Packard 5950A spectrometer using monochromatized Al $K\alpha_{1,2}$ radiation ($h\nu = 1486.7$ eV). PPS powder was pressed in a pellet and deposited on a gold substrate. During the measurements, the surface charging effect was neutralized by an electron "flood-gun". The sample was studied at the normal operating temperature ($\simeq 315$ K). The vacuum in the spectrometer was in the $1.10^{-9}$ Torr range.

The XPS spectra calibration was achieved by the gold decoration technique : thin film or islands of gold were evapored on the polymer surface and the line Au $4f_{7/2}$ was set at 84.0 eV.

## CORE LEVELS ANALYSIS

Table I lists the C1s and S2p$_{3/2}$ core levels binding energies of PPS, two related molecules PS and BS and elemental α-cyclooctasulfur.

The C1s level binding energy is constant and identical to the value reported for hydrocarbons (lowest acenes [10], poly-p-phenyls [9] series). In the case of BS, the methyl signal cannot be distinguished from the benzenic one. As observed in the acenes series, the C1s signal if followed by a weak shake-up satellite resulting from a valence π → π* excitation. The energy separation between this satellite and the main photoelectron peak is around 6.4 eV for each compound. A complete analysis of the shake-up lines observed in numerous organic sulfur compounds is in progress and will be published elsewhere [11].

The S2p$_{3/2}$ level shows small but reproducible and significant variations and a small negative charge on the sulfur atom in PPS can be deduced. By using the relation $\Delta E_B$ (Sulfur)/$\Delta q$ (Sulfur) = 5.3 eV/electron established by Siegbahn et al [12], we estimate, by comparison with cyclooctasulfur, this atomic charge to 0.08 electron. So, we assume that the final state effects, namely the molecular relaxation, similarly affect the photoemission energy balance and that the observed differences between the core levels binding energies reflect the differences in the polarity of the molecules before the photoelectric perturbation.

Table 1. C1s, S2p and S3p Lone Pair Binding Energies in Poly-p-phenylene Sulfide and Related Molecules (all values in eV).

| COMPOUND | C1s | S2p | S3p lone pair (a) |
|---|---|---|---|
| PPS (b) | 284.5 | 163.8 | 2.6 - 2.8 |
| PS (c) | 284.5 | 164.1 | 2.7 |
| BS (d) | 284.5 | 163.7 | 3.3 |
| CYCLOOCTASULFUR | —— | 164.2(5) | 3.6 |

(a) Centroïd of the highest occupied band (peak F in Figs. 1 and 2)
(b) PPS : poly-p-phenylene sulfide
(c) PS : phenyl sulfide
(d) BS : benzyl sulfide

## VALENCE LEVELS ANALYSIS

The XPS valence band of PPS is compared in Figure 1 with the PPP spectrum and in Figure 2 with that of related molecular spectra.

The 22 to 11 eV energy region (Figure 1) is strongly similar for the two polymers and mainly marked by the fingerprint of the C2s molecular orbitals derived from benzenic levels : $2a_{1g}$ (peak A), $2e_{1u}$ (peak B), $2e_{2g}$ (peak C) and $2b_{1u}$ (peak D) (Figure 3).

A bonding admixture occurs between the S3s atomic orbital and deepest C2s molecular orbitals, $2a_{1g}$ and $2e_{1u}$ (a), having and electron density on the carbon atom linked to the chalcogen.

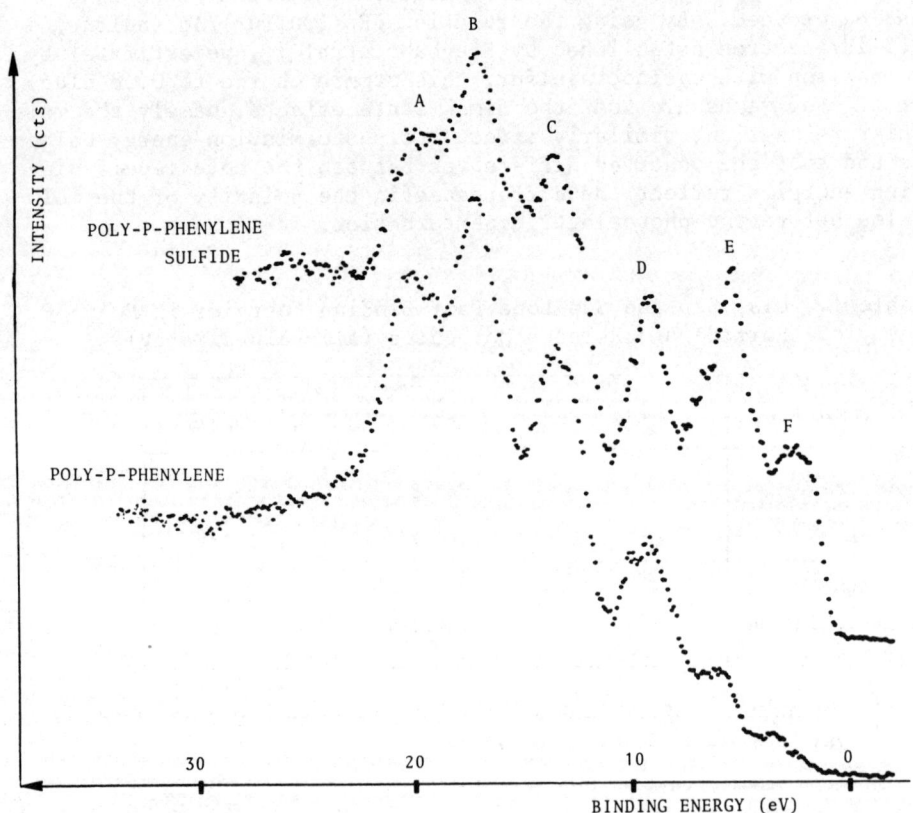

Figure 1. X-ray photoelectron valence band spectra of poly-p-phenylene sulfide and poly-p-phenylene.

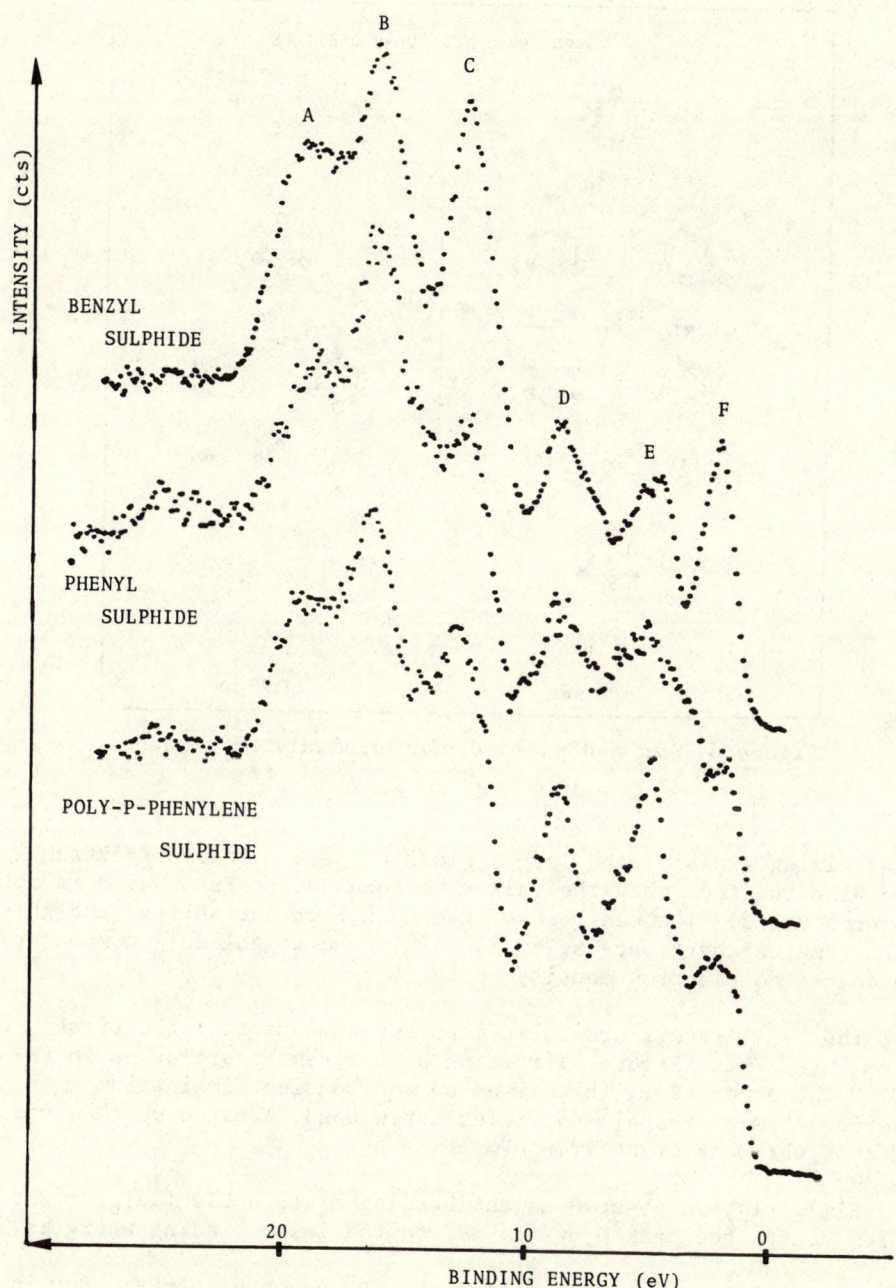

Figure 2. X-ray photoelectron valence band spectra of poly-p-phenylene sulfide and two related molecules.

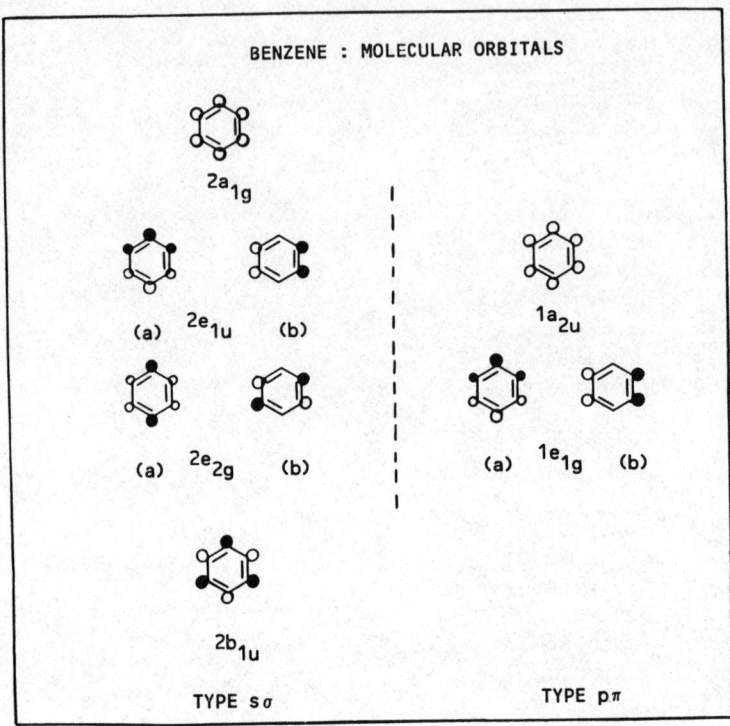

Figure 3. sσ and pπ molecular orbitals of benzene.

That mixing explains the larger width of peak A in PPS spectrum and also the fact that the valley between the peaks A and B is not so pronounced. The centroïd of the peak B is not shifted and this peak remains sharp because it is mainly due to the $2e_{1u}$ component which has no electron density on the $C_\alpha$.

The $2e_{1g}$ levels are split into two components : the first ($2e_{1g}$ (a) in Figure 3) remains at the same position as in PPP (C). The other, $2e_{1g}$ (b), gives an antibonding combination with the S3s atomic orbital and yields a new band, located on the left side of the peak C, at 12.6 eV.

Similarly, we observe an antibonding mixture C2s ($2b_{1u}$) - S3s which shifts the peak D by 0.5 eV to the lower binding energies.

These mixtures appear also on the XPS spectrum of PS. But in BS, a methyl group is inserted between the phenyl and the sulfur atom. Therefore, the interactions between the molecular orbitals of the aromatic cycle and the atomic orbitals of the chalcogen are hindered and the above described effects are less marked :

the shoulder on the left side of the peak C is not so apparent and the shift of the peak D does not occur (Figure 2).

The peak E contains a lot of molecular orbitals : $\sigma$ (C2p - S3p), $\sigma$ (C2p - H1s), $\pi$ C2p (1a$_{2u}$ and 1e$_{1g}$ (b), Figure 3) and the $\pi$ C2p (1e$_{1g}$ (a)) - S3p (lone pair) bonding combination.

The highest occupied band in PPS(F), is due to the photoionization of the S3p lone pair. The intensity of this peak is similar in PS spectrum and reflects an appreciable antibonding $\pi$ overlap between the 1e$_{1g}$ (b) and the S3p orbitals. Indeed if the chalcogen lone pair is strongly localized on the sulfur atom as in BS, the molecular orbital has essentially an S3p atomic orbital character and therefore its cross-section is high. A lowering in the intensity of this peak translates an admixture of C2p atomic orbital character (low cross-section). The antibonding combination gives rise in the polymer to a band located at 2.6 - 2.8 eV.

The highest occupied band in PPS is very close to the position of the HOMO in PS (2.7 eV) and 0.6 eV lower than in BS (3.3 eV). In PPP we have located the centroïd of the highest peak at 2.0 - 2.2 eV [9]. That means that the energy gap is approximatively 1.2 eV larger in PPS than in PPP. This result is related to the $\pi$ electron delocalization, which is hindered in PPS by the non planar conformation of the polymeric chain.

## CONCLUDING REMARKS

From XPS core valence levels spectra, the C-S bond appears very similar in both (di)phenyl sulfide and poly-p-phenylene sulfide. The S lone pair of PPS interacts with the $\pi$ electronic aromatic system and the C-S bond has some double character, as observed in aromatic sulfides or disulfides [13]. So, there is in the polymer a potential delocalization which is strongly hindered by the orthogonal arrangement of the phenyl rings. These results agree with the observation of a chemical modification in the doped material : breaking of the van der Waals forces to reach the planar conformation allowing the $\pi$ electron delocalization [6]. Therefore, the S3d orbitals involvement do not have to be invoked to explain the electrical properties of doped PPS.

## ACKNOWLEDGMENTS

J.P.B. acknowledges I.R.S.I.A. for a doctoral fellowship and J.J.P. is very grateful to the F.N.R.S. for financial support.

## REFERENCES

1. V.V. Walatka Jr., M.M. Labes and J.H. Perlstein, Phys. Rev. Lett. 31, 1139 (1973).
2. G.B. Street and T.C. Clarke, IBM J. Res. Develop. 25, 51 (1981)
3. C.B. Duke in "Extended Linear Chain Compounds", J.S. Miller, Editor, Plenum Press, New York, 1980.
4. A.G. Mac Diarmid and A.J. Heeger, Synthetic Metals, 1, 101 (1979/80).
5. a) D.M. Ivory, G.G. Miller, J.M. Sowa, L.W. Shacklette, R.R. Chance and R.H. Baughman, J. Chem. Phys. 71, 1506 (1979)
   b) L.W. Shacklette, R.R. Chance, D.M. Ivory, G.G. Miller and R.H. Baughman, Synthetic Metals 1, 307 (1979).
6. a) J.L. Brédas, R.R. Chance, R.H. Baughman and R. Silbey, Int. J. Quantum Chemistry, S15, 231 (1981).
   b) L.W. Schacklette, R.L. Elsenbaumer, R.R. Chance, H. Eckhardt, J.R. Frommer and R.H. Baughman, submitted for publication.
7. a) J.F. Rabolt, T.C. Clarke, K.K. Kanazawa, J.R. Reynolds and G.B. Street, J.C.S. Chem. Comm., p. 347 (1980)
   b) R.R. Chance, L.W. Schacklette, G.G. Miller, D.M. Ivory, J.M. Sowa, R.L. Elsenbaumer and R.H. Baughman, J.C.S. Chem. Comm., p. 348 (1980).
8. B.J. Tabor, E.P. Magre and J. Boon, European Polymer J. 7, 1127 (1971).
9. J. Riga, J.J. Pireaux, J.P. Boutique, R. Caudano, J. Verbist and Y. Gobillon, Synthetic Metals, 4, 99 (1981).
10. J. Riga, J.J. Pireaux, R. Caudano and J. Verbist, Physica Scripta 16, 346 (1977).
11. J.P. Boutique, J. Riga, J.J. Verbist (1982), unpublished data.
12. K. Siegbahn, C. Nordling, A. Fahlman, R. Nordberg, K. Hamrin, J. Hedman, G. Johansson, T. Bergmark, S.-E. Karlson, I. Lindgren, B. Lindberg, "ESCA, Atomic Molecular and Solid State Structure Studied by Means of Electron Spectroscopy", Almqvist and Wiksells, Uppsala, Sweden, 1967.
13. J. Riga, Ph. D. Thesis, University of Namur, Namur, Belgium, 1978, available from University Microfilms, Publ. N° 7970007.

# XPS ANALYSIS OF FLUOROCARBON FILMS PRODUCED BY SPUTTERING OF A PTFE BULK CATHODE[*]

J.J. Pireaux[§‡], J.P. Delrue[◊‡], A. Hecq[☆], J.P. Dauchot[☆]

‡ Facultés Universitaires, Laboratoire de Spectroscopie Electronique
   61, rue de Bruxelles, 5000 NAMUR - Belgium

☆ Université de l'Etat, Laboratoire de Chimie Inorganique
   23, Avenue Maistriau, 7000 MONS - Belgium

Polymer films were deposited on stainless steel substrates by RF sputtering from a bulk polytetrafluoroethylene cathode in various deposition conditions (gas pressure, gas composition, cathode self-bias voltage and cathode to specimen distance) : the surfaces of these sputtered films were then characterized by X-ray photoelectron spectroscopy in order to deduce information on their electronic structure and composition, as well as on the sputtering-deposition mechanism of the films.

This first systematic XPS analysis has shown that it is possible to sputter-deposit compounds whose surface properties (fluorine to carbon ratio, cross-linking and branching) can be varied over a large range of values, as it has already been observed for polymer films prepared by plasma polymerization. We note here that a film very similar to polytetrafluoroethylene

---

[*] Work performed under the auspices of the IRIS (Institute for Research in Interface Science) programme sponsored by the Belgian Ministry for Science Policy.
[§] Research Associate of the National Fund for Scientific Research (Belgium).
[◊] Postdoctoral IRIS fellow.

has been grown at large distance from the discharge axis, and that all the prepared polymers were free of any oxygen contamination.

## INTRODUCTION

During the last few years, the XPS (X-Ray Photoelectron Spectroscopy) technique has frequently been used and has proved to be the most successful spectroscopy to characterize surfaces of fluorine containing polymers. Indeed, the method provides valuable information concerning the different types of carbon present at or near the surfaces of these compounds[1]. To our knowledge, all the data available in the literature are related to fluoropolymers prepared by discharge in various gaseous atmospheres (plasma polymerization) or to stereoregular homopolymers[2-13]. RF sputtering from a bulk fluoropolymer cathode is another preparation technique which can also be used to deposit polymeric material uniformly over large surface areas. These films were already studied by conventional techniques (Infra-Red, UV absorption edge, contact angle measurements[14-16]). These sputtered polymers present interesting properties : perfect inertness, good adherence, electrical and mechanical properties like those of bulk PTFE (polytetrafluoroethylene or Teflon)†.

Therefore, the surface characterization of these sputtered polymers, and the investigation of the effects of selected sputtering conditions (gas pressure, gas composition, cathode self-bias voltage and cathode to specimen distance) on the structure and composition of these films were performed for the first time. Some information is also deduced on the mechanism of the sputtering-deposition of the films.

Teflon films are actually used in synthesis of selective membranes which could serve in desalination [18], gas separation[19] and gas sensors applications[20].

## EXPERIMENT

RF sputtering depositions have been done in a high-vacuum chamber (Figure 1) evacuated with a turbomolecular pumping unit. Background pressure prior to the introduction of argon, $CF_4$ or Ar-$CF_4$ mixtures was lower than $1.10^{-6}$ Torr (1 Torr = 133 Pa). In

---

† Note added in proof:
  Since this paper presentation, Dilks and Kay[17] suggested briefly that a sputtered PTFE film is almost identical with plasma PTFE.

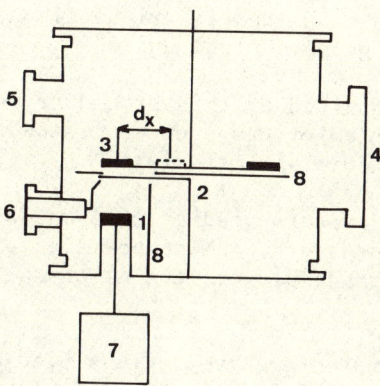

Fig. 1 Sputtering equipment : 1 - PTFE cathode; 2- shutter; 3- substrate; 4 - pumping direction; 5 - gas preparation vessel and gas admission; 6 - quartz crystal oscillator; 7 - RF generator; 8 - horizontal and vertical masks. $d_x$ is the substrate distance with regard to its normal position in front of the PTFE cathode.

this experiment, the 4 cm diameter PTFE bulk cathode bonded to a water-cooled copper backing was connected to a 9.0 MHz power supply. The target-substrate distance was 3.5 cm, but some substrates were also positioned at a distance $d_x$ from the discharge axis (see Figure 1) in order to study the energetic species bombardment influence of the freshly deposited layers. During such deposition the horizontal and vertical masks (n° 8 on Figure 1) were removed

Thin films were deposited on stainless steel substrates (area ≃ 1 cm² for the XPS measurements) or on quartz (for the optical measurements). According to the design of the uncooled sample holder, it was possible to prepare a maximum of five samples without breaking the vacuum. The deposition rate was followed in situ with a quartz crystal oscillator calibrated against the optical Tolansky method. For the XPS analysis, the thickness of the sputtered films was measured to be in the range of 500 to 3000 Å. The sputtering time required to deposit these films varied between 6 and 90 minutes, depending on the discharge parameters.

Mixtures of gases were made in a separate vessel after bakeout. Argon and $CF_4$ were admitted to a total pressure of 1000 to 1500 Torr. The ratio of partial pressures was checked with Pirani and Bourdon gauges.

The gas mixture was then introduced into the sputtering chamber through a regulating magnetic valve. The working pressure

($1.10^{-2}$ to $1.10^{-1}$ Torr) of the electrical discharge was determined with an MKS Baratron gauge control unit.

Polymer films deposited by this procedure were transparent or yellowish : this last color has been understood previously as indicating fluorine deficient materials[14].

The discharge parameters studied were the self-bias of the cathode $V_R$, the gas pressure, the composition of the gas phase, and finally, the distance between the discharge axis and the substrate.

A list of the samples prepared, classified according to the sputtering parameters, is given in Table I. As each series was prepared in a same run, we are confident in discovering general relative trend(s) - if any ! - for each set of polymers...

The XPS technique has already been extensively described in its application to the study of polymer surfaces[1]. Some useful information on the present work is nevertheless given here.

The XPS core level spectra of fluorine, carbon, oxygen and argon 1s levels were recorded using monochromatized $AlK_\alpha$ radiation on a Hewlett-Packard 5950A-ESCA spectrometer. All these data were measured in a low $10^{-9}$ Torr range with an X-ray source power of 800 W, keeping the samples at room temperature.

The fluoropolymers being studied as thin films on conducting substrates, the charging effect was reduced and homogeneous : it was easily neutralized with the conventional electron flood-gun technique.[A hot filament expells low energy electrons very close to the sample in order to compensate the positive electrostatic charging of the polymer surface due to the photoelectrons departure.]As a consequence, no problem was encountered during the calibration of the binding energy scale of the spectra; it was obtained for the homogeneous series of studied PTFE films by internal referencing : literature data recorded for the F1s level were used to assign the energy scale of the C1s structure. More details are given below.

As the X-ray photoelectron spectroscopy technique is very surface sensitive, the fluoropolymers were carefully transferred - through the air - from the sputtering vessel to the ESCA spectrometer, and analyzed without further chemical or mechanical treatment. All the studied samples were stable in air (see below) and in the spectrometer during the analysis,i.e. no degradation could be detected.

The XPS spectra were stored via a mini-computer on floppy disks for further data handling. This treatment consisted of fit-

Table I. Characteristics of the Preparation Procedure for the Studied Samples.

| Series Code | Ar-CF$_4$ % composition | p gas pressure | Self bias Voltage $V_R$* |
|---|---|---|---|
| A-1 | 100-0 | 2.10$^{-2}$ Torr | 800 |
| 2 | 100-0 | 2.10$^{-2}$ Torr | 1200 |
| 3 | 100-0 | 2.10$^{-2}$ Torr | 1400 |
| 4 | 100-0 | 2.10$^{-2}$ Torr | 1500 |
| B-1 | 100-0 | 1.5 10$^{-2}$ Torr | 1200 |
| 2 | 94-6 | 1.5 10$^{-2}$ Torr | 1200 |
| 3 | 75-25 | 1.5 10$^{-2}$ Torr | 1200 |
| 4 | 60-40 | 1.5 10$^{-2}$ Torr | 1200 |
| 5 | 0-100 | 1.5 10$^{-2}$ Torr | 1200 |
| C-1 | 85-15 | 1.5 10$^{-2}$ Torr | 800 |
| 2 | 85-15 | 1.5 10$^{-2}$ Torr | 1200 |
| 3 | 85-15 | 1.5 10$^{-2}$ Torr | 1400 |
| D-1 | 85-15 | 7.0 10$^{-2}$ Torr | 500 |
| 2 | 85-15 | 7.0 10$^{-2}$ Torr | 800 |
| 3 | 85-15 | 7.0 10$^{-2}$ Torr | 1200 |
| 4 | 85-15 | 7.0 10$^{-2}$ Torr | 1400 |
| E-1 | 85-15 | 1.0 10$^{-2}$ Torr | 1200 |
| 2 | 85-15 | 3.0 10$^{-2}$ Torr | 1200 |
| 3 | 85-15 | 5.0 10$^{-2}$ Torr | 1200 |
| 4 | 85-15 | 8.5 10$^{-2}$ Torr | 1200 |
| F-1 | | Same as B-5 with $d_x$ = | 0 cm |
| 2 | | | 4 cm |
| 3 | | | 7 cm |
| 4 | | | 10 cm |

* negative potential

ting gaussian peaks above a linear background via a least square method using the SIMPLEX algorithm, a procedure that was successfully implemented at the CERN Center[21]. In order to keep the data analysis within a reasonable time scale, we fixed a unique full width at half maximum (FWHM) for the Cls components -spectral analyses using all FWHM's as free parameters did show on sample data that this procedure did not modify in a sensible way the calculated binding energies and relative intensities-. To begin the data analysis of the Cls spectra, the same input parameters were injected into the program: as it is shown on Figure 2, we had to suppose the existence of eight Cls components (Authors in ref. 6 and 3 recorded a minimum of seven forms of carbon for TFE polymers, and surface fluorinated polyethylene respectively). Therefore, the same number of eight components was used to fit <u>all</u> the Cls spectra; but in our further discussion of the deconvolution data, we will group peaks together, according to their functionality.

## XPS RESULTS

### Binding Energy Calibration

As the present paper is the first one to deal with a systematic XPS characterization of polymer films prepared by RF sputtering from a bulk PTFE cathode, we have to base our interpretation on results obtained from the study of compounds prepared by plasma polymerization, or from the study of pure stereoregular homopolymers [2-11,17] : we suppose a priori that "the chemistry" of the polymer formation will be similar.

Whereas different chemical environments as a function of the existing first and second neighbours cause quite large shifts of the Cls lines, it has been shown that the Fls peak is recorded at a more constant binding energy; and in the case of polymers close to polytetrafluoroethylene, various authors did measure a quite comparable Fls binding energy and therefore a similar Fls-Cls (for $CF_2$-type carbon) binding energy difference (Table II). Our present series of spectra did show only one distinct peak in the Fls region with a constant FWHM $\simeq$ 1.6 ± 0.1 eV (Table VI); to ensure consistency with our previous work we chose a Fls binding energy at 689.9 eV to serve as a reference, inferring that at $\simeq$ 292.2 eV, we should record the carbon signal characteristic of a $-CF_2-$group (PTFE type)[7].

### Cls Peaks Assignment

In the study of fluorosubstituted polymers, simple and direct relation cannot a priori be established between the binding ener-

Table II. Examples of F1s Binding Energies (in electron volt) and Binding Energy Differences ΔE Measured Between the F1s and C1s ($CF_2$) type Core Level Peaks for Tetrafluoroethylene Polymers.

| Sample | Preparation | F1s (eV) | ΔE (eV) | Ref. |
|---|---|---|---|---|
| PTFE | Standard * | 690.3 | 397.3 | (2) |
| TFE | Plasma polymerization | 690.3 | 397.2-397.7 | |
| FEP teflon | Standard | 690.2 | 398.4 | (3) |
| FEP teflon | Standard | 689.1 | 397.3 | (4) |
| PTFE | Standard | 690.2 | 398.0 | (5) |
| TFE polymer | Plasma polymerization | 689.85 | 397.05 | (6) |
| PTFE | Standard | 689.9 | 397.7 | (7) |
| PTFE | Standard | ≃ 690 eV | 397 | (8) |

* Ionic or radical polymerization.

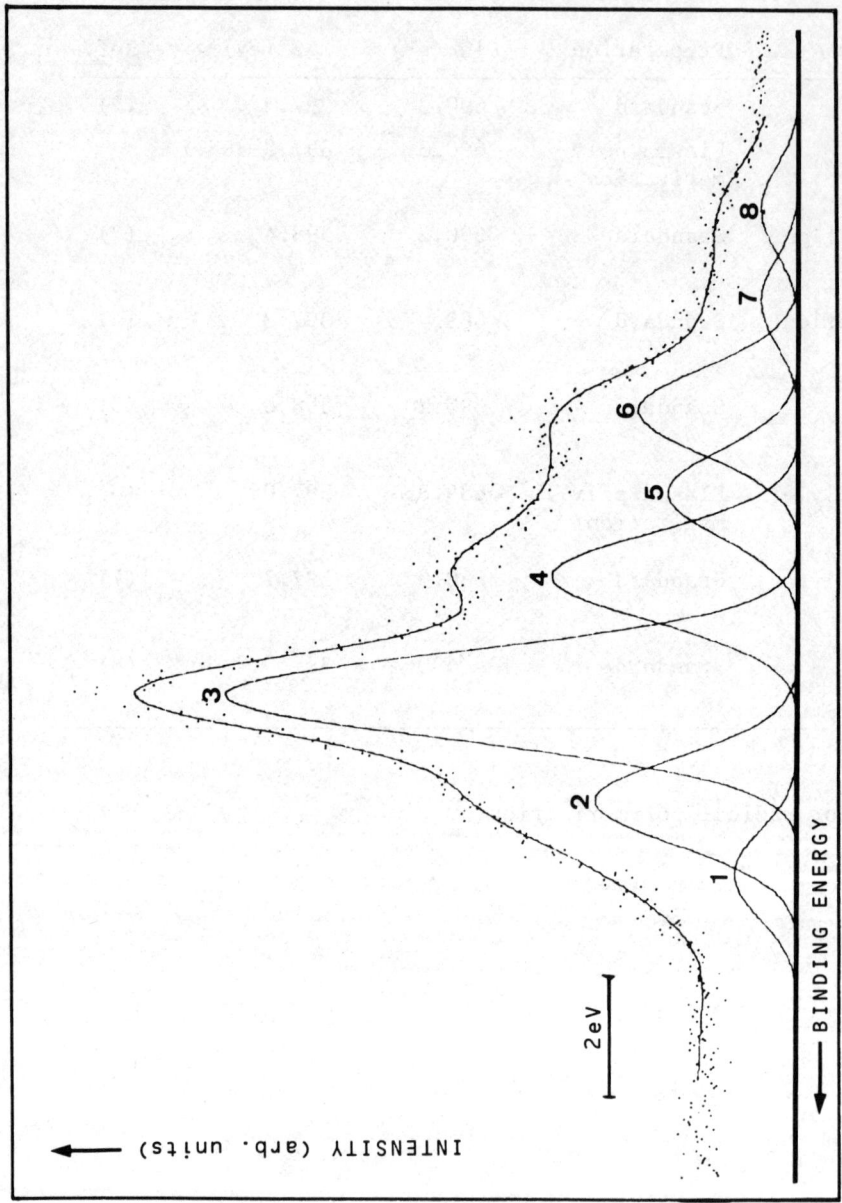

Figure 2. Representative result of a deconvolution analysis for the C1s structures recorded for the fluoropolymer film labelled E-1.

gy and the first neighbour-limited chemical environment cf the atom considered : assignments restricted to an analysis based on the sole first substituent effect are not unambiguous; Table III shows e.g. that various authors attribute a binding energy range of at least ± 0.5 eV to different carbon groups. Indeed, the fluoropolymers are characterized by quite large chemical shifts of the Cls level, and e.g. the second substituent also contributes significantly to the binding energy. We chose to base our assignement on a consistent binding energy scale deduced from a study of a well defined series of stereoregular polymers; that scale was shown to be consistent with the conventional potential model calculations (Table IV).

Taking into account the data displayed in Tables III and IV, the eight structures that were observed in the Carbon ls region can be given the most probable assignement of Table V. The effective error in the binding energy reported in column 3 is for a 99 % confidence interval : the effective energy range ± 0.1 eV measured for each Cls peak reflects slight modification in the chemical environment of the carbon atoms considered. Noteworthy is the well defined binding energy 291.98 ± 0.03 eV measured for the "Teflon-like" carbon : the binding energy difference of 0.2 eV measured in this work and for pure PTFE (Table IV, ref. 7) is an estimation of the <u>absolute</u> precision of our measurements.

For discussion purposes in the following chapters, we will group peaks and attribute them as given in the heading of Table VI. The hydrogen content of our polymer is expected to be very low, as the sole hydrogen contamination is from residual $H_2$ and $H_2O$ gases in the preparation vessel ($p \simeq 10^{-6}$ Torr). We propose therefore that peaks 4 to 6 correspond to carbon atoms with the given fluorine substituent, the other bonds corresponding to cross linking, branching, or unsaturated bonds.

Another possible assignment for peak # 1 at highest binding energy would be a $\pi \rightarrow \pi^*$ 7 eV shake-up transition originating from photoelectron peak # 6 at 287.3 eV. That supposes that part or all of the electrons of type # 6 are in fact CF vinyl radicals with a double bond. The intensity analysis of the 8 carbon components (see Table VI below) completely rules out that interpretation; peak 6 is not intense enough to be accompanied by such an intense satellite (peak 1).

We would like here to stress the fact that the polymers studied were almost free of "any contamination". Indeed, the Cls signal attributed to "saturated hydrocarbon" has an intensity almost always negligible. We point out also that <u>all</u> compounds studied were absolutely free of any trace of oxygen and argon. This absence of oxygen is in good agreement with the amounts of oxygen functionalities which is less than 0.1 atomic % [22]. This implies

Table III. Literature Data for the Binding Energy of Different Fluorine Substituted Carbon Atoms as Measured for Polymers Prepared by Plasma Polymerization and for Model Systems.

| $-CF_2-CF_3$ | $-CF_3$ | $-CF_2-$ | $>CF(II)$ | $>CF(I)$ | $-C-CH_2$ | $CH_2$ | Reference |
|---|---|---|---|---|---|---|---|
| 295.1<br>295.3 | – | 293.1<br>292.6 | – | – | – | – | 2 |
| – | 293.5 | 291.5 | 289.5 | – | 286.8 | 284<br>285 | 4 |
| – | 292.9<br>292.8 | 290.5<br>291.0 | 289.0 | 287.8 | 285.8<br>286.3 | – | |
| 294.8 | – | 292.8 | 290.6 | 288.6 | – | 285 | 6 |
| 295.6 | 293.7 | 291.3 | 289.1 | 288.3 | 286.7 | 285.0 | 9 |
| – | 293.8 | 291.6 | 289.2 | 287.2 | 287.2 | 285.1 | 10 |
| – | 293.8 | 291.5 | 289.5 | 288.2 | 286.9 | 285.5 | 11 |

Table IV. Typical Binding Energies for the C1s Peaks of Fluorinated Stereoregular Homopolymers, versus the Nature and Number of First and Second Neighbour (adapted from ref. 7).

| Binding energy (eV) | Polymer* | Atomic neighbourhood | Number of fluorine atom(s) as 1st neighbour | Number of fluorine atom(s) as 2nd neighbour |
|---|---|---|---|---|
| 284.6 | PE | $-CH_2 - \overset{*}{C}H_2 - CH_2 -$ | 0 | 0 |
| 285.5 | PVF | $-CHF - \overset{*}{C}H_2 - CHF -$ | 0 | 1 |
| 286.4 | $PVF_2$ | $-CF_2 - \overset{*}{C}H_2 - CF_2 -$ | 0 | 2 |
| 287.7 | PVF | $-CH_2 - \overset{*}{C}HF - CH_2 -$ | 1 | 0 |
| 289.4 | $PVF_3$ | $-CF_2 - \overset{*}{C}HF - CF_2 -$ | 1 | 2 |
| 290.9 | $PVF_2$ | $-CH_2 - \overset{*}{C}F_2 - CH_2 -$ | 2 | 0 |
| 291.6 | $PVF_3$ | $-CHF - \overset{*}{C}F_2 - CHF -$ | 2 | 1 |
| 292.2 | PTFE | $-CF_2 - \overset{*}{C}F_2 - CF_2 -$ | 2 | 2 |

* PE = polyethylene
PVF = Polyvinylfluoride, $PVF_2$ = polyvinylidene fluoride, $PVF_3$ = polytrifluoroethylene,
PTFE = polytetrafluoroethylene.

Table V. Assignment of the Cls-peaks Deconvoluted from the Spectra Recorded for the Sputtered Polymer Films.

| Peak number | Calibrated binding energy | Statistical error E* | Assignment model polymer chain |
|---|---|---|---|
| # 1 | 294.62 | ± 0.15 | $\underset{*}{C}F_3 - CF_2 -$ |
| # 2 | 293.70 | ± 0.14 | $\underset{*}{C}F_3 - C -$ |
| # 3 | 291.98 | ± 0.03 | $- CF_2 - \underset{*}{C}F_2 - CF_2$ |
| # 4 | 290.03 | ± 0.04 | $\begin{cases} - CF_2 - \underset{*}{C}FH - CF_2 - \\ - CH_2 - \underset{*}{C}F_2 - CH_2 \end{cases}$ |
| # 5 | 288.53 | ± 0.09 | $- CHF - \underset{*}{C}HF - CHF$ |
| # 6 | 287.35 | ± 0.14 | $- CH_2 - \underset{*}{C}HF - CH_2 -$ |
| # 7 | 285.33 | ± 0.16 | $- CHF - \underset{*}{C}H_2 - CHF -$ |
| # 8 | 283.65 | ± 0.27 | "graphite" decomposition |

* for a 99 % confidence interval, $E = \pm 2.58 \sqrt{\frac{\Sigma (x-\bar{x})^2}{n(n-1)}}$ where

$\bar{x}$ = average of the binding energies measured

$n$ = number of measurements.

a very small amount of free radical sites. We will come back to this point.

### C1s Peak Width.

Stereoregular homopolymers in the fluorinated series are characterized by small FWHM's, ranging from 1.0 to 1.4 eV for the C1s peaks, and of about 1.7 eV for the F1s peak[4-7]. The bulk cathode used in the present experiment produced F1s and C1s peaks with a FWHM of 1.9 and 1.45 eV respectively (the nominal resolution of the spectrometer used is 0.6 eV).

The FWHM's of the peaks recorded for the PTFE sputtered films are generally larger than these values (see Table VI, column 2). This is another indication of "modified" chemical environment of a given atom, e.g. second and 3rd neighbour effect, or cross-linking, or unsaturared bonds ... contribute to broaden the photoelectric lines.

### C1s Intensity Analysis and Overall Compounds Stoichiometry

The computer components analysis of the multipeak carbon 1s region produced the results - expressed as relative intensity in percent - presented in Table VI. As we propose in the previous chapter to group peaks according to their assignment, we give also in italics the total intensity corresponding to the same carbon functionality.

The sensitivity factor of our spectrometer is known from the study of standard samples : the measured relative cross section for the F1s/C1s peak intensities is $1.0/0.20$[2,4,6]. Therefore, by normalizing the observed intensities for an equal data accumulation time, it is possible to calculate the F/C stoichiometry of the fluoropolymer films, and to evaluate an average $CF_x$ formula (two last columns of Table VI). It is very satisfying to note that our bulk PTFE cathode standard gave a value of F/C = 10.20, and thus a $CF_{2.04}$ formula, in agreement with other authors[2,4,6].

The carbon 1s spectra could also be used to determine from the areas of the individual deconvoluted peaks a F/C ratio that should be comparable to the value computed from the total F1s/C1s intensities. In fact, such a calculation reproduces the trends given for the $CF_x$ formula (last column of Table VI), but the absolute values differ systematically, pointing out :
1. the relative accuracy (estimated to a few percent) of the peak intensity results and/or
2. the relative accuracy of the F1s/C1s relative cross section used.

Table VI. Intensity Analysis of the Components in the C1s Region and Overall Polymeric Film Stoichiometry.

| Series Code | F1s FWHM | C1s FWHM | #1 CF$_3$ | #2 | #3 CF$_2$ | #4 | #5 | #6 CF | #7 | #8 | XPS F/C intensity ratio | CF$_x$ formula |
|---|---|---|---|---|---|---|---|---|---|---|---|---|
| A1 | 2.3 | 1.6 | 9  | 23 | 40 | 21 | 8  | 16 | 1 | 0 | 7.37 | 1.47 |
| 2  | 2.2 | 1.6 | 12 | 23 | 45 | 18 | 7  | 13 | 1 | 0 | 7.50 | 1.50 |
| 3  | 2.4 | 1.7 | 8  | 19 | 47 | 15 | 8  | 17 | 2 | 0 | 7.27 | 1.45 |
| 4  | 2.2 | 1.6 | 8  | 17 | 38 | 16 | 12 | 24 | 5 | 0 | 6.83 | 1.37 |
| B1 | 2.1 | 1.5 | 3  | 21 | 48 | 20 | 4  | 9  | 1 | 0 | 7.73 | 1.55 |
| 2  | 2.2 | 1.6 | 4  | 21 | 46 | 20 | 6  | 11 | 1 | 1 | 7.77 | 1.53 |
| 3  | 2.2 | 1.6 | 4  | 22 | 42 | 21 | 6  | 13 | 1 | 1 | 7.30 | 1.46 |
| 4  | 2.3 | 1.8 | 4  | 18 | 41 | 20 | 8  | 20 | 1 | 1 | 6.50 | 1.30 |
| 5  | 2.3 | 1.7 | 5  | 18 | 39 | 18 | 10 | 21 | 3 | 0 | 6.30 | 1.26 |
| C1 | 2.4 | 1.8 | 14 | 28 | 33 | 17 | 10 | 19 | 2 | 1 | 7.00 | 1.40 |
| 2  | 2.2 | 1.7 | 9  | 21 | 39 | 19 | 8  | 20 | 2 | 1 | 6.70 | 1.34 |
| 3  | 2.3 | 1.7 | 7  | 16 | 45 | 17 | 9  | 19 | 2 | 1 | 6.73 | 1.35 |
| D1 | 1.9 | 1.3 | 1  | 19 | 46 | 17 | 3  | 11 | 5 | 6 | 7.70 | 1.54 |
| 2  | 2.3 | 1.7 | 2  | 22 | 40 | 19 | 6  | 17 | 1 | 0 | 7.33 | 1.47 |
| 3  | 2.1 | 1.6 | 2  | 22 | 40 | 19 | 6  | 18 | 1 | 0 | 7.23 | 1.46 |
| 4  | 2.2 | 1.6 | 2  | 15 | 49 | 16 | 7  | 19 | 1 | 1 | 7.00 | 1.40 |
| E1 | 2.3 | 1.7 | 4  | 18 | 39 | 17 | 9  | 20 | 3 | 3 | 6.3  | 1.26 |
| 2  | 2.2 | 1.6 | 3  | 21 | 46 | 19 | 5  | 12 | 1 | 0 | 7.20 | 1.44 |
| 3  | 2.3 | 1.7 | 4  | 22 | 44 | 29 | 7  | 14 | 1 | 1 | 7.47 | 1.49 |
| 4  | 2.2 | 1.5 | 4  | 23 | 43 | 20 | 6  | 12 | 1 | 1 | 7.50 | 1.50 |

# XPS ANALYSIS OF FLUOROCARBON FILMS

| | | | | | | | | | | | | | |
|---|---|---|---|---|---|---|---|---|---|---|---|---|---|
| F1 | 2.3 | 1.7 | 5 | 18 | 13 | 39 | 18 | 10 | 21 | 11 | 3 | 0 | 6.30 | 1.26 |
| 2 | 2.1 | 1.6 | 3 | 18 | 15 | 47 | 17 | 8 | 17 | 9 | 2 | 0 | 7.20 | 1.44 |
| 3 | 2.3 | 1.6 | 5 | 18 | 15 | 49 | 16 | 7 | 14 | 7 | 1 | 0 | 7.33 | 1.47 |
| 4 | 1.9 | 1.5 | 0 | 18 | 15 | 61 | 13 | 6 | 9 | 3 | 1 | 1 | 9.90 | 1.98 |
| Heated sample | 2.2 | 1.6 | 3 | | 15 | 41 | 21 | 9 | 9 | | 1 | 0 | 7.10 | 1.42 |
| Bulk cathode | 1.9 | 1.45 | — | | — | 78 | — | — | — | | — | 22 | 10.20 | 2.04 |
| Average over all the sputtered films | 2.2 | 1.6 | 5.1 | | 15 | 43.8 | 18.0 | 7.2 | 8.6 | | 1.7 | 0.7 | | |

## DISCUSSION

It is certainly not advisable in this XPS characterization approach to the sputtered fluoropolymers, with so many varying parameters in the sample preparation, to discuss every detail of the figures and tables. So we will only
(1) extract from our analysis (Figs. 3 to 8, and Table VI) the main tendencies to try to understand the relation between the six series of preparation parameters that were used (Table I) and the XPS results;
(2) compare those trends with the conclusion reported for fluoropolymers produced by other methods, mainly the plasma polymerization technique.

We must here warn the reader that, for their presentation (Figures 3 to 8) we have normalized each Cls spectrum intensity relatively to the largest components (structure # 3 attributed to the $CF_2$ peak).

### Sample Composition and Structure

<u>Effect of the sputtering bias voltage.</u> Figure 3 presents the Cls spectra recorded for the polymers prepared using varying sputtering self-bias voltage in a pure argon atmosphere (samples Series A, according to Table I).

With increasing bias voltage (and therefore power) of the sputtering minor changes are observed for the first three samples
- (Figure 3), at normalize $CF_2$ (teflon carbon like) fraction, the content in $CF_3$, CF and CH (to a lesser extent) species are diminishing in relative percentage;
- (Table VI) the tendency in $CF_2$ species (teflon like) is still to increase with the power, what is counterbalanced by a decrease of peak # 4. The $CF_3$ groups,(peaks 1 and 2) are slightly diminishing. Sample # 4 marked off from the others by a much lower $CF_2$ content, accompanied by an increase in the CF (# 5,6) moieties.

<u>Effect of the gas composition.</u> When operated at a constant medium bias voltage and pressure, with an increasing $CF_4$ fraction in the gas mixture (series B), the fluoropolymers films produced show a continuous decrease in $CF_2$ (Teflon type) carbon content, at the profit of the less fluorinated CF species (peaks # 5 and 6). One observes that the peak II ($CF_3$ carbon) also follows the trend observed for the Teflon component.

According to Lehman et al.[16], in an effort to produce PTFE films with a better stoichiometry (i.e. a F/C atomic ratio closer to 2.0), we have used various mixtures of Ar-$CF_4$ as discharge gas. (The proposed mechanism is that the growing polymer film will be

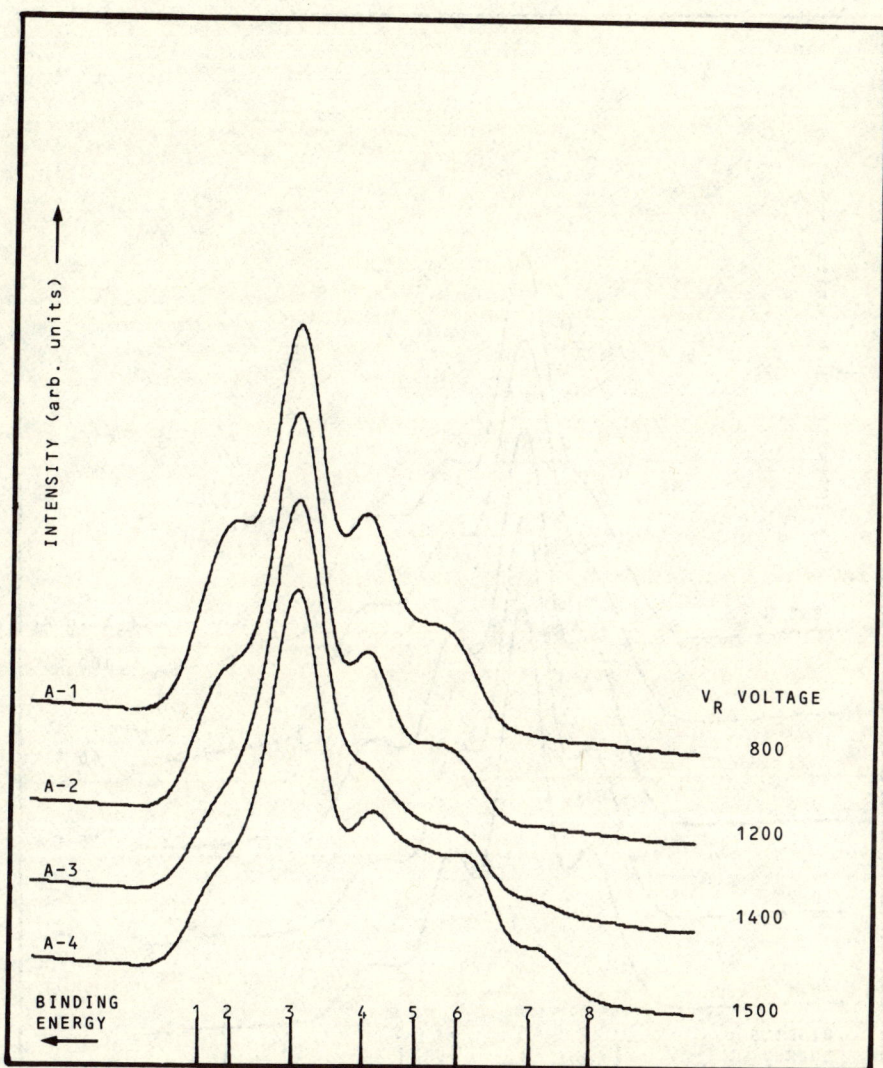

Figure 3. C1s spectra recorded for polymers of the series A (see Table I). All the intensities of peak III ($CF_2$ PTFE like) have been normalized to the same arbitrary value.

bombarded by F* radicals in the Ar-$CF_4$ discharge[16]). Nevertheless, we observe that the fluorine to carbon content decreases progressively with increasing $CF_4$ concentrations in the present study (Figure 5).

One should then consider the fact that the fluorine deficiency can be caused by a fluorine scavenger; at high $CF_4$ concentration,

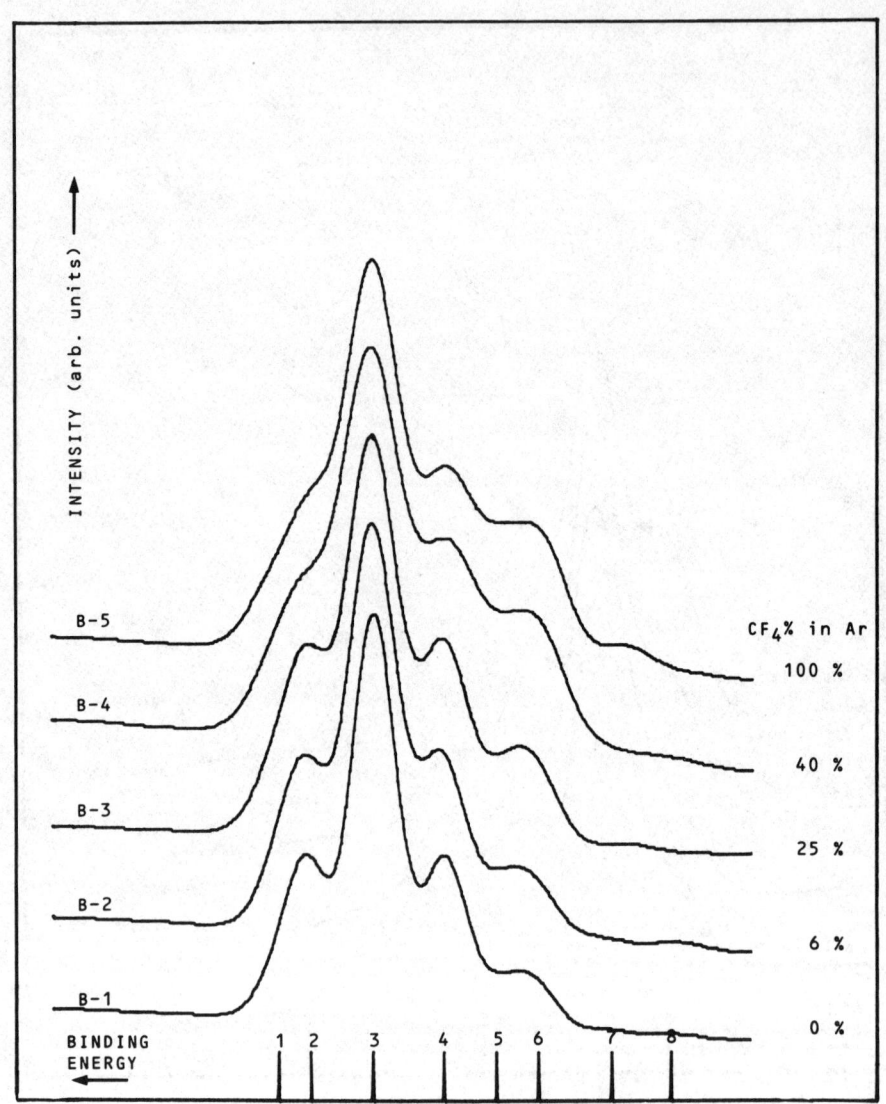

Figure 4. C1s spectra recorded for polymers of the series B. Cf. Figure 3 for further comments.

there is competition between film deposition and etching of the substrate but also of the growing film. Kay and Coburn[23] report that in a pure $CF_4$ discharge the carbon which is deposited on a silicon surface must be removed, most probably by reforming $CF_4$. Thus, a fraction of the F which would be available will be consumed to remove C. Electrodes made of silicon, metal, or Teflon can play the role of fluorine sca-

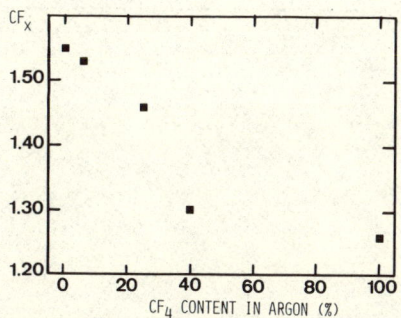

Figure 5. Fluorine to carbon ratio versus percentage of $CF_4$ in Argon atmosphere measured for polymers sputtered at $1.5 \, 10^{-2}$ Torr, with $V_R = -1200$ V (Series B).

venger[12,23-24]. If stoichiometric films are desired, one should use a compound with a F/C ratio such that the etching rate is weak and the polymerization rate is significant.

Effect of self-bias voltage, gas composition and pressure. Series C (Figure 6). Referring to series A, films prepared at a same base pressure but with a sputtering gas containing 15 % $CF_4$, the tendency is similar (a decrease of $CF_3$ species is observed); but some other features should be also noted (Table VI) : Peaks 1 and 2 ($CF_3$ type) start with higher intensity but then decrease simultaneously, while component # 3 ($CF_2$-Teflon like peak) increases very substantially from a very low percentage. The average F/C ratio is almost constant for these films.

Series D (Figure 7). With the same gas composition, but a higher pressure than in series C, it is seen from Figure 7 that at equal $CF_2$ (Teflon like) content, the amounts of other types of carbon atoms increase with the power of the discharge up to $V_a = -1200$ V. (The polymer prepared at $V_R = -1400$ Volt does not follow that overall tendency). On the average the total polymer composition is reduced in fluorine content (F/C ratio decreasing) with $V_R$ potential from 1.54 to 1.40. Noteworhty is the particular shape of the spectrum recorded for polymer D-1 : all the components of the Cls region are especially well resolved; if we add that the deposition rate for this film was slow, and that the shape of spectrum D-1 can be related to a more "regular" (stoichiometric) nature of the film we found some correlation with the work of Lehman[16] who noted a relation between deposition rate and stoichiometry.

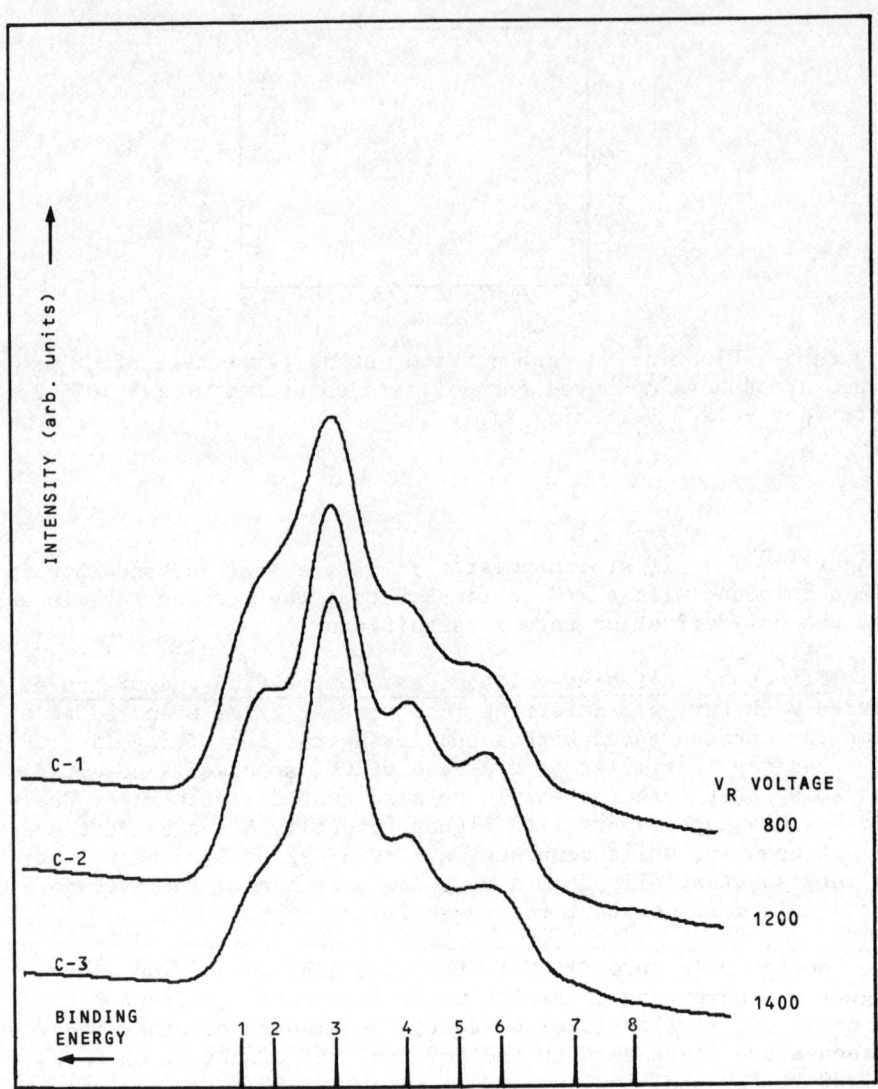

Figure 6. C1s spectra recorded for polymers of the series C. Cf. Figure 3 for further comments.

Effect of gas pressure. With an increasing gas pressure in the RF discharge (series E, Figure 8), it is seen that, at equal $CF_2$ Teflon content, the relative CF and $CH_2$ intensities decrease whereas the $CF_3$ species (peaks 1 and 2) increase. But the pressure increase raises also substantially the $CF_2$ amount, resulting in polymers with F/C content rising from 1.26 to 1.50.

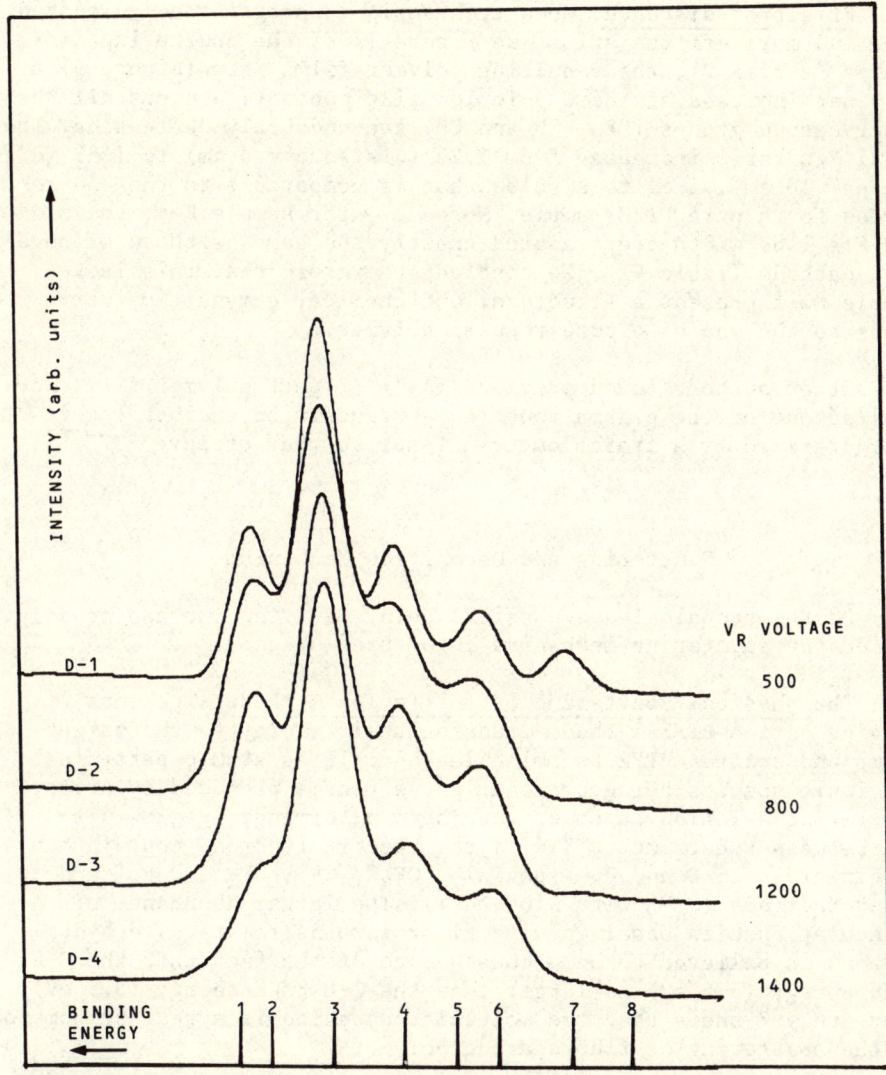

Figure 7. C1s spectra recorded for polymers of the series D. Cf. Figure 3 for further comments.

At low pressure (spectrum E-1), the intensity of peaks # 7 and 8 attributed to $CH_2$ and "graphite" carbons is higher: this is explained by the fact that the ratio between the residual gas pressure (mostly water at $\simeq 10^{-6}$ Torr), and the $C_xF_y$ "monomers" (sputtered from the cathode) pressure is higher with a lower sputtering pressure.

Effect of distance. When the sample is progressively located more and more off the axis (see Figure 1) of the sputtering discharge (series F), the resulting polymer films show (Figure 9) a very net increase of the $CF_2$-Teflon like content, whereas all the other carbon groups ($CF_2$, CF and CH) are undoubtly decreasing. The total F/C ratio increases from 1.26 (distance = 0 cm) to 1.98 (distance = 10 cm), i.e. to a value that is comparable to the one recorded for a pure PTFE sample. Moreover, for sample F-4, the Cls and Fls line width are measured exactly the same as those of pure PTFE cathode (Table VI). We conclude therefore that this last sample must present a structural and chemical composition very close to the one of a pure regular polymer.

Other authors found previously[2,4,6,9] that polymer films deposited out of the plasma zone (in a "deposition chamber") were characterized by a stoichiometry closer to that of pure PTFE.

## Sputtering and Deposition Mechanism

On the technical and physical point of view, we can crudely divide the sputtering mechanism into three stages :

The physical sputtering of the Teflon cathode. The ions impinging on the PTFE cathode transfer their energy to the target atoms and sputter PTFE as molecules as well as atomic particles. The ionic species being present in the plasma produced by argon sputtering a Teflon cathode, have been determined by glow discharge mass spectrometry[16,29] : the spectra recorded contain many combinations of C and F, especially $CF_3^+$, $CF^+$, $CF_2^+$, $C_2F_4^+$ and other radicals ($C_nF_y$ with n up to 7). The larger abundance of molecular species observed with fluorocarbons relative to hydrocarbons is believed to be a consequence of the fact that the C-F bond energy (5.5 eV) is larger than the C-H bond energy (3.5 eV). That study[29] shows that the molecular ejection is a real phenomenon in the sputtering of fluorocarbon polymers.

The transfer of the sputtered excited species from the cathode to the substrate. During the travel in the gas phase, the sputtered species can react between themselves and with the discharge gas atoms .From the results of plasma polymerization, it is generally recognized that the gas phase reaction plays an important part in the polymerization of fluorocarbons, in the pressure regime where the collisional frequency between species able to polymerize is sufficiently high[12]. Although $CF_4$ has been observed to weakly polymerize in some plasmas[16,23,25] it is only for fluorocarbons with $F/C \leq 2$ that a significant rate of polymerization is observed under normal conditions.

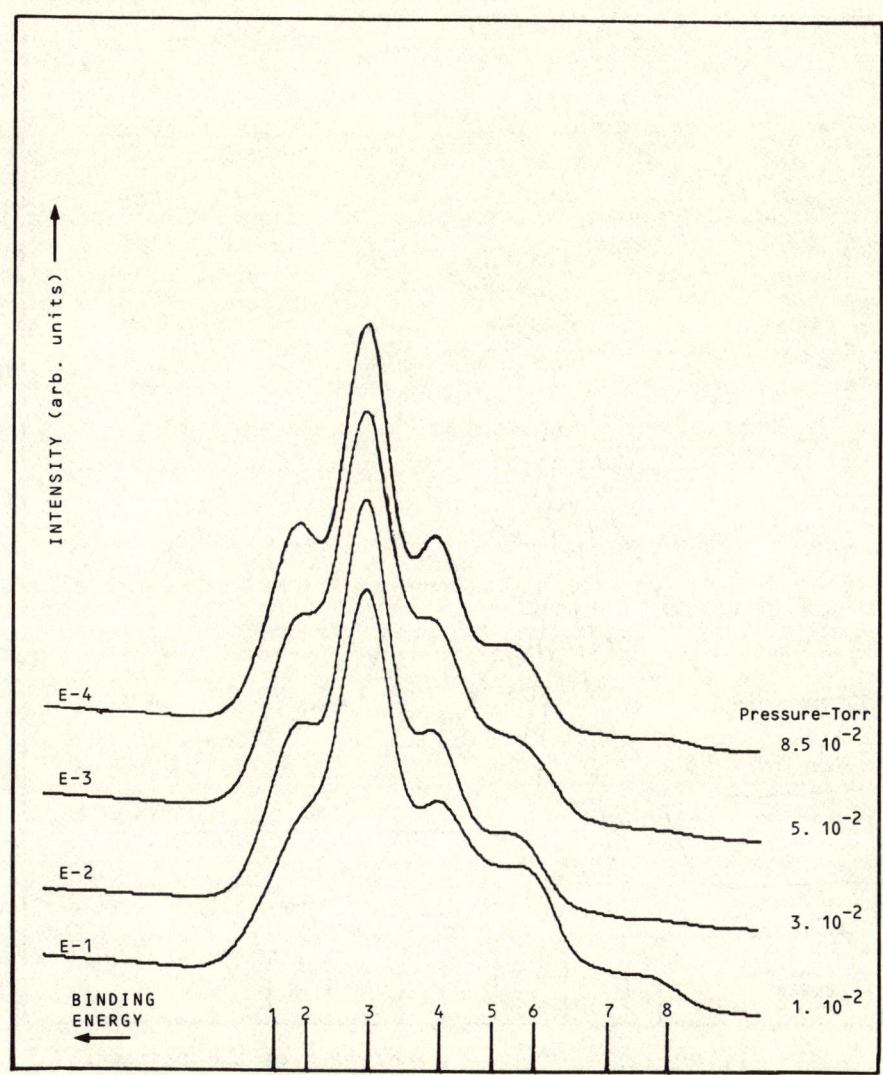

Figure 8. C1s spectra recorded for polymers of the series E. Cf. Figure 3 for further comments.

The film condensation. The ions, electrons and energetic neutrals induce chemical and structural modifications of the growing sputtered films. This fact has been shown recently in the case of rare gas interactions on ethylene-tetrafluoroethylene copolymers[26]. During the deposition these interactions eliminate fluorine atoms and also small fragments to form cross links, unsaturated and free radicals in the deposited polymer. These free radicals and un-

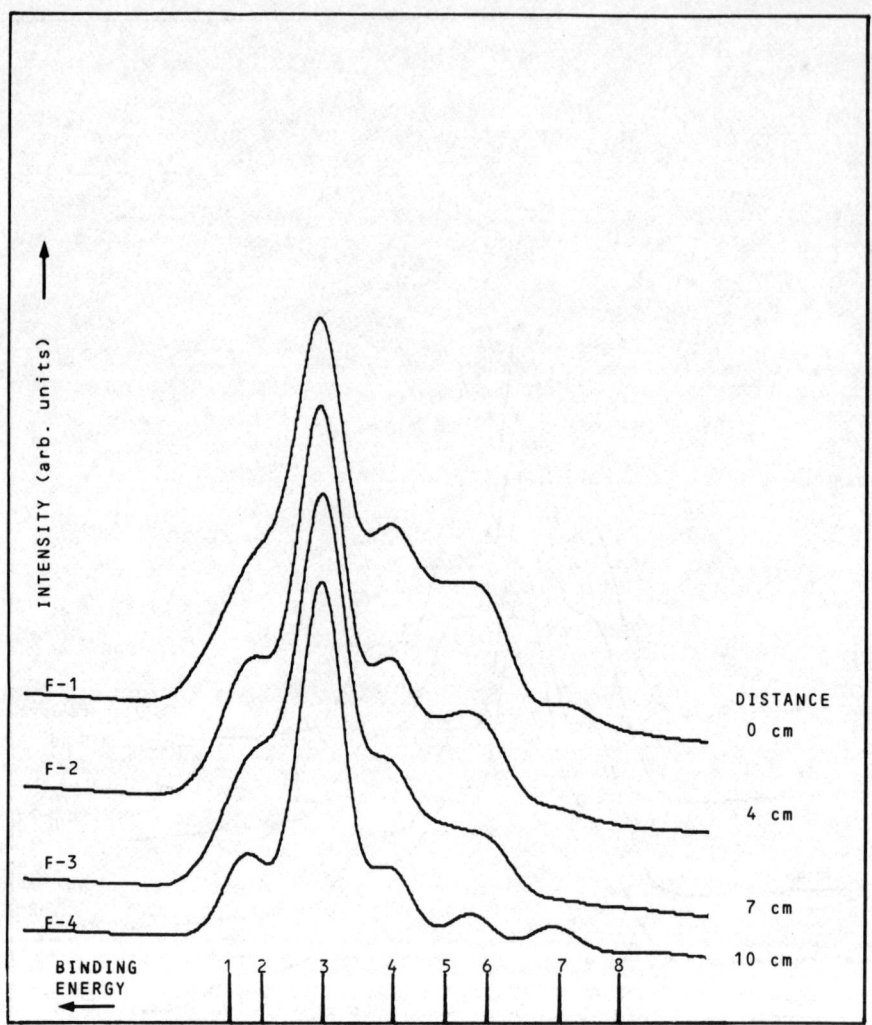

Figure 9. C1s spectra recorded for polymers of the series F. Cf. Figure 3 for further comments.

saturated carbon sites can also react with free radicals ($CF_3$) or with impurities (H,O,...). In the $CF_4$ discharge, etching of the growing film may also occur. The decrease of the fluorine to carbon content with increasing $CF_4$ concentration in the gas phase suggests the etching of the surface of the growing PTFE thin film. Further, the idea that the mechanism III plays an important role is supported by investigations involving the effect of the distance dx from the discharge axis. In the series F, the sputtering of the cathode and the interactions in the plasma do not change but the larger is the distance dx, the smaller is the bombardment of the growing film by energetic species. In agreement with the results of previous works[2,4,6,9], films with a composition closer to PTFE (F/C = 2) are deposited at larger distance from the cathode axis. In this case, the deposition rate decreases rapidly with increasing dx.

From the XPS measurements, we detect three main kinds of carbon groups : the $CF_2$ species we associate with the polymer network, the $CF_3$ groups we imagine terminating the chains, and the $CF_{1.0}$ components indicating branching or cross-linking points.

From all our results (Table VI), it appears that intensity of the $CF_3$ (chain terminators) decreases as the discharge power increases and as the pressure decreases. This fact is in good agreement with other work[27] which reports that the degree of polymerization increases with an increase in the power-to-pressure ratio (W/P).

The highest $CF_2$ (Teflon like) content is measured, in our series of experiment, for higher bias voltage, lowest $CF_4$ content in the gas, in the high pressure range, and especially at larger distance from the discharge axis.

## Sputtering versus Plasma Polymerization

If we compare stoichiometry and composition of fluoropolymer films prepared by RF sputtering (this work) and by plasma polymerization, we note that by the latter method polymers with high[2,4] or low[6,10,11] fluorine to carbon ratio could be prepared : integrated peak intensities corresponding to $CF_x$ formula of X - 2.06 to $2.36^2$, X = 0.96 to $2.18^6$, X = 0.94 to $1.3^{10}$ and X = 0.75 to $1.43^{11}$ were measured. Polymer films prepared by sputtering did show an average $CF_x$ value of 1.42 (the extremes being 1.26 and 1.98).

When these data are available, we compare also the trends in the relative C1s intensities attributed to various groups; except for ref. 2, the $CF_3$, $CF_2$, CF and CH species count generally for about 20 % each of the total C1s intensity; the authors interpreted this as an indication of a high degree of branching and cross linking. Table VI shows that the sputtered polymeric films contain an average of more than 40 % $CF_2$-Teflon like species and about 20 % $CF_3$ moieties. These sputtered-grown polymers contain undoubtly more $CF_2$ species, attributable to carbon atoms forming the polymer network; as a consequence, the amount of branching and cross linking should be reduced.

Our work points out that the sputtering is a powerful method for depositing films of polymer with stoichiometry and degrees of polymerization, branching and cross linking varying progressively. Films are synthetized in a reproductible way controlling every parameter of the discharge. Films like Teflon (F/C = 1.98) are produced at large distance $d_x$ of the discharge axis, whereas films with ratio of F/C=1.26 are deposited in a $CF_4$ discharge.

## Polymer Stability

As mentioned above, the polymer films studied were absolutely free of any oxygen contamination, despite the fact that they have been transported in air between the sputtering vessel and the XPS spectrometer. One sample (E.2 code) was, after the XPS analysis, submitted for three days to a heating at 150°C in air and reanalyzed by the XPS technique. It showed a slight O1s signal estimated to correspond to an effective chemical composition O/C = 1/13, i.e., about 8 % of the carbon atoms were "oxidized". No new signal appeared in the C1s region, but some intensity loss is observed from the $CF_3$ and $CF_2$ (Teflon) regions, together with an increase of the $CF_2$, CF ones (see Table VI).

## Valence Band Information

Polymers can not only be studied by the XPS technique through their core level spectra, but it has also been shown that the valence band spectra contain a lot of information which is unique to characterize some compounds[28]. But the accumulation of these valence band spectra is very difficult, and the interpretation of these data can actually be made only for well defined compounds, as it rests on theoretical calculations and on XPS study of model molecules. Figure 10 shows nevertheless as an example that such a valence band spectrum recorded for the polymeric film E-1 is profoundly different from that of pure PTFE. This suggests that in the future some additional information would perhaps be gained from the analysis and interpretation of this kind of data.

## CONCLUSIONS

In this first systematic XPS analysis of polymeric films grown by RF sputtering from a bulk polytetrafluoroethylene cathode, we have shown that it is possible to synthetize compounds whose characteristics - at the surface level - (chemical composition, fluorine to carbon stoichiometry, cross linking and branching) can be varied over a wide range of values, depending on the preparation parameters (discharge power, gas pressure, gas composition, cathode to substrate distance) used. Noteworthy is the fact that films similar to PTFE have been grown at large distance from the discharge axis. All the prepared polymers are free of any oxygen

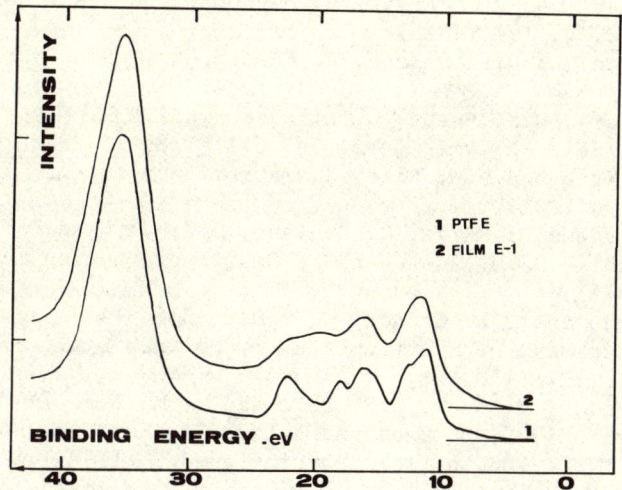

Figure 10. Valence band spectra of polytetrafluoroethylene and of sample E1.

(and argon) contamination.

As for the preparation mechanism, we have shown that our analysis is compatible with the following : the cathode is sputtered as $C_xF_y$ species which interact between themselves but also with the discharge gas. As a consequence, polymerization and dissociation compete, depending on the pressure range, gas composition, and discharge power. Finally, the interaction between energetic species occurs mainly on the substrate where fluorine loss (and thus cross linking and branching) is caused by the bombardment of the growing film with energetic species.

ACKNOWLEDGEMENT

One of us (J.J.P.) is grateful to the Belgian NFSR for financial support.

REFERENCES

1. J.J. Verbist, "Quantum Theory of Polymers", J.M. André, J. Delhalle and J. Ladik, Editors, p. 31, Reidel Publ. Co., Dordrecht, 1978;
2. H.R. Anderson, Jr., F.M. Fowkes and F.H. Hielscher, J. Polymer Sci. Polymer Phys. Ed., 14, 879 (1976).

3. D.T. Clark and W.J. Feast, J. Macromol. Sci. Rev. Macromol. Chem. 12, 2, 191 (1975).
4. D.W. Rice and D.F. O'Kane, J. Electrochem. Soc. 123, 1308 (1976).
5. D.T. Clark, W.J. Feast, D. Kilcast and W.K.R. Musgrave, J. Polymer Sci. Polymer Chem. Ed., 11, 389 (1973).
6. D.F. O'Kane and D.W. Rice, J. Macrom. Sci. Chem., A10 (3), 567 (1976).
7. J.J. Pireaux, J. Riga, R. Caudano, J.J. Verbist, J. Delhalle, S. Delhalle, J.M. André and Y. Gobillon, Physica Scripta 16, 329 (1977).
8. P. Cadman and G.M. Gossedge, J. Mat. Sci. 14, 2672 (1979).
9. D.T. Clark and D. Shuttleworth, J. Polymer Sci. Polymer Chem. Ed., 18, 27 (1980).
10. M.R. Pender, M. Shen, A.T. Bell and M. Millard in "Plasma Polymerization", M. Shen and A.T. Bell, Editors, ACS Symposium Series, Vol. 108, p. 147, Washington D.C. (1979).
11. K. Nakajima, A.T. Bell, M. Shen and M.M. Millard, J. Appl. Polymer Sci., 23, 2627 (1979).
12. E. Kay, J. Coburn and A. Dilks, IBM Res. Rep. RJ 2584 - 7/18/79.
13. G.C.S. Collins, A.C. Lowe and D. Nicholas. Europ. Polymer J. 9, 1173 (1973).
14. I.H. Pratt and T.C. Lausman, Thin Solid Films, 10, 151 (1972); M. White, Thin Solid Films, 18, 157 (1973); W. De Wilde, Thin Solid Films, 24, 101 (1974).
15. L. Holland, H. Biederman and S.M. Ojha, Thin Solid Films, 35, L19 (1976).
16. H.W. Lehmann, K. Frick, R. Widmer, J.L. Vossen and E. James, Thin Solid Films, 52, 231 (1978).
17. A. Dilks and E. Kay, Macromolecules 14, 855 (1981).
18. H. Yasuda, Appl. Polymer. Symp. 22, 241 (1973).
19. A.F. Stancell and A.T. Spencer, J. Appl. Polymer. Sci. 16, 1505 (1972)
20. J.P. Dauchot, A. Hecq and M.T. Grogna in"Proc. 3rd European Hybrid Microelectronic Conference, Avignon, France 1981" (Intern. Society for Hybrid Microelectronics - Europe).
21. W. Murray, "Numerical Methods for Unconstrained Optimization," Academic Press, New York, 1972; F. James and M. Roos, Computer Physics Comm. 10, 343 (1975).
22. A. Dilks and E. Kay in "Plasma polymerization". M. Shen and A.T. Bell, Editors. ACS Symposium series, Vol. 108, Washington D.C. (1979).
23. J.W. Coburn and E. Kay in"Proceed. 7th Intern. Vac. Congr. 3rd Intern. Conf. Solid Surfaces", Vienna 1977, 1257.
24. S. Matsuo and Y. Takehara, Jpn. J. Appl. Phys. 16, 175 (1977).
25. M. Biederman, S. Ojha and L. Holland, Thin Solid Films, 41, 329 (1977).
26. D.T. Clark and A. Dilks, J. Polymer. Sci. Polymer. Chem. Ed. 16, 936 (1978).

27. J. Castonguay and A. Theoret, Thin Solid Films, 69, 85 (1980).
28. J.J. Pireaux, J. Riga, R. Caudano and J. Verbist in "Photon, Electron and Ion Probes of Polymer Structure and Properties," D.W. Dwight, T.J. Fabish and H.R. Thomas, Editors, ACS Symposium Series No. 162, p. 169, Washington D.C., 1981.
29. J.W. Coburn, L.W. Eckstein and E. Kay, J. Vac. Sci. Technol. 12, 151 (1975).

LOCALIZED AUGER STATES IN POLYETHYLENE

J. A. Kelber, R. R. Rye, D. R. Jennison and
J. C. Houston
Sandia National Laboratories
P. O. Box 5800
Albuquerque, New Mexico  87185

The final states of a core-valence-valence Auger decay involve two valence level holes which may remain localized on one atomic site or functional group or move as independent particles through the system. The x-ray excited Auger spectrum of polyethylene indicates the existence of both localized and independent particle-like Auger final states. A comparison of the polyethylene Auger spectrum to those of the gas phase normal alkanes shows that certain spectral features have a final state binding energy which is independent of chain length, indicating the presence of localized states. A broadening of the alkane spectra with increasing chain length indicates, however, that some final states exhibit independent particle behavior. The width of the polyethylene Auger spectrum is significantly wider than the self-fold of the valence band density of states, demonstrating the partial breakdown of the independent particle model and providing additional evidence for the existence of localized states in polyethylene.

## INTRODUCTION

An Auger core-valence-valence (CVV) decay (denoted as KVV when the core hole is in the K shell) involves a transition from an initial state with a single hole in a core electron energy level to a final state with (at least) two valence level holes. In one limit, these two holes may move in an uncorrelated manner in ground state-like molecular orbitals (i.e., the independent particle approximation). Alternatively, in the limit of strong correlation the two holes may be localized on the same atomic site for times long compared to single-hole hopping times, although they may move together from site to site (a localized final state). While the question of final state hole-hole interaction is of interest to molecular physics, the issue is also of practical importance to the interpretation of Auger spectra. Auger electron spectroscopy has proven to be an effective tool for obtaining local chemical information especially by analysis of the Auger spectral lineshape.[1,2] Recent work has shown that the Auger spectra of polyethylene and poly(ethylene oxide) can be interpreted using the "fingerprint" spectra of analogous gas phase species.[3] More detailed information concerning the local valence electronic structure may in principle be obtained from a comparison of experimental and theoretical Auger spectra. Accurate calculations, and therefore accurate assignments of Auger transitions require an adequate knowledge of final state hole-hole interaction. Because of the large number of theoretical and experimental investigations of the electronic structure of polyethylene,[4] we concentrate on this system and in this report summarize the existing experimental evidence for correlated (localized) two-hole final states in polyethylene; that is, two-hole states which are localized as a specific result of hole-hole correlation, even though the one-hole spectrum is well given by independent particle theory. A more detailed account of this work will be given elsewhere.[5]

The nature of final state hole-hole interactions may be examined by comparing the experimentally determined hole-hole interaction ($U_{eff}$) with that predicted by independent particle calculations. In particular, one can compare the experimental Auger lineshape to theoretical predictions for a series of systems in which electronic properties are systematically varied. In the independent particle approximation, $U_{eff}$ is defined, for gas phase systems, by the relationship[2b]

$$E_k = I_c - I_j - I_k - U_{eff} \tag{1}$$

where $E_k$ is the kinetic energy of the Auger electron, $I_c$ is the core electron ionization energy; and $I_j$ and $I_k$ are valence electron ionization energies. Within this approximation the use of experimental values for $I_c$, $I_j$ and $I_k$ (Equation (1)) allows a straightforward experimental determination of $U_{eff}$. Since one-electron relaxation effects are contained in the experimental ionization potentials,[2b] $U_{eff}$ should be a measure of the deviation from independent particle behavior. For uncorrelated final state hole motion, $U_{eff}$ should decrease with increasing molecular size.[2,6] As will be shown below, no such decrease is observed for the gas-phase linear alkanes or polyethylene.

## RESULTS

The electronic properties of the alkanes vary with increasing chain length in such a way that for orbitals of corresponding symmetry in different molecules, $I_c - I_j - I_k$ (Equation (1)) is varied by no more than ∼3% from methane through pentane.[7,8] This observation, coupled with the fact that major features in the Auger spectra of the alkanes occur at kinetic energies which are constant with increasing chain length, implies a value of $U_{eff}$ which is also constant with increasing chain length[9] (Equation (1)). This has been cited as evidence for the existence of localized two-hole states in the normal alkanes.[9]

The Auger spectrum of polyethylene, corrected for Auger electron energy losses to the solid,[3,5,10] is an extrapolation of those trends observed in the alkane series from methane through hexane.[3,5] The "raw" and loss-corrected polyethylene Auger spectra are compared in Figure 1. As shown in Figure 2, the loss-corrected spectrum is now directly comparable to the spectra from gas phase alkanes. The spectra in Figure 2 are plotted on a common two-hole binding energy (BE) scale, where

$$BE = I_c - E_k = I_j + I_k + U_{eff}. \qquad (2)$$

Such a scale allows one to directly compare transitions to corresponding final states in different systems. The polyethylene spectrum corresponds closely to those of the alkanes in terms of spectral lineshape and energy, and this similarity strongly suggests that correlated two-hole final states also exist in the polymer as well as in the alkanes. For example, the methane, neopentane and n-pentane spectra (Figure 2) display a dominant feature at ∼40 eV binding energy and two smaller features at ∼54 eV and ∼63 eV. The polyethylene spectrum (Figure 2) shows similar features shifted by ∼3 eV due to the dielectric response of the solid.[8] The electronic properties of the alkanes are such that the occurrence of a spectral feature at a constant binding energy with increasing chain length implies a $U_{eff}$ which is constant with

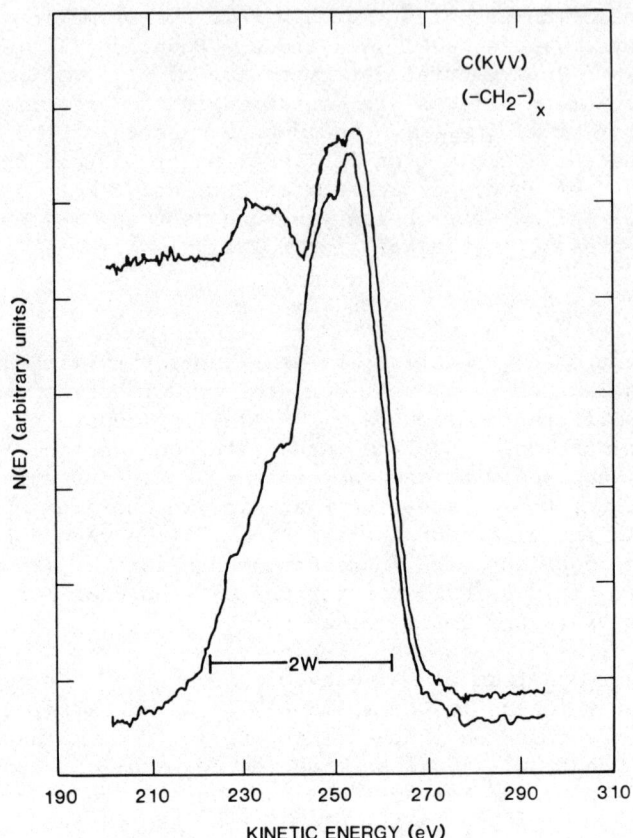

Figure 1. X-ray excited Auger spectrum of polyethylene. Top trace: 'raw' data with secondary electron background subtracted (Ref. 3,5). Bottom trace: Auger spectrum corrected for electron energy loss functions to the solid (Ref. 10). The corrected spectrum has a width of ~50 eV, compared to the 40 eV predicted by independent particle theory (see text).

increasing molecular size, in contradiction to the predictions of the independent particle model.[2,6] For independent particle-like behavior, the spectral features in Figure 2 should shift steadily to lower binding energy with increasing molecular size. For polyethylene the shift should be ~14 eV relative to methane.[2,6,9] The observed shift is only 3 eV and is due to other causes.[8] In Figure 2, all the spectra except methane display a feature just

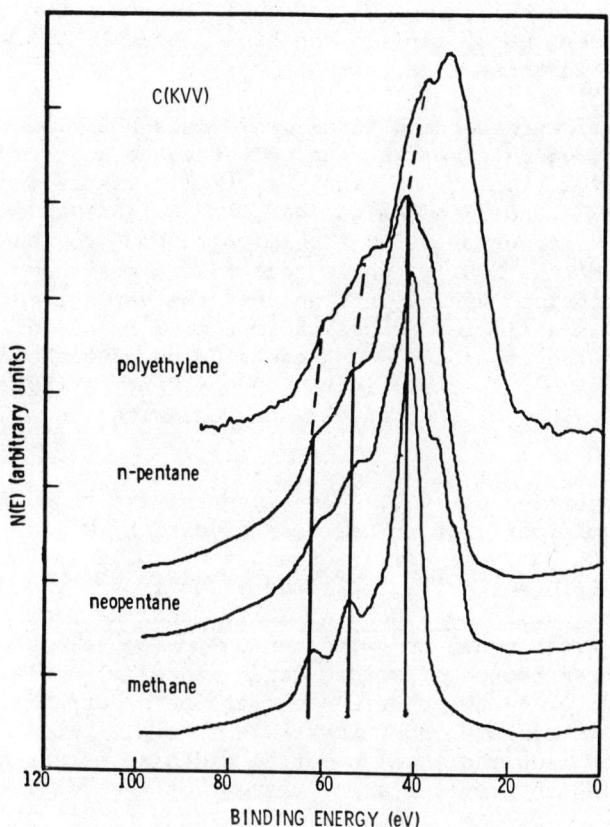

Figure 2. Comparison of the loss-corrected Auger spectrum of polyethylene with the spectra of gas-phase methane, neopentane and n-pentane. The polyethylene spectrum is similar in lineshape and energy to those of the gas phase alkanes. The ∿3 eV shift of features in the spectrum relative to corresponding features in the alkane spectra, emphasized by the vertical solid and slanting dashed lines, is due to the dielectric response of the solid (see Ref. 3).

to the right (lower binding energy) of the feature at ∿40 eV. The intensity of this feature increases relative to that of the 40 eV feature as the alkane chain length increases.[9] In polyethylene, this feature dominates the spectrum. This is consistent with independent particle-like behavior.[5,9] Thus a comparison of the alkane and polyethylene spectra strongly suggests the presence of both independent particle-like and highly correlated two hole final states in these materials.

Information concerning final state hole-hole interactions may also be obtained by comparing the polyethylene Auger spectrum to the self-convolution of the single-particle valence density of states (DOS). Lander suggested that, in the independent particle limit, the CVV Auger spectrum should correspond to the self-convolution of the DOS.[11] While recent work has shown that a close comparison requires taking into account the variation of Auger transition probabilities across the spectrum[12] and initial state screening of the core hole,[13] these studies have shown that the width of the Auger spectrum in the independent particle limit is equal to the width of the convolution, or twice the width of the DOS.

The distortion of a CVV Auger spectral lineshape due to final state hole-hole correlation has been studied by Cini[14] and Sawatzky and Lenselink.[15] These workers found that for a sufficiently large local interaction, a localized state is split off from the bandlike portion of the spectrum. This bandlike region, however, still has a width equal to twice the DOS or one-electron bandwidth, so the presence of strong final state hole-hole correlation leads to an overall two-particle excitation spectrum wider than that predicted by independent particle theory. "Width" is understood here to mean the total spectral width as measured from the background, which provides a measurement of the total single particle (or 2- particle) DOS.[12,13] A large number of overlapping final states, as in polyethylene,[16] may prevent the resolution of individual transitions to localized states, but unless all Auger amplitude goes into the localized state one would still have a spectrum of greater overall width than that predicted by independent particle theory. The valence band photoemission spectrum of polyethylene has been obtained by several groups who have used x-ray[17,18] and ultraviolet[19] radiation as an excitation source. These results are in good agreement with each other and with semi-empirical[17] and ab initio[3,17] calculations, and show the width W of the valence band spectrum to be $20 \pm \sim 1$ eV wide. If all the final state holes in polyethylene are uncorrelated, the experimental Auger spectrum should then be $2W \sim 40$ eV wide (see Figure 1). As can be seen in Figure 1, the actual width is at least 50 eV, or twenty-five percent wider than the 40 eV prediction of independent particle theory. In view of the agreement between the experimental spectral width and the self-convolution of the DOS for other systems,[12,13] this large discrepancy cannot be ascribed to greater vibrational broadening in the two-hole relative to the one-hole spectrum.

## SUMMARY AND CONCLUSION

Though the presence of localized two-hole final states in polyethylene is indicated by both the shape and width of the Auger spectrum, many questions remain to be answered concerning the detailed nature of such states (e.g., lifetime, nature of the localization site, etc.). A motivation for further work in this area is the possibility that such states, which localize a large amount of electronic excitation energy on a specific site in the system, may provide a detailed mechanism for radiation induced desorption and dissociation in covalent materials analogous to the Knotek-Feibelman mechanism[20] in ionic solids. If this is the case, ion emission thresholds for electron and photon-stimulated desorption should occur at energies corresponding to excitation of the two-hole localized state via shake-up, direct excitation or decay from a core hole.

## REFERENCES

1. R. R. Rye, T. E. Madey, J. E. Houston and P. H. Holloway, J. Chem. Phys., 69, 1504 (1978); R. R. Rye, J. E. Houston, D. R. Jennison, T. E. Madey and P. H. Holloway, Ind. Eng. Chem. Prod. Res. Dev., 18, 2 (1978).
2. a) D. R. Jennison, Phys. Rev., A23, 1215 (1981)
   b) D. R. Jennison, Chem. Phys. Lett., 69, 435 (1981).
3. J. A. Kelber, R. R. Rye and G. C. Nelson, J. Vac. Sci. Tech., 18, 712 (1981).
4. D. Bloor, Chem. Phys. Lett., 40, 323 (1976).
5. J. A. Kelber, R. R. Rye, G. C. Nelson and J. E. Houston, Surf. Sci., 116, 148 (1982). The experimental work referred to in this paper was first presented at the International Symposium on Physicochemical Aspects of Polymer Surfaces (New York, August, 1981). A paper which includes a brief account of this experimental work plus a theoretical explanation has subsequently been published; J. A. Kelber and D. R. Jennison, J. Vac. Sci. and Tech., 20, 848 (1982).
6. D. R. Jennison, J. Vac. Sci. Tech., 17, 172 (1980).
7. A. D. Baker, D. Betteridge, N. R. Kemp and R.E. Kirby, J. Mol. Structure, 8, 75 (1971).
8. J. J. Pireaux, R. Caudano, S. Svensson, E. Basilier, P. A. Malquist, V. Gelius and K. Siegbahn, J. de Physique, 38, 1221 (1977); J. N. Murrel and W. Schmidt, Faraday Trans. II, 68, 1709 (1972).
9. R. R. Rye, D. R. Jennison and J. E. Houston, J. Chem. Phys., 73, 4867 (1980).
10. H. H. Madden and J. E. Houston, J. Appl. Phys., 44, 3071 (1976).

11. J. J. Lander, Phys. Rev., 91, 1382 (1953).
12. P. J. Feibelman, E. J. McGuire and K. C. Pandey, Phys. Rev., B15, 2202 (1977); D. R. Jennison, Phys. Rev., B18, 6865 (1978).
13. D. R. Jennison, H. H. Madden and D. M. Zehner, Phys. Rev., B21, 430 (1980).
14. M. Cini, Surf. Sci., 87, 483 (1979):
    Sol. State Commun., 24, 681 (1977); 20, 605 (1978).
15. G. A. Sawatzky, Phys. Rev. Lett., 39, 504 (1977);
    G. A. Sawatzky and A. Lenselink, Phys. Rev., B21, 1790 (1980).
16. The number of two-hole final states increases rapidly with alkane chain length. See Ref. 9.
17. J. Delhalle, J. M. Andre, S. Delhalle, J. J. Pireaux, R. Caudano and J. J. Verbist, J. Chem. Phys. 60, 595 (1974).
18. M. H. Wood, M. Barber, I. H. Hillier and J. M. Thomas, 56, 1788 (1972).
19. K. Seki, S. Hashimoto, N. Sato, Y. Harada, K. Ishii, H. Inokuchi and J. Kanbe, J. Chem. Phys., 66, 3644 (1977).
20. D. R. Jennison, Phys. Rev., B18, 6865 (1978).
21. M. L. Knotek and P. J. Feibelman, Surf. Sci., 90, 78 (1979).

THE INFLUENCE OF A SUBSTRATE ON THE SURFACE CHARACTERISTICS OF SILANE LAYERS

Hatsuo Ishida, Somsak Naviroj, and Jack L. Koenig

Department of Macromolecular Science
Case Western Reserve University
Cleveland, Ohio 44106

We have studied the structure of γ-methacryloxypropyltrimethoxysilane (γ-MPS) interphase as well as the glass/γ-MPS interface using FT-IR. Multilayer formation of γ-MPS on the surface of E-glass fibers was reconfirmed and the results are in good agreement with the work using $^{14}$C-labeled γ-MPS. The amount of γ-MPS adsorbed increases linearly with concentration in solution. There is a sudden change in the amount of adsorbed γ-MPS at around 0.4% by weight. This transition is associated with the onset of incomplete dispersion of the coupling agent in the solution. There is a significant amount of γ-MPS in the outermost layers that can be washed away by organic solvents. The more tightly bound layers of γ-MPS undergo, in addition to the siloxane reaction, surface induced homopolymerization of acrylic polymer and loss of most of the organofunctionality. The surface induced homopolymerization arises due to the surface catalytic effect of the acid centers that strongly interact with the carbonyl groups of the γ-MPS. The area occupied by a single γ-MPS molecule at the surface is experimentally determined to be 48Å$^2$ which is in fair agreement with the expected size of the molecule.

## INTRODUCTION

Reinforcement mechanisms of fiberglass reinforced plastics (FRP) have been in dispute for many years. Molecular aspects of the mechanisms have been reviewed[1] and, among many theories, the chemical bonding theory appears to be the most useful concept that explains a wide variety of experimental observations. When thermosetting plastics are used as matrices, chemically reactive silane coupling agents generally produce FRP superior to that made from unreactive silanes.[2] The effect of compatibility between silane coupling agents and matrices is minimal in thermosets as compared to the case of thermoplastics where compatibility seems to be one of the important reinforcement factors. Although the possibility of chemical reaction between silane coupling agents and thermoplastics has been postulated,[3-5] the concentration of chemically bonded silanes may be very small. In general, the usefulness of coupling agents is better realized in thermosetting resins.

A question remains as to the number of bonds necessary to produce a sufficient reinforcement effect. Also, the nature of the effective chemical bond remains undetermined. Recently, interpenetrating networks in the silane interphase have been proposed to be one of the effective reinforcement mechanisms.[6] The function of interpenetrating networks may be similar to that of chemical bonding in that molecular interlocking provides an effective connection between the silane layers and the matrix resin. However, no experimental evidence of interpenetrating networks has yet been reported.

One of the useful coupling agents for unsaturated polyesters is γ-methacryloxypropyltrimethoxysilane (γ-MPS). We have reported that γ-MPS can homopolymerize prior to the polymerization with the matrix resin. The degree of polymerization depends on the thermal history of the silane adsorbed on the glass surfaces. Upon homopolymerization of the γ-MPS, it loses the functional $C=C$ groups that can copolymerize with the matrix. Formation of chemical bonds between the silane and the matrix resin is not possible at normal conditions when homopolymerization of the silane is complete. It has been observed that the major part of the $C=C$ groups of γ-MPS on E-glass fibers polymerize in air at room temperature in one month without the matrix resin.[7] Nevertheless, the glass fiber/γ-MPS system that has been stored in air for many months produces an FRP similar to freshly treated glass fibers.[8] Small concentrations of the residual $C=C$ bonds of the γ-MPS adds further strength to reinforcement by chemical bonds.

Another observation concerning the γ-MPS interphase is that the outermost layers can be washed away by organic solvents indicating that only physically adsorbed layers exist.[9] An obvious consequence is that the dispersed oligomers or microgel particles in the matrix phase near the silane layers will modify the properties of the matrix, which forms the interphase. The modes of polymerization of the matrix resin are either a graft copolymerization at the surface of the particles, a random copolymerization with the silane, or interpenetrating network formation. Similar considerations can be extended to the more tightly bound silane layers.

It is the purpose of this study to elucidate detailed structures of the coupling agent interphase, especially the interphase made of γ-MPS on E-glass fibers. The carbonyl group of γ-MPS has very high specific absorptivity as well as a frequency sensitivity to its environments. Therefore, studies of γ-MPS on the glass fiber-silane interface are feasible by utilizing Fourier transform infrared spectroscopy.

## EXPERIMENTAL

A mat made of E-glass microfilaments (Crane glass, grade 50-01) was kindly supplied by Dr. F.J. Crane, Jr., Crane & Co., Massachusetts, and heat cleaned in air at 500°C for 24 hours prior to use. The coupling agent, γ-methacryloxypropyltrimethoxysilane (γ-MPS) was purchased from Petrarch Systems Inc., Levittown, Pennsylvania, and used as received.

Deionized distilled water was adjusted to pH = 3.5 by acetic acid and used to hydrolyze γ-MPS. The hydrolysis times were between one and three hours depending on the concentration of the silane. A slightly hazy appearance was noticed for the hydrolyzed γ-MPS solution of concentration above 0.5% by weight. The E-glass mat was immersed into the silane solution for one minute and the solution was then filtered out by an aspirator in order to improve reproducibility of silane adsorption experiments. This procedure was necessary because the solution remaining between the filaments deposits extra silane as water evaporates. The powder samples were centrifuged after silane treatment for one minute and the solution decanted. Silane-treated samples were stored in air at $21 \pm 1°C$ with relative humidity of $50 \pm 10\%$.

A Fourier transform infrared spectrophotometer (Digilab FTS-14®) was used at a resolution of 3 cm$^{-1}$ throughout the spectral range (3800 - 450 cm$^{-1}$) and with a dry air purge. Co-addition of 300 scans was a common condition except for the fibers with very little silane. In this case, 600 scans with dry nitrogen purge were used in order to improve the signal-to-noise ratio. All spectra are shown in the absorbance mode and the difference between the maximum and minimum absorbances are shown as ΔA in absorbance unit.

## RESULTS

Silane coupling agents adsorb onto the surface of glass fibers as multilayers.[7,10] The amount of adsorbed silane is a function of the concentration of silane treating solutions. Since the mechanical properties of FRP have some correlations with the amount of coupling agent present, it is important to determine the concentration dependence of the silane adsorption. Johannson $et\ al$[10] reported the first correlation using $^{14}$C-labeled γ-MPS.

We have collected nearly 40 data points and obtained the correlation as shown in Figure 1. The amount of γ-MPS on glass fibers was calculated from the integrated absorbance of the C=O group at 1720 cm$^{-1}$ using the measured integrated specific absorptivity of the C=O stretching mode at 1720 cm$^{-1}$ which was $(1.92 \pm 0.03)\ 10^7$ cm/mole for the unhydrolyzed γ-MPS. There is a transition around 0.4% by weight which coincides with the observation of the onset of a hazy silane solution. Above 1.0% by weight, the reproducibility of data becomes increasingly poor. A closer look at lower concentrations in range 0 - 0.1% by weight, which is shown in Figure 2, reveals another feature. At very low concentrations, the linearity is lost and the amount of adsorbed silane becomes almost independent of concentration. As shown later, these concentrations correspond to nearly a monolayer equivalent of silane on the surface.

The data reported by Johannson $et\ al$[10] for the $^{14}$C-labeled γ-MPS on E-glass fibers is replotted in Figure 3 where the dotted line is the result obtained by FT-IR. Considering the difference in the form of glass fibers as well as other factors such as sample purity and hydrolysis time, the agreement between the two techniques is excellent. The difference is certainly within the experimental variation of the data.

In earlier studies[9] it was observed that some of the coupling agent can be washed away by organic solvents. We have studied the amount of silane remaining on the surface of glass fibers as a function of the concentration of the silane treating solution and the results are shown in Figure 4 along with the result, which is illustrated as a dotted line, obtained in Figure 1. An interesting fact is that there is a relatively large scatter in the data for the washed sample reflecting extreme sensitivity of the process. The reproducibility below 1.0% by weight is generally good. At low concentrations, e.g., 0.2% by weight, very little silane can be washed away.

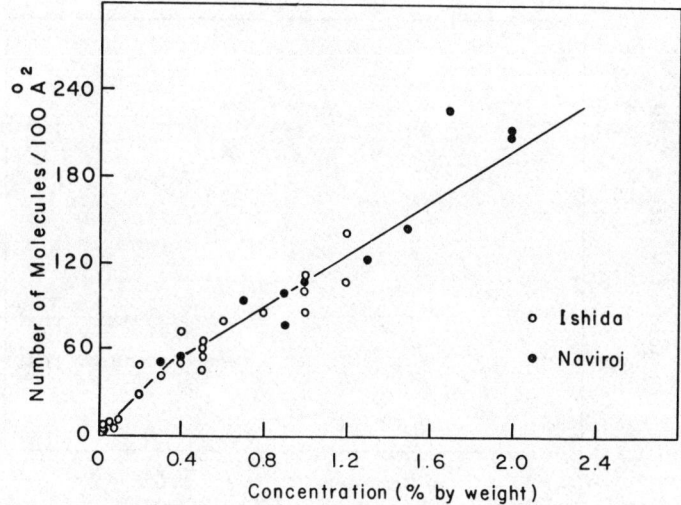

Figure 1. The number of γ-MPS molecules per 100Å² of E-glass surface as a function of the concentration of silane treating solution. The open and closed circles are the results obtained independently by two researchers (Ishida and Naviroj) using the same sample preparation procedure.

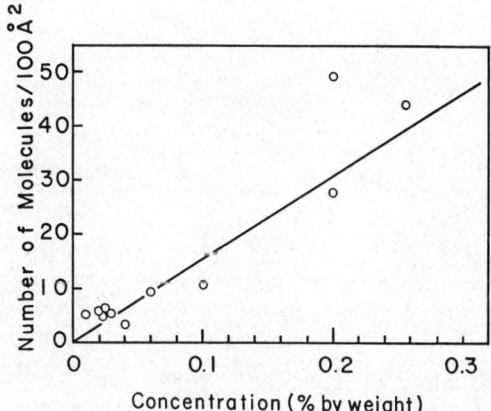

Figure 2. Magnified view of Figure 1 at lower concentrations. The solid line is drawn according to the determination in Figure 1.

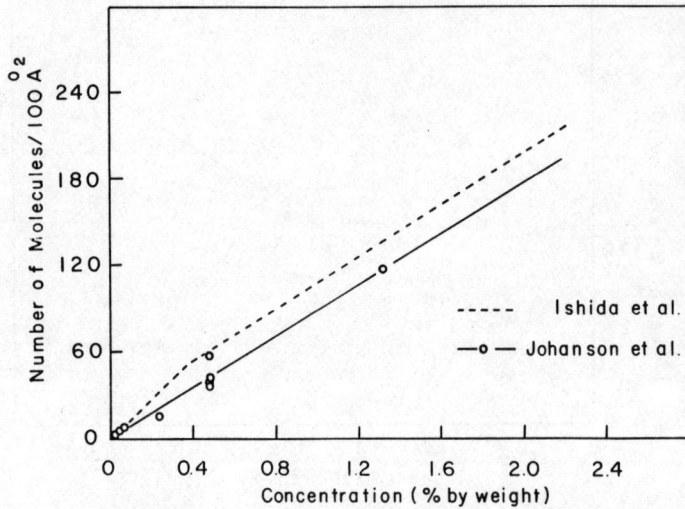

Figure 3. Comparison of the results on the adsorption of γ-MPS on E-glass fibers by radioisotope labeling technique and FT-IR. The solid line was drawn based on the radioisotope results which are shown as open circles, and the dotted line is the result by FT-IR shown in Figure 1.

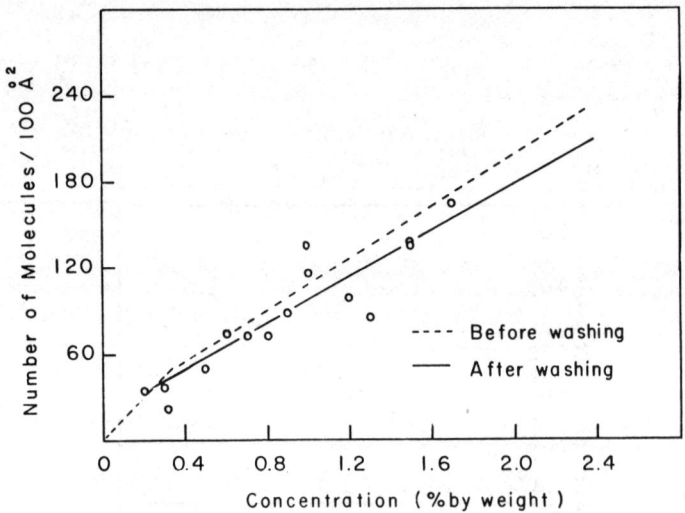

Figure 4. The amount of residual γ-MPS on E-glass fibers after washing with Tetrahydrofuran (THF) for five minutes. The dotted line is the result for the unwashed samples. The same samples used for Figure 1 were used for the measurements.

# SURFACE CHARACTERISTICS OF SILANE LAYERS

The FT-IR difference spectra, from which the absorbance contributions of E-glass fibers are excluded, are shown in Figure 5 in the region of the C=O stretching mode. The concentration of the treating silane solution is below 0.1% by weight, hence the poor signal-to-noise ratio. Three bands at 1719, 1699, and 1670 $cm^{-1}$ are observed. These bands are all attributed to the C=O groups with different environments. Previously, the bands at 1719 and 1699 $cm^{-1}$ were assigned to the non-hydrogen bonded carbonyls, respectively.[7] The band at 1670 $cm^{-1}$ is probably due to the C=O groups interacting strongly with the surface. Below 0.02% by weight, only the band at 1670 $cm^{-1}$ is observed and the frequency of the band decreased to 1660 $cm^{-1}$ for the 0.01% by weight sample. Also, the band at 1670 $cm^{-1}$ of the 0.02% by weight sample shifted to 1660 $cm^{-1}$ when the sample was dried an additional 24 hours in air at room temperature.

If the band at 1670 $cm^{-1}$ is unique to the C=O groups interacting with the glass surface, thicker layers should show either bands at 1719 or 1699 $cm^{-1}$. In other words, the carbonyl group can be used as a molecular probe to determine the concentration of the treating solution that yields a monolayer of silane on the surface.

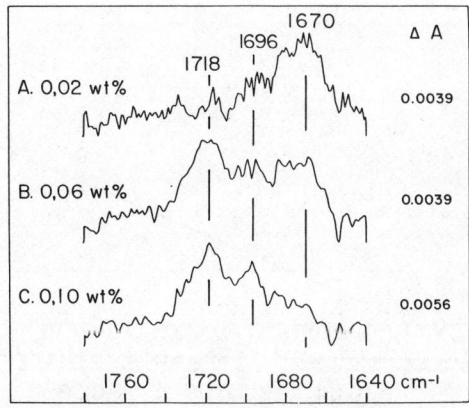

Figure 5. The FT-IR difference spectra of γ-MPS on E-glass fibers at very low concentrations. The spectra A, B, and C are obtained from the E-glass fibers treated with γ-MPS solutions at 0.02, 0.06, and 0.10% by weight.

The relative absorbances 1719 cm$^{-1}$/1670 cm$^{-1}$ are plotted in Figure 6 as a function of the concentration of the silane solution. When all carbonyl groups are interacting with the surface, the ratio 1719 cm$^{-1}$/1670 cm$^{-1}$ is zero. If the first layer covers the glass surface completely and the additional silane is only to increase the thickness beyond the second layer, the graph 1719 cm$^{-1}$/1670 cm$^{-1}$ versus concentration should result in a linear plot. The figure is approximately linear and the line intersects with the x-axis at the concentration of 0.016% by weight. Hence, this concentration yields a monolayer equivalent silane layer to a first approximation with respect to the C=O interaction with the glass surface.

Polymerization of the C=C group of γ-MPS can be followed by observing the frequency of the C=O vibration. Although the C=C stretching mode provides direct information, the rather low specific absorptivity of the C=C stretching mode and frequency overlap with the bending mode of water at 1635 cm$^{-1}$ make the study difficult. The non-hydrogen bonded carbonyl group at 1719 cm$^{-1}$ shifts to around 1740 cm$^{-1}$ upon polymerization. However, the polymerized γ-MPS gives rise to the hydrogen bonded C=O stretching frequency at around 1720 cm$^{-1}$. In this case, the band width at half height is wider than the unpolymerized one in Figure 7 where reversibility of the C=O frequency upon hydration-dehydration cycle and broadening upon polymerization are seen.

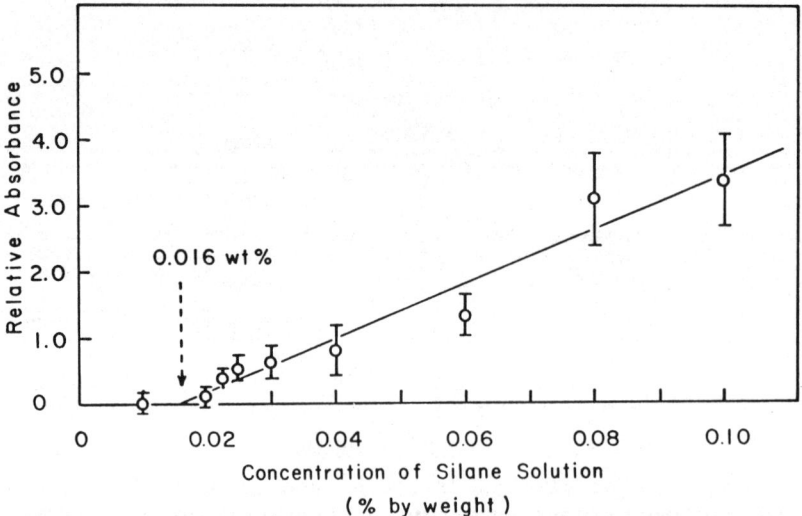

Figure 6. The relative absorbance of the 1718 cm$^{-1}$ band over the 1670 cm$^{-1}$ band as a function of the concentration of silane treating solutions.

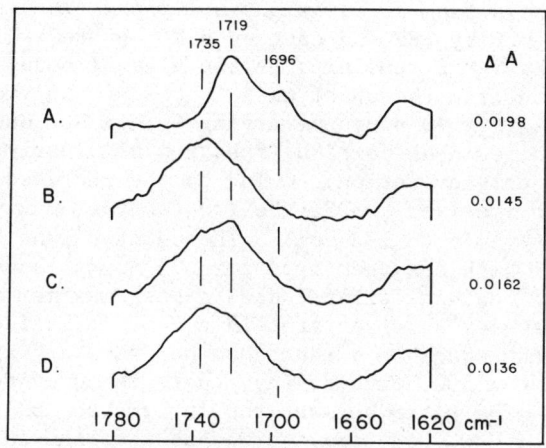

Figure 7. The FT-IR difference spectra of γ-MPS adsorbed from 0.5% by weight solution. A: dried at room temperature for one hour. B: heat treated at 130°C in air for one hour. C: exposed to water vapor at 130°C for 15 minutes. D: heated again at 110°C in air for 30 minutes.

With these facts in mind one can study the state of polymerization of the γ-MPS interphase at various aging times. Illustrated in Figure 8 are the C=O stretching regions of the γ-MPS adsorbed

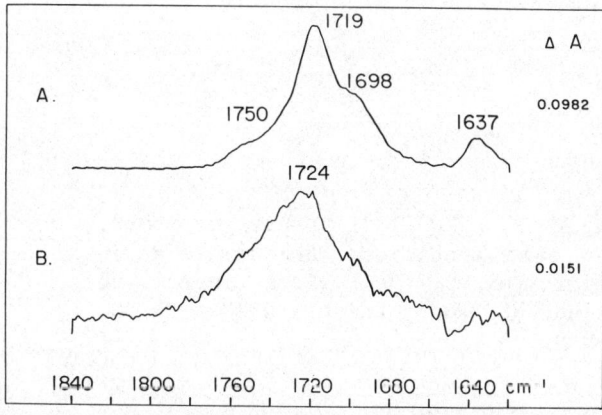

Figure 8. The FT-IR spectra of γ-MPS adsorbed from 1.0% by weight solution and dried at room temperature for 90 hours in air. A: the silane-treated E-glass fibers were washed with THF and the THF-soluble components were cast on a KBr plate which is evacuated at 80°C for one minute. B: the residual silane on E-glass fibers after the THF wash. (Difference spectrum.)

onto E-glass fibers from 1% by weight solution. The spectrum A is the washed γ-MPS by THF and cast on a KBr plate, while the spectrum B is the γ-MPS remaining on the glass fibers after washing. Judging from the spectrum A, almost no polymerization has taken place after 90 hours of drying in air at room temperature whereas the residual portion (spectrum B) clearly shows a sign of partial polymerization. Owing to the increased contribution of the polymerized γ-MPS, the frequency at maximum height shifted from 1719 $cm^{-1}$ to 1724 $cm^{-1}$. The aging effect is clearer when the sample which has been aged for 500 hours is used (Figure 9). Now the residual silane shows a peak maximum at 1730 $cm^{-1}$ with only a weak shoulder at 1719 $cm^{-1}$. It is interesting to notice that the washable silane does not show any aging effect. In fact, even after 500 hours drying, there is no evidence of polymerization. The ratio between the integrated absorbances of the C = O mode and C = C mode for the washable silane is 0.0995, whereas the unwashable silane gives the ratio of 0.0425. It is estimated that approximately 57% of the C = C bonds are polymerized at this stage.

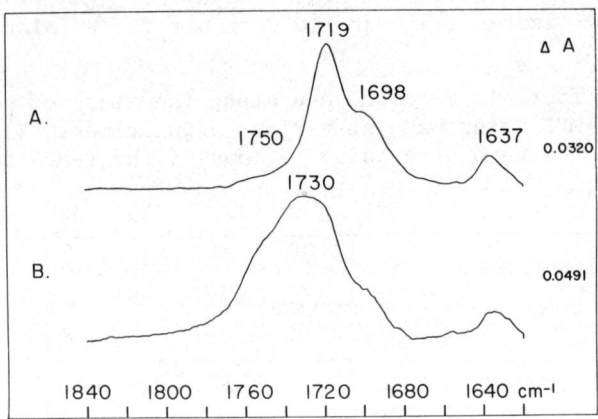

Figure 9. The FT-IR spectra of γ-MPS. The same conditions and sample as in Figure 8 except the E-glass fibers were dried at room temperature for 500 hours.

## DISCUSSION

The amount of silane coupling agent required to yield a well defined monolayer[11] is difficult to determine for γ-MPS due to its conformational flexibility. Nonetheless, the term "monolayer equivalence" is of theoretical and practical interests. One can define a monolayer equivalence as the maximum amount of molecules which adsorb onto the surface with specific interactions that are unique to the substrate-adsorbent interface. From Figure 6, the concentration which yields the monolayer equivalence has been determined to be 0.016% by weight. From Figure 1 (or Figure 2), one can determine the number of molecules per $100\text{Å}^2$ corresponding to this concentration. It is worth mentioning that the specific absorptivity of the carbonyl group may vary when it interacts strongly with the surface, thus the amount calculated for the silane is potentially in error. When the thickness of the silane increases, the relative contribution of the first layer decreases. Extrapolation from higher concentrations will yield results which are not affected by the possible changes in the specific absorptivity since the γ-MPS in the outer layer are the same. An assumption introduced is that the plot at lower concentrations (0.02 - 0.4% by weight) is linear. Since the specific interaction between the silane and the surface does not extend beyond the second layer, this assumption seems reasonable. The relationship obtained by the least squares curve fitting in the concentration range described above gives 2.1 molecules per $100\text{Å}^2$, that is $48\text{Å}^2$ per molecule. Johannson et al[10] assumed the area occupied by one γ-MPS molecule on the glass surface to be $100\text{Å}^2$. However, no basis for this value was reported. If one assumes that all carbonyl groups in the first layer are hydrogen bonded to the glass surface, the conformational freedom is more restricted than for the free chain. A rough estimate of the area occupied by a molecule assuming an all trans conformation in the propyl chain of γ-MPS is approximately $50\text{Å}^2$. The agreement between the observed and calculated values is good. It is important to note that this area may not apply to the γ-MPS beyond the second layer because there are different intermolecular interactions in those layers that may vary the conformational freedom and the area contributed per molecule.

The extrapolated value, 2.1 molecules per $100\text{Å}^2$, is compared to the observed value of 4 - 5 molecules per $100\text{Å}^2$ which seems to be almost constant below 0.015% by weight. The observed value is calculated by using the specific absorptivity of the C = O group in the solution state. It is possible that the integrated specific absorptivity of the adsorbed carbonyl group is different from that in the solution. It is well known that surface species which are strongly interacting with solid surfaces often show an increase in their specific absorptivities. Hence, the apparent number of molecules per unit area calculated by the specific absorptivity which is obtained using solution value would yield higher values. Tensmeyer et al[12] and Hoffmann et al[13] studied specific absorptivities of 2,5-hexanedione and 2,5,8-nonanetrione in solution as well as on calcium montmorilonite surface as a monolayer or less. Both compounds showed increases in specific absorptivity by factors of 1.94 and 1.83 for 2,5-hexanedione and 2,5,8-nonanetrione, respectively, at low surface coverages. If these values are used as correction factors, the previously obtained 4 – 5 molecules per 100Å$^2$ reduces to 2 – 2.5 molecules per 100Å$^2$ which is in good agreement with the extrapolated value for the monolayer equivalence.

Tensmeyer et al[12] and Hoffmann et al[13] also reported that the C = O stretching frequency of the compounds described previously reduced from 1740 cm$^{-1}$ to 1690 cm$^{-1}$ upon adsorption. Reduction of the C = O frequency by about 50 cm$^{-1}$ is similarly observed in the case of γ-MPS ON E-glass fibers. Reduction of C = O stretching frequencies on inorganic oxide surfaces are commonly observed.[14,15] Sheppard and others reported the frequency decrease of the 1724 cm$^{-1}$ band of acetaldehyde to 1689 cm$^{-1}$ upon adsorption.[14] They explained that surface catalyzed keto-enol tautomers exist on the surface and caused the frequency reduction. Although regular keto-enol tautomerism does not exist in the case of γ-MPS due to the lack of C$^5$ hydrogen (C$^1$ being the carbon closest to the silicon), a frequency reduction by strong hydrogen bonding and subsequent stabilization by electron delocalization may be possible.

It should be mentioned that regular hydrogen bonded γ-MPS shows the C = O stretching frequency at 1699 cm$^{-1}$. Hence, it is assumed that a stronger hydrogen bonding structure exists in addition to the regular C = O ···· HOSi$\leq$ type bonding. The Lewis acid centers are possible candidates for the strong hydrogen bonded sites. However, the Lewis acid centers are converted to the Brönsted acid centers since γ-MPS is adsorbed from an aqueous solution. The Brönsted acid centers act as strong proton donors, and the C = O groups are then strongly hydrogen bonded.

The band at 1720 cm$^{-1}$ might arise from the hydrogen bonded, polymerized γ-MPS whose structure is similar to poly(methyl methacrylate). Adsorbed poly(methyl methacrylate) on silica shows a frequency decrease from 1736 cm$^{-1}$ to 1714 cm$^{-1}$ due to hydrogen bonding with the surface silanols.[16]

The surface of E-glass fibers is complex but is believed to have Lewis acid centers and Brönsted acid centers depending upon the history of the surface. One of the useful methods to examine Lewis acid centers is the adsorption of hexachloroacetone.[17] The adsorbed hexachloroacetone shows three carbonyl bands at 1692, 1680, and 1585 cm$^{-1}$. These are assigned to the non-hydrogen bonded and hydrogen bonded C=O groups for the first two bands and the C=O which is interacting with the Lewis acid center for the 1585 cm$^{-1}$ band. Similar measurements were made of hexachloroacetone on aluminosilicate, where the adsorbed C=O shows two bands at 1620 and 1583 cm$^{-1}$.[17]

It has been reported that methyl methacrylate (MMA) can undergo surface induced polymerization.[18,19] A sand ($SiO_2$ content > 97%) and a granulated blast furance slag (CaO, 41%; $SiO_2$, 35%) in the presence of $SO_2$ and water induced polymerization of MMA at 40 - 50°C yielding polymers with yields up to 87.6%. The reaction time was shorter than five hours. Ionic polymerization was probably induced by $SO_2$ in water. In the case of γ-MPS, it polymerizes without $SO_2$. The difference between MMA and γ-MPS should be noted. While the MMA study focused on the bulk polymerization induced by the surface, only the surface species has been under consideration for γ-MPS. Also, polymerized PMMA can be removed from the surface by dissolving into MMA and consequent renewal of the active surface centers. On the other hand, γ-MPS forms cross-linked polymers that cover the surface permanently.

Thus apparently γ-MPS undergoes surface induced polymerization. Extensive loss of the C=C groups leads to an important consequence concerning the reinforcement mechanism of FRP. If a high concentration of chemical bonds is desirable between the coupling agent and the matrix resin, the loss of C=C groups is certainly undesirable to yield good mechanical performance. On the other hand, if interpenetration is the major reinforcement mechanism, then good mechanical performance can be attained even after the loss of the C=C groups. In practice, however, the surface induced polymerization seldom leads to 100% conversion. The residual C=C group could be sufficient to yield good strength without interpenetration. Plueddemann[20] reported that heat treatment of γ-MPS treated fibers at 160°C yields results inferior to that which is dried at lower temperature. Therefore, interpenetration only may not be sufficient since the heat-treated sample at 160°C likely contains very few C=C groups.

## CONCLUSIONS

We have studied the structure of γ-MPS interphase as well as the glass/γ-MPS interface using FT-IR. Multilayer formation of γ-MPS on the surface of E-glass fibers was reconfirmed and the results are in good agreement with the work using $^{14}$C-labeled γ-MPS reported by Johannson et al.[10] There is a sudden change in the amount of adsorbed γ-MPS at around 0.4% by weight when plotted as a function of the concentration of silane treating solutions.

It is associated with the onset of incomplete dispersion of the coupling agent in the solution. There is a significant amount of γ-MPS in the outermost layers that can be washed away by organic solvents. More tightly bound layers of γ-MPS undergo a surface induced homopolymerization of acrylic polymer within a matter of days and loses most of the organofunctionality. The surface induced polymerization is interpreted as a result of the surface catalytic effect of the acid centers that strongly interact with the carbonyl groups of the γ-MPS. The area occupied by a single γ-MPS molecule at the surface is experimentally determined as 48Å$^2$ which is in fair agreement with the expected size of the molecule.

## ACKNOWLEDGMENT

The authors gratefully acknowledge the financial support of the Office of Naval Research under grant no. N00014-80C-0533.

## REFERENCES

1. H. Ishida and J.L. Koenig, Polym. Eng. Sci., 18, 128 (1978).
2. E.P. Plueddemann, J. Paint Technol., 40, 1 (1968).
3. E.P. Plueddemann, Mod. Plast., 43, 121 (August 1966).
4. S. Sterman and J.G. Marsden, Mod. Plast., 43, 133 (July 1966).
5. M.S. Akutin, M.L. Kerber, I.O. Stal'nova, N.L. Grodskaya, and E.E. Alekseev, Mekhanika Polimerov, 8, 1048 (1972).
6. E.P. Plueddemann, in "Proc. 35th Ann. Tech. Conf., Reinf. Plastics/Composites Inst.", SPI, Section 20-B (1980).
7. H. Ishida and J.L. Koenig, J. Colloid Interface Sci., 64, 565 (1978).
8. E.P. Plueddemann, private communication (January 1980).
9. H. Ishida and J.L. Koenig, J. Polym. Sci.-Phys. Edition, 18, 1931 (1980).
10. O.K. Johannson, F.O. Stark, G.E. Vogel, and R.M. Fleishmann, J. Composite Materials, 1, 278 (1967), and AFML-TRI-65-303, Part 1 (1965).
11. J. Sagiv, J. Am. Chem. Soc., 102, 92 (1980).
12. L.G. Tensmeyer, R.W. Hoffmann, and G.W. Brindley, J. Phys. Chem., 64, 1655 (1960).
13. R.W. Hoffmann and G.W. Brindley, ibid, 65, 443 (1961).
14. R.P. Young and N. Sheppard, J. Catal., 7, 223 (1967).
15. R.P. Young and N. Sheppard, Trans. Faraday Soc., 63, 229 (1967).
16. B.J. Fontana and J.R. Thomas, J. Phys. Chem., 65, 480 (1961).
17. M.L. Hair and I.D. Chapman, J. Phys. Chem., 69, 3949 (1965).
18. T. Yamaguchi, T. Ono, and H. Ito, Die Agn. Makromol. Che., 32, 177 (1973).
19. T. Yamaguchi, H. Tanaka, T. Ono, O. Itabashi, and H. Ito, Kobunshi Ronbunshu, 32, 126 (1975).
20. E.P. Plueddemann, "Silane Coupling Agents", Plenum Press, New York (in press).

# Part II
# Contact Angle, Wettability and Surface Energetics

WETTABILITY OF POLYMER SURFACES

M. A. Fortes

Departamento de Metalurgia, Instituto Superior Técnico
Centro de Mecânica e Materiais da Universidade Técnica
Av. Rovisco Pais, 1000 Lisboa, PORTUGAL

The basic theory of wetting is reviewed, covering the concepts of contact angle and polymer surface and interfacial tensions. The necessity of taking into account the interactions between interfaces in the three-phase region is stressed.

Particular emphasis is given to the problem of (equilibrium) contact angle hysteresis. A detailed thermodynamic treatment of hysteresis on solids with axially or cylindrically symmetric rough and heterogenous surfaces is presented, including an analysis of the stability of equilibrium as a function of the location of the triple line on the surface. The equations of Wenzel and of Cassie are criticized and a possible explanation for the drop size effect on contact angle is given. Edges on the solid surface are treated as a particular case of roughness; they are narrow regions of high curvature and of different surface energy. Some edges have a pinning effect on the triple line, allowing a range of stable configurations. Frontiers between two homogeneous patches also have this property, which explains the stick-and-slip movement of the triple line.

Brief considerations on the dynamics of wetting are included.

## 1 - WETTABILITY

The more common situation in wetting involves a solid S, a liquid L, and a second immiscible fluid V. For example, wetting of a polymer surface by water in air. V can simply be the saturated vapor of the liquid. Four-phase wetting[1] may also occur (two fluid phases in contact with the solid and a surrounding fluid that forms an interface with the other two) but shall not be discussed here.

For a closed system with fixed volumes of the three phases S, L, V, the more stable configuration at a given temperature, T, is in general one where the fluids are simply connected (e.g., just one drop of liquid as in Figure 1) and have interfaces with the solid (this may not be true in an externally applied field). Each of the three interfaces is characterized by its interfacial tension under equilibrium conditions at T, respectively $\gamma_{SV}$, $\gamma_{SL}$ and $\gamma$ (of the fluid interface), which for the moment are assumed to have the same value at all points in each interface. This implies, in particular, that the solid surface is homogeneous both chemically and energetically (no edges, for example) but not necessarily planar. If the Helmholtz energy, A, of a closed system S,L,V is minimized, two fundamental equations result (e.g. reference 2):

a) Laplace equation for the equilibrium shape of the fluid interface:

$$\gamma \, \text{div} \, \vec{n} + P_L - P_V = c \tag{1}$$

where c is a constant, $\vec{n}$ is the unit normal to the surface directed from L to V, and $P_L$, $P_V$ are the potential energies per unit volume of the two fluids at a point of the interface. These are due to externally applied fields (gravitational, centrifugal) which are assumed not to alter the values of the interfacial tensions. div $\vec{n}$ is the mean curvature of the interface.

b) Young's equation for the equilibrium contact angle, θ, of the fluid interface with the solid, defined as the angle between the SL and LV interfaces (Figure 1):

$$\cos\theta = \frac{\gamma_{SV} - \gamma_{SL}}{\gamma} \, . \tag{2}$$

This equation can be regarded as a boundary condition for the fluid interface, and is frequently interpreted as an equilibrium of surface tension forces acting on the line of contact of the three interfaces (triple line or triple junction) as shown in Figure 1. The contact angle is independent of geometry and of externally applied fields that do not alter the values of the γ (such as weak gravitational and centrifugal fields). If the second term in Equation (2) is outside the interval [-1,+1], a contact angle cannot be defined. In particular, if that term is larger than unity,

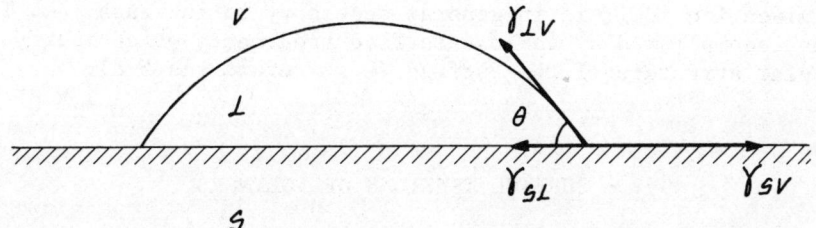

Figure 1. A simple geometry system comprising a solid S with a planar surface wetted by a fluid L surrounded by a second immiscible fluid V. The surface tensions can be regarded as forces acting on the line of contact. The equilibrium contact angle θ of L over S is a property of the three phases.

the liquid will cover completely the solid surface. It is then convenient to define the spreading coefficient, $S_{L/S}$, of L over S in presence of the vapor V of the liquid[3]:

$$S_{L/S} = \gamma_S - \gamma_{LV} - \gamma_{SL} = W_{SL} - W_{LL}. \quad (3)$$

$\gamma_S$ is the surface tension of the solid which is in general different from $\gamma_{SV}$ (see below);

$$W_{SL} = \gamma_S + \gamma_{LV} - \gamma_{SL} \quad (4)$$

is the free energy of adhesion between S and L and

$$W_{LL} = 2\gamma_{LV} \quad (5)$$

is the free energy of cohesion of L. Complete wettability of the solid by the liquid therefore occurs if $S_{L/S} > 0$ or $W_{SL} > W_{LL}$; in this case a contact angle cannot be defined but is loosely taken as zero. The liquid spreads on the solid.

This situation of complete wetting is not often met with polymers because their surface tensions are small (between 10 and 50 mJ.m$^{-2}$; 1 mJ.m$^{-2}$ = 1 erg cm$^{-2}$). Compared with metals and ceramics, for which the surface tensions are up to a few thousand mJ.m$^{-2}$, polymers are therefore poorly wettable solids.

The value of $\theta(0 \leqslant \theta \leqslant 180°)$ is a good measure of the wettability of S by L in presence of V. For a given volume of L, the smaller the θ the larger is the area wetted by a sessile drop of L resting on S (the same applies to a pendent drop[4]). To improve wettability by a given liquid it is in general necessary to increase $\gamma_{SV}$. This can be accomplished by various surface treatments which modify the molecular structure of the surface (e.g. references 5-8).

## 2 - SURFACE TENSIONS OF POLYMERS

Since $\theta$ refers to a particular liquid, it is not a good parameter to characterize the wettability of a surface, irrespective of the liquid. $\gamma_{SV}$, where V is the vapor of the liquid, is a better choice, but it is still dependent on the liquid. The best choice is $\gamma_S$, the surface tension of the solid in a vacuum (or in equilibrium with its vapor). The difference

$$\pi_e = \gamma_S - \gamma_{SV} \tag{6}$$

which is intrinsically positive, is termed the spreading pressure. $\pi_e$ is thought to be negligible for almost all liquids on low-energy, smooth solid surfaces[9].

There is currently no reliable experimental technique for measuring the surface tensions with the solid, $\gamma_S$, $\gamma_{SV}$ and $\gamma_{SL}$[10]. The experimentally accessible quantities are $\gamma$ and $\theta$, although difficulties arise in the proper interpretation of contact angle measurements as will be discussed later. But even these two quantities are not sufficient to calculate $\gamma_S$ because there are two unknowns in Equation (2) (clearly, $\pi_e$ should also be determined). Calorimetric measurements of the free energy of adhesion, $W_{SL}$, which at any rate are difficult, do not contribute to solve the problem of obtaining $\gamma_S$.

Various paths have been used to circumvent the difficulty. The simplest is to obtain $\gamma_S$ by extrapolation of the surface tensions of the polymer melt as a function of temperature[11] (for data on surface and interfacial tensions of polymer melts see reference 12). Calculations based on the parachors have also been undertaken[13]. Other methods have been used to estimate the surface energy of polymers from various molecular properties (e.g., reference 14).

The more important method presently available is the one based on the equation of Good and Girifalco (for recent reviews on these topics see references 15 and 16), which relates the interfacial tension between two condensed phases A and B to the surface tensions of A and B (A and B may be a solid polymer and a liquid):

$$\gamma_{AB} = \gamma_A + \gamma_B - 2\phi(\gamma_A \gamma_B)^{1/2} . \tag{7}$$

This equation is based on the geometric mean rule of molecular interactions. Deviations to this rule are accounted for by the interaction parameter $\phi$ which can in principle be calculated from the properties of the molecules of A and B. If $\phi$ is knwon for a given polymer-liquid pair, it is possible, by combining Equation (7) with Young's equation (2), to obtain $\gamma_{SV}$ (but not $\gamma_S$). Values of $\phi$ were calculated[17] for various polymers using Equation (7) and assuming that $\gamma_S$ coincides with the critical surface tension (see below) which is fairly easy to measure; also, $\pi_e$ was assumed equal to zero. The calculated values of $\phi$ are in several cases appreciably smaller than unity, notably when the liquid molecules are small and polar.

An important contribution in this direction was made by Fowkes[18,19] who suggested that the surface tension of a pure solid or liquid can be decomposed additively in contributions associated with the various types of intermolecular forces. For example:

$$\gamma = \gamma^d + \gamma^p + \gamma^h \qquad (8)$$

where d,p,h designate the London (dispersion), polar and hydrogen bonds. A statistical mechanical basis for this decomposition was recently given[20]. For an interface A, B the various contributions to $\gamma_{AB}$ can be related by a geometric mean rule, each with an interaction parameter $\phi$. For the London interaction, Fowkes[18,21] showed that $\phi \approx 1$. Other rules have been proposed to calculate the contributions to $\gamma_{AB}$ of the various components of $\gamma_A$ and $\gamma_B$[12]. For example, Wu[22] gives the following expression for London and polar contributions:

$$\gamma_{AB} = \gamma_A + \gamma_B - 4\frac{\gamma_A^d \cdot \gamma_B^d}{\gamma_A^d + \gamma_B^d} - 4\frac{\gamma_A^p \cdot \gamma_B^p}{\gamma_A^p + \gamma_B^p}. \qquad (9)$$

This is called the harmonic mean rule.

From contact angle data for judiciously chosen liquids it has been possible to calculate the various components of the surface tension of a large number of polymers (see examples in reference 16). These data are very useful not only to predict wettability by different liquids, but also, at a more fundamental level, to the understanding of molecular interactions across an interface and of their relation to the interactions within each individual phase.

Although this approach gives quite reliable values of $\gamma_S$ for polymers, it is not practical. Frequently one is interested in assessing the wettability of a new polymer or of a commercial polymer blend or even of a surface treated polymer and this is not simple to do with the previous method.

A simple measure of wettability is given by the so-called "critical surface tension of wetting", $\gamma_c$, a concept due to Zisman[23]. To obtain $\gamma_c$, the contact angles, $\theta$, of a polymer with various liquids forming a homologous series (e.g., alkanes or mixtures of two liquids) are measured. The $\cos\theta$ is plotted against $\gamma_{LV}$ and the curve is extrapolated to $\cos\theta=1$. The corresponding value of $\gamma_{LV}$ is the critical surface tension, $\gamma_c$, of the polymer. $\gamma_c$ is the highest surface tension of a liquid (in the series) which completely wets the solid. Although the values of $\gamma_c$ may deviate from one series of liquids to another and extrapolation may lead to some error, the concept proved to be a very useful one.

Turning to Equation (7), setting $\pi_e=0$, and comparing with Young's equation, the following result is obtained[15]

$$\cos\theta = -1 + 2\phi\left(\frac{\gamma_S}{\gamma_{LV}}\right)^{1/2} \tag{10}$$

which shows that a plot of $\cos\theta$ versus $\gamma_{LV}^{-1/2}$ should give a straight line, provided $\phi$ is constant. Also, setting $\cos\theta=1$, the critical surface tension is given by

$$\gamma_c = \phi^2 \gamma_S . \tag{11}$$

These equations show that in order to obtain $\gamma_c$ it may be preferable to plot $\cos\theta$ as a function of $\gamma_{LV}^{-1/2}$ (instead of $\gamma_{LV}$) and explain why $\gamma_c$ may vary from one liquid series to another. The surface tension of a polymer can be identified with the critical surface tension if the solid-liquid system is regular ($\phi=1$). This occurs[24] when the interactions at the interface are exclusively of the London type (in addition to $\pi_e=0$).

Values of $\gamma_c$ for various polymers are given in reference 25. It has been suggested[26] that the value of $\gamma_c$ for a copolymer can be calculated from the sum of the mole fractions multiplied by the $\gamma_c$ of each homopolymer. Deviations to this rule have been reported[27].

A quite different approach to the determination of the surface tensions of polymers was developed recently[28,29] based on the semi-empirical derivation of an additional relation between $\gamma_{SV}$, $\gamma_{SL}$ and $\gamma$. This relation, termed an equation of state, combined with Young's equation, allows the determination of $\gamma_{SV}$ from contact angle measurements.

Empirical methods are also used to measure wettability. One is the wetting tension test[30] which consists of using a range of liquid mixtures with different surface tensions to find the mixture of highest surface tension that does not contract within 2 seconds after brushing it on the surface. Another[31] is based on watching the spreading of a mixture of polar and nonpolar liquids (e.g., ethanol and octane) on the surface of the polymer.

Both $\gamma_c$ and $\gamma_S$ of polymers decrease slightly with increasing temperature. However, systematic studies of the effect of temperature (and also of molecular weight and crystallinity) on $\gamma_c$ and $\gamma_S$ are very scarce. For polymer melts both the experimental and calculated[32] values of $-d\gamma/dT$ are typically 0.04-0.1 mJ.m$^{-2}$.K$^{-1}$. Using the equation of state approach combined with contact angle measurements as a function of temperature, Neumann[29] obtained the value 0.06-0.07 for solid polytetrafluoroethylene. The contact angle for a polymer-liquid pair may increase or decrease with temperature (e.g. reference 29).

## 3 - MODIFICATION OF YOUNG'S EQUATION

The derivation of Young's equation in its form (2) assumes that the solid is homogeneous and the interfacial tensions have the same value at all points in each interface. Even in the absence of curvature effects, interactions between the three interfaces will occur in the region of contact[33-35] and these will modify the interfacial tensions in that region. The unperturbed interfacial tensions should be replaced by their perturbed values in the region of contact. If interface interactions are taken into account, both Laplace and Young equations will be altered. The first takes the form[2]

$$\text{div}(\gamma \vec{n}) + P_L - P_V = c \qquad (12)$$

which should be compared with Equation (1). The generalized Young equation for a variable $\gamma$ has been obtained[2] by treating an axially symmetric system with a non-homogeneous solid; it was admitted that $\gamma$ could also change with the orientation of the fluid interface at a given point. The resulting equation is

$$\bar{\gamma}_{SV} - \bar{\gamma}_{SL} = \bar{\gamma} \cos\theta_m - \sin\theta_m \cdot \frac{\partial \bar{\gamma}}{\partial \sin\phi} \qquad (13)$$

where $\theta_m$ is the microscopic, actual angle of contact, and $\phi$ is the angle between the normal to the fluid interface at the contact line and the symmetry axis. A bar indicates the values of the $\gamma$ at the line of contact. The meaning of the "torque" term in Equation (13) has not been exploited, but may be particularly important when dipole-dipole intercations between the solid and liquid are relevant[2].

It is not certain that Equation (13) will keep the same form if the perturbation in $\gamma_{SL}$ and $\gamma_{SV}$ is taken into account, that is, if they are treated as functions of the distance to the line of contact. However, if the radius of curvature of the line of contact is not too small, it is likely that Equation (13) can still be used, the $\bar{\gamma}$ now indicating the perturbed values at the line of contact.

At any rate it is necessary to distinguish between a microscopic angle of contact given by Equation (13) and the macroscopic angle $\theta_M$ measured by optical or other "macroscopic" means[2,33-38]. The two angles are shown in Figure 2. To indicate the macroscopic angle, the fluid interface profile is extrapolated down to the solid surface. Note that this angle is not given by Young's equation (2).

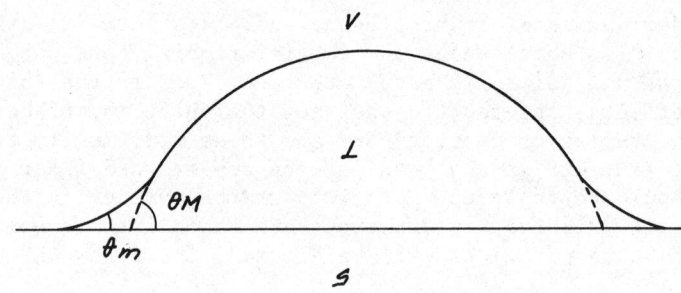

Figure 2. Schematic representation of the microscopic, $\theta_m$, and macroscopic, $\theta_M$, equilibrium contact angles, and of the perturbed interface shape in the contact region.

While the microscopic angle is certainly a property of the three phases in contact (at least if the radius of curvature of the line of contact is not too small), it is not obvious that the macroscopic angle $\theta_M$ should also be independent of geometry and externally applied fields. Both molecular and continuous approaches have been used by various authors to analyse the interactions between interfaces[33-40] but no general conclusions have been drawn on the relation between the microscopic and macroscopic angles. A simple relation between the two angles has been obtained recently[41] in the form

$$\gamma \cos \theta_M = \bar{\gamma} \cos \theta_m \qquad (14)$$

which implies that $\theta_M$ is also a property of the three phases in contact.

Although this is a comforting conclusion, the fact is that $\theta_m$ may not be accessible to measurement; and it is this angle that is directly related to the actual interactions between the three interfaces and therefore reflects the wettability of a surface.

A different, "macroscopic", approach to the problem of interface interaction is based on the concept of a line tension, $\tau$, which is essentially the excess free energy per unit length of the triple junction. This concept is due to Gibbs. If the condition of minimum free energy is introduced, Young's equation is replaced by (see reference 42)

$$\cos \theta_M = \frac{\gamma_{SV} - \gamma_{SL}}{\gamma} - \frac{\tau}{\gamma\rho} \qquad (15)$$

where $\rho$ is the radius of curvature of the line of contact. The angle defined by Equation (15) is obviously a macroscopic angle. $\tau$ may be positive or negative[42,43], but in general the term $\tau/\gamma\rho$ will be negligible except for very large curvatures and for $\theta_M$ not close to 90°. This term is zero for cylindrically symmetric geometries, such as when a plate is partially immersed in a liquid. A general form of the equations of Laplace and Young with the line tension has been obtained by Boruvka and Neumann[44]. The concept of line tension avoids the details of the interactions at the triple junction and does not lead to a distinction between a microscopic and a macroscopic angle of contact. It is nevertheless formally simple.

While we have to await advances in the understanding of the molecular interactions at a triple junction, the operationally more sensible attitude is to keep using the concept of a contact angle as a property of a solid and two fluids in contact, allowing for the possibility that this angle changes from region to region in a non-homogeneous surface, and regard the macroscopic angle as a good parameter to measure wettability. The distinction between the macroscopic and microscopic angles can be avoided in calculations of the free energy, provided only relative energies of different macroscopic configurations are of interest. This is in fact the procedure commonly used.

## 4 - CONTACT ANGLE HYSTERESIS

### 4a - Advancing and Receding Contact Angles

Equilibrium contact angles can be measured directly by optical means on sessile drops (or less frequently on pendent drops) on horizontal flat solid surfaces. Indirect methods rely on the height of rise of the liquid meniscus at a place partially immersed in the liquid, or on the dimensions of sessile and pendent drops on a large plate, or even on the measurement of capillary forces. All these methods give a macroscopic angle of contact, since it is this angle that governs the observable shape of the fluid interface and the capillary forces.

The measured contact angle for a given system is found to be dependent on the way the experiment is made, more specifically on whether the final equilibrium position of the triple junction is reached by advancing on dry solid or by receding on initially wetted solid. The corresponding angles are termed advancing, $\theta_a$, and receding (or retracting), $\theta_r$, contact angles. It is generally found that $\theta_a > \theta_r$, the difference being very large in some cases. This difference is of major importance, considering that contact angle measurements are an essential tool in the study of wettability and in the determination of surface tensions of solids. The fact that the two angles are different and have reproducible values in sucessive advancing-receding cycles is termed contact angle hysteresis. Hysteresis implies that the sequence of equilibrium configurations in advancing and receding are different. The difference is generally attributed to surface heterogeneity and roughness[45-47].

A major assumption that is made in the analysis of hysteresis is that the macroscopic contact angle has a well defined value, $\theta$, at each point of the solid surface, although this value can vary from point to point in a heterogeneous surface. This assumption seems to be thermodynamically sound[44].

Two equations due respectively to Wenzel[48] and to Cassie and Baxter[49,50] are frequently quoted to determine the contact angle on a rough and on a heterogenous surface. Both equations can be derived thermodynamically (e.g. reference 10) by considering average values of the surface tensions. Wenzel's equation relates the observed contact angle, $\theta_w^*$, measured in relation to the mean surface, to the roughness ratio r of the surface (true area divided by projected area):

$$\cos \theta_w^* = r \cos \theta \qquad (16)$$

where $\theta$ is the characteristic contact angle. This equation is not applicable for $|r \cos\theta| > 1$. It predicts that wettability should increase with roughness for $\theta < 90°$; this is not observed experimentally. Cassie's equation applies to an heterogeneous surface with regions of contact angle $\theta_i$ occupying a fraction $f_i$ of the (planar) surface

$$\cos \theta_c^* = \sum_i f_i \cos \theta_i . \qquad (17)$$

The observed angle $\theta_c^*$ is the weighted average of the component $\theta_i$. This equation can also be derived[51] for a liquid contacting a two component air-solid surface. Equations (16) and (17) do not distinguish between $\theta_a$ and $\theta_r$ and the exact meaning of the angles $\theta_w^*$ and $\theta_c^*$ in them is not clear. This question will be discussed later in the paper with the conclusion that no special significance should be attached to the angles $\theta_w^*$ and $\theta_c^*$ since they do not correspond to observable contact angles.

An effect that has been attributed by Good and Koo[52] to surface heterogeneity is the observed dependence of $\theta_a$ and $\theta_r$ on sessile drop size. Both angles were found to decrease for small drops[52]. The variation of $\theta$ from region to region will give rise to a waved contact line and to a fluid interface with convolutions near the solid, like a beer bottle cap (Figure 3). While this description is likely to be correct and is confirmed by direct calculation of the interface shape in simple cases[53], it is doubtful that it can explain the observed size effect on the contact angle. Good and Koo proposed that the increased area of the fluid interface due to the convolutions should be equivalent to a large negative line tension, increasing (in modulus) as the size of the drop decreases. It is not clear why the line tension should be negative instead of positive, considering the increased energy of the triple junction associated with the increased fluid interface area. On the other hand Equation (15) requires a positive line tension for the contact angle to decrease if it is larger than 90°, and the same effect was observed in such a case[52]. A size effect, but not of the type observed experimentally, is predicted in sections 4b and 4c for drops on symmetrical rough or heterogeneous surfaces.

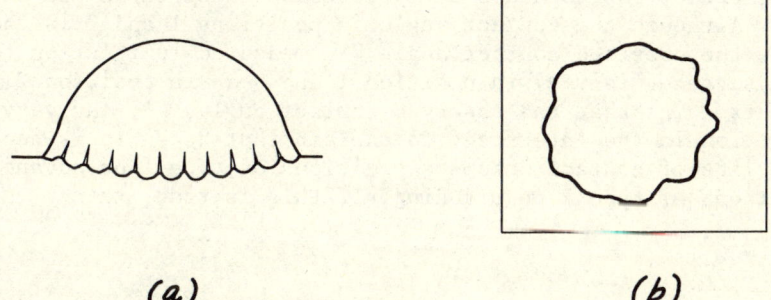

(a)   (b)

Figure 3. a) Convolutions on drop surface due to heterogeneity of the solid surface (from reference 49). The triple line is contorted (b).

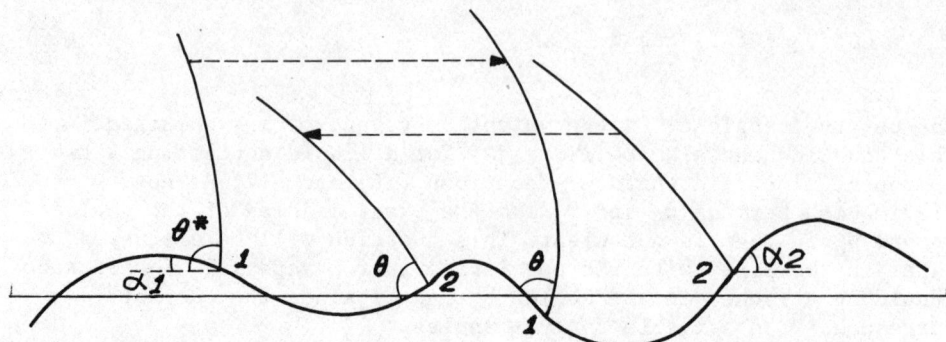

Figure 4. Profile of rough surface showing the expected critical positions of the contact line for the advancing (1) and receding (2) modes. The line of contact jumps (arrows) between the positions indicated as the volume of the drop is made to change. The horizontal line represents the mean surface.

It is now generally accepted[29,46,47] that contact angle hysteresis is related to a multiplicity of equilibrium configurations of the (closed) system at different locations of the contact line on the surface. These multiple configurations can occur in a rough surface and admitedly[46] also on a heterogeneous surface. The triple line "chooses" different locations in advancing and receding. In both cases the triple line makes jumps to adjacent positions as the liquid volume is changed. Such jumps may give rise to the retention of air pockets in advancing while liquid patches may be left on the surface on receding[45].

The effect of roughness on hysteresis can be explained with Figure 4. Although the contact angle in positions 1 and 2 is the same ($\theta$), the observed contact angle $\theta^*$, measured in relation to the mean surface, is $\theta+\alpha_1$ in position 1 and $\theta-\alpha_2$ in position 2. This argument explains that the observed contact angle, $\theta^*$, can vary with position in the interface. To explain that $\theta_a > \theta_r$ it is necessary that the line of contact occupies positions of type 1 on advancing and positions of type 2 on receding. If this is true, then

$$\theta_a = \theta + \alpha_1$$
$$\theta_r = \theta - \alpha_2$$
(18)

where $\alpha_1$ and $\alpha_2$ are the maximum slope angles of the groove faces in the receding and advancing directions. These equations were first proposed by Shuttleworth and Bailey[54]. Subsequently, it was shown[45,47] by analysing a spherical drop resting on a sinusoidal grooved surface, that the line of contact actually moves in the way required by Equation (18), essentially because the volume has maxima and minima at those points. We shall re-examine in section 4b the effect of roughness on contact angle for a similar but more general system.

Hysteresis due to chemical heterogeneity requires that an advancing triple line prefers the contact with the less wettable regions, the reverse being true for a receding triple line. This behavior will be explained in section 4c. The combined effect of roughness and heterogeneity will be considered in section 4d. In section 4e the effect of edges and, in general, of line imperfections on the surface is discussed as an extension of the previous analysis. Finally, in section 4f we briefly discuss hysteresis on random surfaces.

### 4b - Rough Surfaces

In this section we extend previous studies[45,47] of surfaces with sinusoidal concentric grooves to the more general case of an axially or cylindrically symmetric rough surface of arbitrary profile. Gravity is neglected. The entire system has the same symmetry as the solid surface and in the case of cylindrical symmetry there is a plane of symmetry such that the triple line contacts identical regions of the solid on the two sides.

The main hypotheses that will be made in the following analysis of the contact angle of a drop resting on the rough surface are:
a) The fluid interface is always a solution of Laplace equation with a constant $\gamma$.
b) The local contact angle with the surface is the same everywhere (homogeneous surface).
c) A non-reversible jump to another equilibrium configuration occurs whenever the triple line reaches an unstable position as it moves on the surface due to the imposed change in the drop volume. The special positions of the triple line, separating stable and unstable regions, will be termed critical positions or configurations.
d) A transition between two stable equilibrium configurations can only occur if the Helmholtz free energy of the system decreases (for a fixed volume of liquid).

Huh and Mason[47] replaced condition c) by a condition dealing with the maxima and minima of the volume. As we will show the two conditions are equivalent. Condition d) is necessary at least if the change in drop volume is made slowly with no energy transferred to the liquid. Eick et al.[55] in their analysis of the roughness effect discarded condition b) and assumed instead that the observed angle of contact was "chosen" by the system to minimize the free energy. However this was made by imposing that the fluid interface would contact a specific line on the surface, a condition which should not be introduced a priori.

The profile of the solid surface is defined by a continuous function z(x) in both cases, with x measuring the distance to the symmetry axis in axial symmetry or to the symmetry plane in cylindrical symmetry (see Figure 5); the z axis is directed into the liquid. The surface of the drop has constant curvature, that is, it is either spherical or cylindrical. When the contact line is at a position x with slope $\alpha$ (positive or negative)

$$\operatorname{tg} \alpha = z' = \frac{dz}{dx} \tag{19}$$

the observed contact angle $\theta^*$ is (condition b)

$$\theta^* = \theta - \alpha \tag{20}$$

where $\theta$ is the actual macroscopic contact angle. The Helmholtz energy A of the system is calculated in the way used by Johnson and Dettre[45]

$$\frac{A}{\gamma} = \Omega^{LV} - \Omega^{SL} \cos\theta \tag{21}$$

where the $\Omega$ are the areas of the interfaces. This equation is adequate to compare energies of different configurations at constant volume and temperature.

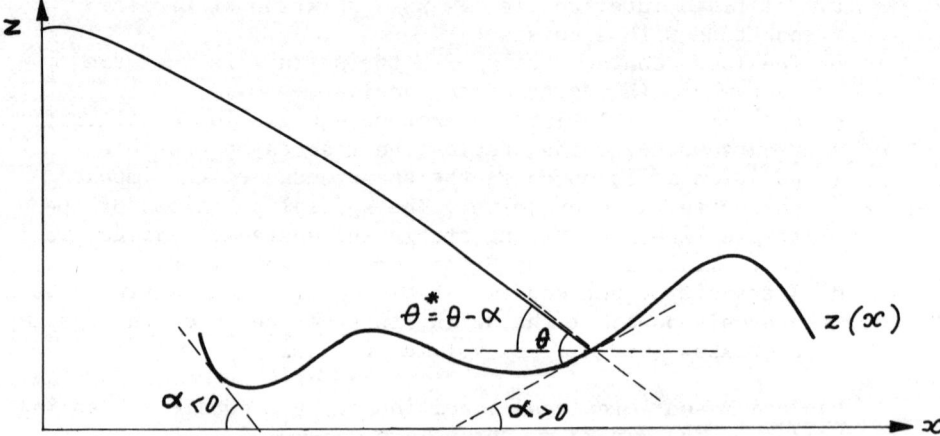

Figure 5. Coordinate system for axially or cylindrically symmetric geometry of a rough wetted solid surface with profile z(x) of any tipe. z is on a symmetry plane of the system. $\alpha$ is the slope at each point.

# WETTABILITY OF POLYMER SURFACES

The revelant equations for the volume V and Helmholtz energy A are as follows (for cylindrical drops V and A are the quantities per unit length of the system):

Spherical drop (22a)

$$\frac{V}{\pi} = \frac{x^3}{3} f(\theta^*) + \int_0^x x^2 z' \, dx \; ; \quad f(\theta^*) = \frac{(1-\cos\theta^*)(2+\cos\theta^*)}{\sin\theta^* \, (1+\cos\theta^*)}$$

$$\frac{A}{2\pi\gamma} = \frac{x^2}{1+\cos\theta^*} - \cos\theta \int_0^x x \sqrt{1+z'^2} \, dx \qquad (22b)$$

Cylindrical drop

$$V = x^2 g(\theta^*) + 2 \int_0^x x \, z' \, dx \; ; \quad g(\theta^*) = \frac{\theta^* - \sin\theta^* \cos\theta^*}{\sin^2\theta^*} \qquad (23a)$$

$$\frac{A}{2\gamma} = \frac{\theta^*}{\sin\theta^*} x - \cos\theta \int_0^x \sqrt{1+z'^2} \, dx \; . \qquad (23b)$$

We first calculate the derivatives of V and A with respect to x.

Spherical

$$\frac{1}{\pi x^2} \frac{dV}{dx} = f(\theta^*) + z' - x \frac{z''}{(1+\cos\theta^*)^2 (1+z'^2)} \qquad (24a)$$

$$\frac{1}{2\pi\gamma x} \frac{dA}{dx} = \frac{2}{1+\cos\theta^*} - \cos\theta \sqrt{1+z'^2} - \frac{x \sin\theta^* \, z''}{(1+\cos\theta^*)^2 (1+z'^2)} \qquad (24b)$$

Cylindrical

$$\frac{1}{2x} \frac{dV}{dx} = g(\theta^*) + z' - x \frac{z''(\sin\theta^* - \theta^* \cos\theta^*)}{\sin^3\theta^* (1+z'^2)} \qquad (25a)$$

$$\frac{1}{2\gamma} \frac{dA}{dx} = \frac{\theta^*}{\sin\theta^*} - \cos\theta \sqrt{1+z'^2} - \frac{x \, z''(\sin\theta^* - \theta^* \cos\theta^*)}{\sin^2\theta^* (1+z'^2)} \qquad (25b)$$

Next we investigate the equilibrium of the configuration with the line of contact at x. To do this we allow x and $\theta^*$ to change slightly at constant volume and with the liquid drop keeping a constant curvature and find the sign of $d^2A/dx^2$. Clearly, a more general change of shape should be considered to assess stability but would considerably complicate the mathematical treatment. We have

Spherical

$$\frac{1}{\pi\gamma} \frac{dA}{dx} = x \frac{\cos(\theta^*+\alpha) - \cos\theta}{\cos\alpha} \qquad (dV=0) \qquad (26a)$$

Cylindrical

$$\frac{1}{2\gamma}\frac{dA}{dx} = \frac{\cos(\theta^*+\alpha)-\cos\theta}{\cos\alpha} \qquad (dV=0) \qquad (26b)$$

These equations indicate that the equilibrium configurations are those for which $\theta^*=\theta-\alpha$, as already admitted when we wrote Equation (20). The value of $d^2A/dx^2$ for $\theta^*=\theta-\alpha$ can now be calculated; it turns out that this value can be simply related to the corresponding value of $dV/dx$.

Spherical

$$\frac{1}{\pi\gamma}\frac{d^2A}{dx^2} = \frac{\sin\theta(1+\cos\theta^*)^2}{\cos\alpha}\frac{1}{x^2}\frac{dV}{dx} \qquad (27a)$$

Cylindrical $\qquad (dV=0 \ ; \ \theta^*+\alpha=\theta)$

$$\frac{1}{2\gamma}\frac{d^2A}{dx} = \frac{\sin\theta\sin^3\theta^*}{\cos\alpha(\sin\theta^*-\theta^*\cos\theta^*)}\frac{1}{2x}\frac{dV}{dx} \qquad (27b)$$

Since the coefficients of $\frac{dV}{dx}$ in these equations are positive, it is concluded that the stable configurations are those for which $dV/dx>0$. The critical configurations occur when $\frac{dV}{dx}=0$, that is, at the maxima and minima of the volume. For large values of x, stability implies that $z''<0$, see Equations (24a) and (25a); the critical positions are at the inflection points of the profile, the maxima of the volume corresponding to the inflection points on the descending side of the groove. For smaller values of x, the critical positions can be either above (larger z) or below the inflection points, depending on the contact angle and type of profile. Figure 6 shows, for a particular sinusoidal profile and for $\theta=60°$, the location of the critical points. If $\alpha_i$ is the absolute value of the slope at the inflection points, the range of stability is larger or smaller than for $x\to\infty$ depending on whether $f(\theta^*)-tg\alpha_i$ is positive or negative. For small values of $\theta$ the range is smaller. As the size of the drop decreases, the values of $\theta^*$ at the maxima of V decrease while those at the minima of V increase. It is also interesting to note that for very small values of x (or $z''\to 0$) there may be drops which are necessarily unstable, that is, no stable configuration is possible. This occurs if the triple line is on regions for which $f(\theta^*)+tg\alpha$ or $g(\theta^*)+tg\alpha$ is always negative (which implies $\alpha<0$). Figure 7 shows the values of $\theta$ and $\alpha$ for which $\frac{dV}{dx}<0$ as $x\to 0$. Such drops, of dimensions comparable with the wave length of the roughness, must separate into smaller drops. Line tension effects may become important in this context.

In general the volume varies with x in the way shown in Figure 8 with successive maxima and minima. The stable regions, for large drops, are shown in heavy lines, between the inflection points.

WETTABILITY OF POLYMER SURFACES 123

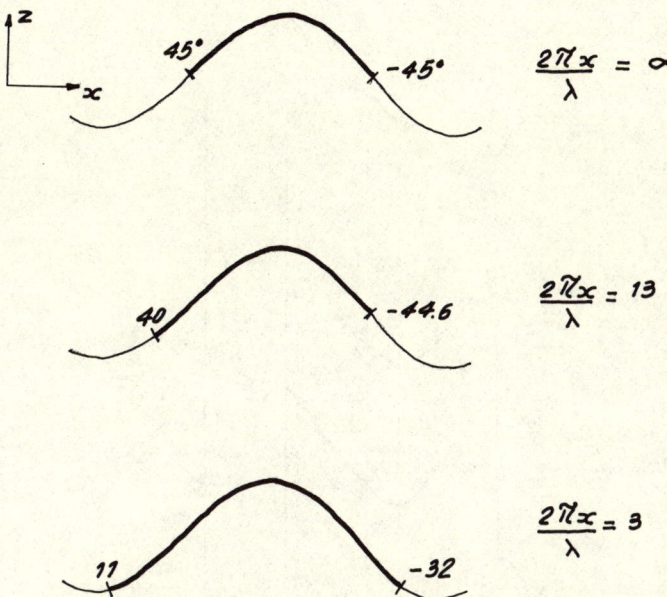

Figure 6. Critical positions on an axially symmetric sinusoidal surface of wavelength $\lambda$ and maximum slope 45° for various drop sizes and for $\theta=60°$. The stable regions are shown heavier. The slope angles at the critical positions are indicated. x is the radius of the contact line.

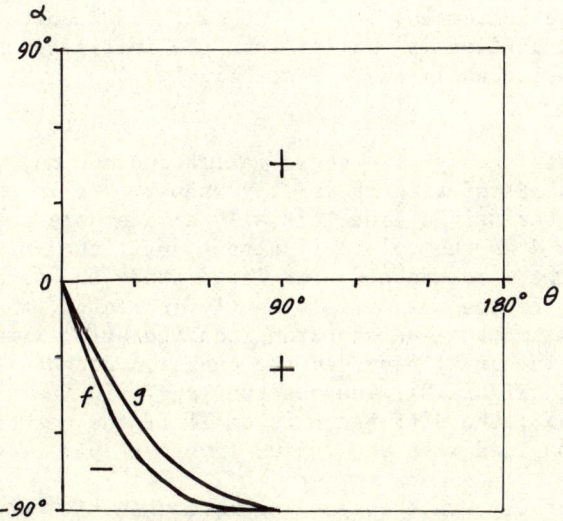

Figure 7. The sign of $f(\theta^*)+tg\alpha$ and $g(\theta^*)+tg\alpha$ as a function of $\theta$ and $\alpha$. The sign is negative below the curves marked f and g, respectively.

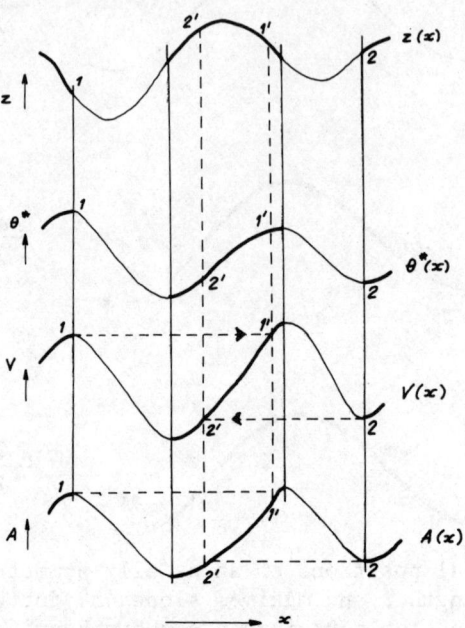

Figure 8. Rough surface profile z(x) showing the regions of stability (heavy) for large drops. The other curves give $\theta^*$, V and A as a function of position of the triple line. The symmetry plane of the drop is to the left. Jumps 1→1' in advancing and 2→2' in receding are indicated; in both cases the Helmholtz energy of the final configuration is smaller than the initial energy, at the critical positions 1 and 2, respectively.

The Helmholtz energy also shows maxima and minima, which occur for the same values of x as those of V when $x \to \infty$. In the advancing mode, with the triple line initially at a stable position, the line moves forward as the volume increases until the critical position with negative $\alpha$ is reached. For large drops the observed angle is then $\theta + \alpha_i$. If the volume is slightly increased, the triple line may make a jump to a neighbouring configuration with the same volume and larger x, provided the energy A decreases (such as the jump 1→1' in Figure 8). The observed angle in the new configuration is below $\theta + \alpha_i$; the difference is small if the new position is not far away compared with the radius of the triple line.

Conversely, in the receding mode when the triple line reaches the critical position with positive $\alpha(\theta^* = \theta - \alpha_i$ for large drops) it may jump to another position with slightly larger $\theta^*$ (such as the jump 2→2' in Figure 8). For large drops the advancing angle is therefore near $\theta+\alpha_i$ and the receding angle is slightly larger than $\theta-\alpha_i$. For small drops the observed contact angles should be smaller for advancing drops and larger for receding drops. This is not the type of size effect observed experimentally for receding drops[52].

It is now necessary to find whether A decreases or increases in the jumps. To do this we use approximate expressions for V and A, neglecting the small variation in V due to the roughness and assuming that the area of the solid-liquid interface can be obtained from the projected area $\pi x^2$ by multiplication by the roughness ratio r, which is taken as a constant. For axial drops we then have

$$V = \frac{\pi}{3} x^3 f(\theta^*)$$
$$\frac{A}{\gamma} = 2\pi x^2 \left(\frac{1}{1+\cos\theta^*} - \frac{r\cos\theta}{2}\right) \quad . \tag{28}$$

These approximate expressions neglect the detail of the rough surface and do not lead to maxima and minima in V and A. They simply give the trend in the variation of V and A with x or $\theta^*$. Eliminating x between Equations (28) we find

$$A = K\left(\frac{1}{1+\cos\theta^*} - \frac{r\cos\theta}{2}\right)\frac{1}{f(\theta^*)^{1/3}} \tag{29}$$

where K is a constant, and

$$\frac{dA}{d\theta^*} = N(\theta^*)(r\cos\theta - \cos\theta^*) \tag{30}$$

where $N(\theta^*)$ is a positive function of $\theta^*$. An analogous result can be obtained for cylindrical drops. In both cases A has a minimum for $\theta^*_w$ given by

$$\cos\theta^*_w = r\cos\theta \quad . \tag{31}$$

This is Wenzel's equation. For $r\cos\theta>1$, A increases monotonically with $\theta^*$ and for $r\cos\theta<1$ it decreases with $\theta^*$.

We may now analyse the changes in A for jumps at constant V. In advancing $\theta^*$ decreases in the jump, starting with a value $\theta+\alpha_i$ for large drops. In receding $\theta^*$ increases in the jump, starting with a value $\theta-\alpha_i$ for large drops. Therefore, for the energy to decrease in the jumps the following conditions must be met

$$\cos(\theta+\alpha_i) < r\cos\theta \tag{32a}$$

$$\cos(\theta-\alpha_i) > r\cos\theta \ . \tag{32b}$$

Since r is the average value of $\sec\alpha$, it follows that $r\cos\alpha_i<1$. Conditions (32) can fail in some cases. The first may fail for $\theta>90°$ and is never satisfied for $r\cos\theta<-1$. The second may fail for $\theta<90°$ and is never satisfied for $r\cos\theta>1$. In such cases it is not clear what the sequence of shapes will be when the volume is slightly varied from the critical values. The detachement of a small drop (toroidal or cylindrical) is a possibility that should be investigated.

In Figure 8. the curve A(x) is for a situation where A decreases in the jumps. Even in such cases the formation of un-connected configurations cannot be ruled out. The problem is a dynamic one and no definite conclusions can be drawn from the equilibrium analysis.

When $|r\cos\theta|<1$ the more stable configuration for $\theta^*>\theta_w^*$ (advancing) is the one with smaller $\theta^*$, that is, with the larger x compatible with the drop volume; and conversely for $\theta^*<\theta_w^*$ (receding). The angle $\theta_w^*$ is never observed and simply indicates the absolute minimum in the Helmholtz energy curve.

In the previous discussion it was assumed that at the critical configurations the angle $\theta^*$ is in the interval $0, \pi$. If $\theta\pm\alpha_i$ is outside this interval it is expected that the critical configurations will occur for $\theta^*=0,\pi$. The observed contact angles should then be near 0 or $\pi$, respectively.

Finally, at a region of constant slope, $z''=0$, the stability will depend on the sign of $f(\theta^*)+tg\alpha$ (or $g(\theta^*)+tg\alpha$) according to Equations (24a) and (25a). The values of $\theta$ and $\alpha$ for which these quantities are positive or negative are shown in Figure 7. The presence of regions of constant slope does not alter the conclusions drawn previously on hysteresis: the observed angles are near $\theta\pm\alpha_i$, where $\alpha_i$ is again the absolute value of the slope of the groove profile.

## 4c – Heterogeneous Surfaces

We consider now a planar surface with regions of different contact angle. Only axially and cylindrically symmetric geometries will be considered; that is, $\theta=\theta(x)$, using the same notation as previously. Hypotheses a), c) and d) of section 4b will be maintained. The observed contact angle at position x is $\theta(x)$. The basic equations are:

Spherical

$$\frac{V}{\pi} = \frac{1}{3} x^3 f(\theta) \tag{33a}$$

$$\frac{A}{2\pi\gamma} = \frac{x^2}{1+\cos\theta} - \int_0^x x\cos\theta(x)dx \tag{33b}$$

Cylindrical

$$V = x^2 g(\theta) \tag{34a}$$

$$\frac{A}{2\gamma} = \frac{\theta}{\sin\theta} x - \int_0^x \cos\theta(x)dx \tag{34b}$$

where f and g are defined by Equations (22a) and (23a). We have

Spherical

$$\frac{1}{\pi}\frac{dV}{dx} = x^2 \left[ f(\theta) - \frac{x}{\sin\theta(1+\cos\theta)^2} \frac{d\cos\theta}{dx} \right] \tag{35a}$$

$$\frac{1}{\gamma}\frac{dA}{dx} = \frac{2\sin\theta}{x}\frac{dV}{dx} \tag{35b}$$

Cylindrical

$$\frac{1}{2}\frac{dV}{dx} = x\left[g(\theta) - \frac{x(\sin\theta - \theta\cos\theta)}{\sin^2\theta}\frac{d\cos\theta}{dx}\right] \tag{36a}$$

$$\frac{1}{\gamma}\frac{dA}{dx} = \frac{\sin\theta}{x}\frac{dV}{dx} \tag{36b}$$

For large x, the maxima and minima of V occur where $\theta$ is maximum or minimum respectively, and dV/dx is positive where $\theta$ increases with x (or $\cos\theta$ decreases with x). For smaller values of x, the interval where dV/dx>0 includes regions where $d\theta/dx<0$ and is therefore larger than for large x. The maxima and minima of A coincide with those of V.

We shall now discuss the stability of axial drops (a similar procedure can be used for cylindrical drops). Suppose the spherical drop makes an angle w with the surface. Then, at constant V, we obtain

$$\frac{1}{2\pi\gamma}\frac{dA}{dx} = x(\cos w - \cos\theta) \qquad (dV=0) \tag{37}$$

and for w=$\theta$ (equilibrium configurations)

$$\frac{1}{2\gamma}\frac{d^2A}{dx^2} = \sin\theta (1+\cos\theta)^2 \frac{1}{x^2}\frac{dV}{dx} \qquad (dV=0 \ ; \ w=\theta) \tag{38}$$

For cylindrical drops the final result is

$$\frac{1}{\gamma}\frac{d^2A}{dx^2} = \frac{\sin^4\theta}{\sin\theta - \theta\cos\theta}\frac{1}{x^2}\frac{dV}{dx} \qquad (dV=0 \; ; \; w=\theta) \qquad (39)$$

In both cases the stable configurations are those for which dV/dx>0. As concluded above, for large drops these are the regions where θ increases with x. Figure 9 applies to this case.

Contact angle hysteresis on a heterogeneous surface can now be easily explained. As the drop volume is made to increase, a critical configuration is reached where V is a maximum (θ maximum for large drops): for example, the critical position 1 in Figure 9. The drop then makes a jump to another configuration with the same volume and a slightly smaller θ (if the new position is not far away), such as position 1 in Figure 9. For a receding drop the critical configuration is for θ minimum for large drops. An advancing drop contacts regions of larger θ while a receding drop contacts the more wettable regions. The advancing angle decreases and the receding angle increases as the drop size decreases.

As regards the change of energy in the jumps we use a similar procedure as for rough surfaces. We introduce an average contact angle $\theta_c^*$ defined by

$$\cos\theta_c^* = \frac{1}{S}\int\cos\theta \; dS \qquad (40)$$

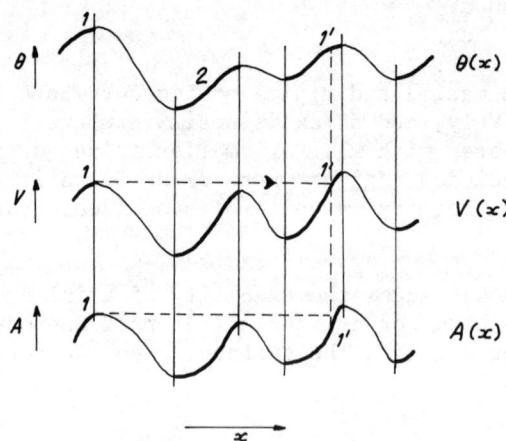

Figure 9. Heterogeneous planar surface, with contact angle changing as indicated by θ(x). The other curves give V and A as a function of position of the triple line. The symmetry plane of the drop is to the left. A jump 1→1' in advancing is indicated; the energy decreases in the jump. Note that no position with the same volume as 1 occurs in the stable region 2.

# WETTABILITY OF POLYMER SURFACES

which we assume to be independent of x. This angle is identical with the one given by Cassie's equation (17). We write, for axial drops,

$$\frac{V}{\pi} = \frac{1}{3} x^3 f(\theta) \tag{41}$$

$$\frac{A}{2\pi\gamma} = x^2 \left(\frac{1}{1+\cos\theta} - \frac{\cos\theta_c^*}{2}\right) . \tag{42}$$

At constant V we obtain

$$\frac{dA}{d\theta} = K \frac{f(\theta)^{-2/3}}{\sin\theta(2+\cos\theta)} (\cos\theta_c^* - \cos\theta) \tag{43}$$

where K is a positive constant. There is a minimum of A for $\cos\theta = \cos\theta_c^*$. This minimum always exists. In advancing $\theta > \theta_c^*$ and $\theta$ decreases in the jump: the Helmholtz energy therefore decreases (as in the jump 1→1' of Figure 9). The same is true in receding.

These conclusions also apply to cylindrical drops. Contrary to what happens on rough surfaces, the energy always decreases in the jumps.

### 4d - Roughness and Heterogeneity. Other Geometries

The same type of analysis can easily be generalized to a rough and heterogeneous solid surface with axial or cylindrical symmetry. As expected from the previous results, the stable configurations are those for which V increases with x. For large drops this implies that $\theta^* = \theta - \alpha$ should increase with x (or $\alpha - \theta$ decrease with x). It is possible to construct from the actual profile an equivalent homogeneous profile with a slope $\alpha_e = \alpha - \theta +$ constant. The stable regions are those where $\alpha_e$ decreases with x and the critical positions for large drops are at the maxima and minima of $\alpha_e$. Not all critical positions of one type (maxima or minima) will be reached on advancing or receding, respectively. Only those where $\theta - \alpha$ have the largest and smallest values will be "visited" by the triple line as the volume of the drop is changed. For example, on a rough homogeneous surface, small crests on the surface will in general not be touched by the triple line since the corresponding configurations will not have the required volume. Figure 9 shows a similar example for a heterogeneous surface: the line jumps 1→1' ignoring position 2.

In advancing the contact angle is near the largest maxima of $\theta - \alpha$ and in receding near the smaller minima. It is interesting to note that for a rough but homogeneous surface the difference $\theta_a - \theta_r$ should be the same for drops (of the same volume) of different liquids (except if the surface contains regions for which $\theta - \alpha$ for some liquids is outside the interval $0, \pi$). Heterogeneity of the surface will produce a different $\theta(x)$ profile for different liquids and the difference $\theta_a - \theta_r$ will in general change from liquid to liquid. It should in principle be possible to assess heterogeneity of a surface by measuring $\theta_a - \theta_r$ for different liquids. Another conclusion is that the critical surface tension method described in section 2 probably gives the $\gamma_c$ of the more wettable regions if the contact angles are measured in receding and the $\gamma_c$ of the more repellent regions if the advancing angles are measured.

Other geometries of rough and/or heterogeneous surfaces with axial or cylindrical symmetry have been investigated using the same procedure as for drops. For example, a liquid in a closed tube or a liquid between parallel plates at a given distance and with a closed end. The results are similar to those obtained for drops in that the stable positions of the triple line are those for which V increases with x, where x defines the position of the triple line (in advancing x increases). The critical positions for large radii of curvature of the fluid interface are at the maxima and minima of $\theta - \alpha$. Some differences from drop behavior occur at small radii of curvature: for a liquid in a tube or between plates all positions are stable so that no hysteresis is expected. For example, for a liquid in a tube with a profile z(x), where z is the distance to the tube axis and x defines the position of the triple line relative to the closed end of the tube, it is

$$\frac{1}{z^2}\frac{dV}{dx} = 1 - f(\phi)z' + \frac{z}{(1+\cos\phi)^2}\frac{z''}{1+z'} \qquad (44)$$

where $\phi = \pi/2 - \theta + \alpha$ and $tg\alpha = z'$. The observed angle of contact $\theta^* = \pi/2 - \phi$. Since $1 - f(\phi)z'$ is always positive all configurations are stable as $z \to 0$.

### 4e – Edge Effects

It is well known that a sharp macroscopic edge can stop the spreading of a liquid. Gibbs[56] derived a purely geometrical condition for the equilibrium of a drop bound by an edge such as the circular edge shown in Figure 10a. The liquid will leave the edge and wet the inclined region when the angle $\theta^*$ reaches a critical value

$$\theta^* = \theta - \alpha \qquad (45)$$

where $\alpha$ is taken as negative. The stopping effect of the edge for $\alpha < 0$ (Figure 10a) was recently re-analysed by Oliver et al.[57] for an axially symmetric drop in the absence of gravity. Both their analysis and experimental observations[57,58] confirm that Gibbs equation is correct, even for $\theta = 0$. That surface edges may have a major role on hysteresis was suggested by Huh and Mason[47]; more recently, Oliver and Mason[59] reported scanning microscope observations on the effect of micro-edges on liquid spreading.

# WETTABILITY OF POLYMER SURFACES 131

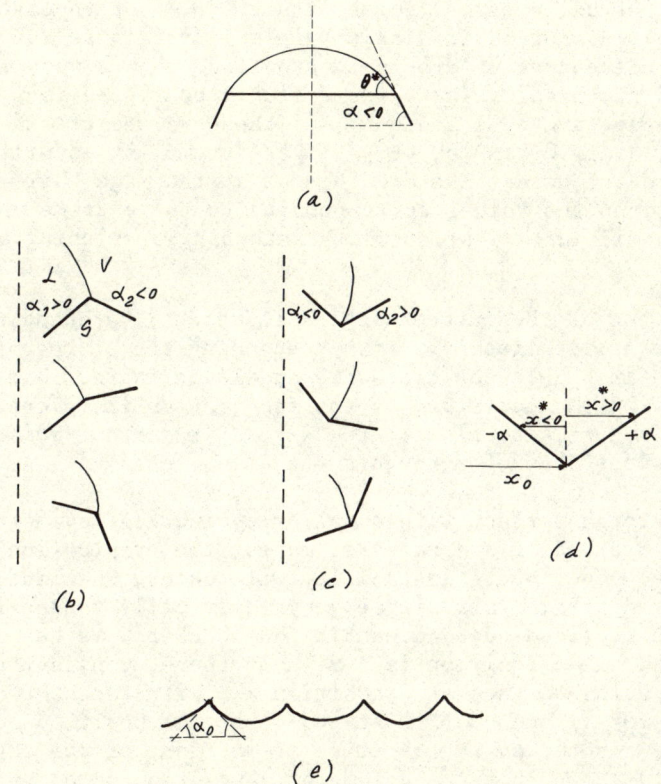

Figure 10. a) Macroscopic edge on a solid. The drop leaves the edge to the inclined region when $\theta^* = \theta - \alpha$. b) Stable edges (protruding edges): the slope decreases as the distance to the symmetry plane (dotted) of the drop increases. c) Unstable edges (recessed edges). d) A symmetric recessed edge at a position $x_0$; $x^*$ is the coordinate relative to the edge. e) Rough surface with stable edge configurations; all other positions are unstable. $\alpha_0$ is the slope angle at the edge.

We shall in this section provide a thermodynamic basis for the pinning effect of edges, by extending the results of the previous analysis of rough surfaces. An edge is conveniently described as a region of the surface with large curvature. Using the notation of the previous sections, $z'' \to \pm \infty$ at an edge. This region of large curvature contacts the "faces" of the edge with slopes $\alpha_1$ and $\alpha_2$ at the edge, $\alpha_1$ being the slope in the near face, that is, the one closer to the symmetry axis or plane in spherical or cylindrical drops, respectively; $\alpha_2$ is the slope in the far face. As shown in section 4b, stability at an edge will occur if $z'' \to -\infty$. Edges of this type have $\alpha_1 > \alpha_2$ and examples are given in Figure 10b; they will be termed protruding edges. Edges with $z'' \to +\infty$, or recessed edges (Figure 10c), do not admit stable triple lines. These results are of course independent of drop size provided $|z''|$ is very large; this is expected at an edge. The extreme stable configurations at a protruding edge are $\theta-\alpha_1$ and $\theta-\alpha_2$, if these values are in the interval $0$, $\pi$. If $\theta-\alpha_1<0$ or $\theta-\alpha_2>\pi$ the extreme angles will be $0$ or $\pi$ according to the cases. The drop hinges at the edge while the volume changes between the values corresponding to the extreme angles. This is a thermodynamic justification of the stopping or pinning effect of an edge.

When one of the extreme angles is reached by changing the volume, the triple line may start descending the corresponding face of the edge if stable positions are available there. This will depend essentially on the curvature of the face, or on its slope if the curvature is zero, and also on the size of the drop, according to the conclusions drawn for rough surfaces.

If a rough surface with axial or cylindrical symmetry contains successive protruding and recessed edges, the triple line may reach the recessed edges where instability will occur. The line may then jump to an adjacent protruding edge or eventually to the far or near face of the recessed edge (depending on whether V is being increased or decreased) if a position is available there. Consider a recessed edge in an axial surface at a position $x_0$, with faces of constant and equal slope $\alpha$ and $-\alpha$ (Figure 10d). Let the position relative to the edge be denoted by $x^*$ (positive if in front of the edge). Then it is easily shown that the volume at a position $x^*$ on the face, with angle $\theta^*$, is

$$\frac{V}{\pi x_0^3} = \text{constant} + \frac{(1+X)^3}{3} f(\theta^*) \pm \text{tg}|\alpha| X(1+X+\frac{X^2}{2}) \qquad (46)$$

where $X = \frac{x^*}{x_0}$. For $X>0$, the $+$ sign should be used and for $X<0$, the $-$ sign. In receding, the recessed edge is reached with $\theta^*=\theta-|\alpha|$ and in advancing with $\theta^*=\theta+|\alpha|$. The position on the faces of the edge, with the same volume, are defined by a X which, for $X\ll 1$, is given by

$$|X| = \frac{f(\theta+|\alpha|)-f(\theta-|\alpha|)}{3[f(\theta\pm|\alpha|)\pm\text{tg}|\alpha|]} \qquad (47)$$

The top signs are for an advancing drop. If the faces are not wide enough, no position is available there and the triple line will move to the protruding edge. This will in general happen for large drops.

In special cases, such as the one shown in Figure 10e, only

(protruding) edge configurations are stable and the triple line moves between such edges either in advancing (with $\theta^*$ near $\theta+|\alpha_o|$) or receding (with $\theta^*$ near $\theta-|\alpha_o|$), where $|\alpha_o|$ is the angle indicated in Figure 10e.

As in rough surfaces, some of the jumps may require an increase in energy. In this case, but possibly also if the energy decreases, the drop may give rise to un-connected shapes.

So far it has been assumed that the energy of an edge is the same as that of the smooth (low curvature) regions of the surface. Because of the high curvature of the edge this is unlikely to be so. We then assign a positive line energy $\varepsilon$ to the edge[60] and distinguish between a "dry" and a "wet" edge. The corresponding line energies will be denoted by $\varepsilon_d$ and $\varepsilon_w$ respectively. When the triple line is on the edge, the energy is probably between $\varepsilon_d$ and $\varepsilon_w$ and we shall assume that $\varepsilon$ varies continuously from $\varepsilon_d$ to $\varepsilon_w$ as shown in Figure 11. It is expected that $\varepsilon_w < \varepsilon_d$ for $\theta<90°$ and $\varepsilon_w>\varepsilon_d$ for $\theta>90°$. Let us see how the range of stability of the edge is affected by this effect. The complete analysis can be made by introducing the edge energy term in the equation for A. This term is $2\pi x_o \varepsilon(x)$ for axial symmetry, where $x_o$ is the radius of the edge. The range of stability is not affected for large drops, but for small drops the extreme stable values of $\theta^*$ will change. For example, if $\varepsilon_w<\varepsilon_d$ ($\theta<90°$) the receding $\theta^*$ will be smaller while the advancing $\theta^*$ will be larger than in the absence of edge energy effects.

The same type of analysis can be applied to sharp changes in the contact angle, such as at the boundaries between two homogeneous patches on the surface. Such boundaries also pin the triple line and provide a range of stable configurations with $\theta^*$ values between $\theta_1$ and $\theta_2$, the contact angles in the two homogeneous patches. This is true if $\theta$ decreases with x. Otherwise the position at the discontinuity in $\theta$ is unstable.

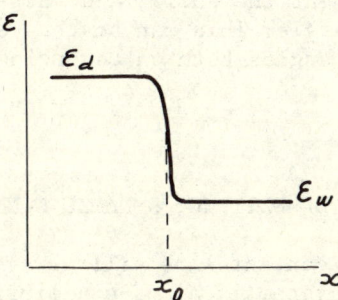

Figure 11. The line energy $\varepsilon$ of an edge changes from $\varepsilon_d$ to $\varepsilon_w$ as the triple line changes position. $\varepsilon_d$ is the energy for the dry edge and $\varepsilon_w$ that for the wet edge.

In general, the pinning effect and the corresponding stick-and-slip movement of the triple line can be attributed to singular lines on the surface where there is a rapid change in $\theta-\alpha$, such that $\theta-\alpha$ decreases as x increases.

### 4f - Random Rough and Heterogeneous Surfaces

When rugosity and heterogeneity are not such that the entire system has either axial or cylindrical symmetry the analysis of hysteresis is considerably complicated by the fact that the shape of the fluid interface, even in the absence of gravity, is very difficult to determine. A perturbation treatment was used by Huh and Mason[47] to determine the fluid interface shape in contact with a random rough surface, but unfortunately the method does not allow the treatment of hysteresis.

The hysteresis behavior on an arbitrary surface can however be predicted qualitatively from the results obtained for simple geometries. The main complication is that the triple line will be undulated and correspondingly the fluid interface will show convolutions. On advancing it is likely that "most" of the triple line will be on regions of high $\theta-\alpha$ or pinned on protruding edges near the maximum stable $\theta*$, while in receding it will prefer places of low $\theta-\alpha$ or be pinned on protruding edges near the minimum stable $\theta*$.

The drop size effect on the observed contact angle is probably related to the path of the triple line on the surface (that is, to its undulations), but not in the way proposed by Good and Koo[52]. It is likely that, for small drops, the corrugations of the triple line are smaller than for large drops, because of the relatively larger increase in the area of the fluid interface that will occur in small drops for the same amplitude and wavelength of the corrugations. On the other hand, corrugations increase the volume of the liquid region near the solid, and therefore tend to increase the observed contact angle. This can be the reason why small drops exhibit lower contact angles both on advancing and receding.

## 5 - KINETICS OF WETTING; DYNAMIC CONTACT ANGLES

Movement of the contact line will always occur if the initial configuration of the system is a non-equilibrium one or if the system is acted by external stresses, such as when a drop is squeezed between two plates. In the absence of such stresses, movement of the liquid front may be regarded as produced by surface tension and gravity forces. Inertial and viscous forces have to be considered as well. The complexity of the problem is appreciable, specially on rough polymer surfaces[61].

The contact angle measured during movement of the liquid is termed the dynamic contact angle. It should be noted that dynamic surface tensions, particularly of the fluid interface[62], should also be considered. These are in general variable with time and distinct from the equilibrium surface tensions (for example, due to adsorption in liquid solutions).

In a very crude approach to the problem, let us consider a sessile drop on a flat plate advancing slowly to its equilibrium configuration and assume that the driving force for the displacement of the contact line is due to the surface tensions regarded as forces acting on this line (Figure 1). It is then necessary that the components of such forces parallel to the plate have a resultant directed in the advancing direction. This implies that

$$\cos \bar{\theta}_a > \cos\theta_a \qquad (48)$$

where $\bar{\theta}_a$ is the advancing dynamic contact angle and $\theta_a$ is the equilibrium advancing contact angle. If the drop recedes, the condition is

$$\cos \bar{\theta}_r < \cos\theta_r . \qquad (49)$$

For example, the dynamic advancing angle is larger than the equilibrium angle for $\theta_a < \pi/2$ but smaller for $\theta_a > \pi/2$. If the surface tension forces are opposed by a friction force proportional to the velocity v of the line of contact, with a proportionality constant $\lambda$, then

$$\cos \bar{\theta} - \cos\theta = (\lambda/\gamma)v \qquad (50)$$

both for the advancing and receding modes. $\lambda$ should depend on the liquid viscosity and also on some property or properties of the solid-liquid interface. If this equation is correct then the dynamic angles should be constant for constant v, and deviate more and more from the equilibrium values as v increases. Equation (50) also shows that a correlation between v and contact angles should be tried with ($\cos \bar{\theta} - \cos \theta$) rather than with $\cos \theta/\cos \theta$ as attempted by Schonhorn et al.[62] to interpret their experimental results on wetting by polymer melts.

In spite of the complexity of the problem some progress has been achieved in various theoretical studies[64-71], while careful experimental observations have also been undertaken[1,61,72,73] at low and high wetting speeds.

When the liquid spreads on the solid ($S_{L/S} > 0$, see Equation (3)) the problem may be even more complicated. There is in this case a film of liquid that spreads ahead of the bulk[74,75], but very little is known on the mechanism of formation and propagation of this film.

## 6 - CONCLUDING REMARKS

The enormous practical importance of wettability of solids explains the large effort that has been put in recent years on the understanding of the mechanism and factors that govern wetting. Nevertheless, a large number of problems on wetting and spreading have not yet been solved. This is not surprising considering the complexity of the phenomena (which is related to the intrinsic complexity of solid surfaces, particularly those of polymers) and in spite of the powerful experimental techniques which are now available to study solid surfaces.

The study of the molecular mechanism of wetting is still in its initial stage and can hardly provide, in the near future, a general explanation for the complexities observed in wetting. The macroscopic thermodynamics of interfaces should be examined more carefully and probably the concepts of line tension or, alternatively, of perturbed interfacial tensions, should be included in the routine theoretical analysis of wetting. Contact angles are central to the theoretical and experimental treatment of wetting, but unfortunately an enormous variety of them can, and have to, be defined. It is important to realize that the simple indication of a contact angle value, without clearly defining the type of angle that was measured, will not tell much about the wettability of a surface.

Although the essential of contact angle hysteresis seems to be understood, it is likely that the determination of the equilibrium shape of liquid surfaces in contact with more "real" solid surfaces and a better understanding of the dynamic aspects of spreading, together with a more careful characterization of polymer surfaces (topography, chemical heterogeneity, etc.), will be decisive in this context.

## REFERENCES

1. M. C. Wilkinson, R. Ellis, M. P. Aronson, J. W. Vanderholf and A. C. Zettlemoyer, J. Colloid Interface Sci., $\underline{68}$, 545 (1979)
2. M. A. Fortes, Phys. Chem. Liq., $\underline{9}$, 285 (1980)
3. A. W. Adamson, "Physical Chemistry of Surfaces", Wiley, New York, 1976
4. E. A. Boucher, M. J. Evans and H. J. Kent, Proc. Royal Soc. $\underline{A349}$, 81 (1976)

5.  D. T. Clark, A. Dilks and D. Shuttleworth, in "Polymer Surfaces", D. T. Clark and W. J. Feast, Editors, Chap. IX, John Wiley, 1978
6.  H. Schonhorn, ibid., Chap. X
7.  F. Galembeck, J. Polym.Sci., Polym. Lett. 15, 107 (1977)
8.  T. E. Nowlin and D. Foss Smith, Jr., J. Appl. Polym. Sci., 25, 1619 (1980)
9.  R. J. Good, J. Colloid Interface Sci., 52, 308 (1975)
10. B. W. Cherry, "Polymer Surfaces", Cambridge Univ. Press, 1981
11. S. Wu, Polymer Reprints, 11, 1921 (1970)
12. J. D. Andrade, S. M. Ma, R. N. King and D. E. Gregonis, J. Colloid Interface Sci., 72, 488 (1979)
13. D. W. von Krevelen, "Properties of Polymers - Correlations with Chemical Structure", Elsevier, New York, 1976
14. B. D. Davis, J. Colloid Interface Sci., 59, 420 (1977)
15. R. J. Good, J. Colloid Interface Sci., 59, 398 (1977)
16. A. J. Kinloch, J. Materials Sci., 15, 2141 (1980)
17. P. Becher, J. Colloid Interface Sci., 59, 3 (1977)
18. F. M. Fowkes, J. Phys. Chem., 66, 682, 1863 (1962)
19. F. M. Fowkes, J. Phys. Chem., 67, 2538 (1963)
20. G. Navascués, J. Colloid Interface Sci., 72, 150 (1979)
21. F. M. Fowkes, Ind. Eng. Chem., 74, 3305 (1970)
22. S. Wu, J. Polym. Sci., Part C, No.34, 19 (1971)
23. W. A. Zisman, in "Contact Angle, Wettability and Adhesion", Advan. Chem. Ser., 43 p. 1, Amer. Chem. Soc., Washington, 1964
24. G. Navascués, Rep. Prog. Phys., 42, 1131 (1979)
25. E. G. Shafrin, in "Polymer Handbook", J. Brandrup and E. H. Immergut, Editors, John Wiley, 1975.
26. L. H. Lee, J. Polym. Sci., A25, 1103 (1967)
27. M. Toyama, A. Watanabe and T. Ito, J. Colloid Interface Sci., 47, 802 (1974)
28. A. W. Neumann, R. J. Good, C. J. Hope and M. Sejpal, J. Colloid Interface Sci. 49, 291 (1974)
29. A. W. Neumann, Adv. Colloid Interface Sci., 4, 105 (1974)
30. L. K. Sharples, Plast. Polym., 37, 135 (1969)
31. A. Marmur, M. Narkis and W. Woogen, J. Appl. Polym. Sci., 25, 1253 (1980)
32. Y. Oh and M. S. John, J. Colloid Interface Sci., 73, 467 (1980)
33. M. V. Berry, J. Phys. A, 7, 231 (1974)
34. G. J. Jameson and M. C. G. del Cerro, J. C. S. Faraday Trans. I, 72, 883 (1976)
35. L. R. White, J. C. S. Faraday Trans. I, 73, 390 (1977)
36. G. Saville, J. C. S. Faraday Trans. II, 8, 1182 (1977)
37. J. A. de Feijter and A. Vrij, Electroanal. Chem. Interfac. Electrochem., 37, 9 (1972)
38. G. Navascués and M. V. Berry, Molecular Phys., 34, 649 (1977)
39. D. E. Sullivan, J. Chem. Phys., 74, 2604 (1981)
40. P. C. Wayner, Jr., J. Colloid Interface Sci., 77, 495 (1980)
41. M. A. Fortes, J. C. S. Faraday Trans. I, 78, 101 (1982)

42. B. A. Pethica, J. Colloid Interface Sci. 62, 567 (1977)
43. P. Tarazona and G. Navascués, J. Chem. Phys., in press (1981)
44. L. Boruvka and A. W. Neumann, J. Phys. Chem. 66, 5464 (1977)
45. R. E. Johnson, Jr. and R. H. Dettre, in "Contact Angle, Wettability and Adhesion", Advanc. Chem. Ser. No. 43, p. 112 (1964); "Surface and Colloid Science", E. Matijevic, Editor, Vol. 2, p. 85, Interscience, New York, 1969
46. R. E. Johnson Jr., R. H. Dettre and D. A. B. Brandreth, J. Colloid Interface Sci., 62, 205 (1977)
47. C. Huh and S. G. Mason, J. Colloid Interface Sci., 60, 11 (1977)
48. R. N. Wenzel, Ind. Eng. Chem., 28, 988 (1936); J. Phys. Colloid Chem., 53, 1466 (1949)
49. S. Baxter and A. B. D. Cassie, Textile Inst. 36, T67 (1945)
50. A. B. D. Cassie, Discuss. Faraday Soc. 3, 11 (1948)
51. R. H. Dettre and R. E. Johnson, in "Wetting", p. 144, Soc. Chem. Industry, Univ. of Bristol, 1967
52. R. J. Good and M. N. Koo, J. Colloid Interface Sci., 71, 283 (1979)
53. L. Boruvka and A. W. Neumann, J. Colloid Interface Sci., 65, 315 (1978)
54. R. Shuttleworth and G. L. J. Bailey, Disc. Farad. Soc., 3, 16 (1948)
55. J. D. Eick, R. J. Good and A. W. Neumann, J. Colloid Interface Sci., 53, 235 (1975)
56. J. W. Gibbs, "Scientific Papers" vol. I, p. 326 (1906)
57. J. F. Oliver, C. Huh and S. G. Mason, J. Colloid Interface Sci., 59, 568 (1977)
58. E. Bayramli and S. G. Mason, J. Colloid Interface Sci., 66, 200 (1978)
59. J. F. Oliver and S. G. Mason, J. Colloid Interface Sci., 60, 480 (1977)
60. J. M. Blakely, "Introduction to the Properties of Crystal Surface p. 83, Pergamon Press, 1973
61. J. F. Oliver, C. Huh and S. G. Mason, Colloids and Surfaces, 1, 79 (1980)
62. M. Ronay, J. Colloid Interface Sci., 66, 55 (1978)
63. H. Schonhorn, H. L. Frisch and T. K. Kwei, J. Appl. Phys., 37, 4967 (1966)
64. E. B. Dussan V., J. Fluid Mech. 77, 665 (1976)
65. E. B. Dussan V. and S. H. Davis, J. Fluid Mech. 65, 71 (1974)
66. L. M. Hocking, J. Fluid Mech. 76, 801 (1976)
67. J. Lopez, C. A. Miller and E. Ruckenstein, J. Colloid Interface Sci., 56, 460 (1976)
68. E. Ruckenstein and C. S. Dunn, J. Colloid Interface Sci., 59, 135 (1977)
69. C. Huh and S. G. Mason, J. Fluid Mech. 81, 401 (1977)
70. H. P. Greenspan, J. Fluid Mech. 84, 125 (1978)
71. J. Lowndes, J. Fluid Mech. 101, 631 (1980)

72. T. A. Elliot and D. M. Ford, J. C. S. Faraday Trans. I, 68, 1814 (1972)
73. A. M. Schwartz and S. B. Tejada, J. Colloid Interface Sci., 38, 359 (1972)
74. D. H. Banghan and Z. Saweris, Trans. Faraday Soc., 34, 554 (1938)
75. A. Marmur and M. D. Lelah, J. Colloid Interface Sci., 78, 262 (1980)

NOVEL METHODS OF STUDYING POLYMER SURFACES BY EMPLOYING CONTACT ANGLE GONIOMETRY

Frank J. Holly

Department of Ophthalmology and Visual Sciences
Texas Tech University Health Sciences Center
Lubbock, Texas 79430

Highly sophisticated instrumentations are available today for studying solid polymeric surfaces. Nevertheless, the relatively modest technique of contact angle goniometry has a disproportionately high potential for providing information on surface energetics and surface transformation that are largely unexploited. Such techniques employing one- and two-condensed-phase contact angle measurements are especially useful for studying the interfacial properties of polymeric surfaces, both electrically charged and neutral, that exhibit variability in surface composition and thus in surface properties, due to segmental mobility. Plasma-treated polymer surfaces, hydrogel surfaces, tissue and cellular interfaces are ordinarily prone to rapid and drastic surface changes depending on the polarity of the adjacent fluid phase or solid surface. Relations that have proven to be useful include adhesion tension and the relative contact angle hysteresis both as a function of the advancing contact angle. Hence, information can be obtained on surface hydrophillicity, thin film stability as related to disjoining pressure, and water-solid interfacial tension, all which play an important role in such widely diverse phenomena as the biocompatibility of contact lenses and other biomaterials, surgical trauma, tear film stability, cellular interaction, and bioadhesion.

## INTRODUCTION

The affinity of solid polymeric surfaces for water is of great practical importance in industry and even more so in biology and the medical sciences. The water wettability of solid surfaces, the stability of thin aqueous films over solids, and the magnitude of water-solid interfacial tensions are important characteristics that have been recognized in such seemingly diverse phenomena as the biocompatibility of prostheses, contact lens tolerance, surgical trauma resulting from contact adhesion between tissue and foreign surfaces, tissue adhesion, and cellular interaction. Despite the importance of surface hydrophilicity, a great deal of confusion and contradiction over the term exists in the literature. Even the term is often poorly defined or is derived from bulk properties.

During the past decade, we have been involved in studying the water wettability of polymeric surfaces including those of hydrogels as well as the wettability of tissue boundaries primarily in the eye. Using mostly contact angle goniometry, we discovered interesting and revealing interrelations among some of the quantities measured that yielded insight into processes occurring at surfaces and interfaces.

### Various Surface Aspects Studied

<u>Surface Hydrophilicity</u>. Usually the hydrophilic nature of a solid surface is described in terms of wetting. Since the phenomenon of wetting can be described in various ways, it often becomes a matter of controversy whether a solid surface is hydrophilic.

Wetting can be considered as a <u>spreading phenomenon</u>. The solid can be classified as hydrophilic if water spontaneously spreads on its surface. This implies that the spreading coefficient of water is positive and thus the advancing contact angle of water is zero. By this criterion, only a few solid surfaces in practice can be classified as hydrophilic. In a clean state, such a surface has a high surface free energy and thus can readily become contaminated. Such a surface would also have a fairly high interfacial tension against water despite their water wettability.

Wetting can also be considered as a <u>capillary phenomenon</u>. The inner wall of the capillary may be defined as hydrophilic if the capillary rise of water is positive. This implies that the advancing contact angle of water is less than $90°$. This criterion is much less stringent than the previous one, and hydrophilicity would imply that the solid-vapor interfacial tension is greater than the solid-water interfacial tension.

Another criterion for surface hydrophilicity requires that the solid surface exhibit a <u>preference toward water</u> in the presence of a nonpolar liquid. When the two-phase contact angle measured across the water phase is less than $90°$, the surface is defined as hydrophilic. By carefully choosing the nonpolar liquid so that its surface tension is about the same as the dispersion force component of the water surface tension (21.8 mN/m), then the dispersion force components will cancel and the cosine of the two-phase contact angle will be proportional to the magnitude of the polar interaction across the interface.[1] It is known that the magnitude of molecular interaction across the interface directly affects the solid-water interfacial tension, therefore, this approach enables one to rank various solid surfaces according to the relative magnitude of their interfacial tensions.[2]

The critical surface tension of Zisman[3] has been widely used to characterize solid surfaces with low surface free energy. Since the method uses hydrophobic liquids to determine contact angles, the quantity obtained is characteristic of the dispersion force field component of the solid surface free energy.[4,5] Only for a completely apolar solid surface would the method yield reliable information as to its hydrophilicity. Otherwise, the magnitude of the critical surface tension could be completely misleading. For example, the critical surface tension of polyethylene consisting mostly of $-CH_2-$ groups is about the same as that of a poly(acrylamide) gel containing 78% water at equilibrium-hydration. The water contact angle on polyethylene is $94°$ while it is only about $10°$ on poly(acrylamide) (cf. Table I).

In our experience, the water-in-air advancing contact angles are also poor indicators of surface hydrophilicity, even if one uses the criterion of the capillary rise, i.e. the surface is hydrophilic if the advancing contact angle is less than $90°$. Poly(methylmethacrylate) [PMMA], a solid polymer that can only absorb about 1.5% water at equilibrium hydration exhibits an advancing contact angle of about $73°$. Poly(2-hydroxyethyl methacrylate) (PHEMA), a hydrogel with an equilibrium water content of 38-42% can also exhibit an advancing contact angle of similar magnitude even when its surface is fully hydrated (such as measured by the captive-bubble technique).

We found the criterion of preferential wetting to be the most realistic for deciding the hydrophilic character of solid and gel surfaces. Table I contains the water-in-octane contact angle for polyethylene, PMMA, PHEMA, poly(2-hydroxyethyl acrylate) [PHEA], poly(glyceryl methacrylate) [PGMA], and poly(acrylamide) [PAA]. Using the $90°$ as the dividing value, we can see that by this definition polyethylene and PMMA have hydro-

Table I. Critical Surface Tension and Contact Angle Values for Polymeric Solids and Hydrogels.

| Gel or Solid | Water Content | Critical Surface Tension (dyne/cm) | Advancing Contact Angle Water/Air | Advancing Contact Angle Water/Octane |
|---|---|---|---|---|
| Polyethylene | 0% | 32.1 | 94° | 153° |
| PMMA | 1.5% | 38.5 | 73° | 120° |
| PHEMA | 42.0% | 36.0 | 60° | 88° |
| PHEA | 72.% | 34.1 | 44° | 79° |
| PGMA | 73.% | 37.6 | 41° | 63° |
| PAA | 78.% | 32.9 | 10° | 12° |

Note: For abbreviations see text.

phobic while the hydrogels have hydrophilic surfaces.

Variance of Surface Properties. When describing a solid surface, one often presumes or implies that the surface characteristics are well-defined, inherent, and invariant properties of the surface that will not change unless the composition of the solid is changed. Unfortunately, this assumption is commonly untrue. Often the method of measurement, such as the liquid used in contact angle goniometry will effect surface changes that will result in completely different surface characteristics. One indication of such changes could be the often-observed contact angle hysteresis, i.e. the dependence of the contact angle on the manner of the drop formation.

One well-known cause of contact angle hysteresis is surface roughness. If the advancing contact angle is less than $90°$ on such a surface, then the water will penetrate the valleys and crevices of the surface where the drop is in contact with the solid. When the volume of the droplet is decreased, the drop edge will have to recede over a composite surface that partially consists of water. Naturally this will decrease the value of the receding contact angle.

Adam[6] has listed two additional causes of contact angle hysteresis. One is physical interaction between the fluid and the solid. In the case of water, this would take the form of hydration, dissolution, or possible deposition of soluble contaminants. The other cause is chemical interaction between the liquid and solid. Thus, with water it could be hydrolysis or water-induced decomposition. In both cases, the solid surface in contact with water changes so that the receding contact angle is now defined by a different solid-vapor interface.

By forming hydrogels with an optically smooth surface that were equilibrated with water for prolonged periods of time, we could eliminate the three possible causes of contact angle hysteresis that were discussed above. Nevertheless, the contact angle hysteresis on hydrogels such as PHEMA, PHEA, PGMA, and PAA[7] was found to be quite pronounced. This led us to postulate another cause of contact angle hysteresis; the stereo-chemical interaction, i.e. conformational and/or orientational changes in the surface polymeric chains and segments due to the change in the polarity of the adjacent fluid phase (Figure 1).

We have shown[2] that the water wettability of hydrogels as measured by the advancing contact angle of water-in-air does not increase with increasing water content. In fact, the wettability often decreases with increasing water content for a given type of polymer matrix. We explained this anomaly by pointing out

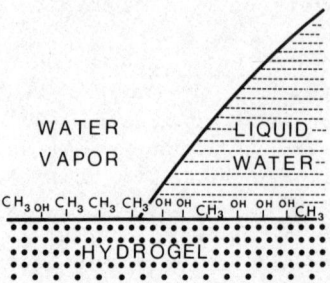

Figure 1. Schematic view of the stereochemical changes at the water-gel and vapor-gel interfaces (from ref. 7).

that increasing equilibrium water content is achieved by decreasing the crosslink density in the matrix, and this would allow greater mobility in the surface polymer chains and segments. The hydrophilicity of the polymer matrix, the surface polymeric chain mobility, and the relative density of the hydrophilic and hydrophobic groups on the polymer backbone (amphipathic character of the segments) determine both water wettability and the extent of contact angle hysteresis.

More recently Yasuda and Sharma have demonstrated[8] that polymeric solids having little or no water content also exhibit variability of surface properties due to short-range rotational mobility of the surface polymer molecules.[9] The same has been found to be true for plasma-treated surfaces and for surface of hydrophilic polymers grafted onto a hydrophobic substrate polymer[10] by using the sophisticated ESCA technique.

We defined relative contact angle hysteresis ($H_R$) as the difference between the advancing and receding contact angle values divided by the advancing contact angle. Table II contains the equilibrium water content, the water-in-air advancing contact angle, and the relative contact angle hysteresis for polyethylene, PMMA, and the four hydrogels. It is apparent from the data that polyethylene has negligible hysteresis. The $H_R$ for PMMA, while considerable, is still lower than the hysteresis on the hydrogels. Among the hydrogels, the highest degree of hysteresis is exhibited by the PGMA gel, which has two hydroxyl groups and one methyl group on its side chains capable of stereochemical changes. PHEMA has one hydroxyl group and a methyl group, so its hysteresis value is still quite high. Both PHEA and PAA gels are missing the hydrophobic methyl group and

Table II. Water Wettability and Contact Angle Hysteresis of Polymers and Hydrogels.

| Solid or Gel | Water Content | Advancing Contact Angle in degrees* | Relative Contact Angle Hysteresis |
|---|---|---|---|
| Polyethylene | 0% | 94° | 0.02 |
| PMMA | 1.5% | 73° | 0.33 |
| PHEMA | 42.0% | 60° | 0.72 |
| PHEA | 72.% | 44° | 0.59 |
| PGMA | 73.% | 41° | 0.76 |
| PAA | 78.% | 10° | 0.60 |

*water-in-air

Note: For the abbreviations see text.

thus both exhibit less hysteresis than the previous two gels.

An interesting and useful relationship is found when the relative contact angle hysteresis is plotted against the advancing contact angle for the hydrogels and for a copolymer of dimethyl siloxane and vinyl pyrrolidone (SIL-PVP) (see Figure 2). For a given hydrogel, the value of $H_R$ appears to increase linearly with the advancing contact angle, i.e. with diminishing wettability. As one can see, the slope is greater for the hydrogels where the polymer matrix does not contain the highly hydrophobic methyl side-groups, i.e. for PAA and PHEA, indicating that even high segmental mobility at the surface cannot make the surface more hydrophobic above a given level. The positive slope for all the hydrogels indicate that the higher segmental mobility of the amphipathic polymeric segments will lead to both decreased wettability when exposed to water-vapor saturated air, and higher contact angle hysteresis.

It is clear from the figure that the copolymer SIL-PVP exhibits a behavior just the opposite of that of the hydrogels. The reason for this seemingly anomalous behavior is that only the PVP component imparts hydrophilicity to the surface and only this component possesses amphipathic character provided that sufficient chain mobility is present. Thus increasing the PVP content will increase both wettability and relative hysteresis, hence the negative slope observed. This technique can be quite useful for evaluating polymer surfaces that had been exposed to glow-discharge treatment and where the surface composition and stability are usually unknown.[11]

<u>Effect of Adsorbed Solutes at Polymer-Water Interfaces</u>. It is well known that solid-liquid interfacial tension cannot be directly measured. It is possible, however, to determine the film pressure of the solute(s) adsorbed at the interface, i.e. the decrease in interfacial tension due to adsorption, provided that the solid-vapor interfacial free energy has not been affected appreciably by the presence of the solute in the solution. The solute film pressure at the interface ($\Pi_i$) then is simply given by the difference between the adhesion tension of the solution and that of the solvent[12]:

$$\Pi_i = \gamma_{sol'n} \cos\theta_{sol'n} - \gamma_w \cos\theta_w$$

We have measured the interfacial film pressure of various water-soluble polymers at several hydrogel-water interfaces and at the polyethylene-water interface.[13] For the latter, $\Pi_i$ was positive and approximately equal to the surface tension depression of the solvent ($\Pi_o$) at the solution-air interface. When the soluble polymer was more hydrophilic than the gel

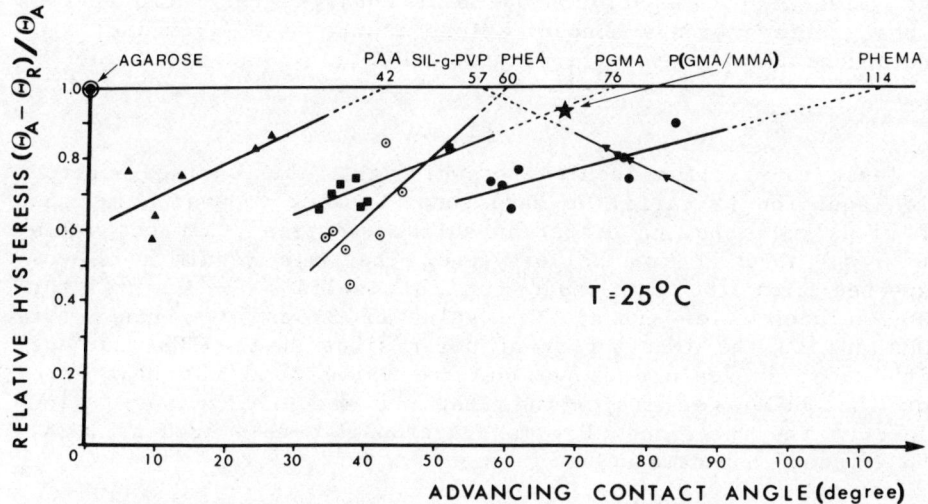

Figure 2. Relative contact angle hysteresis for various hydrogels and dimethyl siloxane - vinyl pyrrolidone copolymers (SIL-g-PVP). The value obtained for agarose gels containing more than 90% water is also induced (from ref. 7).

polymer matrix, the film pressure of the adsorbed polymer was positive. When the polymeric adsorbate was the same as the polymer in the gel matrix, the value of $\Pi_i$ was approximately zero or even less. When the polymeric adsorbate was less hydrophilic than the gel matrix, the value of $\Pi_i$ was negative possibly indicating an increase in interfacial tension due to solute adsorption most likely entropy-driven.

<u>Adhesion Tension of Solutions to Gels and Solids</u>. It is known for many systems that the adhesion tension of the solution to the solid defined as the difference between the solid-vapor surface tension and the solid-solution interfacial tension;

$$W_T = \gamma_{sv} - \gamma_{sl} - \gamma_{lv} \cos\theta$$

is a linear function of the solution surface tension $\gamma_{lv}$. In a $W_T$ versus $\gamma_{lv}$ plot, the zero contact angle (complete wetting) is represented by a straight line with a slope of 1, while the rectilinear behavior of the solutions can be described as

$$W_T = A - B\gamma_{lv}$$

Wolfram[14] and later Lucassen-Reynders[15] demonstrated that the value of the negative slope B is equal to the ratio of the Gibbs' surface excess concentrations at the solid-liquid ($\Gamma_{sl}$) and at the liquid-vapor ($\Gamma_{lv}$) interfaces:

$$B = \Gamma_{sl}/\Gamma_{lv}$$

at least for small molecules for which the Gibbs surface adsorption equation is valid. We have found[13] using aqueous solutions of biopolymers having different surface tensions that the value of B is about 1 for polyethylene. The same result could be expected from the fact that for this solid $\Pi_i = \Pi_o$ for many solutes (vide supra). The value of B was less than 1 for PMMA due to the interaction of polar sites on the PMMA surface with water lessening adsorption. The value of B was near zero for the PHEMA gel indicating that not much change was taking place at the interface. For more hydrophilic gels such as PHEA, the value of B became negative.

Especially with biopolymeric solutes, the value of B is affected not only by the type of solid used but, to a lesser extent, by the polymeric solutes as well. Furthermore, as the solutions became so dilute that their surface tension was higher than 60 dyne/cm, the linearity of the $W_T$ versus $\Pi_{lv}$ relation was lost. Despite these shortcomings, the Wolfram plot could be used to determine the approximate surface tension of body fluids and gland secretions available only in microliter quantities. First the advancing contact angle of the drop is measured. Then it is serially diluted to volumes high enough so both the surface tension and the contact angle could be determined. These latter data are employed to determine the value of A and B in Equation 3. Then the surface tension of the undiluted droplet could be calculated from the equation:

$$\Pi_{sol'n} = A/(\cos\theta+B)$$

The same drop could also be used to determine the receding contact angle and the water wettability of the biopolymer adsorbed on the solid. We employed glossy high-density polyethylene slabs as the solid substrate.

<u>Contact Angle as an Indicator of Surface Polarization</u>. It is possible to provide certain polymers with various levels of electric charge density by plasma treatment or by subjecting a polymer film to an electric bias of sufficient strength.[16] The resulting electrets are often characterized by their net charge density as measured in air by appropriate techniques. The value thus obtained is the algebraic sum of the homocharge and the heterocharge densities, i.e. it is a small difference between two, relatively large numbers. It is reasonable to assume that

interfacial phenomena occurring at a charged polymer – water interface will depend more on the magnitude of electrical polarization at the solid surface produced during the formation of the electret rather than the net charge density due to the short range of the electrical forces in an aqueous, electrolyte-type of medium. Such a surface polarization would affect the contact angle of sessile droplets or captive gas bubbles as they are known to exhibit systematic variation with surface potential and surface charge density.[1]

We have shown[18] that the difference between charge densities located at the solid-liquid and solid-vapor interfaces is given by the variation in adhesion tension, or more specifically of the cosine of the contact angle, with respect to the potential difference across the interface:

$$\gamma_{lv}\{\partial(\cos\theta)/\partial(\Delta\Phi)\}_{T,p,\mu} = \sigma_{sl} - \sigma_{sv}$$

For nonconducting surfaces such as electrified polymers, the zeta potential could be used as the independent variable, since it would be proportional to the surface potential at a given electrolyte concentration in the absence of potential-determining ions. We found[18] for a model system that the adhesion tension of water was proportional to the second power of the interfacial potential, i.e. that the charge density and the potential were linearly related. Furthermore, the charge effect was localized almost entirely at the solid-vapor interface resulting in a decrease in adhesion tension upon increasing charge density. The magnitude of the adhesion tension variation with surface potential indicated that contact angle changes in the range of 5 to 10 degrees can be expected for a nonconductive polymer surface having a surface charge density of $10^2 - 10^3$ stat-C.cm$^{-2}$ provided that the charge effects dominate at only one of the interfaces.

Contact angle measurements using water droplets on poly(vinylidene chloride – vinyl chloride) and poly(fluoroethylene-propylene) copolymer sheets, cleaned with 1% aqueous solution of Micro in an ultrasonic cleaner, then rinsed repeatedly in double-distilled water, laboratory detergent (International Products Corp., Trenton, N.J.) were made. The contact angle consistently decreased 3 – 6 degrees upon electrification. Thus the charge effect for electrets appears to dominate the electret-water interface. The net charge density for these polymers as measured by the vibrating condenser technique was less than 10 stat-C.cm$^{-2}$; less than one tenth of the charge density estimated from contact angle variation.[18]

## DISCUSSION

Water-solid interactions are especially important in biological systems, since life processes occur almost exclusively at interfaces in an aqueous medium. It is important to realize that not only the water structure adjacent to the interface becomes different[19] but that the solid surface itself undergoes changes that lead to drastically altered interfacial properties. Thus, it is usually impossible to predict interfacial properties from surface properties determined in air or from the bulk properties of the material.

The stability of a thinning aqueous film at tissue boundaries and cellular interfaces is especially important since it determines whether the water will act as a lubricant and as an abhesive, or whether contact adhesion will take. There are abundant examples in medical practice where the rupture of a thin layer of bodily fluids takes place due to changes in the surface properties of tissue boundaries induced by the proximity of a hydrophobic interface. For example, during the surgical implant of PMMA intraocular lenses in the eye, the delicate and irreplaceable endothelial layer on the posterior surface of the cornea is often traumatized due to hydrophobic conversion of the endothelium and the resulting contact adhesion to the lens.[20] Surgeons also observed that in abdominal surgery the trauma to the intestines increased when hydrophobic rubber gloves were used.[21] During tonography, when the intraocular pressure of the eye is monitored by a plastic footplate pressed against the anesthetized cornea, considerable tissue damage can occur when the cornea is not protected by a hydrophilic polymeric layer incapable of hydrophobic transition.[22] The biocompatibility of contact lenses, whether hard (PMMA, cellulose acetate-butyrate) or soft (hydrogel or silicone), depends mostly on the interfacial properties of the lens after it had equilibrated with the tears.[23]

In summary, contact angle goniometry can be employed to estimate water-solid interfacial tension, which in case of prosthesis determines biocompatibility. Contact angle measurements can also yield a clue whether hydrophobic-hydrophilic transformation take place at certain boundaries. Two-phase contact angle measurements can decide whether a solid surface is truly hydrophilic irrespective of the water content of the solid. Receding contact angles can also shed light on thin fluid film stability that is capable of determining the sign of the disjoining pressure between two solid interfaces leading to abhesion or adhesion. Contact angle variation can even be used to estimate the surface charge density of electrically charged solids.

## CONCLUSIONS

The importance of the properties of water-solid interfaces and their variability for biomaterials and tissue boundaries are just now beginning to be recognized. On the cellular level, it is reasonable to assume that many aspects of cellular interactions are also controlled by the interfacial properties of the cell membranes and by the changes induced in the vicinal water layer by metabolically controlled changes in the cell membrane as well as changes induced by external factors. Even in the investigation of such complex processes the humble but classic technique of contact angle goniometry, when properly executed and interpreted, can be a highly useful tool.

## ACKNOWLEDGEMENT

This study was supported in part by PHS Grant No. EY-03002 from the National Eye Institute, National Institutes of Health and in part by Research to Prevent Blindness, Inc., N.Y.

## REFERENCES

1. F.J. Holly and M.F. Rofojo, in "Colloid and Interface Science" M. Kerker, Editor, Vol. III, p. 321, Academic Press, New York, 1976.
2. F.J. Holly and M.F. Refojo, in "Proc. Inter. Conf. Colloid Surface Sci.", E. Wolfram and T. Szekrenyessy, Editors, Vol. 2, p. 159, Akademiai Kiado, Budapest, Hungary, 1976.
3. W.A. Zisman in "Contact Angle, Wettability and Adhesion", F.M. Fowkes, Editor, Advances in Chemistry Series, No. 43, p. 15. American Chemical Society, Washington, D.C., 1964.
4. K.L. Mittal, Polym. Eng. Sci., $\underline{17}$, 467 (1977).
5. K.L. Mittal, in "Adhesion Science and Technology", L.H. Lee Editor, Vol. 9A, pp. 129-167, Plenum Press, N.Y., 1975.
6. N.K. Adam, "Physics and Chemistry of Surfaces", p. 79, Clarendon Press, Oxford, England, 1930.
7. F.J. Holly and M.F. Refojo, in "Hydrogels for Medical and Related Applications", J.D. Andrade Editor. ACS Symposium Series No.31, p. 252, American Chemical Society, Washington, D.C., 1976.
8. H. Yasuda and A.K. Sharma, J. Polym. Sci.. Polym. Phys. Ed. $\underline{19}$, 1285 (1981).
9. H. Yasuda, H.C. Marsh, S. Brandt, and C.N. Reilly, J. Polym. Sci. Polym. Chem. Ed., $\underline{15}$, 991 (1977).
10. B.D. Ratner, P.K. Weathersby, A.S. Hoffman, M.A. Kelly, and L.H. Scharpen, J. Appl. Polym. Sci., $\underline{22}$, 643 (1978).
11. F.J. Holly (1976), unpublished data.
12. W.D. Harkins, in "Recent Advances in Surface Chemistry and Chemical Physics", F.R. Moulton, Editor, p. 19, The Science

Press, Lancaster, PA 1939.
13. F.J. Holly and M.F. Refojo, in "Hydrogels for Medical and Related Applications", J.D. Andrade Editor, ACS Symposium Series No.31, p.267, American Chemical Society, Washington, D.C., 1976.
14. E.Wolfram, Plaste und Kautschuk, 12, 604 (1962).
15. E.H. Lucassen-Reynders, J. Phys. Chem., 67, 969 (1963).
16. M.M. Perlman and J.L. Meunier, J. Appl. Phys., 36(2), 420 (1965).
17. H.B. Moller, Z. Phys. Chem., 65, 226 (1909).
18. F.J. Holly, J. Colloid Interface Sci., 65, 435 (1977).
19. W. Drost-Hansen, in "Chemistry of the Cell Interface", H.D. Brown Editor, Vol. B, Chap. 6, p. 1, Academic Press New York, 1971.
20. H.E. Kaufman and J.I. Katz, Invest. Ophthalmol., 15, 996 (1976).
21. E. Goldberg, (1978), personal communication.
22. Y Do, D.W. Lamberts, and F.J. Holly, unpublished data.
23. F.J. Holly, Am. J. Optometry and Physiol. Optics, 58, 324 (1981).

# THE SOLIDIFICATION FRONT TECHNIQUE:

# ITS SCOPE AND USE TO DETERMINE INTERFACIAL TENSIONS

R.P. Smith, S.N. Omenyi and A.W. Neumann
Department of Mechanical Engineering
University of Toronto, Toronto
Canada  M5S 1A4

Experiments with particle engulfment and rejection by an advancing solidification front have shown that these phenomena are governed by the interfacial tensions operative in these systems. The results have been used to develop a technique for determining interfacial tensions, particle wettability and contact angles from the observation of particle pushing by a freezing front. An outline of this technique is presented. To test its internal consistency, particle-vapor surface tension values for polymer particles were determined and checked by independent means. Solid-vapor surface tensions and liquid-vapor surface tension of organic matrix materials were also determined, using this technique. The results presented demonstrate the potential of the solidification front technique for determining otherwise difficult to obtain quantities such as the particle-vapor surface tension of small particles or the solid-vapor surface tension of the matrix material.

## INTRODUCTION

The interaction of small particles with solidification fronts is a phenomenon of interest from several points of view. A particle, initially embedded in a melt, will encounter the freezing front upon solidification of this melt. At relatively low rates of solidification, the particle may either be engulfed by the solidification front or else be swept along. Such phenomena are of interest in process metallurgy, the study of perma frost problems and as a model for phagocytic engulfment[1].

At low rates of solidification and in the absence of elctrostatic interactions, inert particles may be expected to be engulfed if the free energy of engulfing

$$\Delta F^{eng} = \gamma_{PS} - \gamma_{PL} \tag{1}$$

$\gamma_{PS}$: Particle - solid interfacial tension
$\gamma_{PL}$: Particle - liquid interfacial tension

is negative. If it is positive, rejection and pushing of the particle is expected. As the determination of $\gamma_{PS}$ and $\gamma_{PL}$ is a difficult and controversial proposition, observation of particle engulfment and rejection is therefore a means of testing the validity of methods to determine these quantitites. This proposition was studied extensively with a number of polymers as particles and organic materials as matrices[2]. The main result of this study is the fact that our equation of state approach[3] for the determination of interfacial tensions of solids is strongly supported by the microscopic observations of particle engulfment and rejection.

The procedure used to study particle/solidification front interactions is as follows. Matrix material and a small quantity of the test particles are placed in a groove machined in the upper face of a rectangular copper plate. Then the matrix material is melted. By selectively heating or cooling each end of the copper cell by means of Peltier units, the solidification rate of the matrix material is controlled. The interaction of particles with the advancing solidification front is then observed through a microscope[2].

Next, a more detailed study of those cases was undertaken where particles are rejected by the solidification front at low rates of solidification. As the rate of solidification is increased, viscous drag on the particle will increase and eventually lead to engulfment against the resistance due to the positive free energy of engulfing. The latter may also be interpreted as a repulsive van der Waals interaction[4]. In such situations, engulfing occurs at a limiting rate of solidification, called the critical velocity of engulfment, $V_c$. From a large body of data, giving $V_c$ as a function of particle size, the free energy of adhesion

$$\Delta F^{adh} = \gamma_{PS} - \gamma_{PL} - \gamma_{SL} \quad (2)$$

$\gamma_{SL}$: Solid - melt interfacial tension

was related to the critical velocity, $V_c$, through a scheme of dimensional analysis. Using the technique of dimensional analysis the following equation was formulated to describe particle behaviour at a solidification front[5]:

$$Re = h \hat{F}^\ell Q^m \quad (3)$$

In Equation (3)

$$Re = \frac{\rho_L V_c D}{\mu} \quad (4)$$

is the Reynolds number,

$$\hat{F} = \frac{\Delta F^{adh}}{\rho_P C_P DT} \quad (5)$$

$$Q = \frac{\Delta F^{adh}}{(\mu k_P T)^{\frac{1}{2}}} \quad (6)$$

The individual terms in Equations (4), (5) and (6) are:

- $\rho_P$ : particle density
- $\rho_L$ : liquid matrix density
- $C_P$ : heat capacity of the particle per unit mass
- $k_P$ : thermal conductivity of the particle
- $\mu$ : viscosity of the liquid matrix
- $T$ : melting temperature of the matrix material
- $D$ : particle diameter

The constant h and the exponents $\ell$ and m were determined using experimental critical velocity data in systems were all parameters were known. Having determined h, $\ell$ and m, it is possible to produce the following equation[5] from Equation (3):

$$\Delta F^{adh} = 2.64 \times 10^5 \frac{\rho_L^{0.847} T^{0.280} k_P^{0.720}}{\mu^{0.127} (\rho_P C_P)^{0.441}} D^{0.407} V_c^{0.847} \qquad (7)$$

where the individual terms are in S.I. units.

Initially, Equation (7) was produced for polymer particles from 10 μm to 100 μm in diameter, in organic matrix materials[5]. Further studies indicate that this equation may be applied to particles up to 200 μm in diameter. Equation (7) has also been used for certain biological cells, in an ice/water matrix. These results are given in Reference (6) and are in agreement with results obtained by other means.

Equation (7) provides a novel means of determining the free energy of adhesion of small particles to solidification fronts, provided this free energy is positive, i.e. provided there is particle pushing at very low rates of solidification. This possibility was explored with the aim of determining particle-vapor surface tension and wettability of small particles. It was found that from the critical velocity of engulfment of polymethyl methacrylate (PMMA) particles, particle-vapor interfacial tensions and contact angles on PMMA could be predicted from the equation of state approach[3] which were in close agreement with contact angles measured on smooth and extended surfaces of PMMA by conventional means[7]. The technique has since been used to study the wettability of particles as different as coal powders[8] and biological cells[6]. It is the purpose of this paper to review, illustrate and extend the solidification front technique. This will be done in four parts.

As there is a need for more matrix materials than are presently available, thymol will be introduced for this purpose. This requires the determination of the viscosity of thymol and its liquid and solid surface tension, $\gamma_{LV}$ and $\gamma_{SV}$, respectively.

In the second part, thymol will be used as a matrix material to determine the surface tension, $\gamma_{PV}$, of small particles of polyvinyl chloride (PVC) and PMMA.

In the third section, two additional matrix materials are introduced, benzophenone and bibenzyl. However, this is not done through measurements of the temperature dependence of contact angles as in the first section above, but through the solidification front technique itself, using particles of known surface tension as probes.

In the fourth and final part, a test of the internal consistency of the procedure is presented. The thermodynamic analysis in conjunction with the equation of state approach stipulates that a plot of the free energy of adhesion versus the surface tension of particles yields a straight line at least for relatively small values of $\Delta F^{adh}$ and that this line intersects the $\gamma_{PV}$ axis at $\gamma_{PV} = \gamma_{LV}$, independent of the value of $\gamma_{SV}$. This prediction is tested for benzophenone as matrix material.

## DETERMINATION OF THE MELT VISCOSITY AND SOLID-VAPOR AND LIQUID-VAPOR INTERFACIAL TENSIONS OF THYMOL

In order to use thymol in quantitative freezing front experiments, we require data for all the parameters on the right hand side of Equation (7). As literature data on the viscosity of thymol are not available, this quantity will have to be measured. Furthermore, as the determination of the critical velocity of engulfment, $V_c$, yields only the free energy of adhesion $\Delta F^{adh}$, cf. Equation (7), it will be necessary to determine independently the solid-vapor and the liquid-vapor interfacial tensions in order to obtain, through the equation of state approach[3], the particle-vapor interfacial tension and the contact angle between particle and other liquids of known surface tension.

The viscosity of thymol and all other matrix materials presented in this paper were measured with an Ostwald viscometer placed in a liquid bath at the melting point of the matrix material. The surface tension, $\gamma_{LV}$, for thymol at its melting point was determined from temperature dependent surface tension measurements at and above the melting point using the Wilhelmy plate method[9]. These data are given in Figure 1. The surface tension at the melting point of thymol was taken from the curve fit of the data in Figure 1 using linear regression. The results of these viscosity and surface tension measurements are given in Table I along with the liquid density and the melting temperature of the matrix materials obtained from the literature.

The solid-vapor surface tension, $\gamma_{SV}$, of thymol at the melting point was obtained from the extrapolation of temperature dependent contact angles measured on thymol in the solid state. Smooth surfaces of thymol were prepared by solvent casting[10]. Using the vertical plate method[11,12], capillary rise measurements, h, with glycerol were taken for a range of temperatures. These measurements were converted into contact angles, θ,[11,12] and then via the equation of state approach to surface tensions. A line fit to the data using linear regression was extrapolated to the melting point to obtain $\gamma_{SV}$. These data are given in Figure 2. The value of $\gamma_{SV}$ obtained for thymol at its melting point is 29.4 mJ/m².

Having measured the required physical properties for thymol, it is now possible to determine particle surface tensions, $\gamma_{PV}$, from critical velocity measurements using this matrix.

### DETERMINATION OF THE SURFACE TENSION OF PVC AND PMMA PARTICLES FROM CRITICAL VELOCITY MEASUREMENTS IN THYMOL

To illustrate the procedure for determining $\gamma_{PV}$ from freezing front experiments, measurements with PMMA and PVC

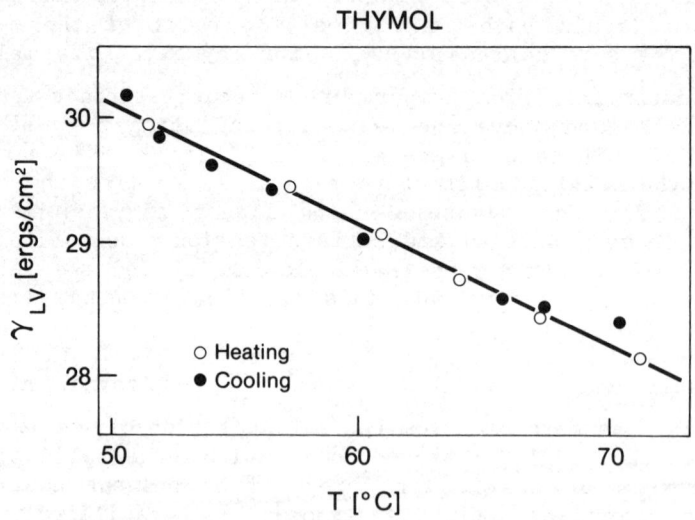

Figure 1. Temperature dependence of the liquid-vapor surface tension, $\gamma_{LV}$, for thymol at and above its melting point, 51.5°C.

Table I. Physical Properties of Matrix Materials.

| Material | $\rho_L$ (Kg/m³) | T (°C) | μ (N·s/m²) | $\gamma_{LV}$ (mJ/m²) |
|---|---|---|---|---|
| Thymol | 925 | 51.5 | 0.00397 | 29.9 |
| Benzophenone | 1140 | 48.0 | 0.00515 | 39.9 |
| Bibenzyl | 958 | 52.0 | 0.00246 | 24.9 |

particles in thymol will be used to calculate the surface tension of PVC and PMMA. These surface tensions will then be compared with values obtained from direct contact angle measurements.

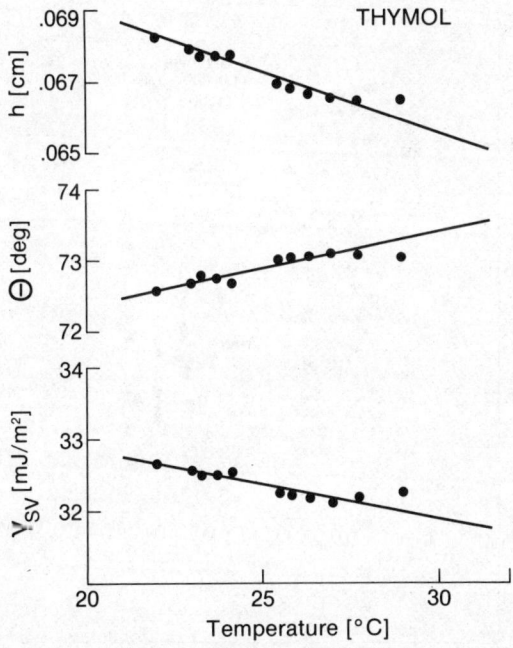

Figure 2. Temperature dependence of the contact angle, θ, of glycerol on solid thymol, determined from the capillary rise, h, on a flat vertical surface, and the temperature dependence of the resulting surface tension, $\gamma_{SV}$, of thymol in the solid state.

Solidification front experiments with PVC and PMMA particles must first be performed in order to obtain $V_c$ data. Figure 3 shows the experimental observations of pushing, engulfment and transition, i.e. momentary pushing followed by engulfment for PVC particles in thymol; see Reference 1 for more details. The critical velocity curve for this system is then obtained from the center line of the shaded band in Figure 3. Critical velocity curves produced in this manner for PVC and PMMA in thymol are given in Figure 4. Points on each curve for a range of diameters were selected and $\Delta F^{adh}$ was calculated from the physical properties of PVC and PMMA given in Table II. The surface tension, $\gamma_{PV}$, was calculated for each diameter from $\Delta F^{adh}$ using the previously determined values of $\gamma_{LV}$ and $\gamma_{SV}$ for thymol. The results are summarized in Table III. The slight variation in $\gamma_{PV}$ for the different diameters is thought to be due to experimental error. An average value of $\gamma_{PV}$ for the various particle sizes was calculated.

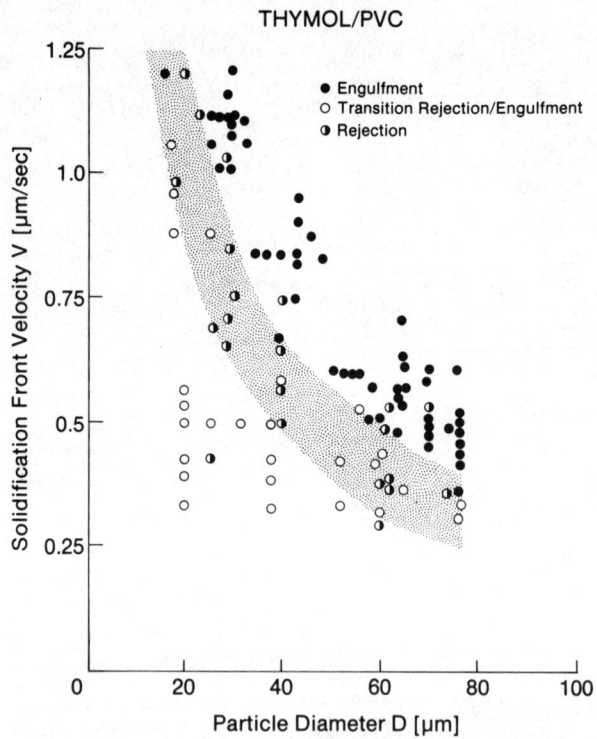

Figure 3. Solidification front velocity, V, of thymol as a function of particle diameter, D, for PVC particles.

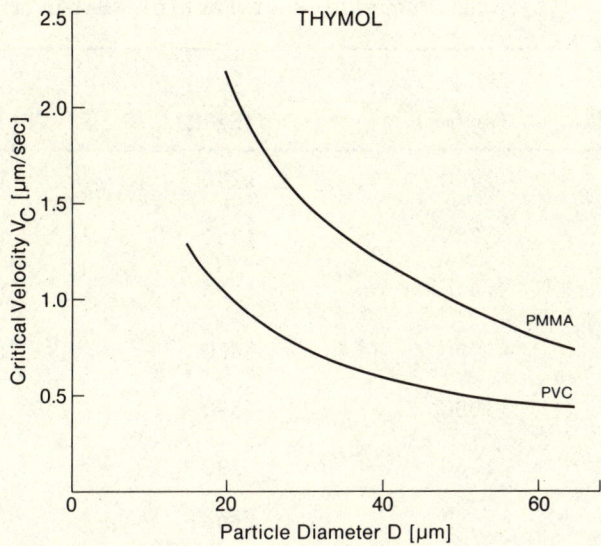

Figure 4. Critical velocity, $V_c$, of thymol solidification front as a function of particle diameter, D, obtained from the mean of the shaded band of plots of the type given in Figure 3.

As a check on the $\gamma_{PV}$ values obtained from $V_c$ measurements, the surface tensions of PVC and PMMA were determined from contact angle measurements. Since these measurements were performed at 20°C, a value of $d\gamma_{PV}/dT = -0.075$ mJ/m$^2$ was assumed[10] for both PVC and PMMA in order to obtain surface tension values at 51.5°C. These results are also given in Table III. It is apparent that there is good agreement between the two experimental strategies, supporting our contention that the freezing front technique is a valid tool to characterize surface properties of small particles.

## DETERMINATION OF THE SOLID-VAPOR INTERFACIAL TENSION OF BENZOPHENONE AND BIBENZYL USING THE FREEZING FRONT TECHNIQUE

In the preceding section, particle-vapor interfacial tensions were obtained from $V_c$ in situations where the matrix properties $\gamma_{SV}$ and $\gamma_{LV}$ were known. It stands to reason that it should be possible to determine the surface tension of the matrix material, $\gamma_{SV}$, in the solid state if its surface tension

Table II. Physical Properties of Particle Materials.

| Material | $\rho_P$ (Kg/m$^3$) | $C_P$ (J/Kg°K) | $k_P$ (J/sec.m°K) |
|---|---|---|---|
| PVC | 1400 | 1250 | 0.163 |
| PMMA | 1250 | 1380 | 0.188 |
| Acetal | 1430 | 1520 | 0.260 |
| Nylon-6 | 1090 | 1600 | 0.260 |
| Nylon-6,6 | 1080 | 1590 | 0.260 |
| Nylon-12 | 1060 | 1590 | 0.260 |
| Nylon-6,12 | 1040 | 1590 | 0.260 |

Table III. Particle Surface Tension, $\gamma_{PV}$, Determined from Critical Velocity Measurements in Thymol and Compared with $\gamma_{PV}$ Values Obtained from Direct Contact Angle Measurements. $\gamma_{PV}$ Given for 51.5°C

| Particle Material | Diameter D (μm) | $V_c$ (μm/sec) | $\gamma_{PV}$ (mJ/m$^2$) | $\gamma_{PV}$ (average) (mJ/m$^2$) | $\gamma_{PV}$, from $\theta$ measurements (mJ/m$^2$) |
|---|---|---|---|---|---|
| PVC | 20 | 1.10 | 33.2 | | |
|  | 40 | 0.60 | 32.5 | 32.7 | 31.8 |
|  | 60 | 0.45 | 32.3 | | |
| PMMA | 20 | 2.2 | 36.4 | | |
|  | 40 | 1.1 | 34.8 | 35.3 | 36.8 |
|  | 60 | 0.8 | 34.8 | | |

$\gamma_{LV}$ in the liquid state as well as the surface tension $\gamma_{PV}$ of the particles are known. This possibility may be of practical interest as it is easier to perform freezing front experiments than precise temperature dependent contact angle measurements.

Surface tensions of the melt of benzophenone and bibenzyl as well as viscosities were determined as described above. Freezing front experiments were performed with particles of acetal and Nylon, the surface tensions of which are known[2]. The results for acetal particles in benzophenone are shown in Figure 5. Again, the center line of the "transition" band was established as the critical velocity $V_c$ and plotted in Figure 6, together with the corresponding curves obtained from measurements with Nylon particles. Corresponding curves for bibenzyl were similarly obtained and are given in Figure 7. $\Delta F^{adh}$ values for each particle-matrix system were then calculated as before, using the data in Tables I and II. Table IV shows the results of these calculations for particles in benzophenone. In the case of bibenzyl, only average $\Delta F^{adh}$ values are reported (see Table VI).

Finally, to obtain $\gamma_{SV}$ for these materials, it is necessary to use the equation of state approach in conjunction with the known values of $\gamma_{PV}$, $\gamma_{LV}$ and $\Delta F^{adh}$. The surface tension values, $\gamma_{SV}$, obtained from the measurements with each type of polymer particle are given in Tables V and VI, respectively. Final average $\gamma_{SV}$ values are also given. The consistency of the results obtained with the different particle materials is remarkable.

In order to test the accuracy and reliability of these results, contact angles which a liquid of known surface tension, i.e. glycerol with $\gamma_{LV}$ = 63.4 mJ/m$^2$ would form on benzophenone and bibenzyl were predicted and compared with actual contact angle measurements. In order to avoid the complexities of the capillary rise measurements at elevated temperatures, contact angle measurements on solid benzophenone and bibenzyl were simply performed at room temperature. However, the $\gamma_{SV}$ values in Tables V and VI refer to the respective melting points of the two matrix materials. Therefore the $\gamma_{SV}$ for 20°C were calculated assuming a temperature coefficient $\frac{d\gamma_{SV}}{dT}$ = -0.075 mJ/m$^2$°C. Contact angles which glycerol would form on surfaces of benzophenone and bibenzyl were then predicted from the equation of state approach. Next, actual contact angle measurements with glycerol on layers of benzophenone and bibenzyl were performed.

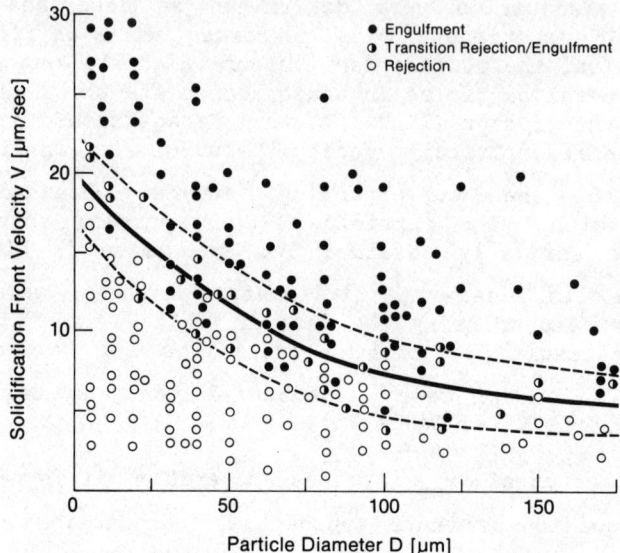

Figure 5. Solidification front velocity, V, of benzophenone as a function of particle diameter, D, for acetal particles.

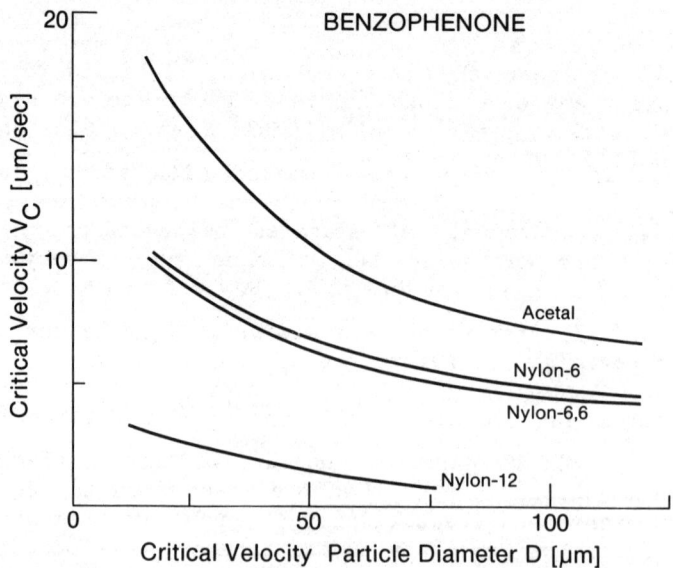

Figure 6. Critical velocity, $V_c$, of benzophenone solidification front as a function of particle diameter, D, obtained from the mean of the band on plots of the type given in Figure 5.

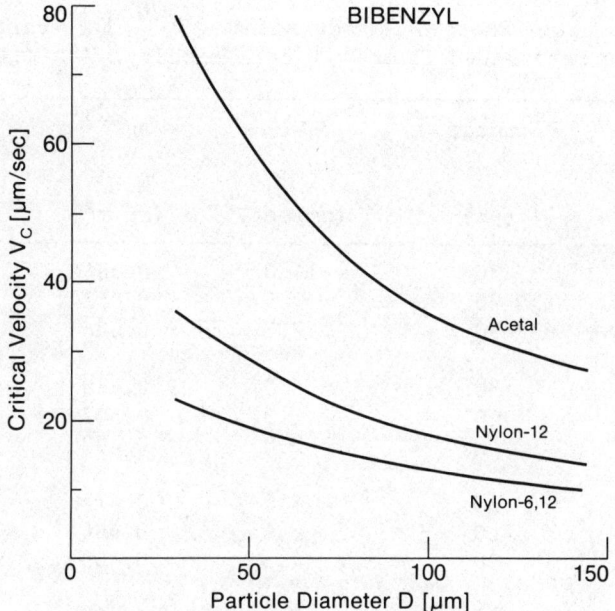

Figure 7. Critical velocity, $V_c$, of bibenzyl solidification front as a function of particle diameter, D, obtained from the mean of the band on plots of the type given in Figure 5.

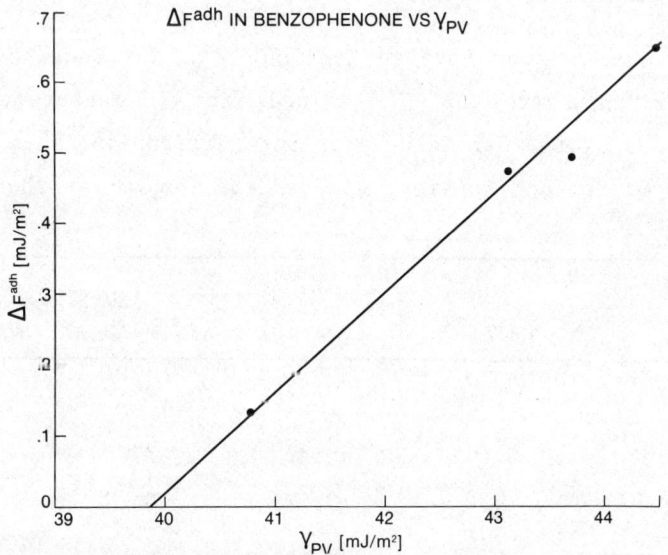

Figure 8. Free energy of adhesion, $\Delta F^{adh}$, obtained from $V_c$ measurements in benzophenone as a function of particle surface tension of, in order of decreasing surface tension, acetal, Nylon-6, Nylon-6,6 and Nylon-12 particles.

Table IV. Free Energy of Adhesion, $\Delta F^{adh}$, for Particles in Benzophenone Determined from Critical Velocity, $V_c$, Measurements.

| Particle Material | Diameter D (μm) | $V_c$ (μm/sec) | $\Delta F^{adh}$ (mJ/m$^2$) | $\Delta F^{adh}$ (average) (mJ/m$^2$) |
|---|---|---|---|---|
| Acetal | 30 | 14.0 | 0.684 | |
|  | 60 | 9.5 | 0.652 | 0.655 |
|  | 90 | 7.3 | 0.628 | |
| Nylon-6 | 30 | 8.5 | 0.488 | |
|  | 60 | 6.3 | 0.507 | 0.496 |
|  | 90 | 5.0 | 0.492 | |
| Nylon-6,6 | 30 | 8.1 | 0.477 | |
|  | 60 | 5.9 | 0.484 | 0.475 |
|  | 90 | 4.6 | 0.463 | |
| Nylon-12 | 10 | 2.9 | 0.130 | |
|  | 20 | 2.2 | 0.135 | 0.131 |
|  | 30 | 1.7 | 0.129 | |

Table V. Solid-Vapor Surface Tension, $\gamma_{SV}$, for Benzophenone at 48°C, Calculated from $\Delta F^{adh}$ Obtained from $V_c$ Measurements and Known Values of the Particle-Vapor Surface Tension, $\gamma_{PV}$, and the Liquid-Vapor Surface Tension, $\gamma_{LV}$ ($\gamma_{LV}$=39.9 mJ/m$^2$ at the melting point).

| Particle Material | $\gamma_{PV}$ (mJ/m$^2$) | $\Delta F^{adh}$ (mJ/m$^2$) | $\gamma_{SV}$ (mJ/m$^2$) | $\gamma_{SL}$ (mJ/m$^2$) |
|---|---|---|---|---|
| Acetal | 44.5 | 0.655 | 34.5 | 0.381 |
| Nylon-6 | 43.7 | 0.496 | 34.9 | 0.326 |
| Nylon-6,6 | 43.1 | 0.475 | 34.3 | 0.410 |
| Nylon-12 | 40.8 | 0.131 | 34.3 | 0.410 |
| Average Values |  |  | 34.5 | 0.381 |

These solid layers were obtained by solvent casting. Clean glass slides were dipped in a bath containing a 1% solution of benzophenone or bibenzyl dissolved in a suitable solvent. These glass slides were then stored in an evacuated desiccator for at least four hours to remove excess solvent. Contact angle measurements were performed using the vertical plate method[11,12]. The results of these measurements, together with the predictions from the critical velocity measurements are given in Table VII. The agreement is excellent, indicating that the surface tensions, $\gamma_{SV}$, obtained for benzophenone and bibenzyl from freezing front experiments are reliable. Thus, the two materials benzophenone and bibenzyl may be used in future quantitative freezing front work.

## A TEST FOR INTERNAL CONSISTENCY OF THE FREEZING FRONT TECHNIQUE

Calculations with the equation of state approach[3] indicate that a plot of the free energy of adhesion, $\Delta F^{adh}$, versus the particle-vapor interfacial tension, $\gamma_{PV}$, is approximately linear for values of $\gamma_{PV}$ which are only a few mJ/m$^2$ larger than the liquid-vapor surface tension, $\gamma_{LV}$, of the melt, independent of the solid-vapor interfacial tension, $\gamma_{SV}$. For benzophenone at its melting point, $\gamma_{LV}$ was found to be $\gamma_{LV} = 39.9$ mJ/m$^2$ by direct measurement. To test the above prediction, $\Delta F^{adh}$ values were obtained from freezing front experiments with acetal and Nylon particles, and benzophenone as matrix material. For each particle, $\Delta F^{adh}$ was plotted against the particle surface tension determined from contact angle measurements. The results are shown in Figure 8. The straight line obtained by linear regression intersects the $\gamma_{PV}$ axis at $\gamma_{PV} = 39.9$ mJ/m$^2$, in excellent agreement with the prediction.

Overall, we conclude that the freezing front technique may be used to determine any one of the interfacial tensions $\gamma_{SV}$, $\gamma_{LV}$ and $\gamma_{PV}$. For the determination of either $\gamma_{SV}$ or $\gamma_{PV}$, the other two interfacial tensions must be known. The determination of $\gamma_{LV}$ requires only information on $\gamma_{PV}$ of several particles. While the determination of $\gamma_{LV}$ from freezing front experiments may not be of practical importance, the determination of $\gamma_{SV}$ and $\gamma_{PV}$ certainly is. The determination of $\gamma_{PV}$ entails the possibility of predicting contact angle values on these particles.

Table VI. Solid-Vapor Surface Tension, $\gamma_{SV}$, for Bibenzyl at 52°C, Calculated from $\Delta F^{adh}$ Obtained from $V_c$ Measurements and Known Values of the Particle-Vapor Surface Tension, $\gamma_{PV}$, and the Liquid-Vapor Surface Tension, $\gamma_{LV}$ ($\gamma_{LV}$=24.9 mJ/m² at the melting point).

| Particle Material | $\gamma_{PV}$ (mJ/m²) | $\Delta F^{adh}$ (mJ/m²) | $\gamma_{SV}$ (mJ/m²) | $\gamma_{SL}$ (mJ/m²) |
|---|---|---|---|---|
| Acetal | 44.3 | 2.59 | 20.3 | 0.354 |
| Nylon-12 | 40.6 | 1.54 | 21.5 | 0.191 |
| Nylon-6,12 | 34.0 | 1.14 | 20.7 | 0.294 |
| Average Values | | | 20.8 | 0.280 |

Table VII. Contact Angle θ of Glycerol ($\gamma_{LV}$=63.4 mJ/m²) on Solid Matrix Material at 20°C.

| Matrix Material | θ from $V_c$ measurements (deg) | θ from direct measurements (deg) |
|---|---|---|
| Benzophenone | 64.0 | 64.9 |
| Bibenzyl | 88.3 | 86.5 |

## REFERENCES

1. S.N. Omenyi and A.W. Neumann, J. Appl. Phys. <u>47</u>, 3956 (1976).
2. S.N. Omenyi, A.W. Neumann and C.J. van Oss, J. Appl. Phys. <u>52</u>, 789 (1981).
3. A.W. Neumann, R.J. Good, C.J. Hope and M.J. Sejpal, J. Colloid Interface Sci. <u>49</u>, 291 (1974).
4. A.W. Neumann, S.N. Omenyi and C.J. van Oss, J. Phys. Chem. <u>86</u>, 1267, (1982).

5. S. N. Omenyi, A. W. Neumann, W.W. Martin, G.M. Lespinard and R.P. Smith, J. Appl. Phys., $\underline{52}$, 796 (1981).
6. J. K. Spelt, D. R. Absolom, W. Zingg, C.J. van Oss and A. W. Neumann, Cell Biophysics, accepted for publication, 1982.
7. S.N. Omenyi, R.P. Smith and A.W. Neumann, J. Colloid Interface Sci. $\underline{75}$, 117 (1980).
8. E.I. Vargha-Butler, M.R. Soulard, A.W. Neumann and H.A. Hamza, C.I.M. Bulletin $\underline{74}$, 54 (1981).
9. A.W. Neumann, R.J. Good, P. Ehrlich, K. Basu and G.J. Johnston, J. Macromol. Sci. Phys. $\underline{37}$, 525 (1973).
10. S.N. Omenyi, Ph.D. thesis, University of Toronto, Toronto, Canada (1978).
11. A.W. Neumann, Z. Phys. Chem. (Frankfurt am Main) $\underline{41}$, 339 (1964).
12. A.W. Neumann and W. Tanner, J. Colloid Interface Sci. $\underline{34}$, 1 (1970).

# SURFACE THERMODYNAMICS OF LIQUID POLYMERS: THEORY

I. C. Sanchez and C. I. Poser*
Center for Materials Science
Polymer Science and Standards Division
National Bureau of Standards
Washington, D.C. 20234

A generalized density gradient theory of interfaces has been combined with a compressible lattice theory of polymers. This yields a unified theory of bulk and surface thermodynamic properties. A unique feature of this theory is that it is parameterless. The only parameters required to calculate a surface tension are obtained from pure component thermodynamic properties. Since the theory is a mean field theory, it is only applicable to non-polar and slightly polar liquids. For such systems, surface tensions can be accurately calculated. The temperature and molecular weight dependence of the surface tension for the n-alkanes is correctly predicted and is shown to be related to the dependence of liquid density on chain length. Polymer liquid surface tensions satisfy a corresponding states principle and can be estimated for many polymers with an error of less than 10%. This method of estimating polymer surface tensions appears to be the most accurate that is available.

---

*Present address: 3M Co. St. Paul, Minnesota 55101

## INTRODUCTION

Recently we proposed a theory of interfaces for low molecular weight and polymer liquids.[1-3] This theory was developed by combining the lattice fluid (LF) model[4-7] with the density gradient theory[8-11] of inhomogeneous fluids. The main objective of this paper is to illustrate that this method of estimating polymer melt surface tensions is superior to existing empirical or semi-empirical methods.

## THEORY

In the LF model, three parameters, either the molecular parameters $\varepsilon^*$, $v^*$, and $r$ or the equation of state parameters $T^*$, $P^*$, and $\rho^*$, serve to characterize a pure fluid. These parameters are related by the following definitions:

$$\tilde{T} \equiv T/T^* \qquad T^* \equiv \varepsilon^*/k \qquad (1)$$

$$\tilde{P} \equiv P/P^* \qquad P^* \equiv \varepsilon^*/v^* \qquad (2)$$

$$\tilde{v} \equiv 1/\tilde{\rho} \equiv V/V^* \qquad V^* \equiv N(rv^*) \equiv N(M/\rho^*) \qquad (3)$$

where $\varepsilon^*$ is the total interaction energy per mer, $v^*$ is the close-packed mer volume, $V^*$ is the close-packed volume of the N r-mers, $\rho^*$ is the close-packed mass density, and M is the molecular weight. $\tilde{T}$, $\tilde{P}$, $\tilde{v}$, and $\tilde{\rho}$ are the reduced temperature, pressure, volume, and density and T, P, and V are the temperature, pressure, and volume. The three parameters required to describe a fluid can be readily calculated using PVT data and have been tabulated for a large number of low molecular weight fluids[4] and a number of common polymers.[6,7]

The reduced chemical potential, $\tilde{\mu}$, of a pure fluid is given by

$$\tilde{\mu} \equiv \mu/(Nr\varepsilon^*) = -\tilde{\rho} + \tilde{P}\tilde{v} + \tilde{T}\left[(\tilde{v}-1)\ln(1-\tilde{\rho}) + \frac{1}{r}\ln\tilde{\rho}\right] \qquad (4)$$

to within an additive constant. At equilibrium the reduced density (or reduced volume) satisfies the equation of state:

$$\tilde{\rho}^2 + \tilde{P} + \tilde{T}\left[\ln(1-\tilde{\rho}) + (1-1/r)\tilde{\rho}\right] = 0 \qquad (5)$$

The interfacial tension, $\gamma$, of a pure liquid in contact with its vapor can be written in reduced form as[1]

$$\tilde{\gamma} \equiv \gamma/\gamma^* = 2\int_{\tilde{\rho}_g}^{\tilde{\rho}_\ell} (\tilde{\kappa}\Delta\tilde{a})^{\frac{1}{2}} d\tilde{\rho} \qquad (6)$$

where $\gamma^* \equiv \epsilon^*/v^{*2/3}$ is the characteristic surface tension and $\tilde{\rho}_g$ and $\tilde{\rho}_\ell$ are the reduced equilibrium gas and liquid densities which are determined by Equation (5). $\tilde{\kappa}$ is a dimensionless parameter related to the second moment of the intermolecualr potential. For a Sutherland type $(m,\infty)$ potential it becomes

$$\tilde{\kappa} = \frac{1}{6} \frac{m-3}{m-5} \qquad (7)$$

For pure dispersion interactions, we would expect $m=6$ and $\kappa=\frac{1}{2}$. $\Delta a$ is a dimensionless excess Helmholtz free energy density defined as

$$\Delta\tilde{a} \equiv \tilde{\rho} \, [\tilde{\mu}(\tilde{\rho},\tilde{P}_e) - \tilde{\mu}_e] \qquad (8)$$

where $\tilde{\mu}_e$ is the equilibrium chemical potential,

$\tilde{\mu}_e \equiv \tilde{\mu}(\tilde{\rho}_g,\tilde{P}_e) = \tilde{\mu}(\tilde{\rho}_\ell,\tilde{P}_e)$, calculated by Equation (4) at the equilibrium vapor pressure $\tilde{P}_e$. Since Equation (5) defines the minima in Equation (4) at $\tilde{\rho}_\ell$ and $\tilde{\rho}_g$, $\tilde{\mu}(\tilde{\rho},\tilde{P}_e) > \tilde{\mu}(\tilde{\rho}_\ell,\tilde{P}_e) = \tilde{\mu}(\tilde{\rho}_g,\tilde{P}_e)$, and thus, $\Delta\tilde{a} \geq 0$.

## RESULTS

It is well-known that the surface tension of a homologous series of liquids such as the n-alkanes is chain length dependent. LeGrand and Gaines[12] have shown that the following empirical relation correlates surface tension with molecular weight very well:

$$\gamma = \gamma_\infty - \frac{K_e}{M^{2/3}} \qquad (9)$$

where $\gamma_\infty$ is the surface tension at infinite molecular weight and $K_e$ is an empirical constant characteristic of a given series of chain liquids. Other empirical relations have also been shown to correlate data equally well.[13] In Figure 1 surface tension data for the n-alkanes[14] (pentane through eicosane) are compared with theoretical lines at two different temperatures. The theoretical lines are least squares lines for the calculated values of the surface tension of n-pentane through n-heptadecane. Equation of state parameters for these n-alkanes are available.[1,4] A value of $\tilde{\kappa} = 0.61$, or equivalently, an exponent of $m = 5.75$ was used in all of the calculations. This particular value of $\tilde{\kappa}$ has been shown to give the best fit to surface tension data for the n-alkanes.[1]

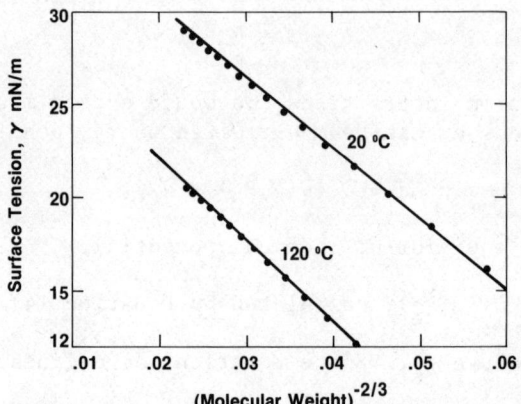

Figure 1. Surface tension (solid circles) vs. molecular weight to the -2/3 power for the n-alkanes. Theoretical lines were calculated from a linear regression analysis of values obtained from Equation (6). Experimental points taken from Reference (14).

One of the results of the LF model is that for high polymers ($r \to \infty$), the equations approach limiting forms that suggest a corresponding states principle. Figure 2 shows a plot of $\tilde{\gamma}$ for $r \to \infty$ as a function of reduced temperature $\tilde{T}$ compared with experimental data for six polymers. A value of $\tilde{\kappa} = 0.55$, which corresponds to $m = 5.9$, yields a good fit to the data. The data were taken between 413 K and 453 K;[15,16] absolute values of $\gamma$ range from a high of 32.1 mN/m for polystyrene to a low of 12.1 mN/m for poly(dimethysiloxane) (1mN/m = 1 dyne/cm). The maximum error between experiment and theory in Figure 2 is about 10%. Equation of state parameters used to reduce the experimental data are shown in Table I.

Table I. Equation of State Parameters.

|  | T* K | P* MN/m$^2$ | v* cm$^3$/mol | γ* mN/m |
|---|---|---|---|---|
| poly(dimethyl siloxane) | 476 | 302 | 13.1 | 84.3 |
| Poly(vinyl acetate) | 590 | 509 | 9.6 | 128 |
| poly(isobutylene) | 643 | 354 | 15.1 | 104 |
| linear polyethylene | 649 | 425 | 12.7 | 118 |
| branched polyethylene | 673 | 359 | 15.6 | 106 |
| atactic polystyrene | 735 | 357 | 17.1 | 109 |

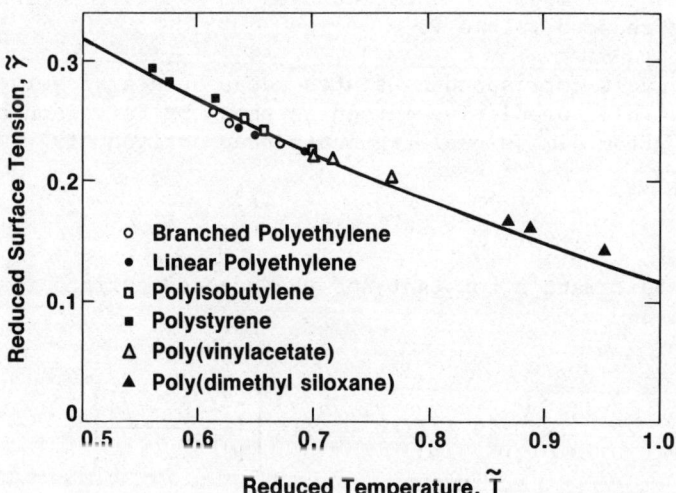

Figure 2. Reduced surface tension vs. reduced temperature of polymers. The theoretical line was calculated for infinite molecular weight (r→∞) and $\widetilde{K}$ = 0.55. Experimental data were taken from References 15 and 16.

Three other methods have been widely used to estimate polymer melt surface tensions. The oldest of these methods is based on the parachor method first proposed for small molecules by Sugden.[17] The parachor, $\mathcal{P}$, is a constant and for a given liquid is given by

$$\mathcal{P} = \gamma^{1/4} M/(\rho_\ell - \rho_g) \tag{10}$$

where M is the molecular weight. When applied to polymers $\mathcal{P}$ and M refer to the repeat unit and $\rho_g = 0$. An extensive tabulation of atomic parameters has been compiled by Quayle[18] and can be used to calculate $\mathcal{P}$, and thus, $\gamma$.

Another method for predicting polymer surface tensions is based on the empirical relation between the tension and solubility parameters, $\delta$, observed by Hildebrand and Scott[19] for low molecular weight liquids. Siow and Patterson[20] have adapted this empirical relation to polymers through the use of the Prigogine corresponding states principle[21] to obtain

$$\gamma = 0.095 \, P*^{2/3} \, T*^{2.1}/\tilde{v} \tag{11}$$

where T* and P* are the characteristic temperature and pressure and are expressed in °K and cal/cm³, respectively. The Flory equation of state model[22-25] can be used to evaluate T* and P* as well as the reduced volume $\tilde{v}$.

Prigogine's corresponding states ideas have also served as the basis for a third predictive method as shown by Patterson and Rastogi.[26] From dimensional arguments they arrived at

$$\gamma = \tilde{\gamma} \, P*^{2/3} (T*/k)^{1/3} \tag{12}$$

where k is Boltzmann's constant and the reduced surface tension, $\tilde{\gamma}$, is given by

$$\tilde{\gamma}\tilde{v}^{5/3} = 0.29 - (1-\tilde{v}^{-1/3})\ln\left[\frac{\tilde{v}^{1/3} - 1/2}{\tilde{v}^{1/3} - 1}\right] \tag{13}$$

T*, P*, and $\tilde{v}$ can be determined from the Flory equation of state theory.

Figures 3 and 4 compare calculated surface tensions of the present theory, Equation (6), with those of the parachor method, Equation (10), the modified solubility parameter method, Equation 11, the corresponding states relation, Equation 12, and with

experimental data for poly(dimethyl siloxane) and polyisobutylene.
As can be readily seen, the present theory is superior to the other
three methods of estimating polymer surface tensions.

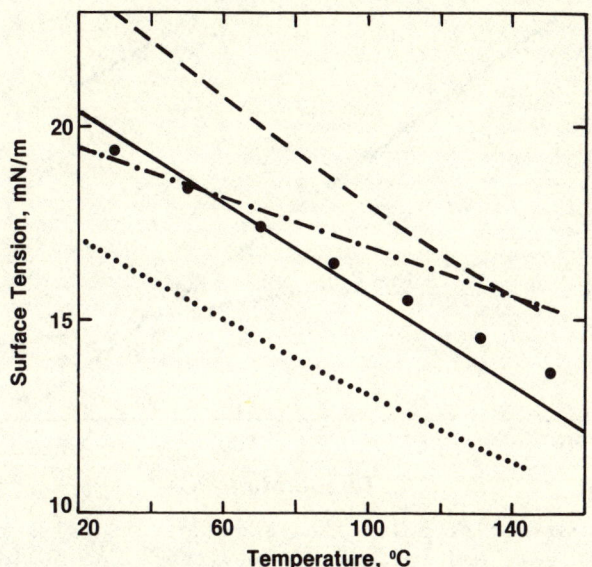

Figure 3. Comparison, of calculated and experimental surface
tensions (solid circles) for poly(dimethyl siloxane). The solid
line is the present theory, Equation (6), with $\tilde{\kappa} = 0.55$. The
dotted line is Equation (10), the dot-dashed line is Equation (11),
and the dashed line is Equation (12). Experimental values were
taken from Reference 16.

## DISCUSSION

Figure 1 illustrates that the lattice fluid/density gradient
theory of interfaces adequately describes the molecular weight
(MW) and temperature dependence of the surface tension of the
n-alkanes. The inherent property of the theory responsible for
this excellent agreement originates with the ability of the LF
model to describe the variation of liquid density with MW and temperature.

Surface tensions are inherently MW dependent because liquid
densities are inherently MW dependent. The density of a polymeric
liquid may be 10-20% larger than the density of a low MW homologue
at the same temperature. This occurs because the free energy cost
of forming a vacancy in a polymeric liquid is larger than in a low
MW homologue. Energetically the cost is approximately the same,
but entropically there is an appreciable difference. The creation

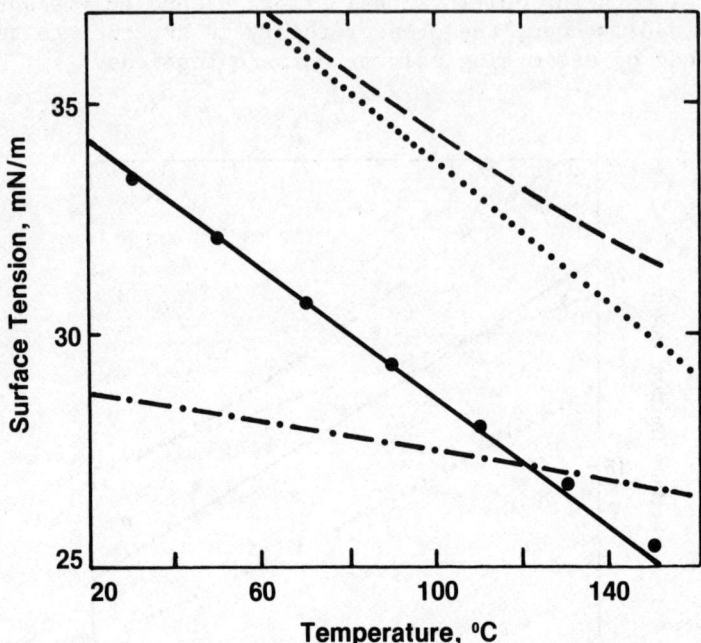

Figure 4. Comparison of calculated and experimental surface tensions (solid circles) for polyisobutylene. The solid line is the present theory, Equation (6), with $\tilde{\kappa} = 0.55$. The dotted line is Equation (10), the dot-dashed line is Equation (11), and the dashed line is Equation (12). Experimental values were taken from Reference 16.

of a vacancy in a polymeric liquid results in a smaller entropy increase than occurs in a low MW homologue. The increase in entropy associated with an incremental volume increase depends on the ability of the system to utilize the extra volume. Because more monomer units are chemically connected to one another in a polymeric liquid, the increase in configurational entropy is smaller for the polymeric liquid. Therefore, at equilibrium, a polymeric liquid will possess a smaller number of vacancies, and thus will be denser, than a low MW homologue. A density difference of 20% can result in a 100% difference in surface tension.

Experimentally the surface entropy, $-d\gamma/dT$, of a polymeric liquid is smaller than for a low MW homologue. The surface entropy of a polymeric liquid may typically be 50% to 100% smaller than a corresponding low MW homologue.[1] The primary reason for this phenomenon can again be related to density variations. A polymeric liquid has a smaller thermal expansion coefficient than a low MW homologue because entropy increases associated with volume expansion are smaller for a polymeric liquid. Since the polymer liquid density varies more slowly with temperature, $-d\gamma/dT$, is smaller for the polymeric liquid.

A second reason why the surface entropy might be smaller is that the surface restricts the number of polymer configurations. But we believe, as others do[20,27,28], that this effect plays a relatively minor role.

## REFERENCES

1. C. I. Poser and I. C. Sanchez, J. Colloid Interface Sci., 69, 539 (1979).
2. I. C. Sanchez, J. Macromol. Sci. - Phys., B17 (3), 565 (1980).
3. C. I. Poser and I. C. Sanchez, Macromolecules, 14, 361 (1981).
4. I. C. Sanchez and R. H. Lacombe, J. Phys. Chem., 80, 2352 (1976).
5. R. H. Lacombe and I. C. Sanchez, J. Phys. Chem., 80, 2568 (1976).
6. I. C. Sanchez and R. H. Lacombe, J. Polym. Sci. Lett. Ed., 15, 71 (1977).
7. I. C. Sanchez and R. H. Lacombe, Macromolecules, 11, 1145 (1978).
8. J. D. van der Waals, J. Phys. Chem. 13, 657 (1894).
9. J. W. Cahn and J. E. Hilliard, J. Chem. Phys., 28, 258 (1958).
10. J. W. Cahn, J. Chem. Phys., 30, 1121 (1959).
11. V. Bongiorno, L. Scriven, and H. T. Davis, J. Colloid Interface Sci., 57, 462 (1976).
12. D. C. LeGrand and G. L. Gaines, Jr., J. Colloid Interface Sci., 31, 162 (1969).
13. S. Wu, in "Polymer Blends", D. R. Paul and S. Newman, Editors, Vol. I, pp. 244-295, Academic Press, New York, 1978.
14. F. D. Rossini, "Selected Values of Properties of Hydrocarbons and related Compounds." Research Project 44, American Petroleum Research Institute, Carnegie Press, Pittsburgh,1953.
15. G. L. Gaines, Jr., Poly. Eng. Sci., 12, 1 (1972).
16. S. Wu, J. Macromol. Sci., C10, 1 (1974).
17. S. Sugden, J. Chem. Soc., 125, 32 (1924).
18. O. R. Quayle, Chem. Rev., 53, 439 (1953).
19. J. H. Hildebrand and R. L. Scott, "Solubility of Non-Electrolytes," Reinhold Publishing Corp., New York, NY, 1950.
20. K. S. Siow and D. Patterson, Macromolecules, 4, 26 (1971).
21. I. Prigogine (with A. Bellemans and V. Mathot), "The Molecular Theory of Solutions," North-Holland Publishing Co., Amsterdam, and Interscience, New York, NY, 1957, Ch. 16.
22. P. J. Flory, R. A. Orwoll, and A. Vrij, J. Am. Chem. Soc., 86, 3515 (1964).
23. P. J. Flory, J. Am. Chem. Soc., 87, 1833 (1965).

24. B. E. Eichinger and P. J. Flory, Trans. Faraday Soc., 64, 2035 (1968).
25. P. J. Flory, Discuss. Faraday Soc., 49, 7 (1970).
26. D. Patterson and A. K. Rastogi, J. Phys. Chem., 74, 1067 (1970).
27. R. J. Roe, J. Phys. Chem., 69, 2809 (1965).
28. K. M. Hong and J. Noolandi, Macromolecules, 14, 1229 (1981).

# CHARACTERIZATION OF HIGH SURFACE AREA POLYESTER FILAMENTS BY MEANS OF WETTING FORCE MEASUREMENTS

B. Miller, H.-D. Weigmann, and D. Simonetti

Textile Research Institute

Princeton, New Jersey  08540

By briefly exposing smooth, undrawn polyester filaments to an interactive solvent just prior to drawing, bimorphic filaments have been produced that have a fibrillar core, which maintains mechanical properties, and a cavitated skin, which provides the high surface area desirable for many applications. A methodology for studying the wetting of single filaments has been found suitable for characterizing the surface cavitation of these experimental filaments. Detailed and reproducible wetting force profiles, obtained by moving a high surface tension liquid up and down a filament suspended from a microbalance, appear to reflect accurately physical features seen in scanning electron micrographs, chiefly grooves. This correspondence between features of the force curve and of the filament surface arises because of the dependence of wetting force on orientation of the filament surface with respect to the vertical at the point of liquid contact. The force scans can be interpreted to give the number of cavities per unit scan length, the range of cavity shapes, and the transverse length and the width of the cavities encountered during a scan.

# INTRODUCTION

The usually smooth surface of a polyester monofilament can be converted to a highly cavitated form by a sequential application of two processes: (1) solvent-induced crystallization (SINC) at room temperature, followed by (2) thermal/stress crystallization at temperatures ranging from 90 to 120°C.[1] If the first step is limited by a very short period of contact between solvent and filament, the result is a radially-differentiated structure having a fibrillar inner core with a high degree of orientation and crystallinity and a surface made up of deformed spherulites. For many end uses, such a bimorphic combination would be quite desirable, since it provides a high surface area while maintaining most of the tensile strength of the filament.

The apparatus for such a sequential process is illustrated schematically in Figure 1. An important point to note is that heating under stress occurs immediately after exposure to the solvent. This causes the cavitating effect of the solvent to be enhanced by the subsequent thermal stressing. In contrast, if the filament is drawn first and then exposed to the solvent, hardly any surface cavitation occurs. Examples of bimorphic (BM) filaments, drawn (D) filaments, and drawn followed by solvent treatment (DS) filaments are shown in Figures 2-4, respectively. The latter two have been used as controls for the investigation described in this paper.

The object of this study was to see if the technique of continuous scanning wetting force measurement could be used to evaluate the extent of surface cavitation of the bimorphic filaments. This technique allows one to follow the interaction (i.e., wetting force) between the surface of a liquid and a single filament suspended vertically and partially immersed in the liquid.[2]

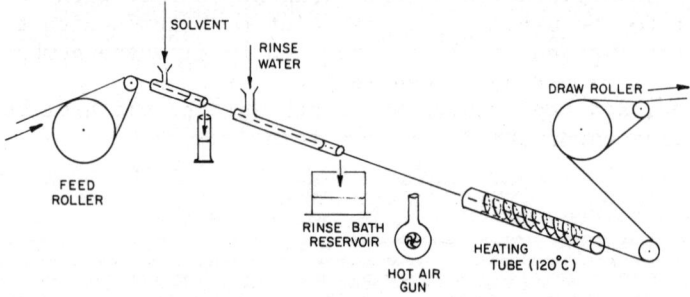

Figure 1. TRI apparatus for continuous sequential SINC and drawing.

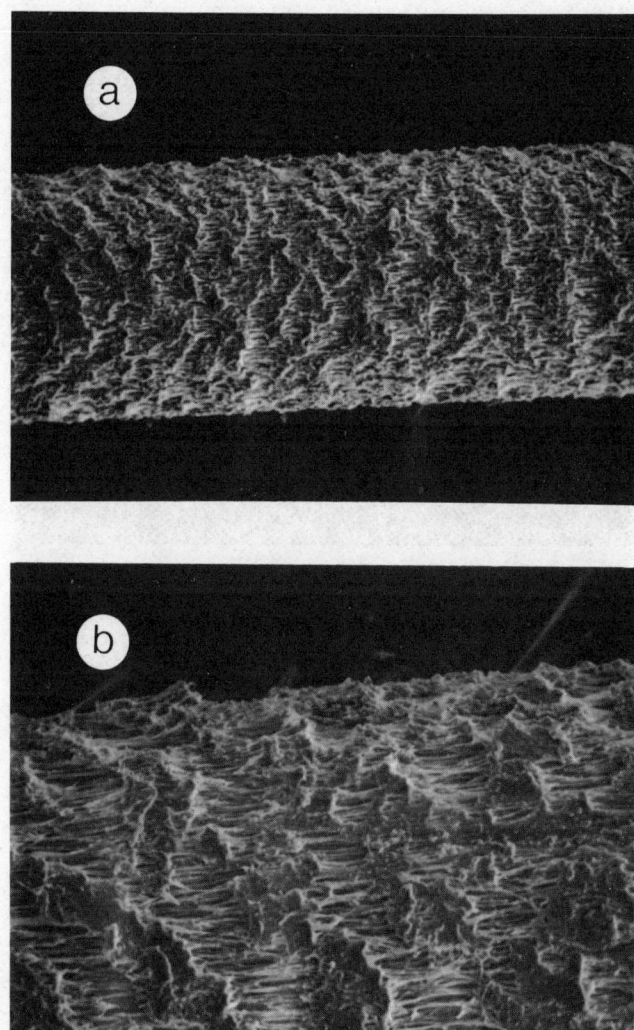

Figure 2. Scanning electron micrographs of bimorphic (BM) polyester monofilaments produced by sequential SINC and thermal/stress drawing. a) 500X. b) 1000X.

Figure 3. Micrograph of drawn (D) monofilament. 400X.

Figure 4. Micrograph of drawn and then solvent-treated (DS) monofilament. 350X.

The filament is attached to the hang-down wire of a recording microbalance, and, if provision can be made to move the liquid reservoir up and down, the result is a scanning of the filament surface in the longitudinal direction by the liquid surface. The general arrangement of the apparatus is shown in Figure 5.

The measured force (after correction for buoyancy) is described by the Wilhelmy equation,

$$F_w = P \gamma_{LV} \cos \theta_w,$$

where  $P$ = perimeter of the solid at the line of contact,
$\gamma_{LV}$ = surface tension of the liquid,
$\theta_w$ = effective contact angle.

The effective contact angle is equal to the thermodynamic (Young) contact angle when the surface of the solid at the three-phase interface is homogeneous and vertically oriented. If any point of contact is not at a vertical face, the wetting force will be altered accordingly. This is illustrated in Figure 6. Since the measured wetting force is proportional to the vertical component of the surface tension vector, any inclination of the surface from the vertical causes the force to become larger or smaller depending on the topographic change. The rationale for this investigation was that if reproducible perturbations of wetting force could be obtained during scanning of a filament surface, the results could be used to describe the extent of cavitation. Considerable experience with this type of scanning has shown that because of chemical heterogeneity all polymeric surfaces show some degree of scanning noise with any liquid,[3] and, because of the extreme sensitivity of present day microbalances, the use of a gear-driven elevator to move the liquid adds mechanical noise. The latter problem has been effectively eliminated through the use of an "Inchworm" linear translator* which provides smooth movement at very low velocities. The possible interference of surface chemical heterogeneity was discounted after comparing the results for the bimorphic filaments with the two control samples.

---

*Burleigh Instruments, Inc., Burleigh Park, Fishers, NY   14453.

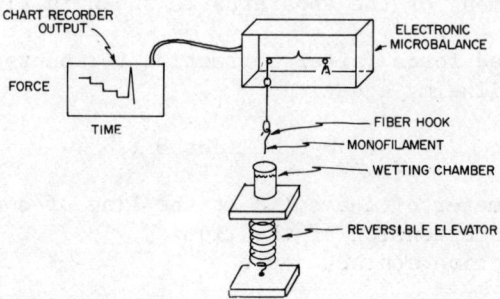

Figure 5. Apparatus for Wilhelmy wetting force measurements on single filaments.

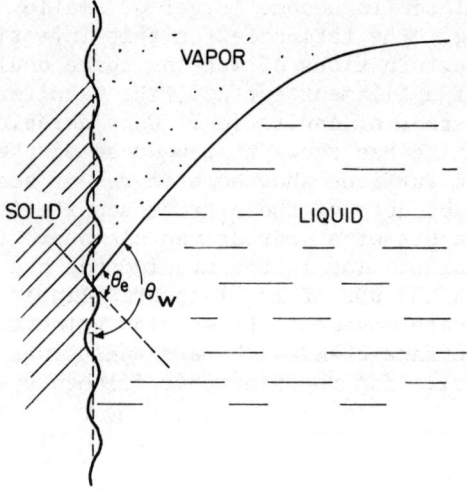

Figure 6. Difference between the thermodynamic contact angle $\theta_e$ and the effective contact angle $\theta_w$ when the solid surface is not vertical.

## RESULTS

At least ten specimens of each sample were scanned with water and three different aqueous butanol solutions to reveal wetting behavior as a function of liquid surface tension. The average wetting force for each scan was obtained over about a 2 mm length. The results for both advancing and receding scans are shown in Figure 7. The observed average wetting forces in both directions were about the same for the BM, D and DS samples when the surface tension of the liquid was low (resulting in contact angles of less than 30°). At the higher values of $\gamma_{LV}$, the average advancing wetting force on the BM sample was lower and the receding force higher than for the two controls (D and DS). It is unlikely that such contrasting behavior results from differences in intrinsic surface wettability or effective perimeter. Instead, the differences are more likely to be the result of the drastically altered surface contours of the BM sample.

The contrast between the BM filament surface and the others is most obvious for the advancing water scans shown in Figure 8. The scanning profile of the D sample shows the typical noise one finds with most polymer surfaces when the contact angle is high. (With water, the advancing contact angle $\theta_a$ was greater than 75° for all the samples). The DS surface shows somewhat larger variations, but these do not come near the size or frequency of the variations in the BM profile. The latter has a recurring distinctive sawtooth pattern, the wetting force repeatedly dropping gradually and then rising abruptly. This type of pattern was never found with the other two materials, nor was it found when the BM sample was scanned with a low surface tension liquid. Receding scans for all three types show much less variation; however, those for the BM sample do show an analogous sawtooth pattern, a gradual rise in force followed by a sharp drop. The scanning profiles for each BM specimen were very reproducible, as illustrated in Figure 9. The amount of detail obtained is slightly increased when the scan rate is decreased (Figure 10).

The explanation for the nature of the scanning profile of the BM sample lies in the aforementioned dependence of the wetting force on the alignment of the solid surface relative to the vertical. We can consider what will happen when a liquid front which has been moving up a vertical surface encounters a protrusion or a cavity.

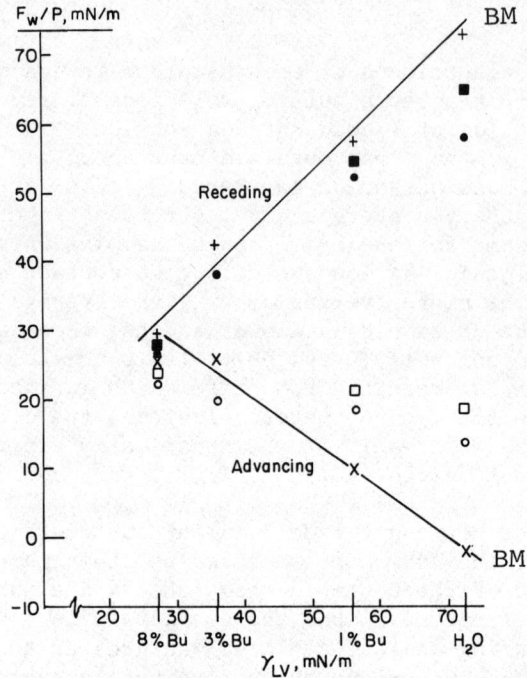

Figure 7. Average advancing and receding wetting forces on the three monofilament samples as a function of liquid surface tension. x,+ - Bimorphic. ○,● - Drawn. □,■ - Drawn and solvent-treated.

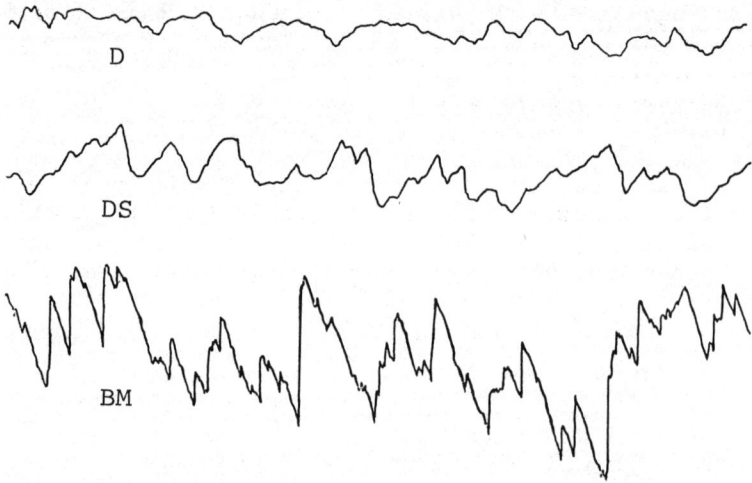

Figure 8. Advancing scans for water on BM, DS and D monofilaments.

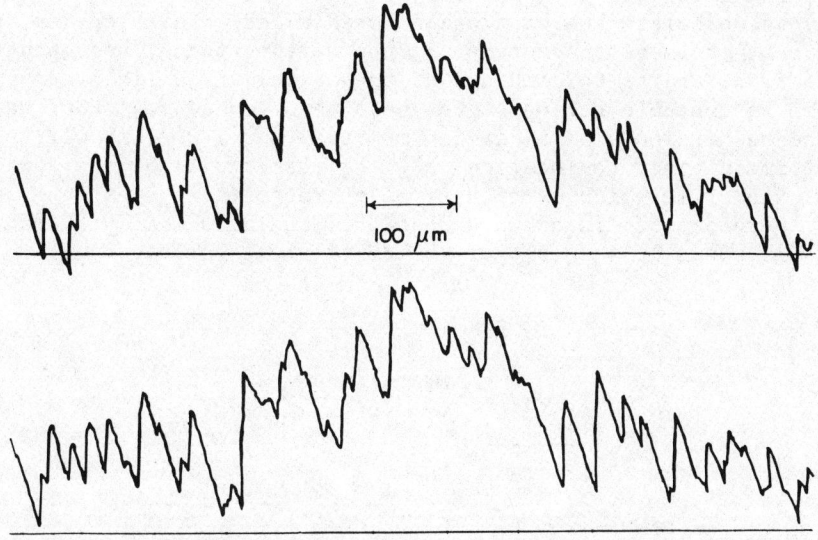

Figure 9. Reproducibility of wetting scans on BM monofilaments.

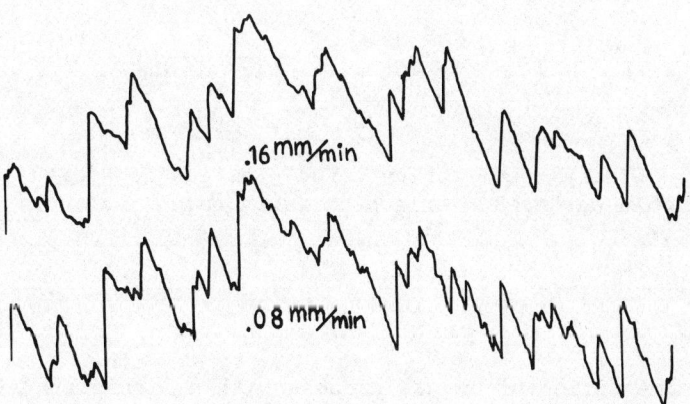

Figure 10. Effect of scan rate on wetting profiles.

Figure 11 shows the contrasting consequences for these two possibilities when the true thermodynamic contact angle is high. For a protrusion, the first contact at B causes an increase in the wetting force which can then decrease to less than the original (vertical) value before the protrusion is engulfed. For a cavity, the first contact causes a decrease in the wetting force, and when the liquid front starts to come out of the cavity it is likely to pass through an unstable region where, because of conservation of mass and the requirements of LaPlacian meniscus curvature, it will skip almost immediately to location D.[4] The result will be the sawtooth scanning profile which we find to be characteristic of the BM sample. Therefore, it is assumed that the liquid is mainly encountering cavities as it passes over the BM filament surface.

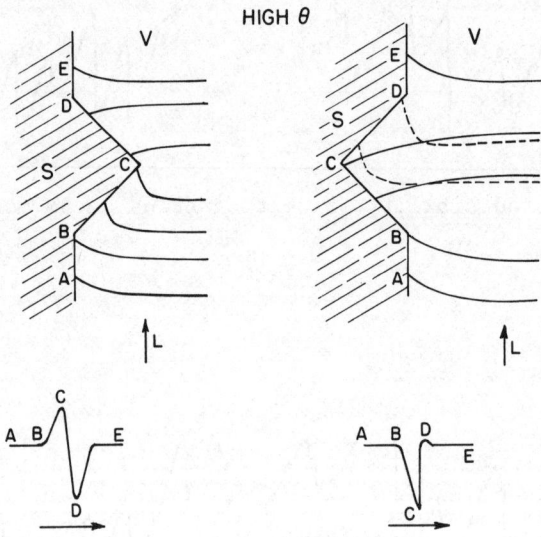

Figure 11. Generalized wetting force profiles as a high surface tension liquid advances over a protrusion and a cavity on a filament surface.

When the same series of events happens with a liquid that forms a small advancing contact angle with the polymer surface, the force variation profile for a protrusion will be the same as in the previous case (Figure 12); however, the unstable region for the scanning of a cavity will be reduced or possibly nonexistent. This would explain why the sawtooth perturbations are not found with a low surface tension liquid.

Scans such as those shown in Figures 9 and 10 can be used to determine the number of cavities detected per unit length of scan. This is done simply by counting the number of abrupt increases in force (excluding any that are less than 1% of the full chart range). The length of each of these jumps can also be measured; however, interpretation of the physical significance of these force change magnitudes is somewhat more complicated.

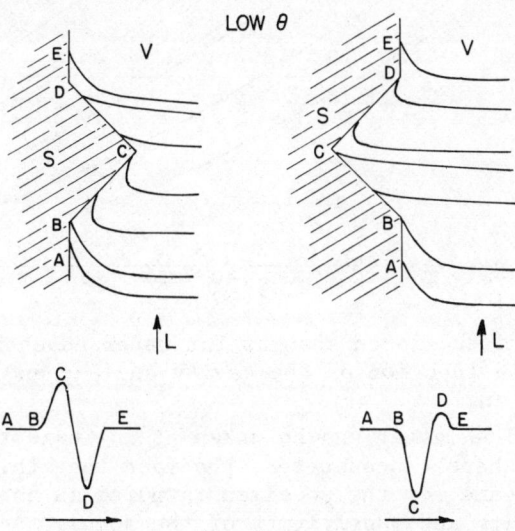

Figure 12. Generalized wetting force profiles as a low surface tension liquid advances over a protrusion and a cavity on a filament surface.

Figure 13 shows the predicted changes for water passing over a model cavity as a function of the cavity angle $\phi$. It is seen that the magnitude of the force change is essentially linearly dependent on $\phi$ or, in other words, the size of a jump is proportional to the steepness of the corresponding cavity, not its depth. The jump size also depends on the fraction of the contact perimeter represented by the cavity. Another interesting point illustrated in Figure 13 is that the effect of cavity steepness is not influenced by contact angle over a wide range, since the curves for contact angles between 30° and 90° are very close to parallel.

The results of surface analyses for a collection of seven BM filament specimens are shown in Table I. The frequency of cavitation as determined by scanning only a few millimeters of each specimen is quite consistent from specimen to specimen. The average size of the vertical jumps for each scanned length is also quite consistent, as is the coefficient of variation(C.V.) for these changes in

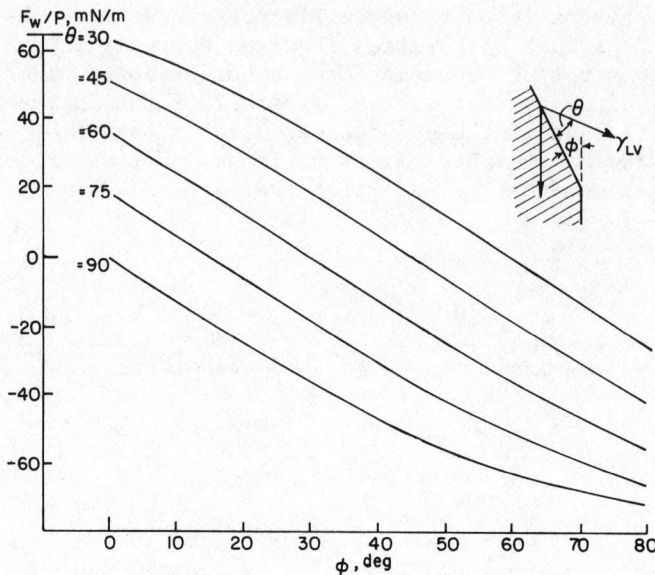

Figure 13. Predicted force changes for water advancing over a model cavity as a function of the cavity angle $\phi$ and the thermodynamic contact angle $\theta$.

wetting force. The latter can be taken as an indication of the range of cavity shapes encountered; the fact that this range is practically the same for the specimens studied is an indication of the reproducibility and sensitivity of the scans.

Table I. Surface Cavitation Data for BM Specimens.

| Specimen | Length scanned, mm | Cavities per mm | Average $\Delta F_w$, µg | Coefficient of Variation |
|---|---|---|---|---|
| 1 | 2.74 | 50 | $109 \pm 85$ | 0.78 |
| 2 | 2.03 | 48 | $115 \pm 104$ | 0.91 |
| 3 | 2.28 | 54 | $81 \pm 68$ | 0.85 |
| 4 | 1.52 | 49 | $73 \pm 57$ | 0.79 |
| 5 | 2.02 | 37 | $169 \pm 162$ | 0.96 |
| 6 | 2.28 | 44 | $106 \pm 98$ | 0.83 |
| 7 | 3.93 | 47 | $105 \pm 98$ | 0.93 |
| | Average | $47 \pm 5.4$ | | |

The scanning electron micrograph shown in Figure 2b, which represents about a 0.1 mm length of BM filament surface, appears to show many individual protrusions; on the other hand, the cavities seem to be connected so as to form extended transverse grooves. According to the results shown in Table I, this length of filament should have about five cavities, which is about the number of grooves one can see in the micrograph.

The general concept of what happens during an advancing scan is that each time a portion of the liquid front comes out of one of these cavity bands, the wetting force will abruptly increase. The size of this increase will depend on the steepness of the cavity and also on the fraction of the total three-phase interface that is emerging from the cavity. Of course, if two separate cavities are encountered at the same time, they would appear as one. The slower the scan rate, the less likely this is to happen, which explains why more jumps can be detected at slower speeds.

If one could assume that the cavity angles were all about the same, then the size of each jump could be used to estimate the relative transverse length of each cavity. Cavity width could also be estimated by the scanning length during the gradual drop. These equivalences are illustrated in Figure 14.

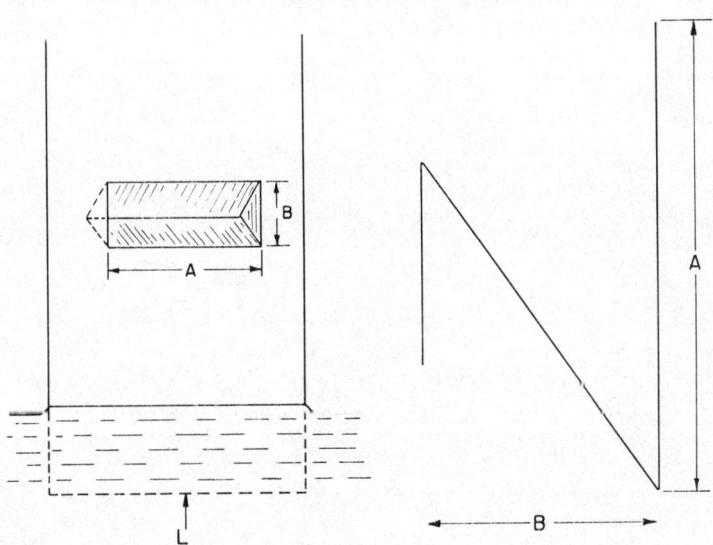

Figure 14. Illustration of the correlations between cavity dimensions and wetting force profile for advancing liquid when cavity steepness is constant.

## ACKNOWLEDGMENTS

These surface characterization studies were carried out as part of the Textile Research Institute project "Surface-Controlled Properties of Fibrous Structures," supported by a group of Corporate TRI Participants. The bimorphic polyester filaments were developed under a grant (No-DMR-7905980) to TRI from the National Science Foundation.

## REFERENCES

1. E. A. Gerold, L. Rebenfeld, M. G. Scott, and H.-D. Weigmann, Textile Res. J., 49, 652 (1979).
2. B. Miller, in "Surface Characteristics of Fibers and Textiles", Vol. II, M. J. Schick, Editor, Marcel Dekker, Inc., New York, 1977.
3. L. S. Penn and B. Miller, J. Colloid Interface Sci., 78, 238 (1980).
4. S. Hedvat, "Capillary Flow Studies on Fibrous Substrates", Doctoral Dissertation, Textile Research Institute and Princeton University Department of Chemical Engineering, October 1980, Chap. 5.

# SURFACE FREE ENERGY OF PLASMA-DEPOSITED THIN POLYMER FILMS

A. M. Wróbel

Department of Polymer Physics, Centre of Molecular and Macromolecular Studies, Polish Academy of Sciences Boczna 5, 90-362 Łódź, Poland

The surface properties of plasma polymer films deposited from numerous organosilicon and hydrocarbon monomers were characterized by contact angle measurements using water/methylene iodide test system. The contact angle data were analyzed to evaluate the dispersion, $\gamma_S^d$, and polar, $\gamma_S^p$, components of the surface free energy, $\gamma_S$, assuming a model for low energy surfaces. The surface energy properties were correlated with the surface structure examined by ATR-IR spectroscopy. It was shown that $\gamma_S^d$ depends on the crosslinking in plasma polymer and increases with crosslink density. The $\gamma_S^p$ component appeared to be very sensitive to monomer composition. For example, organosilicon and hydrocarbon nitrogen-containing monomers, in contrast to organosiloxanes and aliphatic hydrocarbons, were found to produce plasma polymers of high $\gamma_S^p$ values. The effect of aging and thermal post-treatment on the surface energy properties of plasma polymers was also studied. Comparison of plasma polymers with their counterparts polymerized by conventional methods revealed significant differences in the surface energy properties. The higher $\gamma_S^d$ and $\gamma_S^p$ values noted for plasma polymers interpreted in terms of their cross-linked structure and increased contribution of the polar groups in the surface region formed by the atmospheric oxidation, respectively. The current work can provide valuable information relative to the wettability and adhesion behavior of plasma-deposited polymeric materials.

## INTRODUCTION

In the recent development of new polymeric materials, plasma polymerization processes have become an important branch of polymer chemistry. The complex plasma reactions produce polymers of different structure and properties from their counterparts polymerized by conventional methods. From an application viewpoint, it is noteworthy that plasma-deposited polymer films are highly crosslinked, adhere tenaciously to a wide range of substrates and their structure and thickness can be controlled closely by the plasma conditions employed.

In many of the potential applications of these materials, such as reverse osmosis membranes[1,2] and protective coatings[3,4], a knowledge of wettability and surface free energy properties can provide valuable information relating to their permeability and adhesion behavior. Although many papers describe the surface energy of conventional polymeric materials; however, no comprehensive study has been performed to characterize this property for plasma polymers. A few papers which have been published[5-10] deal with this problem for a limited number of plasma polymers only. Therefore, the current study was undertaken to estimate the surface free energies of plasma polymer films deposited from numerous organosilicon and hydrocarbon monomers, and in particular, the effect of monomer structure on the dispersion and polar components of their surface energy.

## EXPERIMENTAL

### Plasma Deposition Procedure

Plasma deposition process was carried out in an audio frequency glow discharge operated at a 20 kHz using the electrode system described in detail elsewhere.[11] Polymer films were deposited onto polished surface of chromium electrodes spaced 3 cm apart and each of 50 cm$^2$ in surface area. Plasma was generated at a steady monomer pressure of 0.3 Torr, power density of 0.1 W/cm$^3$, and discharge duration varied from 10 to 40 s. To maintain the same pressure conditions in all experiments, vapors of monomers of low sublimate pressure such as octamethylcyclotetrasilazane and dodecamethylcyclohexasilane were admixed with argon. Thickness of polymer films deposited under these conditions was within the range of 0.1-1 μm.

### Wettability Measurements

Wettability of plasma polymer films was determined by measuring the advancing contact angle of sessile drops of two test liquids: water and methylene iodide. The measurements were carried out at room temperature using a PZO goniometer eyepiece. Each type of plasma polymer was characterized by the contact angle measurements performed for at least three polymer samples deposited in separate experiments. For each liquid an average of 30-60 readings was calculated. Reproducibility of the measurements was better than ±2 deg. in most cases.

## Infrared Spectroscopy

Infrared spectra of the polymer films were recorded on a Perkin-Elmer, Model 457, spectrophotometer, using the attenuated total reflection (ATR) technique.

## Materials

Gaseous monomers of high grade purity were used as supplied. Liquid and solid monomers prior to polymerization were purified by vacuum rectification or vacuum sublimation, respectively. Water and methylene iodide were triply distilled before contact angle measurements.

## RESULTS AND DISCUSSION

### Calculation of the Surface Free Energy Components

To calculate the disperion and polar components of the surface energy of plasma polymer films, the contact angle data were analyzed according to a model for low energy surfaces developed by Owens et al.[12] and Kaelble[13]. This analysis technique is summarized in Equations (1-5).

$$\gamma_L = \gamma_L^d + \gamma_L^p \qquad (1)$$

$$\gamma_S = \gamma_S^d + \gamma_S^p \qquad (2)$$

$$W_a = \gamma_L (1 + \cos\theta) \qquad (3)$$

$$W_a = 2\left[(\gamma_L^d \gamma_S^d)^{\frac{1}{2}} + (\gamma_L^p \gamma_S^p)^{\frac{1}{2}}\right] \qquad (4)$$

where $\gamma_L$=liquid-vapor surface free energy, $\gamma_S$=solid-vapor surface free energy, $\gamma_L^d$, $\gamma_L^p$ = the respective dispersion and polar components of $\gamma_L$; $\gamma_S^d$, $\gamma_S^p$ = the respective dispersion and polar components of $\gamma_S$, $W_a$ = liquid-solid nominal work of adhesion, $\theta$ = liquid-solid contact angle. From Equations (3) and (4) the following relation is derived:

$$1 + \cos \theta = \frac{2}{\gamma_L} \left[ (\gamma_L^d \gamma_S^d)^{\frac{1}{2}} + (\gamma_L^p \gamma_S^p)^{\frac{1}{2}} \right] \qquad (5)$$

By using the experimental values of $\theta$ measured for two different liquids of defined $\gamma_L^d$ and $\gamma_L^p$ in Equation (5), two simultaneous equations are obtained which can be solved for the unknown surface components $\gamma_S^d$ and $\gamma_S^p$. The sum of these components according to Equation (2) should yield a reasonable approximation of the total solid surface energy $\gamma_S$.

In the present study, Equation (5) was applied for the contact angle data of water and methylene iodide which were utilized as test liquids of widely different surface energy properties. The surface energy of liquids at room temperature used for calculations were as follows[13]: water, $\gamma_L^d = 21.8$, $\gamma_L^p = 51.0$ and $\gamma_L = 72.8$ erg/cm$^2$; methylene iodide, $\gamma_L^d = 48.5$, $\gamma_L^p = 2.3$ and $\gamma_L = 50.8$ erg/cm$^2$.

### Plasma Polymers from Organosilicon Monomers

Wettability studies were carried for plasma polymer films deposited from structurally different organosilicaon monomers such as: siloxanes, silazanes and silanes. To minimize the oxidation of polymer film surface in the ambient, the contact angles were measured immediately after deposition process. The contact angle values and calculated surface energy data for these polymers are summarized in Table I, column A. An examination of the dispersion, $\gamma_S^d$, and polar, $\gamma_S^p$, components of the surface energy reveals distinct differences in their values resulting from various structures of monomers. The lowest values of $\gamma_S^d$ are noted for siloxane plasma polymers as compared with those for silazane and silane polymers. In order to explaine the differences in $\gamma_S^d$, the effect of the polymer structure on the dispersion component has to be analyzed.

Since the plasma polymers are known as highly crosslinked materials, so their degree of crosslinking was felt to be an important factor which may influence strongly the dispersion component. This assumption may be considered in terms of nature of the dispersion forces which, in general, depend on electrical properties of the volume elements involved and the distance between them[14]. The volume element here means atom or molecule; and in the case of polymer, it may be structural unit in its linear or crosslinked bulk structure. The potential for the interaction between two volume elements in liquid or solid is given by the Equation:

$$\mu = - \frac{3 \alpha^2 I}{4 r^6} \qquad (6)$$

where α = the polarizability, I = the ionization potential, and r = the distance between the volume elements. By summation of the pair potentials of all the surface volume elements with all of the volume elements below the surface, the following expression for the dispersion force component of the surface free energy is obtained:[14]

$$\gamma^d = -\frac{\pi N^2 \alpha^2 I^2}{8r^2} \tag{7}$$

where N = the number of volume elements per volume unit. The Equation (7) shows that the dispersion component is proportional to the inverse square of the distance between the volume elements. In the case of crosslinked polymer structure, it is obvious that increase in crosslinking degree will decrease this distance thereby increasing $\gamma_S^d$.

Taking into account the above discussion, it is reasonable to assume that the differences noted for $\gamma_S^d$ (Table I, column A) may arise from various crosslink density produced in plasma polymers. In particular, susceptibility of monomer towards plasma crosslinking reactions seems to play the most important role. The crosslinking, however, depends on the monomer functionality which in terms of plasma polymerization means the ability to generate the reactive species by bond breaking reactions initiated in the glow discharge. The methylsilyl groups present in all monomers investigated were found to convert easily under plasma conditions into ≩Si· and ≩SiCH$_2$ species which subsequently combine to form in the polymer disilymethylene (≩Si-CH$_2$-Si≨) and disilyethylene (≩Si-CH$_2$-CH$_2$-Si≨) crosslinkages.[15] The presence of these bonds in polymer films may be documented by the appearance in IR surface spectra (Figure 1) a weak but distinctive absorption band at 1350 and more intense band at 1050-1020 cm$^{-1}$ (Figure 1C, D and E) which originate from $\delta_S(CH_2)$ and $\omega(CH_2)$ vibrations respectively, in disilymethylene and disilyethylene units.[16] The latter band is indistinguishable in the spectra of siloxane polymer (Figure 1A and B) due to overlap with the ≩Si-O-Si≨ strong absorption which also falls in this region. The methylsilanes may crosslink in this way even more readily than siloxanes and silazanes due to higher concentration of ≩Si· radicals produced by plasma scission of ≩Si-Si≨ bonds, as their bond energy (53 kcal/mole [17]) is considerably lower than those for Si-O bonds (108-112 kcal/mole [18]) in siloxanes and Si-N bonds (77-82 kcal/mole [18]) in silazanes. This may also be confirmed by the results of ESR analysis by Yasuda et al.[19] who have shown that the radical concentration in plasma polymer of HMDS was twice as higher than in the polymer of HMDSO. The evidence for the crosslinking type mentioned is appearance in IR spectrum of plasma-polymerized hexamethyldisilane (PP-HMDS)(Figure 1E) a very intense and broad absorption band with a maximum at 1050 cm$^{-1}$, arising from the high content of ≩Si-CH$_2$-Si≨ and ≩Si-CH$_2$-CH$_2$-Si≨ crosslinkages, In

Table I. Contact Angle and Surface Energy Data for Plasma Polymer Films Deposited from Organosilicon Monomers.

| MONOMER | MOLECULAR STRUCTURE | A. BEFORE AGING | | | | | B. AFTER AGING | | | | | C. % CHANGE | | |
|---|---|---|---|---|---|---|---|---|---|---|---|---|---|---|
| | | CONTACT ANGLE (DEG) | | SURFACE ENERGY (ERG/CM$^2$) | | | CONTACT ANGLE (DEG) | | SURFACE ENERGY (ERG/CM$^2$) | | | | | |
| | | WATER | METHYLENE IODIDE | $\gamma_s^d$ | $\gamma_s^p$ | $\gamma_s$ | WATER | METHYLENE IODIDE | $\gamma_s^d$ | $\gamma_s^p$ | $\gamma_s$ | $\gamma_s^d$ | $\gamma_s^p$ | $\gamma_s$ |
| HEXAMETHYLDISILOXANE (HMDSO) | Me$_3$Si-O-SiMe$_3$ | 94 | 59 | 27.4 | 1.7 | 29.1 | 88 | 60 | 25.4 | 3.9 | 29.3 | -7 | +129 | +1 |
| HEXAMETHYLCYCLOTRI-SILOXANE (HMCTSO) | | 95 | 62 | 25.7 | 1.8 | 27.5 | 92 | 62 | 25.0 | 2.7 | 27.7 | -3 | +50 | +1 |
| OCTAMETHYLCYCLOTETRA-SILOXANE (OMCTSO) | | 95 | 62 | 25.7 | 1.8 | 27.5 | 93 | 62 | 25.2 | 2.4 | 27.6 | -2 | +33 | 0 |
| HEXAMETHYLDISILAZANE (HMDSN) | Me$_3$Si-NH-SiMe$_3$ | 84 | 54 | 28.3 | 4.6 | 32.9 | 71 | 56 | 24.2 | 12.6 | 36.8 | -14 | +174 | -12 |
| HEXAMETHYLCYCLOTRI-SILAZANE (HMCTSN) | | 81 | 50 | 30.1 | 5.3 | 35.4 | 78 | 61 | 22.6 | 9.3 | 31.9 | -25 | +75 | -10 |
| OCTAMETHYLCYCLOTETRA-SILAZANE (OMCTSN) | | 72 | 32 | 38.0 | 7.0 | 45.0 | 68 | 39 | 33.4 | 10.4 | 43.8 | -12 | +49 | -3 |

| Compound | Structure | | | | | | | | | | |
|---|---|---|---|---|---|---|---|---|---|---|---|
| BIS(DIMETHYLSILYL)-TETRAMETHYLCYCLO-DISILAZANE (TMCDSN) | Me₂HSi-N(SiMe₂)₂N-SiHMe₂ | 89 | 48 | 33.4 | 2.0 | 35.4 | 82 | 50 | 30.4 | 4.9 | 35.3 | -9 | +145 | 0 |
| HEXAMETHYLDISILANE (HMDS) | Me₃Si-SiMe₃ | 91 | 53 | 30.7 | 1.9 | 32.6 | 81 | 50 | 30.1 | 5.3 | 35.4 | -2 | +179 | +9 |
| DODECAMETHYLCYCLO-HEXASILANE (DMCHS) | (Me₂Si)₆ ring | 74 | 40 | 34.3 | 7.1 | 41.4 | 67 | 48 | 28.1 | 13.1 | 41.2 | -18 | +85 | 0 |
| SILANE | SiH₄ | 73 | 42 | 32.9 | 8.0 | 40.9 | 69 | 46 | 29.7 | 11.3 | 41.0 | -10 | +41 | 0 |

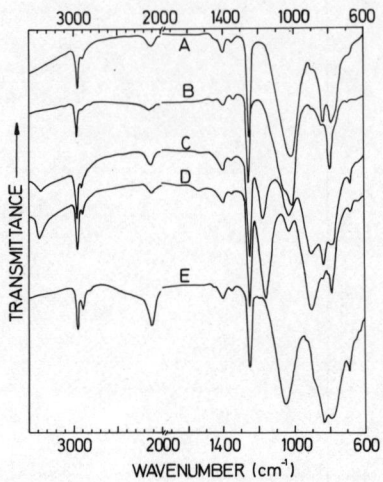

Figure 1. ATR-IR spectra of plasma polymer films deposited from various organosilicon monomers: A, HMDSO, B, HMCTSO, C, HMDSN, D, HMCTSN, and E, HMDS.

contrast, silane ($SiH_4$) may undergo crosslinking by $\equiv$Si-Si$\equiv$ coupling only. In the case of silazanes with $\equiv$Si-NH-Si$\equiv$ units, an additional crosslinking is produced via abstraction of hydrogen from -NH- groups and by the formation of Si-N crosslinkages in reaction with Si· species;[20,21] whereas for cyclodisilazane (TMCDSN), due to tertiary nitrogen, the formation of disilyhydrocarbon links seems to be the prevalent process.

The structural data indicate that silazanes and silanes appear to crosslink more readily than siloxanes, producing plasma polymers of higher degree of crosslinking. This is consistent with a trend observed for the dispersion component in Table I, column A, where for silazane and silane polymer films higher values of $\gamma_S^d$ are noted. The results evidently show that the dispersion component of the surface energy increases with crosslink density in plasma polymers.

The apparent differences in the polar component values noted between particular groups of plasma polymer films (Table I, column A) result from their different surface composition. However, rapid post-plasma reactions of radical species, present in most plasma polymers, with the atmosphere may increase polarity of the film surface, and therefore it is reasonable to assume that the polar components determined are somewhat higher than their real values. The low $\gamma_S^p$ values found for siloxane plasma polymers are due to the presence in their surface structure of $\equiv$Si-H, $\Rightarrow$Si-CH$_2$-Si$\Leftarrow$, $\equiv$Si-CH$_2$-CH$_2$-Si$\equiv$, and $\equiv$Si-CH$_3$ low polar structural units which are identified in IR spectra of these polymers (Figure 1 A and B) by the absorption bands appearing at 2100 cm$^{-1}$, $\nu$(SiH); 1350 cm$^{-1}$, $\delta_s$(CH$_2$); and 1250 cm$^{-1}$, $\delta_s$(CH$_3$); respectively[16]. The higher $\gamma_s^p$ values noted for most silazane plasma polymers may be attributed to nitrogen-containing $\equiv$Si-NH-Si$\equiv$ units whose presence in their structure is confirmed by the appearance of absorption bands at 3390, 1170-1160, and 900 cm$^{-1}$ (Figure 1C and D) corresponding to $\nu$(NH), $\delta$(NH) and $\nu_{as}$(SiNSi) vibrations, respectively.[16] The silane monomers appear to produce plasma polymers of widely different $\gamma_s^p$ values. However, these polymers due to the presence of non-polar $\equiv$Si-CH$_3$, $\equiv$Si-Si$\equiv$, and $\equiv$Si-H units in the respective monomers are expected to be of low polar components as that of PP-HMDS. The rise in $\gamma_s^p$ observed for PP-DMCHS indicates the increased content of polar groups in the surface structure of this polymer. This seems to be associated with the relatively large number of $\equiv$Si-Si$\equiv$ bonds, in DMCHS cyclic monomer which while undergoing plasma scission may produce more $\equiv$Si· radicals as compared with HMDS. The increase in $\equiv$Si· radicals involves also rise in SiCH$_2$· species generated via secondary reaction of silicon radicals with methylsilyl groups.[15] The silicon and carbon radical species formed in plasma are subsequently incorporated into the surface of growing polymer film and on its exposure to the atmosphere these may capture molecular oxygen and water producing rapidly hydroxyl and carbonyl (strongly polar groups) on the surface thereby increasing $\gamma_s^p$. Similarly, high $\gamma_s^p$ found for PP-silane may result from a considerable concentration of $\equiv$Si· species on the surface of this polymer, and such species in contact with the atmosphere are converted easily to $\equiv$Si-OH groups, thus increasing the surface polarity.

The results evidently show that the polar component depends closely on the surface chemistry of plasma polymer films, in particular on the character of monomer structural units incorporated in the surface structure.

### Plasma Polymers from Hydrocarbon Monomers

Contact angles were measured on plasma polymers deposited from numerous hydrocarbon monomers of different structure containing triple bond, olefinic double bonds, aromatic and aliphatic structures. The results of contact angle measurements and evaluated surface energy properties for these polymers are summarized in Table II, column A. The data for plasma polymers from acetylene, ethylene, and hexane indicate that monomer unsaturation does not change substantially the dispersion component but increases the polar component to a considerable extent as in the case of acetylene. This, undoubtedly, is due to the high concentration of radicals in PP-AC and resulting rapid formation of carbonyls and hydroxyls on the film surface on exposure to the atmosphere. The ESR results reported by Yasuda et al.[19] have shown indeed the spin density in this plasma polymer to be a six times greater than that in PP-E.

The data for plasma polymers deposited from double bond-containing monomers reveal strong effect of substituents on carbon atoms on the surface energy properties. Substitution of the hydrogen atoms in ethylene by chlorines (vinylidene chloride) increases $\gamma_S^d$, whereas fluorine substitution (tetrafluoroethylene) leads to a large decrease in this component. Similar trend has been noted for conventional polymers of these monomers[12,13] and the rise in $\gamma_S^d$ for poly(vinylidene chloride) was assumed to derive from the large polarizability of the covalent chlorine atoms, whereas the decrease in $\gamma_S^d$ for polytetrafluoroethylene is not fully understood although it may arise from the larger size of the $-CF_2-$ unit as compared to the $-CH_2-$ unit in polyethylene.[22] However, in the case of plasma polymers the effect of crosslinked structure also has to be considered. Comparison of the average bond energies for C-Cl (81 kcal/mole), C-H (99 kcal/mole), and C-F (116 kcal/mole) bonds indicates that the two former bonds will undergo plasma scission more readily than the latter bonds; and, therefore, in terms of plasma polymerization, VDC is expected to produce polymer of higher crosslinking than TFE. This agrees with the trend noted for the dispersion components of these plasma polymers. Substitution of the ester group methyl methacrylate increases $\gamma_S^p$, whereas $\gamma_S^d$ is seen to decrease slightly. The sharp rise in $\gamma_S^p$ results from the contribution of some ester groups in the polymer structure rather than from the atmospheric oxidation effect since the ester substituent was found to reduce considerably the spin density in plasma polymer.[19] Small decrease in $\gamma_S^d$ may derive from quenching of the radical species in plasma by atomic oxygen formed by cleavage of the ester groups and resulting lower crosslinking produced in the polymer. Phenyl substitution (styrene) increases both components: $\gamma_S^d$ and $\gamma_S^p$. This may arise from the higher crosslink density in this polymer and the oxidation effect, respectively, as the spin density in PP-S was found to be roughly three times higher in comparison to that in PP-E.[19]

The surface energy data for plasma polymers from the aromatic monomers illustrate the effect of substituents in the aromatic ring and monomer heteroaromatic structure. The results for PP-B, PP-T, and PP-AN indicate that the methyl nonpolar substituent does not involve substantial change in the surface energy components, whereas the amine polar substituent increases $\gamma_S^p$. The latter change is evidently caused by the nitrogen incorporation in the polymer. Much stronger effect is noted for PP-P which displays extremely large $\gamma_S^p$ value. This may be attributed to the high content of the nitrogen retained in the polymer structure since a triply bonded nitrogen atom in the heteroaromatic ring seems to be more stable to-

Table II. Contact Angle and Surface Energy Data for Plasma Polymer Films Deposited from Hydrocarbon Monomers.

| MONOMER | MOLECULAR STRUCTURE | A. BEFORE AGING | | | | | | B. AFTER AGING | | | | | | C. % CHANGE | | |
|---|---|---|---|---|---|---|---|---|---|---|---|---|---|---|---|---|
| | | CONTACT ANGLE (DEG) | | SURFACE ENERGY (ERG/CM$^2$) | | | CONTACT ANGLE (DEG) | | SURFACE ENERGY (ERG/CM$^2$) | | | | | | |
| | | WATER | METHYLENE IODIDE | $\gamma_s^d$ | $\gamma_s^p$ | $\gamma_s$ | | WATER | METHYLENE IODIDE | $\gamma_s^d$ | $\gamma_s^p$ | $\gamma_s$ | $\gamma_s^d$ | $\gamma_s^p$ | $\gamma_s$ |
| ACETYLENE (AC) | HC≡CH | 63 | 16 | 41.7 | 10.2 | 51.9 | | 71 | 23 | 41.6 | 6.4 | 48.0 | 0 | -37 | -8 |
| ETHYLENE (E) | H$_2$C=CH$_2$ | 85 | 35 | 40.1 | 2.0 | 42.1 | | 81 | 33 | 40.0 | 3.1 | 43.1 | 0 | +55 | +2 |
| VINYLIDENE CHLORIDE (VDC) | H$_2$C=CCl$_2$ | 79 | 15 | 46.5 | 2.6 | 49.1 | | 72 | 18 | 43.6 | 5.5 | 49.1 | -6 | +112 | 0 |
| TETRAFLUOROETHYLENE (TFE) | F$_2$C=CF$_2$ | 99 | 78 | 16.4 | 2.7 | 19.1 | | 91 | 71 | 19.1 | 4.6 | 23.7 | +16 | +70 | +24 |
| METHYL METHACRYLATE (MMA) | H$_2$C=C(CH$_3$)COOCH$_3$ | 65 | 29 | 37.6 | 10.5 | 48.1 | | 63 | 28 | 37.5 | 11.6 | 49.1 | 0 | +10 | +2 |
| STYRENE (S) | HC=CH$_2$—C$_6$H$_5$ | 77 | 17 | 45.3 | 3.4 | 48.7 | | 69 | 25 | 40.3 | 7.7 | 48.0 | -11 | +126 | -1 |
| BENZENE (B) | C$_6$H$_6$ | 73 | 23 | 42.2 | 5.5 | 47.7 | | 69 | 28 | 39.1 | 8.1 | 47.2 | -7 | +47 | -1 |
| TOLUENE (T) | CH$_3$—C$_6$H$_5$ | 72 | 23 | 41.9 | 6.0 | 47.9 | | 68 | 27 | 39.2 | 8.5 | 47.7 | -6 | +42 | 0 |
| ANILINE (AN) | NH$_2$—C$_6$H$_5$ | 66 | 24 | 39.9 | 9.2 | 49.1 | | 70 | 32 | 37.5 | 8.1 | 45.6 | -6 | -12 | -7 |
| PYRIDINE (P) | C$_5$H$_5$N | 54 | 21 | 38.0 | 16.5 | 54.5 | | 55 | 25 | 36.8 | 16.4 | 53.2 | -3 | 0 | -2 |
| CYCLOHEXANE (CH) | C$_6$H$_{12}$ | 85 | 33 | 41.1 | 1.8 | 42.9 | | 71 | 26 | 40.4 | 6.8 | 47.2 | -2 | +278 | +10 |
| HEXANE (H) | H$_3$C-(CH$_2$)$_4$-CH$_3$ | 86 | 34 | 40.9 | 1.6 | 42.5 | | 78 | 30 | 40.6 | 4.0 | 44.6 | -1 | +150 | +5 |

wards plasma abstraction than that in the amine substituent. On the other hand, it would also explain some decrease in $\gamma_S^d$ noted for PP-P which presumably is due to the reduced crosslinking in this polymer resulting from the lower extent of monomer fragmentation by plasma.

The data for plasma polymers deposited from the aliphatic monomers: cyclohexane and hexane indicate that both structures (cyclic and linear) produce polymers of similar surface energy properties. It is apparent that the lowest values of $\gamma_S^p$ noted for these polymers are associated with the non-polar nature of the aliphatic hydrocarbons. Comparison of the results for polymers from cyclic aliphatic PP-CH and aromatic PP-B monomers shows that the aromatic structure substantially increases $\gamma_S^p$ and only slightly $\gamma_S^d$. The former change may be assigned to the atmospheric oxidation effect arising from the increased spin density level in PP-B which was evaluated to be nearly four times higher than that of PP-CH.[19] The high concentration of radical species formed by plasma fragmentation of aromatic structure may also produce higher crosslinking which would explain some increase in the polar component as noted for this polymer.

In general, the data presented in Tables I and II indicate that plasma polymer films deposited from hydrocarbon monomers (except TFE) display higher total surface energy, $\gamma_S$, than those formed from organosilicon monomers. This is mostly due to the large values of the dispersion component noted for these polymers which may result from higher susceptibility of hydrocarbon monomers towards crosslinking reactions. The results also show that the nature of the surface free energy of plasma polymer film is closely dependent on the chemical structure of the monomer. This seems to be very important, since it opens the possibility of controlling surface energy properties of deposits by suitable selection of monomers.

## Effect of Aging

Since plasma polymer films are well known to be very rich in radicals[23] and the life-times of the radicals are relatively long even in the ambient[19], these are considered to react with oxygen and water forming oxygen-containing polar groups on the surface. This consequently may involve substantial changes in their wettability properties. The estimation of these changes seems to be important from the practical viewpoint.

To evaluate the effect of aging, the contact angles were measured on films after their exposure to the atmosphere for various durations. The most significant changes in wettability were observed during the first two days of aging, and after this period the effect of aging decreased considerably. This is consistent with the kinetics of radicals decay in plasma polymers exposed to the atmosphere and indicates that the decay process is fastest in the initial stages.[19] The contact angle data and surface energy properties evaluated for plasma polymers subjected to 48 hrs. aging process are summarized in Tables I and II, columns B. The relative changes in the surface energy properties resulting from aging effect are illustrated in columns C of Tables I and II. From these data it is evident that aging in most cases involves a marked increase in the polar component whereas the dispersion component appears to decrease. Some deviations from this trend, however, are noted for PP-AC, PP-AN (decrease in $\gamma_S^p$) and PP-TFE (increase in $\gamma_S^d$). The increase in $\gamma_S^p$ is associated with the relatively high level of spin density in plasma polymers which, for most of them, was estimated to be in the range $10^{18}$–$10^{20}$ spins/cm$^3$.[19] Carbon and silicon radicals present in plasma polymers can undergo a variety of oxidation reactions in the atmosphere. The dominant reactions may be described by Equations (8-11),

$$-\overset{|}{\underset{|}{C}}\cdot \; + \; H_2O \; \longrightarrow \; -\overset{|}{\underset{|}{C}}-OH \tag{8}$$

$$-\overset{|}{\underset{|}{C}}\cdot \; + \; O_2 \; \longrightarrow \; -\overset{|}{\underset{|}{C}}-O-O\cdot \; \longrightarrow \; -\overset{|}{\underset{|}{C}}-O-OH \; \rightleftharpoons \; \begin{matrix} -C-OH \\ \overset{|}{C}=O \\ -HC=O \end{matrix} \tag{9}$$

$$-\overset{|}{\underset{|}{Si}}\cdot \; + \; H_2O \; \longrightarrow \; -\overset{|}{\underset{|}{Si}}-OH \tag{10}$$

$$-\overset{|}{\underset{|}{Si}}\cdot \; + \; O_2 \; \longrightarrow \; -\overset{|}{\underset{|}{Si}}-O-O\cdot \; \longrightarrow \; -\overset{|}{\underset{|}{Si}}-O-OH \tag{11}$$

Furthermore, some reactive structural units in organosilicon and hydrocarbon plasma polymers in contact with atmospheric water and oxygen may also produce hydroxyl groups as follows:

$$-\overset{|}{\underset{|}{Si}}-H \; + \; \tfrac{1}{2}O_2 \; \longrightarrow \; -\overset{|}{\underset{|}{Si}}-OH \tag{12}$$

$$-\overset{|}{\underset{|}{Si}}-NH-\overset{|}{\underset{|}{Si}} \; + \; 2H_2O \; \longrightarrow \; 2(-\overset{|}{\underset{|}{Si}}-OH) \; + \; NH_3 \tag{13}$$

$$-\overset{|\;|}{\underset{|\;|}{Si-Si}}- \; + \; 2H_2O \; \longrightarrow \; 2(-\overset{|}{\underset{|}{Si}}-OH) \; + \; H_2 \tag{14}$$

$$-\overset{|\;\;|}{C=C}- \; + \; \tfrac{1}{2}O_2 \; \longrightarrow \; -\underset{\diagdown \diagup}{\overset{|\;\;|}{C-C}}- \; \overset{H_2O}{\longrightarrow} \; -\underset{OH\;\;OH}{\overset{|\;\;|}{C-C}}- \tag{15}$$

$$-\text{C}=\text{C}- \;+\; \text{H}_2\text{O} \longrightarrow -\underset{\underset{\text{OH}}{|}}{\text{C}}-\text{CH}- \tag{16}$$

The contribution of hydroxyl and carbonyl groups in the structure of aged plasma polymers was confirmed by appearance in IR spectra of strong signals in the regions of 3700-3400 cm$^{-1}$ and 1750-1690 cm$^{-1}$ which are assigned to $\geqslant$C-OH and/or $\geqslant$Si-OH, and $\geqslant$C=O, respectively. It is important to note that the signals increased with aging. Similar results have been reported by Yasuda et al.[7] for plasma polymers of numerous hydrocarbon monomers. The presence of oxygen in plasma polymers was also proved by elemental analysis[7,24,25] which indicated 0.3-0.8 oxygen atom per repeating monomer unit.

It is evident that the surface structural changes resulting from reactions (8-16) will increase the polarity of the film surface, but they may also reduce the density in the surface region. This may explain the rise in $\gamma_S^p$ and the decrease in $\gamma_S^d$, as was observed for most aged plasma polymers. However, the deviations from this trend in some of the surface energy components noted for PP-AC, PP-AN, and PP-TFE are rather difficult to elucidate due to a great complexity of the aging process.

## Effect of Thermal Treatment

Of the numerous monomers used in plasma polymerization, the organosilicones are of particular interest. These compounds due to their chemical nature appear to produce polymer films of unique optical properties. Our most recent study[26] confirms the results reported earlier by Tien et al.[27] and indicates that plasma-polymerized organosilicon films display exceptionally low level of light scattering. This stresses their potential use as high quality light-guiding materials in integrated optics. Moreover, we have shown that the chemical structure and properties of plasma-polymerized organosilicon films can be modified selectively by a controlled pyrolytic process used either as a post-treatment[28,29] or combined with the polymerization step, i.e. as a deposition onto heated substrate.[30] Films produced as a result of this process tend to be highly crosslinked glassy materials with increasing inorganic character. These show high thermal stability, strong adhesion to a wide range of metal substrates, and outstanding chemical resistance to aggressive environments. These advantageous properties suggest the possibility of useful applications of such films, for example, as coatings for passivation of metal surface.[4,15]

In view of the evident structural changes produced in organosilicon plasma polymer films as a result of thermal post-treatment[15,29], it was also interesting to establish the effect of this process on the surface energy properties. For this purpose, samples of polymer films were subjected to a vacuum thermal treatment under the following conditions: pressure=$10^{-5}$Torr, temperature= 800°C, and treatment duration=30 min.

Table III. Effect of Vacuum Thermal Treatment for 30 min. at 800°C on Contact Angles and Surface Energy Properties of Plasma Polymer Films Deposited from Organosilicon Monomers.

| Plasma Polymer | Contact Angle (deg) | | Surface Energy (erg/cm$^2$) | | |
|---|---|---|---|---|---|
| | Water | Methylene Iodide | $\gamma_s^d$ | $\gamma_s^p$ | $\gamma_s$ |
| PP-HMDSN | | | | | |
| (U) | 84 | 54 | 28.3 | 4.6 | 32.9 |
| (T) | 73 | 25 | 41.4 | 5.7 | 47.1 |
| PP-HMCTSN | | | | | |
| (U) | 81 | 50 | 30.1 | 5.3 | 35.4 |
| (T) | 70 | 30 | 38.4 | 7.8 | 46.2 |
| PP-HMDS | | | | | |
| (U) | 91 | 53 | 30.7 | 1.9 | 32.6 |
| (T) | 96 | 27 | 47.4 | 0.0 | 47.4 |

(U), Untreated and (T), thermally treated film.

The contact angle and the surface energy data for untreated (U) and thermally treated (T) films from silazane and silane monomers are summarized in Table III. The data clearly show a strong effect of thermal treatment on both the dispersion and polar components. A marked increase in $\gamma_s^d$ noted for each of treated films is due to densification of film structure resulting from thermally induced crosslinking reactions.[15,29] The density value was found to rise from 1.2 g/cm$^3$ for untreated film to nearly 2 g/cm$^3$ for treated film.[15] The increase in $\gamma_s^p$ noted for silazane films is considered to derive mainly from the increased content of nitrogen-containing bonds in the structure of pyrolysed films.[15,29] In contrast to silazane films, however, thermal treatment of silane film eliminates completely the surface polarity, i.e., $\gamma_s^p=0$. This is consistent with IR study carried out for aged silane plasma polymer containing carbonyl and hydroxyl groups. It was shown that thermal treatment involved a decay of the absorption bands characteristic of these structures.

The changes in the surface energy components caused by the thermal treatment appear to be in close relationship to the adhesion behavior of treated films. In contrast to untreated films, thermally treated films demonstrated significant differences in their adhesion. Scratch qualitative test revealed that treated silazane polymer films adhered to the metal substrate so strongly that they could not be removed without destroying the metal surface, whereas treated silane polymer films were easily removed by using Scotch tape. The increase in $\gamma_S^p$ for silazane plasma polymers agrees reasonably well with marked increase in their adhesion to the metal substrate observed after thermal treatment, whereas deterioration in this property noted for treated silane polymer may be accounted for the decrease in $\gamma_S^p$. However, thermal treatment of plasma-polymerized organosiloxanes resulted in a drastic deterioration in their adhesion to the metal substrate and peeling was observed even under milder treatment conditions. This may be due to the low values of the polar component evaluated for this group of plasma polymers (Table I, column A). We realize that the interpretation of adhesion improvement for thermally treated silazane plasma polymer films in terms of surface energy changes, however, may be very approximate only, since the other types of interfacial bonding than van der Waals cannot be excluded here. It is very likely, that under such drastic treatment conditions the chemical bonding may also occur within the polymer/metal interface region.

The results prove that the surface energy properties of plasma polymer films can be modified by using an appropriate thermal posttreatment process which appeared to be particularly useful for the production of strongly adherent protective coatings for metals.[15,29]

### Plasma and Conventional Polymers

Numerous studies have shown that plasma-polymerized materials differ significantly in their structure and properties from those produced by conventional polymerization techniques. The most important difference is that the structure of plasma polymers does not contain long segments with regularly repeating monomer units. In general, plasma polymers are amorphous and highly crosslinked materials of irregular chemical structure.[23] Differences in the structures of plasma and conventional polymers should also be reflected in their surface energy properties.

Table IV exemplifies the surface energy properties of some plasma polymers (data from Tables I and II, columns A) and their conventional counterparts. The surface energy components for conventional polymers were calculated from contact angle data of water/methylene iodide system reported by Shafrin et al.[31] There are also specified the densities of plasma polymers[32-34] and conventional amorphous counterparts,[35] respectively. The data in Table IV clearly indicate that plasma polymers have higher surface energy as compared to their conventional counterparts. This apparently results from the increased dispersion and polar (except PP-VDC) components of their surface energy. The increase in $\gamma_S^d$ noted

Table IV. Surface Energy Properties and Densities of Plasma Polymers and Their Conventional Counterparts.

| Polymer | Surface Energy (erg/cm$^2$) | | | Density (g/cm$^3$) | Ref. |
|---|---|---|---|---|---|
| | $\gamma_S^d$ | $\gamma_S^p$ | $\gamma_S$ | | |
| PP-ethylene | 40.1 | 2.0 | 42.1 | 1.141-1.231 | 32 |
| Polyethylene | 32.1 | 1.1 | 33.2 | 0.855 | 35 |
| PP-(vinylidene chloride) | 46.5 | 2.6 | 49.1 | – | 33 |
| Poly(vinylidene chloride) | 41.6 | 3.1 | 44.7 | – | |
| PP-tetrafluoroethylene | 16.4 | 2.7 | 19.1 | 1.87-2.01 | 33 |
| Polytetrafluorcethylene | 12.3 | 1.5 | 13.8 | – | |
| PP-(methyl methacrylate) | 37.6 | 10.5 | 48.1 | 1.269-1.271 | 33 |
| Poly(methyl methacrylate) | 35.3 | 4.4 | 39.7 | 1.188 | 35 |
| PP-styrene | 45.3 | 3.4 | 48.7 | 1.332-1.408 | 32 |
| Polystyrene | 41.8 | 0.6 | 42.4 | 1.04 | 35 |
| PP-hexamethyldisiloxane | 27.4 | 1.7 | 29.1 | 1.16-1.67 | 34 |
| Polydimethylsiloxane | 21.7 | 1.2 | 22.9 | 0.98 | 35 |

for plasma polymers undoubtedly is due to their crosslinked structure which is manifested by a considerably higher density values found for these polymers (Table IV). This provides strong evidence for the effect of polymer crosslinking on the dispersion component and confirms our previous finding. The higher $\gamma_S^p$ values for plasma polymers are considered to derive mostly from a rapid oxidation of polymer film surface due to reactions of residual radicals with the atmosphere as was described earlier. Somewhat lower $\gamma_S^p$ value for PP-VDC in comparison with that of its conventional counterpart (PVDC) may arise from the lower contribution of chlorine in plasma polymer due to their abstraction resulting in reduced surface polarity.

As a final general comment, the results presented here show clearly that plasma - polymerized materials differ significantly in their surface properties from those polymerized by conventional methods. This accounts for substantial difference in the surface chemistry of plasma and conventional polymers.

## ACKNOWLEDGEMENT

The author wishes to thank Professor M. Kryszewski, Centre of Molecular and Macromolecular Studies, for helpful suggestions and critical reading of the manuscript.

## REFERENCES

1. H. Yasuda and C. Lamaze, J. Appl. Polym. Sci., 17, 201 (1973).
2. A. T. Bell, T. Wydeven, and C.C. Johnson, J. Appl. Polym. Sci., 19, 1911 (1975).
3. H. P. Schreiber, Y. B. Tewari and M. R. Wertheimer, Ind. Eng. Chem. Prod. Res. Dev., 17, 27 (1978).
4. H. P. Schreiber, M. R. Wertheimer and A. M. Wróbel, Thin Solid Films, 72, 487 (1980).
5. P. J. Dynes and D. H. Kaelble, J. Macromol. Sci. Chem., A10, 535 (1976).
6. B. D. Washo, J. Macromol. Sci. Chem., A10, 559 (1976).
7. H. Yasuda, M. O. Bumgarner, H. C. Marsh and N. Morosoff, J. Polym. Sci., Polym. Chem. Ed., 14, 195 (1976).
8. H. Yasuda and T. Hirotsu, J. Polym. Sci., Polym. Chem. Ed., 15, 2749 (1977).
9. T. Smith, D. H. Kaelble and C. L. Hamermesh, Surface Sci., 76, 203 (1978).
10. A. E. Pavlath and A. G. Pittman, in "Plasma Polymerization", M. Shen and A. T. Bell, Editors, ACS Symp. Ser., No. 108, Ch. 11, Am. Chem. Soc., Washington, D. C., 1979.
11. A. M. Wróbel, M. Kryszewski and M. Gazicki, Polymer, 17, 673 (1976).

12. D. K. Owens and R. C. Wendt, J. Appl. Polym. Sci., 13, 1741 (1969).
13. D. H. Kaelble, J. Adhesion, 2, 66 (1970); "Physical Chemistry of Adhesion", Ch. 5, Wiley-Interscience, New York, 1971.
14. F. London, Trans. Faraday Soc., 33, 8 (1937).
15. A. M. Wróbel, J. Kowalski, J. Grebowicz and M. Kryszewski, J. Macromol. Sci. Chem., A17, 433 (1982).
16. D. R. Anderson, in "Analysis of Silicones", Ch.10, A. L. Smith, Editor, Wiley-Interscience, New York, 1974.
17. C. Eaborn, "Organosilicon Compounds", Butterworths, London, 1960.
18. C. T. Mortimer, "Reaction Heats and Bond Strengths", Pergamon Press, New York, 1962.
19. H. Yasuda and T. Hsu, J. Appl. Polym. Sci., 15, 81 (1977).
20. M. Gazicki, A. M. Wróbel and M. Kryszewski, J. Appl. Polym. Sci., 22, 2013 (1977).
21. M. Kryszewski, A. M. Wrobel and J. Tyczkowski, in "Plasma Polymerization", M. Shen and A. T. Bell, Editors, ACS Symp. Ser., No. 108, Ch. 13, Am. Chem. Soc., Washington, D. C., 1979.
22. F. M. Fowkes, Ind. Eng. Chem., 56, 40 (1964).
23. M. Shen and A. T. Bell, in "Plasma Polymerization", M. Shen and A. T. Bell, Editors, ACS Symp. Ser., No. 108, Ch. 1, Am. Chem. Soc., Washington, D. C., 1979.
24. A. R. Westwood, Eur. Polym. J., 7, 377 (1971).
25. A. M. Wróbel, M. Kryszewski and M. Gazicki, Polymer, 17, 678 (1976).
26. A. M. Wróbel, J. Tyczkowski, and M. Kryszewski, (1981), unpublished data.
27. P. K. Tien, G. Smolinsky, and R. J. Martin, Appl. Optics, 11, 637 (1972).
28. A. N. Wróbel and M. Kryszewski, J. Macromol. Sci. Chem., A12, 1041 (1978).
29. A. M. Wróbel and M. Kryszewski in "Plasma Polymerization", M. Shen and A. T. Bell, Editors, ACS Symp. Ser., No. 108, Chap. 14, Am. Chem. Soc., Washington, D.C., 1979.
30. A. M. Wróbel, J. E. Klemberg, M. R. Wertheimer, and H. P. Schreiber, J. Macromol. Sci., Chem., A15, 197 (1981).
31. E. G. Shafrin and W. A. Zisman, "Upper Limits for the Contact Angles of Liquids on Solids", Naval Research Laboratory, Washington, D. C., Report 5985, 1963.
32. W. W. Knickmeyer, B. W. Peace, and K G. Mayhan, J. Appl. Polym. Sci., 18, 301 (1974).
33. W. F. Oberbeck, K. G. Mayhan and W. J. James, J. Appl. Polym. Sci., 22, 2805 (1978).
34. A. M. Wróbel, M. R. Wertheimer, J. Dib and H. P. Schreiber, J. Macromol. Sci. Chem., A14, 321 (1980).
35. J. Brandrup and E. H. Immergut, Editors, "Polymer Handbook", Wiley-Interscience, New York, 1975.

# PROPERTIES OF n-ALKANE FILMS IN THE SYSTEM: TEFLON/n-ALKANE-WATER

E. Chibowski, B. Jańczuk and W. Wójcik

Department of Physical Chemistry, Institute of Chemistry
Maria Curie-Sklodowska University
Lublin, Poland

Although Teflon possesses a very low surface free energy, n-alkane thick liquid films may exist in Teflon/n-alkane-water system. We examined properties of the films by measurements of contact angle, adhesion force of air bubbles and zeta potential for Teflon/n-alkane in water. Series of hydrocarbons ranging from n-hexane to n-hexadecane were tested. The results obtained show that equilibrium values of the measured parameters change nonlineary with chain length of the alkanes. The changes of the parameters correlate between themselves and with melting temperature of the alkanes. For Teflon/n-hexane in water system, the film pressure values were calculated from the experimental relationship between specific film volume and zeta potential. In the extreme case, about 4 statistical monolayers of n-hexane were present on the Teflon surface. If air bubbles are attached and then removed from Teflon surface wetted completely with n-hexane, they take away some amounts of n-hexane. This results in an increase in the adhesion force and decrease in contact angle, both of which reach equilibrium values. About 13 air bubbles are needed to attain the equilibrium, and 13-14 air bubbles remove the film almost completely. Comparison of the results from zeta potential and contact angle measurements leads to the conclusion that the most stable film exists at 1.5 to 2.5 statistical monolayers.

## INTRODUCTION

Polymeric materials are often used as model substances for studying the energetic changes at the solid-air-liquid interfaces. The energetic changes may be caused by the coverage of the polymer surface with n-alkane film.

In the case of Teflon, the calculated value for the surface free energy ranges from ca. 17 to 24 mJ/m$^2$,[1-6] whereas the surface tension of the liquid n-alkanes varies from 16.0 to 27.6 mN/m. Therefore, the conclusions about interactions of Teflon surface with alkanes are not clearly specified, because from the values of the surface tension and surface free energy it is not certain whether a given alkane will or will not spread on Teflon surface.

From the literature[1-7] it may be concluded that the thickness and properties of liquid alkane films on Teflon surface are related to adsorption interactions and condensation of the vapor.

In this paper we have examined the liquid n-alkane films, hexane to hexadecane, on Teflon surface in water by measuring contact angles, detachment force of air bubble from Teflon surface wetted by n-alkanes and zeta potential of such Teflon surface in water.

## EXPERIMENTAL

### Materials

Polytetrafluoroethylene used in present measurements was commercial Teflon sample (from W. Germany). Contact angle measurements were conducted on 1 x 1 cm Teflon plates cut off from a larger Teflon plate. The plates were washed in HCl solution, doubly distilled water, then polished with a flannel, rinsed with doubly distilled water and washed in ultrasonic water bath. Next, they were dried in warm air, cooled to room temperature (20°C) and used for measurements.

For measurements of the detachment force of air bubbles, Teflon cylindrical segments of 0.8568 mm in diameter (as measured with the microscope) and 1 mm in length were prepared by means of a lathe from the same Teflon sample and then cleaned as described above.

Zeta potential measurements were conducted on Teflon powder obtained from grinding Teflon chips (from the same Teflon sample as above) in a coffee mill and sieving 0.25 - 0.385 mm grains. Then it was cleaned with methanol, HCl solution (1 : 10), washed with doubly distilled water (boiling) and dried at 150°C for 10 hr.[8] The BET specific surface of the sample obtained was 0.03 m$^2$/g, as determined chromatographically from the thermal desorption of nitrogen.[9]

The n-alkanes used were chemically pure grade (Reachim, USSR) and were used without further purification. Gas chromatography test showed no detectable polar substances in the alkanes. All measurements were made with doubly distilled water.

## Measurements

Contact angle measurements were performed by the "captive bubble" method with a goniometer-telescope system.[10] Teflon plates were wetted in excess with the n-alkanes tested and dipped into water in the measuring glass cell. Then an air bubble from a microsyringe connected with a micrometer screw was attached to the Teflon surface and after 3-5 min. several readings of the contact angle values were taken. In this period of time changes in the contact angle value did not exceed the error of the method ($\pm 1°$). After that the air bubble was removed and another air bubble was attached to the surface and the contact angle measured. This procedure was repeated until a constant value was achieved for at least three successive air bubbles. Contact angle measurements were made with all n-alkanes studied ranging from hexane to hexadecane in the same way.

The measurements of the detachment force of air bubbles from Teflon surface wetted with n-alkanes (the same series of n-alkanes) were made using the apparatus presented schematically in Figure 1. The details of the apparatus constructed in our Laboratory have been published elsewhere[11], and here only the principle of the method will be given. The cylindrical segments of Teflon are connected by paraffin wax to previously calibrated quartz rod 1 (Figure 1). The quartz rod is calibrated with a cathetometer to establish the departure in relation to the weight applied. The linear departure as a function of the applied weight is found. Then the circular surface of the Teflon segment attached to the quartz rod is wetted with the alkane tested and contacted with the air bubble formed at the top of glass capillary 2 placed in the glass cell 3 filled with doubly distilled water. By the manipulator 4, the quartz rod (with Teflon grain) is strained until the disruption occurs. At the moment of disruption the strain is read on the scale in the ocular of the microscope 5. From the linear departure of the quartz rod and diameter of the Teflon segment (perimeter of the bubble - Teflon contact) the detachment force $F/l$ (mN/m) is calculated. After the bubble - Teflon contact disruption, a new air bubble is formed at the top of the capillary 2 and again the same Teflon segment is adhered to the bubble and the detachment force determined. This is repeated until a constant value of the force $F/l$ is established. Such measurements were made for all

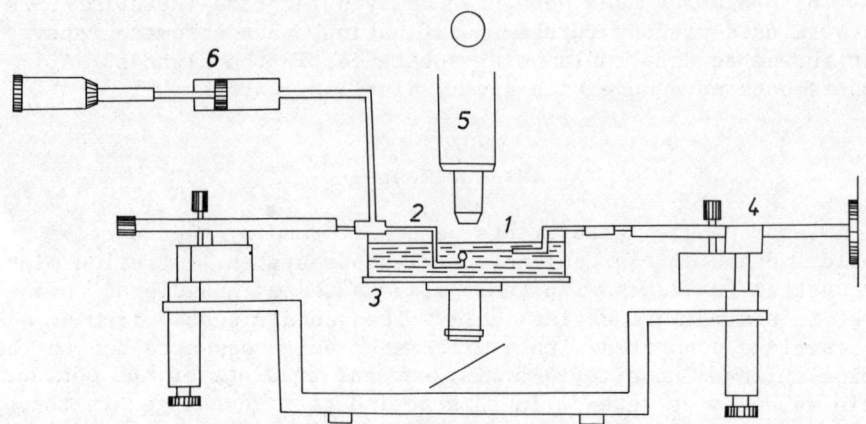

Figure 1. Schematic diagram of the apparatus for measurements of detachment force of an air bubble: 1 - quartz rod, 2 - glass capillary, 3 - glass cell, 4 - manipulator, 5 - microscope, 6 - syringe with micrometer screw.

n-alkanes tested. The accuracy of the measurements is more than 0.1 mN/m (dyn/cm).

Zeta potential determinations were made by the streaming potential method, with the apparatus described by Fuerstenau.[12] Platinum electrodes were employed. The potential difference was measured with a pH meter Elpo N-512 (input $10^{12}$ $\Omega$), and the conductivity with an Elpo RLC-bridge U 915 B. The driving pressure was usually between 4-6 cm Hg. The linearity of streaming potential versus the pressure was established for the range measured. Zeta potential values were calculated from the Smoluchowski equation. In the first series of measurements, the samples of Teflon powder were completely wetted by the n-alkane tested and then rinsed with doubly distilled water under a water pump, and zeta potential was determined in doubly distilled water. In the second series, samples of Teflon powder weighed in glass ampoules were cooled in liquid nitrogen and various amounts of n-hexane were added from a 1 µl microsyringe. Then the ampoules were quickly sealed in a flame, and the closed ampoules were then heated at 110°C for 2 hrs. under constant stirring. After cooling them to 20°C, zeta potential was determined in doubly distilled water. Some of the results obtained have already been published.[8]

## RESULTS AND DISCUSSION

The results of the contact angle of an air bubble, detachment force of an air bubble and zeta potential for Teflon surface covered with n-alkane film in water against the number of carbon atoms in the alkane chain are presented in Figure 2. The values of the contact angle (Figure 2A) and detachment force (Figure 2B) are the constant values determined in a series of attached-detached air bubbles. Hence, these values represent equilibrium values. On

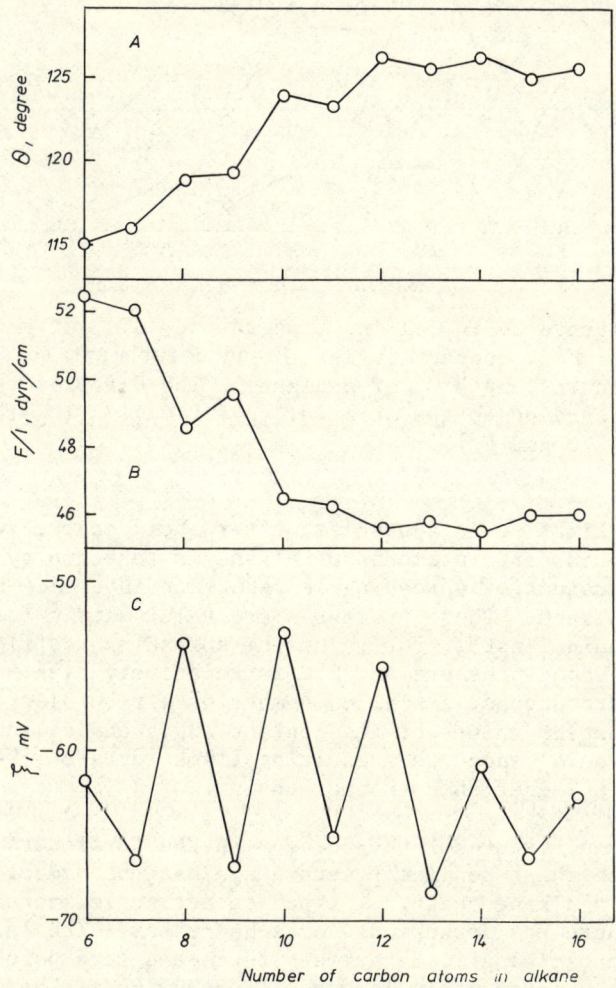

Figure 2. Changes of A - contact angle, B - detachment force, C - zeta potential on Teflon in water as a function of the chain length of n-alkane used for wetting Teflon surface.

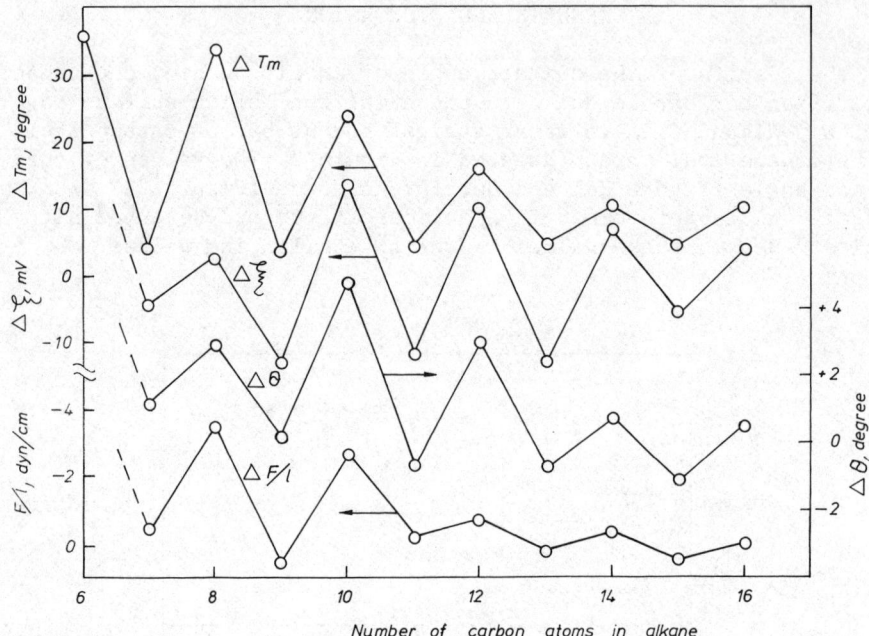

Figure 3. Difference in melting temperature, $\Delta T_m$, of n-alkanes, zeta potential $\Delta\zeta$, contact angle $\Delta\theta$ and detachment force $\Delta F/1$ against the chain length $C_n$ of n-alkane. The differences were calculated according to the formula: $P = P(C_n) - P(C_n-1)$, where P means $T_m$, $\zeta$, $\theta$ and $F/1$.

removing the first few air bubbles, after their contact with n-alkane film, a certain amount of alkane is taken up by the bubbles as a result of spreading at water-air interface and evaporation of the alkane. These processes are valid mainly for alkanes of shorter chain lengths, for which the spreading coefficient is positive and vapor pressure at 20°C is relatively high. Therefore for these hydrocarbons, a smaller number of air bubbles is needed to attain constant values of the contact angle and detachment force. One would expect the resulting films to be thicker for n-alkanes with longer than shorter chains.

The common feature of the changes in the parameters presented in Figure 2 is their nonlinear nature against the number of carbon atoms in the n-alkane chain. As the contact angle increases with the chain length the detachment force decreases. The largest changes in contact angle, detachment force and zeta potential are observed with shorter chain lengths. To show more clearly the common features of the measured parameters, difference of the values from Figure 2 are presented in Figure 3.

The difference of $\Delta\zeta$, $\Delta\theta$ and $\Delta F/l$ were calculated by subtraction of the value determined for $C_n$ alkane from the value for $C_{n-1}$ alkane. Thus calculated differences of the parameters are presented vs. $C_n$ alkane. In the same manner, differences of the melting temperature $\Delta T_m$ of the alkanes were calculated and are also shown in Figure 3. As can be seen the shapes of all the curves are very similar, and "odd" and "even" n-alkanes demonstrate some individuality.

Because $\Delta T_m$, $\Delta\zeta$, $\Delta\theta$ and $\Delta F/l$ correlate between themselves, it may be postulated that the changes in the contact angle, detachment force and zeta potential are related to molecular properties, and thus to the structure of "odd" and "even" n-alkane films.

As the literature data show[5,6], n-alkanes with short chain lengths give detectable film pressure due to adsorption on Teflon surface. Therefore we have examined the energetic changes for Teflon/n-hexane - water system as a function of the amount of hexane present on the surface. In Figure 4A zeta potential values are presented as a function of the number of statistical monolayers of n-hexane assuming vertical orientation of the hexane chains ($18\text{Å}^2$ for cross-sectional area)[8], whereas in Figure 4B and 4C changes of detachment force and contact angle are shown, respectively, as a function of the number of air bubbles attached to the Teflon surface. The surface was previously completely wetted with n-hexane. As it was said earlier the contact angle and detachment force change with the number of bubbles attached and then reach constant values. In the case of hexane, these constant values were obtained after 10-11 air bubbles contacted (Figures 4B and C). The contact angle changes from 160° for the first settled air bubble (in Figure 4C denoted by "0") to a constant value equal to 115°. The decrease in the contact angle value is related to the increase of the detachment force (Figure 4B), which goes through a maximum and then reaches a constant value.

The changes in the contact angle and detachment force (Figure 4B and C) observed in the range of 0-12 attached-detached air bubbles correspond to the range of 0-4 statistical monolayers of hexane, in which zeta potential changes (Figure 4A). Thus it may be suggested that the contact angle of the first settled air bubble and the detachment force of the first contacted air bubble result from the presence of n-hexane film possessing a thickness of 4 statistical monolayers for which zeta potential was determined.

It was discussed elsewhere[8,13,14,15,16] that in nonionogenic solid/n-alkane - water systems, determinations of zeta potential as a function of the film specific volume may be used for studies of energetic changes.

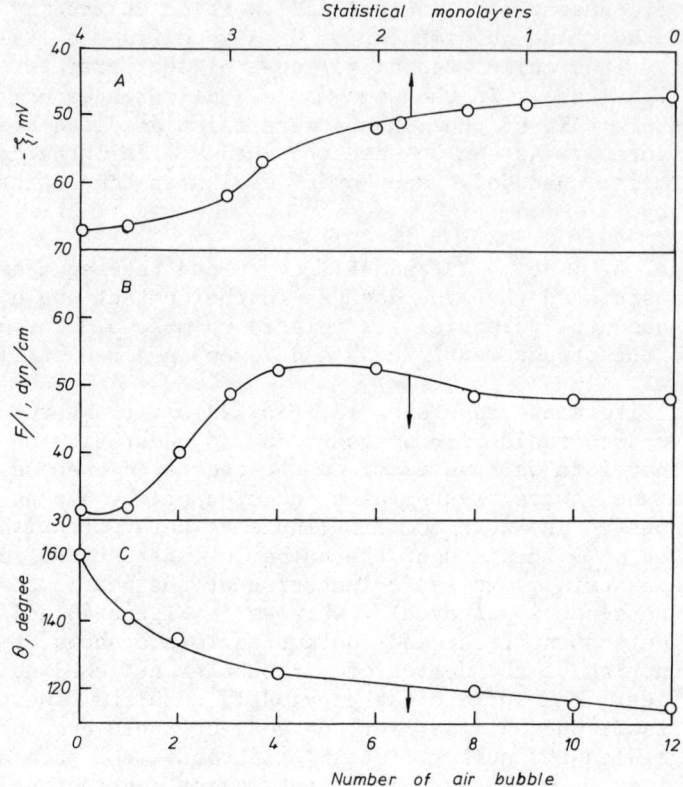

Figure 4. A - zeta potential of Teflon/n-hexane in water vs. the number of statistical monolayers of hexane, B - detachment force of an air bubble, C - contact angle of the air bubble, as a function of the number of air bubbles detached from the surface previously completely wetted with n-hexane.

Thus the experimentally determined relationship between the specific volume of n-hexane film on the Teflon surface and zeta potential value was used[8] to determine the film pressure $\Pi$ from the empirical equation similar to Gibb's adsorption equation, where in place of the vapor pressure zeta potential was introduced, yielding:

$$\Pi = \frac{RT}{V_o A} \int_{\zeta_o}^{\zeta} v \, d\ln\zeta \qquad (1)$$

Where R and T are the gas constant and temperature, respectively, $V_o$ is the molar volume of the alkane, A is the total surface of the sample, $\zeta_o$ is the value of zeta potential for bare surface of the solid (Teflon), and $\zeta$ is the zeta potential value resulting from the deposited film volume v. Some assumptions and verifications of validity of the equation in various systems with hydrophobic (and even polar) solids were presented in papers published earlier.[8,13,14]

Using values from Figure 4A, it is possible to determine n-hexane film pressure $\Pi$ by graphic integration of Equation (1) vs. the number of statistical monolayers.[8] The film pressure is defined as:

$$\Pi = \gamma_T - \gamma_{Tf} \quad (2)$$

where $\gamma_T$ is the surface free energy of bare Teflon, and $\gamma_{Tf}$ is the surface free energy of Teflon covered with hexane film, and it is not equivalent to the value of the free energy of the surface of n-hexane film, as, for the film sufficiently thick, its surface free energy is equal to the surface tension of hexane, i.e. 18.5 mN/m (dyns/cm). Hence the surface free energy of Teflon changes from 19.7 mJ/m$^2$(ergs/cm$^2$)(bare Teflon) to zero for Teflon surface "shaded" with hexane film of a thickness of 4 statistical monolayers. However, the free energy of the surface of the film changes from

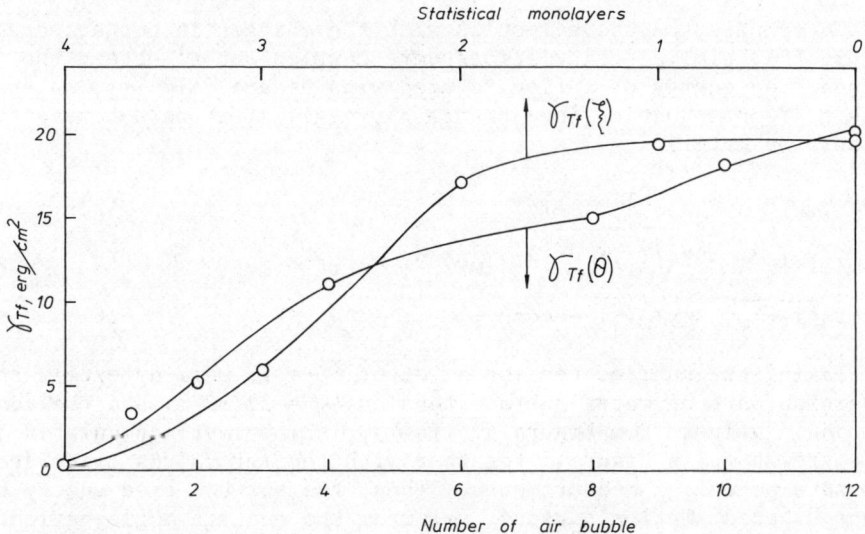

Figure 5. Changes in surface free energy of Teflon surface wetted with n-hexane as a function of the number of statistical monolayers (curve 1, values calculated from zeta potential measurements) and vs. the number of air bubbles detached from the surface (curve 2, values calculated from the contact angle).

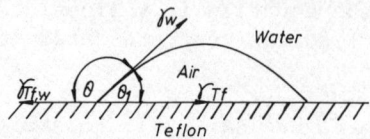

Figure 6. Schematic of the system: Teflon/n-hexane - air - water, for determination of surface free energy changes of Teflon wetted with n-hexane film. Symbols: T - Teflon, f - n-hexane film, w - water.

19.7 mJ/m$^2$ (bare Teflon) to 18.5 mN/m for liquid hexane (4 statistical monolayers)[14]. Then:

$$\gamma_{Tf} = \gamma_T - \Pi \tag{3}$$

If the $\gamma_T$ is known, the $\gamma_{Tf}$ can be calculated. Taking the maximum value of $\Pi$, and considering it as corresponding to the work of immersional wetting, the surface free energy of bare Teflon was determined[8] to be 19.7 mJ/m$^2$ (ergs/cm$^2$). Based on this value and that of $\Pi$'s, the values of $\gamma_{Tf}$ were calculated and are presented in Figure 5 against the number of statistical monolayers of n-hexane.

To compare the above results with the changes in contact angle (Figure 4C), similar calculations were carried out to obtain the surface free energy of Teflon covered with hexane film $\gamma_{Tf}$, on the basis of Young equation (see Figure 6), neglecting the n-hexane film pressure on water.

$$\gamma_{TfW} = \gamma_W \cos\theta_1 + \gamma_{Tf} \tag{4}$$

$$\gamma_{Tf} + \gamma_W - 2\sqrt{\gamma_{Tf}\gamma_W^d} = \gamma_W \cos\theta_1 + \gamma_{Tf} \tag{5}$$

$$2\sqrt{\gamma_{Tf}\gamma_W^d} = \gamma_W(1 - \cos\theta_1) \tag{6}$$

Taking the surface tension of water $\gamma_W$ = 73 mN/m(dyne/cm), and dispersion part of water surface tension $\gamma_W^d$ = 22 mN/m, and the contact angle values from Figure 4C, the $\gamma_{Tf}$ values were calculated and these are shown in Figure 5 together with the $\gamma_{Tf}$ values calculated from zeta potential measurements. Thus the surface free energy of hexane "shaded" Teflon surface $\gamma_{Tf}$ from the contact angle varies from 20.2 mJ/m$^2$ (ergs/cm$^2$) ($\theta$ = 115°, the twelfth contacted air bubble) to 0.22 mJ/m$^2$ (ergs/cm$^2$) ($\theta$ = 160°, the first contacted air bubble). From Figure 5 it can be seen that the changes in $\gamma_{Tf}$ determined by both methods are similar in considerable part of the curves.

So it may be concluded that the properties of n-hexane film on the Teflon surface, as studied by both methods, are similar. On the basis of $\gamma_{Tf}$ changes calculated from the contact angle it can be concluded that the most stable film exists at 1.5 to 2.5 statistical monolayers because the changes in $\gamma_{Tf}$ in this range of thickness are relatively small.

It seems that below 1.5 statistical monolayers of the surface coverage an adsorbed n-hexane film exists, and above two of these monolayers condensation effects predominate.

## REFERENCES

1. S. Wu, in "Polymer Blends", D. R. Paul and S. Newman, Editors, Vol. 1, pp. 243-293, Academic Press, New York, 1978.
2. J. R. Dann, J. Colloid Interface Sci., 32, 302 (1970).
3a. K. L. Mittal, in "Adhesion Science and Technology", L. H. Lee, Editor, Vol. 9A, pp. 129-168, Plenum Publishing Corporation, New York, 1975.
3b. K. L. Mittal, Polymer Eng. Sci., 17, 467 (1977).
4. B. W. Davis, J. Colloid Interface Sci., 59, 420 (1977).
5. W. Hu and A. W. Adamson, J. Colloid Interface Sci., 59, 605 (1977).
6. A. C. Zettlemoyer, in "Hydrophobic Surfaces", pp. 1-27, Academic Press, New York, 1969,
7. J. W. Whalen, in "Hydrophobic Surface", pp. 101-106, Academic Press, New York, 1969.
8. E. Chibowski and A. Waksmundzki, J. Colloid Interface Sci., 66, 213 (1978).
9. A. J. Haley, J. Appl. Chem., 13, 392 (1963).
10. B. Jańczuk, A. Waksmundzki and W. Wójcik, Roczniki Chemii, 51, 985 (1977).
11. J. Barcicki, A. Waksmundzki and E. Maruszak, Chem. Stosowana, 1, 99 (1962).
12. D. W. Fuerstenau, Trans. Amer. Inst. Mining Met. Eng., 206, 834 (1956).
13. E. Chibowski, J. Colloid Interface Sci., 69, 326 (1979).
14. E. Chibowski and L. Holysz, J. Colloid Interface Sci., 81, 8 (1981).
15. E. Chibowski, J. Colloid Interface Sci., 86, 567 (1982).
16. E. Chibowski, B. Biliński, A. Waksmundzki and W. Wójcik, J. Colloid Interface Sci., 86, 559 (1982).

# INTERFACIAL FREE ENERGIES OF CELLS AND POLYMERS IN AQUEOUS MEDIA

Donald F. Gerson*

Basel Institute for Immunology**
Postfach CH-4005
Basel, Switzerland

The interfacial free energy at the surface of cells in aqueous media, $\gamma_{CM}$ determines the free energy of adhesion and the equilibrium constant for adhesion not only between different cells but also between cells and polymer surfaces. Previously described methods for measuring contact angles on cells in polymer containing, biphasic aqueous media (Biochim. Biophys. Acta. 602, 269, 1980; 640, 557, 1980) have been extended to provide values of $\gamma_{CM}$ for cells in physiological media. To do this, an empirical equation of state was developed to describe the relations between contact angles and interfacial free energies in 3 component systems consisting of 2 fluids (gas or liquid) and one solid. In such systems, the bulk phase (gas or liquid) is designated "B", the test droplet (liquid) is designated "D" and the solid phase is designated "S". The contact angle between the drop and the solid when both are immersed in the bulk fluid is given by $\theta$. The equation which was developed thus relates measured contact angles and bulk fluid-drop interfacial free energies, $\gamma_{BD}$ to both of the solid-liquid interfacial free energies $\gamma_{SB}$, and $\gamma_{SD}$. This was used to determine cell-medium interfacial free energies from contact angle data. Cell-medium interfacial free energies in polyethyleneglycol-dextran biphasic media, and the polymer concentrations in these media were extrapolated to give $\gamma_{CM}$ in polymer-free media.

---

*Present address: Genex Corporation, 16020 Industrial Drive, Gaithersburg, MD 20877.
** Founded and supported by F. Hoffmann-La Roche and Company, Limited, Basel, Switzerland.

## INTRODUCTION

Adhesion phenomena are crucial to biological structure and organization. The surfaces involved are complex mixtures of proteins, polysaccharides and lipids; yet the adhesion is largely due to van der Waals interactions. To apply existing knowledge of the physics of adhesion to biological materials requires some modification of the usual concepts and methodologies. Living cells are usually immersed in an aqueous environment, so it is the interfacial free energy at the cell-medium interface which is relevant to cellular adhesion.[1-3] For instance almost all biological polymers and surfaces have an apparent surface energy (against humid air) of 67-72 ergs/cm$^2$, and yet differ widely in their adhesive properties; on the other hand, interfacial free energies at the cell-aqueous interface vary greatly (0.1 to 50.0 x $10^{-3}$ ergs/cm$^2$) among the cell types which have been measured, and these interfacial free energies appear to reflect adhesive properties.[3,4] It is the interfacial free energy between the cell surface and its physiological environment which is relevant to the biophysics of cellular adhesion.

Measurement of the surface energy of hydrophilic polymers has been difficult and estimation of the interfacial free energy at the interface between the polymer and aqueous solutions is even more uncertain. An approach leading to the direct measurement of polymer-aqueous interfacial free energies would appear to be most desirable at this time. The contact angle of a test droplet on a solid surface depends directly on the actual solid-liquid interactions, and contact angle measurement thus provides the best route to the determination of solid-liquid interfacial free energies. Since cell-medium interfacial free energies are quite low ($\sim 10^{-3}$ ergs/cm$^2$) interfacial tension at the interface between the drop and the bulk medium must also be low to obtain measurable contact angles. If the test liquid and the conditions of measurement are physiological, then in principle it is possible to obtain the required interfacial free energy between the living cell and its aqueous environment by the contact angle technique.

Previous attempts to measure the interfacial free energy at the cell surface have also used the contact angle technique. Harvey[6] placed oil droplets inside cells, but found the contact angle to be very high and had to place the cells in a centrifugal field to obtain meaningful measurements. He concluded that the cell-medium interfacial free energy was less than 1 erg/cm$^2$ but could not obtain a precise value. Van Oss, et al.[7] measured the contact angles of drops of physiological saline on air-dried layers of cells. These measurements provided some insight into the role of surface energies in phagocytosis, but, clearly, air-dried cells have been seriously damaged.

Considerable effort has been made recently to develop techniques in which both the cells and the fluids surrounding them are as close as possible to being physiological[1-4,8,9]. An important aspect of these developments was our realization that biphasic mixtures of water-soluble polymers could be used for contact angle measurements[3]. When polyethyleneglycol (PEG) and dextran (DEX), for instance, are mixed at sufficiently high concentrations a biphasic mixture forms spontaneously, and the two phases are as mutually immiscible as are oil and water[9,10]. In this technique, a layer of cells is collected on a membrane filter and immersed in a bath containing the less dense phase. Test droplets of the dense phase are then placed on the cells, and the contact angle is measured. The relation between the interfacial tension between the two aqueous phases and the contact angle is orderly and conforms to a Good-Girifalco Plot, allowing determination of a critical interfacial tension for spreading, $\gamma_C$ on the cells. This is formally identical to Zismann's critical surface tension for spreading[3-4,8,12] and provides a quantitative characterization of the surface energy of the cells.

The $\gamma_C$ for living cells obtained with a dilution series of PEG/DEX biphasic mixtures is clearly only a partial solution to the problem of obtaining measurements of cell surface energy. The difficulty is that $\gamma_C$ refers only to the particular PEG/DEX mixture having a liquid-liquid interfacial tension equal to the interfacial free energy at the surface of cells exposed to the less-dense (bulk) phase of the mixture. This leaves two major problems unsolved:
1. the interfacial free energy at the cell surface is not yet determined for cells in other PEG/DEX mixtures for which contact angle data are available. 2. the interfacial free energy at the cell surface is not yet determined for cells in the normal medium lacking either polymer.

In this paper, an approach to both of these problems is presented. Firstly, a semi-empirical equation of state will be developed which allows calculation of solid-liquid interfacial tension data. Secondly, this empirical equation will be used in conjunction with the phase diagram of the PEG/DEX biphasic system to estimate the surface energy of cells in media which do not contain either polymer.

CALCULATION OF LIQUID-SOLID INTERFACIAL FREE ENERGIES

The contact angle formed by a liquid drop on a plane solid surface depends on the interfacial free energies at each interface: $\gamma_{SB}$, $\gamma_{BD}$ and $\gamma_{SD}$.

Thomas Young originally described the relation between these quantities in 1805 with an enigmatic equation having two quantities which are difficult to measure experimentally ($\gamma_{SB}$ and $\gamma_{SD}$). The general inability to measure or calculate solid-fluid interfacial free energies from the contact angle of a single liquid on a solid has limited our understanding of many problems involving wetting, detergency, emulsion or dispersion stability, and adhesion. Equation 1 is Young's equation.

$$\gamma_{SB} - \gamma_{SD} = \gamma_{BD} \cos\theta \qquad (1)$$

The surface tension of the test droplet ($\gamma_{BD}$) and the contact angle ($\theta$) can be measured easily, but direct measurement of either $\gamma_{SB}$ or $\gamma_{SD}$ is usually difficult. To calculate either of these quantities, a second function of the form of Equation 2 is required.

$$\gamma_{SD} = f(\gamma_{BD}, \gamma_{SB}) \qquad (2)$$

Of the many theoretical attempts to find this function, the most generally accepted result is that derived by Girifalco and Good[11] from the theory of non-electrolyte solutility and Berthelot's geometric mean hypotesis. Equation 3 is the result they obtained.

$$\gamma_{SD} = \gamma_{BD} + \gamma_{SB} - 2\phi\sqrt{\gamma_{BD}\gamma_{SB}} \qquad (3)$$

Unfortunately it introduces a new dimensionless variable $\phi$, the interaction parameter. For hydrophobic surfaces wetted by liquids of low surface tension, $\phi$ is close to 1.0, but the interaction parameter deviates from unity when the ratio of dispersion to polar components of the surface energies of the phases in contact are dissimilar[13].

Zisman and co-workers[12] observed a nearly linear relation between $\cos\theta$ and $\gamma_{BD}$ for a series of liquids on any given solid, up to the point of complete wetting, or critical spreading. This gave an experimental measure of the solid-vapor interfacial free energy, $\gamma_{SB} : \gamma_C$ or the critical surface tension for spreading. Based on this observation, Neumann and co-workers[14] found an empirical linear relation between $\phi$ and $\gamma_{SD}$ which, when combined with Equations 1 and 3 allowed calculation of $\gamma_{SB}$ and $\gamma_{BD}$ for low energy surfaces ($\leqslant 30$ ergs/cm$^2$), but their relation is wildly discontinuous when applied to hydrophilic surfaces (see Fig. 1). They recognized this problem and provided a computer program which filled the gap and calculated numbers which corresponded to existing data. They thus demonstrated that a function of the form of Equation 2 must exist, but did not actually develop such a function. Incorporated into this problem is the relation between the observed $\gamma_C$ and the theoretical free energy of formation of the solid surface in a vacuum. This problem has been carefully reviewed with respect to polymer surfaces by Mittal[15,16].

Inspection and statistical analysis of a large quantity of published measurements (e.g. see references contained in Refs. 4, and 12-16) of contact angles formed by various liquids on solid surfaces in air suggested that $\phi$ was a function of both $\gamma_{SD}$ and $\gamma_{SB}$. Equation 4 is a purely empirical expression of these observations.

$$\phi = \exp\{\gamma_{SD}(a\gamma_{SB}+b)\} \tag{4}$$

Equations 1, 3, and 4 can now be combined to allow calculation of $\gamma_{SB}$ from $\gamma_{BD}$ and $\theta$ (Equation 5), and then $\gamma_{SD}$ from Equation 1. Equation 5 thus combines Equation 3 with a specific empirical version of Equation 2.

$$\frac{\gamma_{BD}(1+\cos\theta)}{2\sqrt{\gamma_{SB}\gamma_{BD}}} = \exp\{(\gamma_{SB} - \gamma_{BD}\cos\theta)(a\cdot\gamma_{SB}+b)\} \tag{5}$$

Solution of Equation 5 for $\gamma_{SB}$ requires numerical evaluation, but this is relatively simple. The best-fit values of the coefficients are:

$$a = 6.5 \pm 1.0 \times 10^{-5}$$

and $\quad b = -0.01 \pm 0.001$

Constant values of a and b in Equation (5) allow one to calculate $\gamma_{SB}$ and $\gamma_{SD}$. Refinements in the fit of Equation 5 to a selected subset of data can be obtained by slight adjustments of the values of a and b.

The fit of Equation 5 to experimental data for the contact angle of water on various solids is shown as the solid line in Figure 1. Measured values of $\theta$ for water ($\gamma_{BD} = 72.8$ ergs/cm$^2$) on solids ranging from fluorocarbons ($\gamma_{SB}<20$ ergs/cm$^2$) to hydrogels ($\gamma_{SB} \simeq 70$ ergs/cm$^2$) are given as data points. The $\gamma_{SB}$ values used are the experimentally determined values of Zisman's critical surface tension for spreading, $\gamma_C$ for each solid. The data were collected from the references listed in References 4, 12-16, and multiple determinations of the same system were averaged. For comparison, calculated values are given in Figure 1 for two previous solutions to this problem, one based on the assumption that $\phi = 1.0$ (dashed line) and the other based on the empirical linear correlation between $\phi$ and $\gamma_{SD}$ of Neumann et al.[14] (dotted line). The calculated values of $\cos\theta$ obtained by solution of Equation 5 clearly fit the data more closely than either of the two previous proposals. The calculated values from Equation 5 also are very close to the values obtained from the computer program which bridges the discontinuity in the equation of Neumann et al.[14]

The importance of Equation 5 is that it allows calculation of interfacial free energies at solid surfaces from measurements of θ for an individual liquid of known $\gamma_{BD}$. In Table I, $\gamma_{SB}$ is listed for several solid materials in air, giving both experimentally determined values of $\gamma_C$ obtained by Zisman's method of critical spreading (see references in Refs. 4 and 12-16; multiple determinations were averaged), and calculated values of $\gamma_{SB}$ obtained from Equation 5.

For each solid, the surface tension of each test liquid and its contact angle on the solid were used in Equation 5 to calculate an estimate of $\gamma_{SB}$. The estimates were averaged, and the average value of $\gamma_{SB}$, and the standard deviation, is given for each solid in Table I.

The surface energies $\gamma_{SB}$ obtained experimentally by Zisman's method are remarkably similar to the average value obtained from Equation 5 throughout the range of $\gamma_{SB}$ regardless of the liquids used. The accuracy of both approaches in the determination of $\gamma_{SB}$ appears to be limited more by the notorious variability of measurements of θ than by other factors.

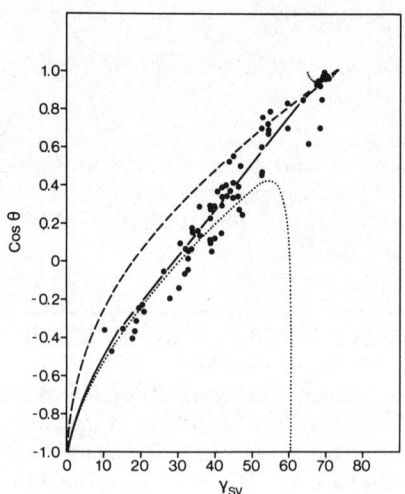

Figure 1. The relation between cosine of the contact angle (cosθ) of water drops on a plane solid and the surface energy ($\gamma_{SV}$) of the solid. The surface energy corresponds to $\gamma_{SB}$ in the text. Solid points (·) are data from the literature and this laboratory; solid line(___), the theoretical expectation calculated from the contact angle (θ) and the surface tension of water ($\gamma_{BD}$ = 72.8 ergs/cm$^2$) using Equation 5 in the text with a = 0.00007 and b = -0.01; dashed line (-----), values calculated assuming φ = 1.0; dotted line (....), values calculated using the equation proposed by Neumann et al.[14]

Table I. Measured values of Zisman's critical surface tension for spreading, $\gamma_C$, culled from the literature (see references contained in Refs. 4 and 12-16) are compared to the average value of $\gamma_{SB}$ obtained from Equation 5 in the text. To calculate the average $\gamma_{SB}$ the measured contact angle and the surface tension of each test liquid were used in Equation 5 to obtain a value of $\gamma_{SB}$, and the values obtained with several liquids on one solid were then averaged. The indicated ranges above are standard deviation.

| Material | Measured $\gamma_C$ | Calculated $\gamma_{SB}$ |
|---|---|---|
| Poly(fluoromethacrylate) | 11.5 | 12.4 ± 1.4 |
| FC-723 (3M) | 12.0 | 13.0 ± 1.0 |
| Poly(tetrafluoroethylene) | 18.5 | 19.0 ± 2.0 |
| Poly(tetrafluoroethylene) poly(ethylene) 1:1 copolymer | 27.0 | 27.3 ± 3.0 |
| Poly(ethylene) | 31.0 | 33.0 ± 3.0 |
| Sulfur (amorphous) | 31.5 | 34.3 ± 0.9 |
| Selenium | 32.0 | 31.2 ± 0.5 |
| Poly(styrene) | 33.0 | 34.0 ± 1.0 |
| Tellurium | 35.5 | 35.1 ± 0.5 |
| Poly(vinylchloride) | 39.5 | 39.4 ± 3.7 |
| Poly(ethyleneterephthalate) | 43.0 | 39.0 ± 4.6 |
| Poly(methylmethacrylate) | 42.8 ± 4.1 | 43.4 ± 1.5 |
| Nylon 6-6 | 46.0 | 44.0 ± 3.0 |
| Poly(hydroxyethylmethacrylate) | 70.2 ± 0.3 | 70.3 ± 0.3 |

The success of Equation 5 in describing the results obtained with one liquid on many solids (Figure 1), and many liquids on many solids (Table I) suggests that it can be used with confidence in situations where only one or a few solid-liquid combinations are possible or relevant. Equation 5 also gives consistent results when the bulk fluid (e.g. air) is replaced by a liquid such as octane or tetradecane, when both fluids are the immiscible phases which form when aqueous solutions of certain polymers are mixed, and when solutions (e.g. water and ethanol) are used as either fluid phase.

The value of Equation 5 in determining the surface energy of cells is that it allows liquid-solid interfacial free energies to be calculated from the observed contact angle and the interfacial tension, $\gamma_{SB}$, between the drop and bulk fluids. In the biphasic mixtures formed with PEG and DEX, $\phi$ is very close to 1.0 since $\gamma_{SD}$ and $\gamma_{SB}$ are both small and close to one-another. Equation 5 also allows the determination of cell surface energy with a hydrophobic, water insoluble liquid (such as hydrocarbon or fluorocarbon) as the test drop, and an unmodified physiological medium as the bulk fluid phase.

## INTERFACIAL FREE ENERGIES OF ADSORBED LAYERS OF PROTEINS IN AQUEOUS MEDIA

To obtain uniform layers of proteins, for contact angle measurements, protein solutions were equilibrated with strips of polymethylpentene. Protein solutions (5 µg/ml in 0.15M HCl) were placed in Costar type 3393, sterile Leighton tubes, which contain a small strip of polymethylpentene. Acid solutions were used to titrate all available carboxyl groups on the proteins. After 24 hours, the strips were removed, washed and placed in a bath of 0.15M HCl. Some samples were air-dried and a drop of water was placed on them; if adsorption had occurred, the contact angle was low or essentially zero, if it had not, the contact angle remained high. Adsorption occurred with all proteins for which data are reported here. Contact angles were measured by placing a small drop (~0.1mm diameter) of a fluorocarbon liquid, FC-48 (3M), on the protein layer in the bath. Boyce et al.[17] have recently demonstrated the applicability of Neumann's computer program to liquid-liquid-solid systems in which one liquid was a fluorocarbon.

The method for measuring contact angles is given by Gerson[4]. Table II gives the contact angles which were observed for each protein under these conditions. The interfacial tension between FC-48 and 0.15M HCl was found to be 51.3 dynes/cm. This value and the contact angle were used in Equation 5 to obtain the interfacial free energy at the protein-aqueous interface, and these values are also given in Table II.

Hydrophobic test fluids have been used to determine the surface free energy of a variety of polymers[18]. Similar measurements have been made on bacterial cells[2,17], mammalian lung[19] and arterial intima[17]. The applicability of both Neumann's computer program[14] and Equation 5 presented above to liquid-liquid-solid systems, in which one liquid is a fluorocarbon, may be due to the low surface tension of fluorocarbon liquids.

Table II. Contact Angles and Interfacial Free Energies of Proteins. Interfacial Free Energies were Calculated from Equation 5.

| Protein | Contact angle of fluorocarbon drop on protein layer in 0.15m HCl | Interfacial free energy at protein-aqueous interface ($\mu N/m^2$) |
|---|---|---|
| insulin (Porcine) | 126° | 4.34 |
| ovalbumin | 127° | 4.21 |
| serum albumin (Bovine) | 133° | 2.79 |
| carbonic anhydrase (Bovine) | 137° | 2.21 |

## INTERFACIAL FREE ENERGIES AT THE CELL SURFACE IN AQUEOUS MEDIA

The low interfacial tension between the phases of biphasic mixtures of polyethyleneglycol and dextran make them ideally suited for measurements of the interfacial free energy at cell surfaces in aqueous media[1-4,8].

In previous studies[2,8], Good-Girifalco plots were constructed using both the contact angle and interfacial tension data obtained with PEG/DEX biphasic systems on a variety of cell types. The resultant critical interfacial tension for spreading applied only to a particular PEG/DEX mixture, and thus lacked generality and strict comparability.

To obtain estimates of the cell surface energies in aqueous media lacking both PEG and DEX, the contact angle and interfacial tension data for each PEG/DEX mixture were used in Equation 5 to obtain values of $\gamma_{SB}$ and $\gamma_{SD}$ for each cell type in each phase of each mixture. These data were then combined with data from the phase diagram of the PEG/DEX mixtures, which gives the concentration of PEG and DEX in each phase. Examples of phase diagrams for PEG/DEX mixtures are given by Albertson[9] and Schurch, et al.[8] A least squares multiple regression analysis of the form of Equation 6 was then performed for each cell type with data from a dilution series of PEG/DEX mixtures. For generality, the cell-liquid interfacial free energy (either $\gamma_{SB}$ or $\gamma_{SD}$) is designated $\gamma_{S*}$ in Equation 6.

$$\sqrt{\gamma_{S*}} = a[PEG] + b[DEX] + \sqrt{\gamma_{CM}} \qquad (6)$$

The best fit value of $\gamma_{CM}$ ($\gamma_{cell-medium}$) is then an estimate of the interfacial free energy at the cell surface in an aqueous medium containing neither PEG nor DEX. Table III lists the values of $\gamma_{CM}$ which were obtained for a variety of cell types. The most hydrophobic cell type, with the highest $\gamma_{CM}$ of those measured was the Human erythrocyte. The physiological significance of this finding is not yet clear, but it may aid in preventing the adhesion of less populous cell types to erythrocytes in the blood stream.

Table III. Cell-Medium Interfacial Free Energies Estimated from Contact Angles in PEG-DEX Biphasic Mixtures.

| Cell type | $\gamma_{cell-medium}$ ($10^{+3}$ ergs/cm$^2$)* | $r^2$ |
|---|---|---|
| Spleen lymphocytes(mouse) | | |
| Balb/c, Control | 12.2 | 0.99 |
| + LPS, t = 1h. | 12.6 | 0.99 |
| + Con A, t = 1h. | 12.3 | 0.99 |
| + A23187, t = 1h. | 11.3 | 0.94 |
| C57BL, Nude | 8.43 | 0.50 |
| Cell lines: | | |
| K | 19.2 | 0.79 |
| BM18-4 | 15.7 | 0.79 |
| ABSL-1 | 16.3 | 0.66 |
| P815 | 19.8 | 0.96 |
| Human erythrocytes | 30.4 | 0.50 |

*Extrapolated by least-squares, from measurements in 4-6 PEG (6000)-DEX (500,000) concentrations; $r^2$ is the correlation coefficient.

The orderliness of the phase diagram for PEG/DEX biphasic mixtures suggests that $\gamma_{CM}$ may be simply related to $\gamma_C$ for a variety of cell types measured with one series of PEG/DEX mixtures. The values of $\gamma_C$ used here have been published previously.[4,8] Figure 2 gives the correlation between $\gamma_{CM}$ and $\gamma_C$ for 10 cell types and demonstrates that this is indeed the case. Some scatter remains, but this is to be expected in an analysis involving so many different types of independent measurements. The overall correlation coefficient ($r^2$) for the relation between $\gamma_C$ and $\gamma_{CM}$ given in Figure 2 is 0.95.

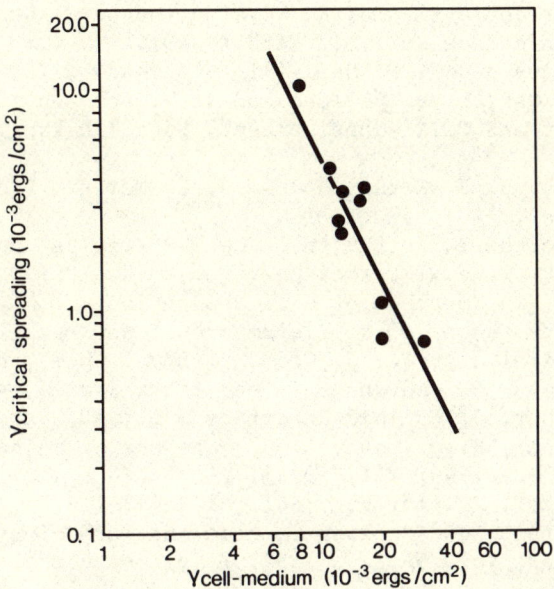

Figure 2. The relation between the critical interfacial tension for spreading, determined with PEG (6000)/DEX(500,000) biphasic mixtures[8] (vertical axis), and the interfacial free energy at the cell surface determined as described in the text (horizontal axis).

## CONCLUSIONS

In conclusion, with the development of an equation-of-state for solid-liquid interfacial free energies, and with the use of contact angle measurements in liquid-liquid-solid systems, it has been possible to estimate the interfacial free energy of biological samples at the aqueous interface.

## REFERENCES

1. D. F. Gerson and J. E. Zajic, Process Biochem., 14, 20 (1979).
2. D. F. Gerson, Biochim. Biophys. Acta., 602, 269 (1980).
3. D. F. Gerson, M. G. Meadows, M. Finkelman and D. B. Walden, in "Advances in Protoplast Research", pp. 447-456, L. Ferenczy and G. L. Farkas, Editors, Akademiai Kiado, Budapest, 1980.
4. D. F. Gerson, in "Immunological Methods", B. Pernis and I. Lefkovits, Editors, Vol. II, pp. 105-138, Academic Press, New York, 1981.

5. J. D. Andrade, S. M. Ma, R. N. King and D. E. Gregonis, J. Colloid Interface Sci., 72, 488 (1979).
6. E. N. Harvey and J. F. Danielli, Biol. Rev., 13, 319 (1938).
7. C. J. van Oss, C. F. Gillman and A. W. Neumann, "Phagocytic Engulfment and Cell Adhesiveness", Marcel Dekker, New York, 1975.
8. S. Schurch, D. F. Gerson and D. J. L. McIver, Biochim. Biophys. Acta., 640, 557 (1981).
9. P. A. Albertsson,"Partition of Cell Particles and Macromolecules", Wiley-Interscience, New York, 1971.
10. H. Walter, in "Methods of Cell Separation", N. Catsimpoolas, Editor, Vol. 1, p. 307, Plenum Press, New York,1977.
11. L. A. Girifalco and R. J. Good, J. Phys. Chem., 61, 904 (1957).
12. W. A. Zisman, ACS Advances in Chemistry Series, 43, 1 (1964).
13. P. J. Becher, J. Colloid Interface Sci., 59, 429 (1977).
14. A. W.Neumann, R. J. Good, C. J. Hope and M. Sejpal, J. Colloid Interface Sci., 49, 291 (1974).
15. K. L. Mittal, Polymer Eng. Sci., 17, 467 (1977).
16. K. L. Mittal, in "Adhesion Science and Technology", L. H. Lee, Editor, Vol. 9A, pp. 129-167, Plenum Press, New York, 1977.
17. J. F. Boyce, S. Schurch and D. J. L. McIver, Atherosclerosis, 37, 361 (1980).
18. J. R. Neufeld, J. E. Zajic and D. F. Gerson, Appl. Env. Microbiol., 39, 511 (1980).
19. S. Schurch, J. Goerke and J. A. Clements, Proc. Natl. Acad. Sci. U.S.A., 73, 4698 (1976).

ADHESION OF ICE TO POLYMERS AND OTHER SURFACES

K. Itagaki

U.S. Army Cold Regions Research and Engineering
Laboratory
Hanover, New Hampshire 03755, USA

A set of simple experiments indicated that water drops can penetrate through a grease layer and make "real" contact with the substrate, then spread over the surface, depending on the surface energy of the substrate, increasing the "real" contact area. Furthermore the ice/substrate bond is stronger than ice itself. The complex problem of ice adhesion may be explainable by combination of these findings in that the "real" contact area multiplied by the strength of ice within the area constitute the apparent adhesive strength. Conceivable effects of various factors are discussed.

## INTRODUCTION

The problem of ice adhesion involves some peculiar features:

1) One of the components (ice) is close to its melting point.

2) Either a) phase transition (water to ice for example) or b) very fast mass transfer on the surface (the so-called liquid-like layer) makes the contact area grow quickly to a size comparable to that of one of the components (ice).

Such conditions are closer to bonding by an adhesive than to adhesion between two solids in contact, in which case the contact area is pretty much limited and considerable internal stress exists within the contact area.

3) The strength of ice is less than that of the structural materials commonly involved.

4) The bond between ice molecules and most substrates is normally stronger than the ice itself.

A lower-energy surface is usually thought to have a weaker bond with ice. The observations reported here indicate that the reason we usually observe lower adhesion of ice on low-energy surfaces is not lower molecular bond strength directly but the smaller "real" contact area caused by low-energy surfaces coated with some contamination.

## EXPERIMENTAL

A circular area on a glass plate was treated with silane ($SiH_4$) vapor to make it hydrophobic. The silane-treated area was outlined on the back side of the glass. Water drops subsequently placed inside the circle assumed hemispherical shapes, while others placed outside it became flatter and more irregular (Figure 1). The same glass plate was later entirely coated with about a 5- to 10-µm-thick layer of grease. Water drops about 3 mm in diameter were then placed in a regular array on the plate. Figure 2 shows the drops 3 minutes after placement. Drops placed outside of the circle became somewhat flatter, while drops placed in the silane-treated area remained hemispherical.

Three kinds of grease were tested: Dow Corning High Vacuum, Primer Industrial Corporation Rotorium Multi-Purpose Blue Grease 35342, and Chesebrough Pond Inc. Vaseline. All gave similar results.

ADHESION OF ICE 243

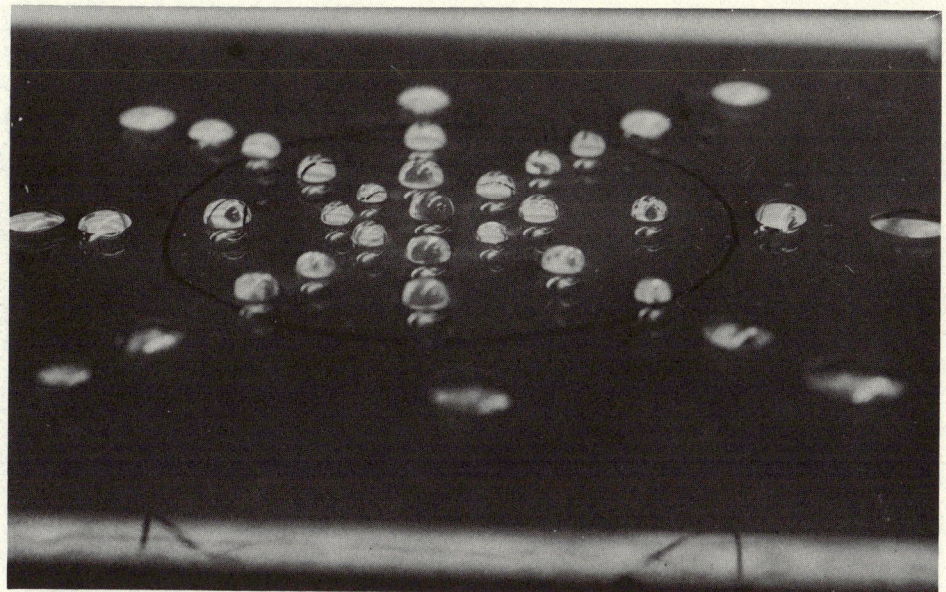

Figure 1. The rounded tops of the water drops inside the circle show that the silane-treated glass surface has a lower surface energy than the untreated area outside the circle.

Figure 2. The same silane-treated glass plate shown in Figure 1, coated with silicone grease. Note that the drops outside the circle are still slightly flatter than those inside.

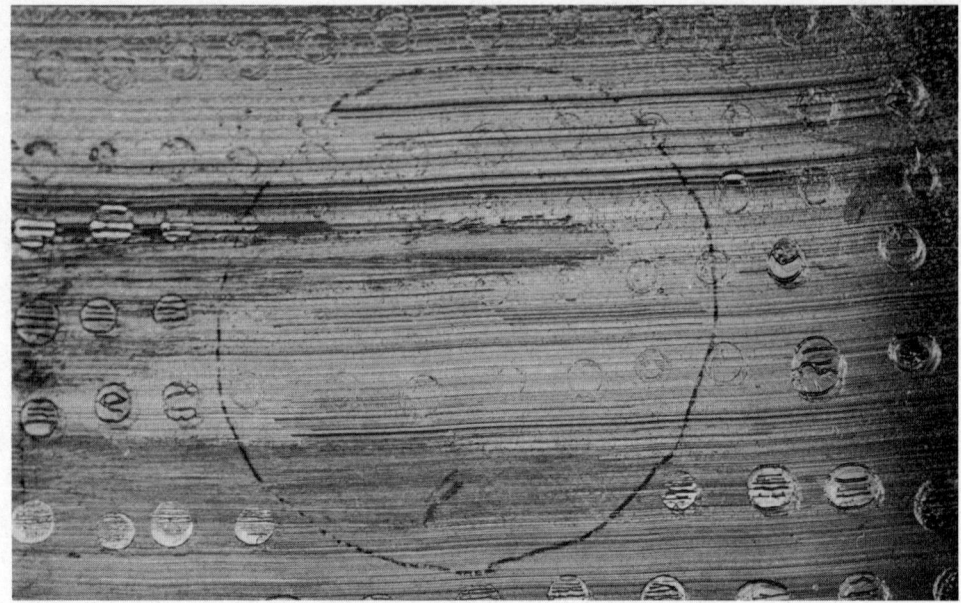

Figure 3. Traces left after the water drops were shaken off the plate. Note the well-defined ring-shaped traces inside the circle in contrast to the wide, grease-free areas outside it.

Ring-shaped traces left on the grease layer after the drops were shaken off from the glass are shown in Figure 3. Sharp, distinct rings were left in the silane-treated area. But in the untreated area, where the grease layer had been removed by the drops and excess grease accumulated toward the centers of the rings, less distinct traces were left.

Apparently, the edges of water drops can attack and penetrate a grease layer by the surface tension effect, until they reach the substrate surface. If the surface energy of the substrate is low, as was that of the silane-treated surface, the contact angle of water is over 90°, as shown in Figure 4a; thus no more spreading occurs. When the surface energy of the substrate is high, the contact angle becomes smaller and the water drops spread, as shown in Figure 4b, until force equilibrium is established. Detailed studies on the penetration process will be discussed elsewhere.

One interesting finding is that the grease layer was displaced at the centers as well as at the edges of the water drops. Figure 5a shows the initial stage of grease removal and Figure 5b shows the final stage. The appearance of the plate indicates that a direct interaction occurred through the thinner part of the grease layer caused by variation in the thickness.

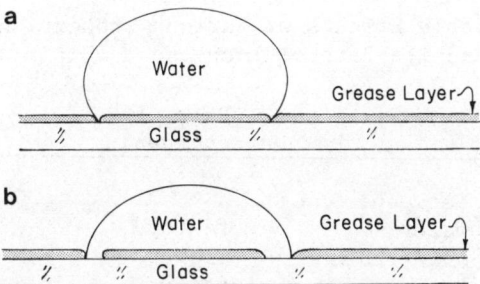

Figure 4a. Water drop on grease-coated low-energy surface keeps its shape. Figure 4b. Water drop spreads by displacing grease layer on high-energy surface.

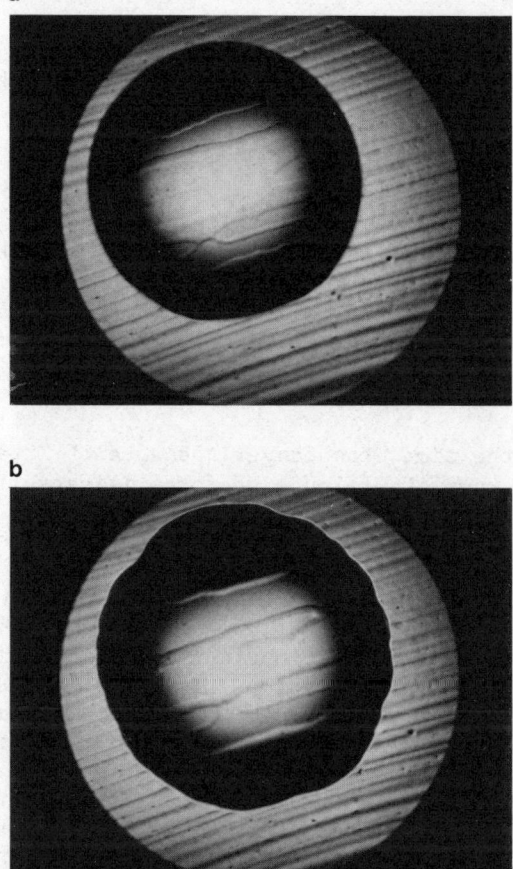

Figure 5a. Grease layer under drop starting to be displaced.
Figure 5b. A considerable portion of the grease removed.

Table I. Conceivable Effects of Various Properties on the Adhesive Strength of Ice and its Rate of Increase.

| Increase in | Rate of strengthening | Final adhesive strength |
|---|---|---|
| Interface energy between: | | |
| substrate and water | faster | stronger |
| substrate and oil/grease | slower | weaker |
| water and oil/grease | faster | stronger |
| Viscosity of oil/grease | slower | unchanged |
| Rest time* | slower | stronger |
| Thickness of oil/grease† | initially faster | weaker |

*Discussed in reference 1.
†Experiments indicated that if the grease is thicker than about 10 μm the edge of the water cannot penetrate through the grease layer.

The area of "real" contact $A_r$ may grow as large as the drop itself when the substrate is hydrophilic and the time of contact is long enough. The adhesive strength when the drop is frozen will, therefore, become higher as the surface energy of the substrate becomes higher, simply because $A_r$ is larger. This growth rate of $A_r$ is controlled by many factors: the surface energy of the substrate, the thickness and viscosity of the oil/grease, the size of the drop, etc.

The longer the time, the larger the area of "real" contact will be, until a complete balance of force is established between the water drop and the substrate surface, resulting in higher adhesive forces as the unfrozen state lasts longer.

The increase in adhesive strength during the "rest time," the length of time drops stay unfrozen on the surface, observed by Baker et al.,[1] supports this notion. Table I indicates the conceivable effects of various properties on the adhesive strength and its rate of increase. The effect of temperature during the rest time may appear through the interfacial energies and the viscosity of the grease/oil.

Another simple experiment is to find the largest water drop that will cling to a vertical low-energy surface. Usually a reasonably clean vertical PTFE (Teflon*) surface is able to hold water drops 3 mm in diameter. The supporting force has to be supplied by the edges of the drops since liquid cannot support

*Trade name: Dupont.

the shear force. The situation can be modeled by a water-filled balloon stuck to flypaper. Only the edge of the balloon touching the flypaper can support the water inside. The equation of the balance of forces for a hemisphere clinging to a vertical surface is then

$$\frac{2}{3} \pi r^3 \rho g \leq 2\pi r\, s\, d \qquad (1)$$

where r is the radius of the drop, $\rho$ is the density of the water, g is the gravitational acceleration, d is the effective width of the surface layer, and s is the average of the real supporting strength S along the perimeter. Equation (1) can be modified to

$$\frac{r^2}{3d} \rho g < s. \qquad (2)$$

If we can assume that d is of the order of the size of a water molecule ($5\times10^{-8}$ cm), then $s > 1.5\times10^8$ dyne/cm$^2$ = $1.5\times10^7$ Pa for a 3-mm-diameter water drop. When this drop is frozen on the PTFE surface, all molecules on the interface would contribute to the support by the real strength S. Therefore the molecular bond strength between substrate and ice would be more than $1.5\times10^7$ Pa. Although the effective width of the surface layer d may be spread over several molecular layers, making s smaller, the real strength S should be much larger than the average value s. If we take into account the fact that s is the results of the difference between advancing and receding contact angles S would further become larger than s.

Generally the shear strength of ice does not exceed $3\times10^6$ Pa.[2,3] Since the bond strength S is much higher than s and the strength of ice is less than one-tenth of S, we could conclude that ice cannot be removed from the PTFE surface without breaking it, insofar as ice is in direct contact with the PTFE.

In order to remove the ice without breaking it we have to find a substrate having a bond strength S lower than the strength of ice $S_i$. Solving Equation (2) for r we obtain

$$r < \sqrt{3\, S_i d/\rho g}. \qquad (3)$$

The generally accepted shear strength of ice at temperatures above $-20°C$ is lower than $2\times10^6$ Pa. Using this value in Equation (3) we obtain $r < 0.55$ mm. In my limited observations I have not yet encountered any such clean, solid surfaces.

Although this discussion may be oversimplified, the rather large adhesive strength between ice and dry Teflon,[3,4] and the adhesive/cohesive nature of the break, indicate that such an estimation remains within the reasonable range. If the Teflon surface is covered by some lubricant the situation is completely different. The specimens used by Raraty and Tabor[4] appear to have been made of solid PTFE, although no description of the method of preparation is given. Usually bulk Teflon exudes some grease-like material. It is possible that their specimen was covered by such a greasy coating, making the adhesive strength much lower than that of a really clean surface.

## POSSIBLE ICEPHOBIC COATINGS

Observations made by Mizuno and Wakahama on the adhesive strength of wet[5] and dry[6] snow give us further insights into the problem of adhesion. They measured adhesive force on various substances including glass and Teflon and found that all showed finite adhesive strength for wet snow at the melting point. With dry snow, however, Teflon shows no adhesive strength. Close observation of the ice-Teflon bond under the microscope revealed that its size decreased with time, while the ice-glass bond kept growing.

One possible explanation is as follows. If we calculate interface tension by the equation of Girifalco and Good,[7]

$$\gamma_{AB} = \gamma_A + \gamma_B - 2\Phi (\gamma_A \gamma_B)^{1/2} \qquad (4)$$

assuming that $\Phi = 1$ for approximation, then the interface tension between water and Teflon is 16.6 dyne/cm, assuming that the surface tension of Teflon is[8] 20 dyne/cm and the surface tension of water is 73 dyne/cm. The interfacial tension is then smaller than the surface tension of Teflon, making spreading of water on the Teflon surface energetically favorable. Since the surface energy of ice of 109 erg/cm$^2$ obtained by Ketcham and Hobbs[9] is much higher than that of water, the interfacial tension becomes 35.6 dyne/cm, which is much higher than the surface tension of Teflon, making a smaller interface energetically favorable. Thus bond size decreases. If this simplified explanation is applicable further, water may not stay on a substrate having surface tension lower than the interfacial tension $\gamma_{AB}$. From Equation (4) the value was calculated to be $\gamma_A/4 = 18.25$ dyne/cm or less. Baker et al.[1] found it difficult to keep water drops on such a lubricated surface because of their high mobility. The mobility is lost when the lubricant film is penetrated by the water drop. If such a low interfacial energy surface was available, water would not stay on the surface, thus no adhesion would be expected.

## DISCUSSION

Since ice is weaker than its bond strength with the substrate, the apparent strength of the adhesive bond is controlled by the area of contact and the strength of the ice. Although the strength of ice is not a unique material constant, but depends on various conditions, the values should remain within a relatively narrow range so that the main factor controlling bond strength is the area of "real" contact.

When some reasonable assumptions are combined with these findings, the rather complex features of ice adhesion seem explainable. The assumptions are:

1. Water can penetrate an oil or grease layer before it freezes.

2. An oil or grease layer usually exists on a substrate unless its surface has been very carefully cleaned.

3. The adhesive strength of the frozen bond is dependent on the "real" contact area.

When a substrate is hydrophobic, the actual contact between a drop of water and its surface is limited to the edge of the drop. "Real" contact between ice and substrate when the drop is frozen, therefore, is also limited to the small area around the edge of the drop. Other areas make indirect contact through the grease layer. The adhesive strength under this condition would be the sum of the strength of the frozen drop edge and the resistance exerted by viscous friction due to the thin grease layer if the ice started to move on the substrate surface. The frictional resistance will be strongly dependent upon loading rate, but will not be present until the drop starts to move.

### Stress intensification

The strength of the bond is controlled by many factors, for instance stress intensification, which is related to the shape and geometry of the bond system. If the shape of the bond system was made to concentrate the stress within the bond, debonding would be easier. A sharp notch near the bond is one example. Bascom et al.[10] observed, by the replica method, that air bubbles were trapped in the interface between a dimethylpolysiloxane coating and ice. Such bubbles and oil droplets concentrated in the interface can reduce the adhesive strength of the bond system, not only by reducing the effective bond area but also by causing stress intensification in the real contact area. The very low shear strength of this bond system can be explained in this way.

Differences in elastic moduli can also cause stress intensification. Frequently we observe that it is much easier to shake off drops frozen on metal foil than drops frozen on plastic film. Since the elastic moduli of metals are generally high, the stress intensity at the edge of the frozen drops becomes very high when the foil is flexed. The elastic moduli of plastics are comparable to or lower than those of ice, and the stress intensity at the drop edges is therefore less.

### Defects

Defects control the strength of materials. In the ice adhesion problem, the dislocation system in ice is the major factor since the ice is the weakest part of the bond system. Bascom et al.[10] observed that the etch pit density of ice touching Teflon-coated foil is much higher than on a cleaved surface of bulk ice. The dislocation systems can modify the mechanical properties, although it is difficult to predict the results. Low-density dislocation systems generally soften the crystal, but interaction between high-density dislocation systems can strengthen the crystal. They found three types of ice separation. Their type B (cohesional, but with the fracture surface close to and parallel with the substrate) indicates that the weak layer exists some distance inside the ice, away from the interface. A difference in the thermal expansion coefficients of the ice and substrate could generate a high-density dislocation loop system near the interface region. This system could strengthen the interface region and the fracture surface could be located somewhat inside the bulk ice, away from the interface.

### Stochastic Processes

The initiation and propagation of cracks are controlled by stochastic processes. The smaller the volume the less chance there is of a defect starting a fracture, and thus there is higher strength. Jellinek[11] showed that logarithms of tensile adhesive strength decreased linearly as logarithms of the volume increased. He also found that the relationship of the strength and the cross-sectional area is linear at constant volume. By assuming that an individual crack is limited to one column within the bundle of columns that constitutes an ice cylinder he was able to explain the relationship of strength with volume and cross-sectional area. If such an assumption is realistic, there must be many microcracks generated before catastrophic fracture occurs. These microcracks could be detected by acoustic emission.[12]

## Relaxation Processes

Various processes of stress intensification may be reduced by relaxation processes, including the dislocation process and liquid-like layers. The nature of the surface (or interface) layer is sometimes described as liquid-like, although the term "liquid-like" is controversial. Valeri and Mantovani[13] mounted silicon strain gauges on the surface of ice by freezing and measured the change in resistance. The thermal expansion coefficient deduced from the resistance measurement showed a sharp change in the direction parallel to the optic axis (C direction) at about -7°C. A rather gradual change, but definitely different results were obtained from the thermal expansion coefficient of ice found by standard measurement perpendicular to the C direction. They attributed the discrepancy to the liquid-like layer, although definite crystal anisotropy is also observed.

Hunsaker et al.[14] and Jellinek[15] found abrupt changes in the shear adhesive strength vs. temperature relationship and the mode of fracture change from adhesive to cohesive at around -12°C for a metal-ice system. Again such features were attributed to the liquid-like layer without reference to any particular properties of the liquid. There is no doubt that some anomalous layer exists on ice near the melting point. However, I have strong reservations about calling the layer liquid-like, since it is anisotropic and a finite stress is required to start sliding relative to the substrate.

The conditions discussed here, and many more, can contribute to the strength of the ice-substrate bond system. Also, certain monolayers can be separated from the substrate surface.[10] In the practical application of ice adhesion control, however, such coatings are effective for only one removal, making the range of application very limited. If ice adhesion is to be controlled by using coatings, probably the best approach is a low energy surface with replenishable lubricant coverage if reasonable shear force is available to overcome the ice-substrate bonding. Mechanical systems that deliberately increase stress intensity or freezing point depressants may find some application, but these are beyond the scope of the present topic and will be discussed elsewhere.

## REFERENCES

1. H.R. Baker, W.D. Bascom and C.R. Singleterry, J. Colloid Sci., 17, 477 (1962).
2. W.F. Weeks and A. Assur, "Fracture of Lake and Sea Ice." USA Cold Regions Research and Engineering Laboratory, Research Report No. 269, 1969 (AD 697750).*
3. T.F. Ford and O.D. Nichols, "Shear Characteristics of Ice in Bulk, at Ice/Solid Interfaces and at Ice/Lubricant/Solid Interfaces in a Laboratory Device." U.S. Naval Research Laboratory, Report 5662, Washington, D.C., 1961.*
4. L.E. Raraty and D. Tabor, Proc. Royal Soc. (London), 245 A, p. 184 (1958).
5. Y. Mizuno and G. Wakahama, Low Temperature Science, Ser. A, 35, 133 (1977).
6. Y. Mizuno and G. Wakahama, Low Temperature Science, Ser. A, 38, 17 (1978).
7. L.A. Girifalco and R.J. Good, J. Phys. Chem., 61, 904 (1957).
8. H.W. Fox and W.A. Zisman, J. Colloid Sci. 5, 514 (1950).
9. W.M. Ketcham and P.V. Hobbs, Philosophical Magazine, 19, 1161 (1969).
10. W.D. Bascom, R.L. Cottington and C.R. Singleterry, J. Adhesion, 1, 246 (1969).
11. H.H.G. Jellinek, "Tensile Strength Properties of Ice Adhering to Stainless Steel". U.S. Army Snow, Ice and Permafrost Research Establishment, Research Report 23, 1957 (AD716663).*
12. W.F. St. Lawrence and D.M. Cole, Cold Regions Sci. Technol. 5, 183 (1982).
13. S. Valeri and S. Mantovani, J. Chem. Phys., 69, 5207 (1978).
14. J.C. Hunsaker, F.G. Keyes, H.G. Houghton, C.A. Johnson and J.V. Dotson, "An Investigation of Methods for the De-icing of Aircraft." Progress Report to National Academy of Sciences, Massachusetts Institute of Technology, Report No. 1, 1940.
15. H.H.G. Jellinek, "Adhesive Properties of Ice." U.S. Army Snow, Ice and Permafrost Research Establishment, Research Report 38, 1957 (AD149061)*.

---

*Technical reports produced by agencies of the United States Government may be obtained from the National Technical Information Service, Springfield, Va. 22161, U.S.A.

# Part III
# Reactions and Interactions at Polymer Surfaces

SURFACES (INTERFACES) AND POLYMER STABILITY


H.H.G. Jellinek

Department of Chemistry
Clarkson College of Technology
Potsdam, New York 13676


An overview of surface degradation and stability of polymers has been presented. Surface thermal, oxidative and photo-oxidative degradation has been discussed and illustrated by examples. The importance of the new surface-instrumental methods has been pointed out. Also, the significance of surface modifications for adhesion has been noted.

## INTRODUCTION

Surfaces of polymers are the locations which are directly exposed to the environment (e.g. sun, oxygen, moisture, pollutants etc.). Degradation such as oxidation, photolysis, photo-oxidation also high energy degradation, corona-discharges, and other intended types of degradation will initially be at a maximum at surfaces. Such changes can be of decisive importance for adhesion. Not only may changes in surface properties influence the adhesive strength of joints, but changes brought about intentionally can improve this strength by increasing surface energy due to, for instance, various controlled oxidations.[1]

Degradation reactions consist of main chain scission, crosslinking, oxygenation or transformation and severance of side groups. Oxidation of chains generate usually hydroperoxide groups first; these groups are formed and eventually decay causing scission of the polymer chains, formation of volatile oxygenated products and oxygenated groups at the polymer chains. Photodegradation will also take place if suitable chromophoric groups are present, eventually leading to chain scission and/or crosslinking.[2]

Another parameter which is of great importance in this context is polymer morphology.[2,3] Quite frequently oxygen cannot or can only diffuse very slowly into crystalline regions of a polymer. This is a matter of polymer density. A polymer can actually become more crystalline during oxidative degradation. Interfacial crystalline/amorphous regions of a polymer are of special interest. Here, the oriented chains bend back parallel to themselves. This creates strain energy of the main chain links in these bends causing main chain link scission to proceed faster here than in the amorphous regions. This has actually been shown.[3] Jellinek,[4] for instance, has elaborated a theory of degradation where main chain links in the amorphous parts scission normally, while those in the crystalline/amorphous interface rupture faster than the normal ones. If this interface is made increasingly larger (e.g. by decreasing average spherulite diameters) then the scission rate constants should become proportional to the reciprocal of average spherulite diameter. This was actually confirmed by Kachi and Jellinek,[5] measuring the chain scission rate constants as a function of average spherulite radius for Nylon 66. The scission rate constant eventually became proportional to the inverse of the average spherulite diameter for small spherulites. Also catalytic degradation reactions can take place on polymer surfaces or in surface layers. Thus it is well known that the oxidative degradation of polyethylene and polypropylene, for instance, is catalytically accelerated by copper or its oxides[6,7] located in the metaloxide/polymer interface. This is of great practical importance as these polymers are used for insulation of copper conductors. Allara and coworkers have studied the catalytic oxidation of polyethylene[6] and Jellinek and coworkers that of isotactic polypropylene.[7] In either case, it was found that copper ions diffuse from the metal/polymer interface into the polymer; carboxylate ions, produced by oxidation, are the anions. (The Cu-ions decompose hydroperoxide groups by a redox reaction. This reaction proceeds catalytically and is much faster than the normal ROOH-decay).

As pointed out above, the field of adhesion is especially concerned with interfaces and surfaces. Here the characteristics (e.g. morphology, oxidation, impurities, monolayers etc.) of these uppermost surfaces or interfaces (they need only to be one or a few monolayers thick) are of prime importance. Unintentional or intentional degradation can take place on surfaces or interfaces, changing their original characteristics leading to impaired or improved adhesion, respectively.

Not only adhesion is of importance but also the opposite phenomenon abhesion i.e. release from substrates is of practical significance (e.g. release coatings for ice). Ice release-coatings of polycarbonate-siloxane co-polymers have been studied by Jellinek and coworkers recently.[8] Not only are the hydrophobic properties of the block polymers in the polymer/metal interface of great significance but also their mechanical and rheological properties have a decisive influence on the magnitude of the decrease in adhesion. In addition Jellinek and Iwaki[9] studied the adsorption of water vapor on these block copolymer coatings in the form of films. The magnitudes of the isosteric heats and the molar differential entropies reflected correctly the abhesion data for the block-copolymer of different siloxane block chain length and content.

It is not intended to give in this brief overview an exhaustive report on surface degradation; it will rather be attempted to highlight important points and to give some experimental examples. The author's and his coworkers' work will quite often be chosen as examples. In addition, the new instrumental methods for surface analysis will be considered briefly.[1] They permit one to investigate thinner surface layers than was hitherto the case; even one molecular thickness can be investigated. These methods have given a great impetus to surface science in all its aspects and have advanced fundamental adhesion work which has threatened to become stagnant. Here again some experimental data will be presented without trying to be exhaustive, especially in view of a number of reviews in the literature[10] and of papers published in these proceedings dealing with these instrumental methods. The main emphasis will again be on the use of these techniques for elucidating surface degradation reactions.

Generally, surfaces can be stabilized as far as thermal and oxidative degradation are concerned by changing the structure and morphology (particularly crystallinity) of polymers. Thus polymers which have aromatic molecules in their backbone, appreciable crystallinity (tacticity), stiffness, crosslinking and ladder polymers, etc. will be thermally more stable and also more stable against oxidative degradation than ordinary thermoplastic polymers. Especially, tertiary H-atoms and unsaturation make polymers very susceptible to thermal-oxidative degradation. Fortunately, polymers show also the cage effect, whereby scissions can be mended especially in solid polymers.[11]

## KINETICS AND MECHANISM

### (Surface Degradation Studies)

<u>Thermal surface degradation</u> can be caused, for instance, by heat irradiation of polymer surfaces. Such a surface will suffer maximal thermal degradation on its very surface (e.g. chain scission may take place). This maximum degradation can be quantitatively expressed by the degree of degradation

$$\alpha_{max,x=0} = \frac{\bar{s}_{max}}{DP_{o,n,x=0}}$$

where $\alpha$ is the so-called degree of degradation at a distance $x = 0$ from the surface, and $\bar{s}$ is the average number of links scissioned in an original chain of length $DP_{o,n}$; $\alpha$ is also the probability of scissioning one main chain link in one original chain at the considered stage of the degradation process.[2] The $\alpha$ value at a distance $x$ from the surface will be a function of the heat conductivity of the polymer. The rate of scissioning at any $x$ will be given by the temperature at $x$. The kinetics of such a process will not be discussed here (for some remarks concerning ablation, see a later section).

<u>Photodegradation</u> of surfaces will be treated in some detail. A polymer sample which has suitable chromophoric groups absorbing monochromatic radiation (usually u.v.) is considered. The radiation impinges on the surface at right angles with its whole intensity $I_o$. As the radiation penetrates the polymer, its intensity will decrease obeying Beer-Lambert's Law. Hence, the surface will suffer maximum degradation. Degradation decreases exponentially with distance $x$ from the surface ($x = 0$). The overall degree of degradation $\bar{\alpha}$ (e.g. for main chain scission) is obtained by determining $\bar{\alpha}$ for the whole film of thickness $\ell$. The polymer is initially homogeneous or has a random distribution i.e. most probable distribution,[12]

$$\bar{\alpha} = \frac{\bar{s}}{DP_{n,o}-1} = \frac{1}{\overline{DP}_{n,t}} - \frac{1}{DP_{n,o}} = \frac{\phi_s kt I_o}{m_1 \ell}(1-e^{-a\ell}) \qquad (1)$$

$$I_o = I'_o/A$$

Here, k is a constant of proportionality, $m_1$ the number of mers in each volume unit (e.g. $cm^3$), 'a' an optical constant and $\phi_s$ the scission quantum-yield. A is the film surface area.

The rate of scissioning links, n, is given by,

$$-\frac{dn}{dt} = \phi_s I_{abs} \tag{2}$$

Where $I_{abs}$ is the total absorbed light intensity by the film. The extent of scissioning is supposed to be relatively small (e.g. <1%, hence for 1% if $DP_{n,o} = 10^3$, $\bar{s} = 10$, then the degree of degradation is $\bar{\alpha} = 10^{-2}$; for such a case the polymer has become useless as far as its mechanical properties are concerned). Integration of Equation (1) yields,

$$\frac{n_o - n_t}{N A x_1} = \bar{s}_t = \frac{\phi_s I_{abs} t}{N A x_1} \tag{3}$$

Here $N = m_1/DP_{n,o}$ is the number of original chains in the film and A the film surface area.

For small light absorption, Equation (1) reduces to

$$\bar{\alpha} = \frac{\bar{s}}{DP_{o,n}} = \frac{kt I_o a}{m_1} \tag{4}$$

The degree of degradation at distance x from the exposed surface (x = 0) is,

$$\alpha_x = \bar{s}_x/DP_{o,n} = \alpha_o e^{-ax} \tag{5}$$

where $\alpha_o = kt I_o a/m_1$ is the degree of degradation at the surface (x = 0) which has the maximum value of all $\alpha$'s.

The size distribution in each layer will change with x; the same is true for the average chain lengths and molecular weights. Space does not allow to present here these distributions and average values.[12] Shultz[13] has derived expressions for initially most probable distributions and has given various elaborate derivations, whereas Jellinek started with a homogeneous sample.[12]

In case of exposure of films to high energy radiation, provided the absorption is weak (i.e. practically homogeneous radiation or constant intensity across the thickness of the film), the random chain scission rate constant $k_{ir}$ is given by,

$$k_{ir} = r'M_1/E_d N_A \tag{6}$$

Here r' is the radiation dose for the time unit, $M_1$ the molecular weight of the monomer, $E_d$ the average energy needed to break one main chain link and $N_A$ Avogadro's number.[14]

Stabilization of polymer surfaces with respect to photolysis has to be carried out similarly as for bulk polymers (i.e. screening, u.v. absorbers, quenching etc.) and also purification will be of help in order to remove chromophoric impurities. The additives must not be volatile as thin films, and surfaces have to be protected. Many polymers absorb in the near u.v. due to surface oxidation during manufacture and subsequent handling. It is not the place here to discuss these topics in detail.[8]

<u>Oxidative Degradation</u> is very important for the stability of polymers and also for their adhesive properties. This reaction can increase the surface energy of polymers thereby modifying (e.g. increasing) their adhesion behavior.[14a,b]

Jellinek and Lipovac[15] studied the oxidative degradation of isotactic polystyrene over a range of temperatures from 249°C to 300°C. The softening point of this polymer is ca. 230°C. Only some important features of this reaction relevant to surfaces will be indicated. Films were exposed to oxygen at constant pressure. The uptake of $O_2$ by films of this polymer was shown to be diffusion controlled. This manifested itself in the small energy of activation of the oxidation process and in the fact that the overall chain scission reaction (i.e. the rate constant) was a function of the

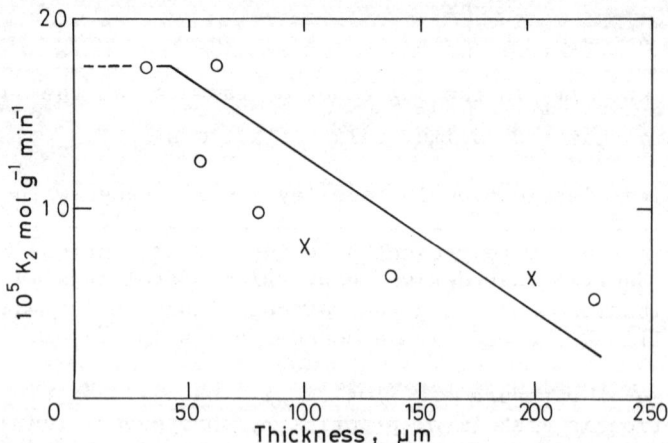

Figure 1. Chain scission rate constants vs. film thickness ($P_{O_2}$ = 176 mm, 280°C).[15]

film thickness for constant film surface area. Figure 1 illustrates this point. The degradation becomes practically independent of film thickness below about 50μ. This indicates the $O_2$-concentration in films $\ell < 50\mu$ is practically homogeneous; in thicker films, however, the $O_2$-concentration decreases with distance x from the surface becoming eventually zero in accordance with the magnitude of the diffusion coefficient of the gas in the polymer.

The oxidation mechanism for all oxidation reactions of polymers passes initially through a stage of hydroperoxide formation. Oxidation can be represented by reactions as indicated below[15] (or by variations thereof),

$$\underline{\text{SLOW}} \quad O_{2,g} \xrightarrow[\text{into film}]{D_{O_2}, \text{Diffusion}} O_{2,f}$$

g and f denote $O_2$ in the gas phase and film respectively. $D_{O_2}$ is the diffusion coefficient for $O_2$ in the film.

<u>Faster</u> reaction steps follow,

$$RH + O_{2,f} \xrightarrow{k_1} R^\cdot + HO_2^\cdot \quad \text{(this step can also be formulated in different ways)}$$

$$R^\cdot + O_{2,f} \xrightarrow{k_2} RO_2^\cdot$$

$$RO_2^\cdot + RH \xrightarrow{k_3} ROOH + R^\cdot$$

$$ROOH + O_{2,f} \xrightarrow[\text{epoxide}]{k_4} R_xOR_y + H_2O + O_2$$

$$\downarrow$$

$$R_{xy} + H_2O$$

unsaturated polymer

$$ROOH \underset{k_6}{\overset{k_5}{\rightleftarrows}} \langle CAGE \rangle_1 \xrightarrow[\text{scission}]{+O_2, \text{chain}, k_m} R_x = O + R_yOH + O_2$$

$$RO_2^\cdot + R^\cdot \underset{k_9}{\overset{k_8}{\rightleftarrows}} \langle CAGE \rangle_2 \xrightarrow{k_{10}} \text{Inert Products}$$

$$+ \frac{\partial [O_2]_g}{\partial t} = D_{O_2} \frac{\partial^2 [O_2]_f}{\partial x^2} - k_2[O_2][R_f^\cdot] = 0 \tag{7}$$

The second term in Equation (7) represents the actual oxidation reaction. The oxygen concentration in the gas is kept constant throughout. The consumption of $O_2$ from the gas phase is practically only due to reaction step 2 of the above mechanism.

It can be shown[15] that for films <50μ thick, the rate of oxygen consumption amounts to,

$$- \frac{d\ O_2, mol_g}{A\ dt} = -D_{O_2} \left( \frac{\partial [O_2]_f}{\partial x} \right)_{x=0} = \frac{K_{r,O_2} V_f}{A} \tag{8}$$

The $O_2$-flux into the film is,

$$- \frac{d\ O_2, mol_g}{A\ dt} = - D_{O_2} \left( \frac{\partial [O_2]_f}{\partial x} \right)_{x=0} \frac{P_{O_2}[O_2]_{f,x=0}}{S_{O_2} \ell} \tag{9}$$

Here $P_{O_2} = D_{O_2} S_{O_2}$ is the permeability constant, and $S_{O_2}$ the solubility constant of $O_2$ in the film (Henry's Law); $\ell$ is the film thickness. x is the distance from the polymer surface, A the surface area, $K_{r,O_2}$ a constant, and $V_f$ the film volume. Equation 8 expresses the fact that the process is diffusion controlled.

The formation and decay of hydroperoxide is given by,

$$\frac{d[ROOH]_f}{dt} = K_{O_2}[RH]_f[O_2]_f - k_4[O_2]_f[ROOH]_f$$

$$= K_{ROOH} - C[ROOH]_f \tag{10}$$

Here, $K_{O_2} = \left\{ \frac{k_1 k_2 k_3 (k_5 + k_{10})}{k_8 k_{10}} \right\}^{\frac{1}{2}}$, $C = k_4[O_2]_f$; $[K_{ROOH}]_f = K_{O_2}[RH]_f[O_2]_f$

and $K_r O_2 = K_{O_2}[O_2]_f[RH]_f$.

If $[O_2]_f$ in the film follows Henry's Law, then the prestationary state for ROOH is,

$$[ROOH]_f = \frac{K_{ROOH}}{C}(1-e^{-Ct}) \tag{11}$$

# SURFACES AND POLYMER STABILITY

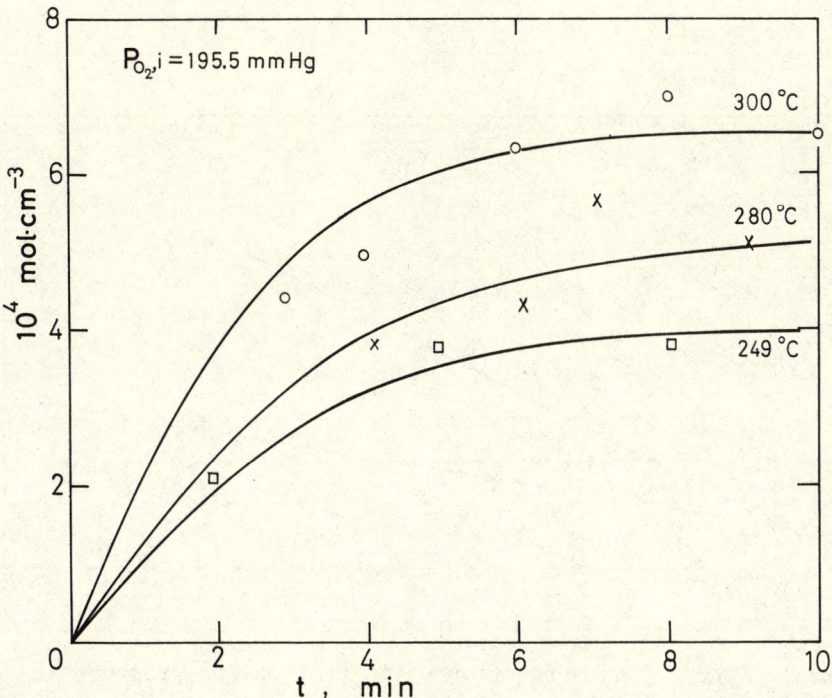

Figure 2. Calculated curves according to Equation (11) and experimental points for initial stages of hydroperoxide formation during thermo-oxidative degradation of isotactic polystyrene.[15]

The steady state for $[ROOH]_f$ is reached when $[ROOH]_f$ reaches a maximum value which is maintained as long as the concentrations $[O_2]_g$ and $[RH]_f$ remain constant. This value is,

$$[ROOH]_{f,max} = \frac{K_{ROOH}}{C}$$

See Figure 2, which shows curves calculated from Equation (11).

Chain scission for small $\bar{s}$ values is given by,

$$\alpha = \frac{1}{\overline{DP}_{n,t}} - \frac{1}{\overline{DP}_{n,0}} = \frac{k_7 k_5}{k_6} \frac{K_{ROOH}[O_2]_f t}{C} = K_{exp}.t \qquad (12)$$

(See Figure 3)

If Henry's Law is followed, then $[O_2]_g = K[O_2]_g$, where K is Henry's constant.

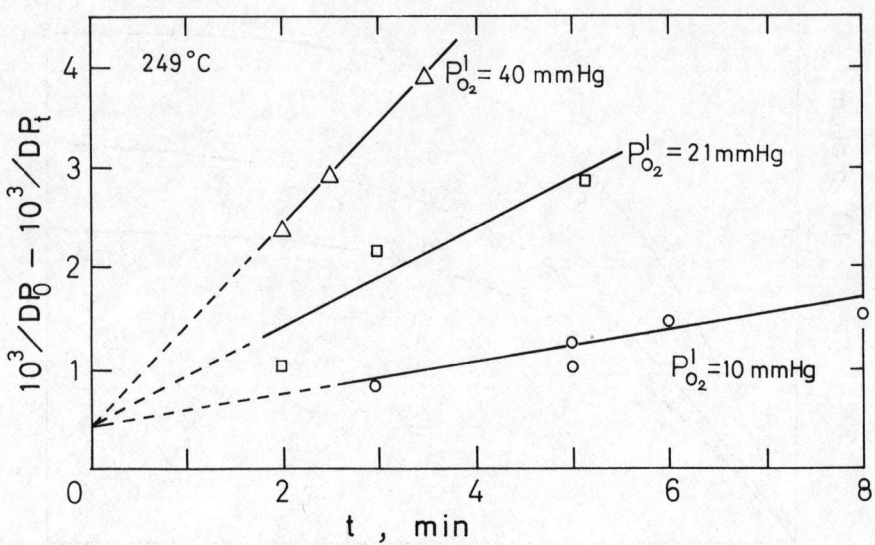

Figure 3. $1/\overline{DP}_{n,t}$ as a function of initial oxygen pressure and time (249°C).[15]

The boundary conditions are;

$d[O_2]_f/dx = 0$ at $x = \ell$ (film thickness) if $\ell > 50\mu$; $[O_2]_{f,x}$ is a function of x (and of $\ell$), and so is $\alpha_x$; $\alpha_{max}$ is located at the film surface.

Work by Jellinek and Igarashi[16] should be noted here. The diffusion of $NO_2$ into atactic polystyrene was studied. As indicator of the progress of diffusion, the degree of degradation as a function of distance x from the polymer surface was chosen. Some data are presented in Figures 4 and 5.

Similar situations as in the case of isotactic polystyrene are encountered during the oxidation of polyethylene and polypropylene films. For the latter case,[17] flakes of different thickness were used. The functional relationship between sample thickness and rate of oxidation was similar as in the previous case. Here, oxygen diffuses from both sides into the flakes (there is no agitation and $[O_2]$ = constant in the film).

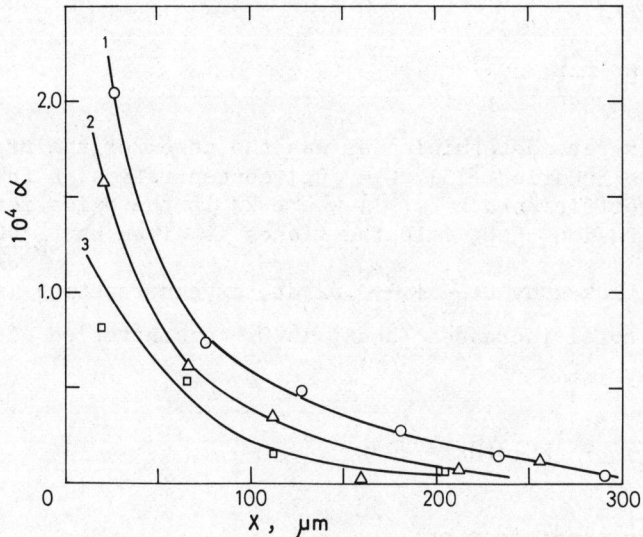

Figure 4. $\alpha_{x,t}$ as a function of distance from film surface for various diffusion times of $NO_2$ into atactic polystyrene. (55°C; $P_{(total)}$= 60 cm; (1) 8 h, (2) 6 h, (3) 4h).[16]

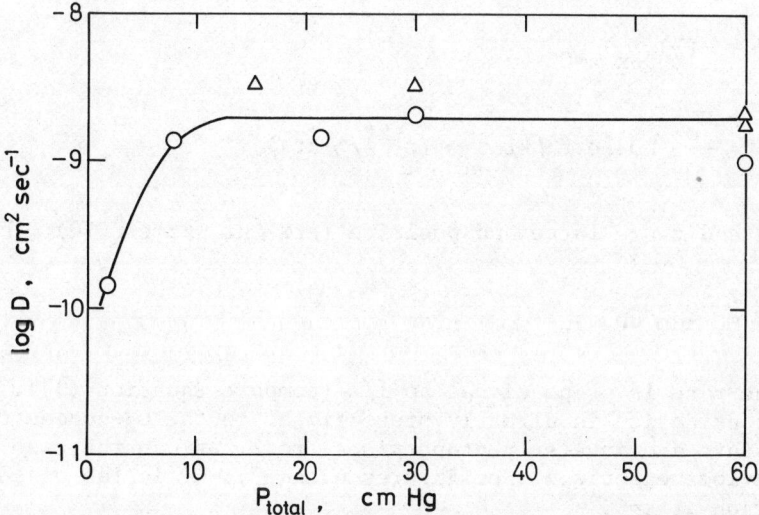

Figure 5. Log of diffusion coefficients D vs. total gas pressure: O penetration, △ sorption, - desorption (45°C).[16]

The rate of oxygen consumption is,

$$\frac{\partial C}{\partial t} = D \frac{\partial^2 C}{\partial x^2} - kC \tag{13}$$

Diffusion is rate determining as was the case for the previous example. In Equation (13), $C = [O_2]$-concentration, $D$ the respective diffusion coefficient, $k=k_1[R\cdot]$ where $k_1$ is the oxidation rate constant. Diffusion of $O_2$ into the flakes is given by $-D(\frac{\partial C}{\partial x})dt$ and out of the flakes by $D(\frac{\partial C}{\partial x} + \frac{\partial^2 C}{\partial x^2} dx)dt$, oxygen reacting is $-kCdtdx$. Hence, the total increases (dCdx) in $O_2$-concentration of the flakes is,

$$[D(\frac{\partial^2 C}{\partial x^2}) - kC]dtdx = dCdx \tag{14}$$

The boundary conditions are,

$C = C_o \quad\quad x = 0 \quad\quad t > 0$

$C = 0 \quad\quad x > 0 \quad\quad t = 0$

$C = 0 \quad\quad x = 0 \quad\quad t > 0$

Then, the rate of oxidation per volume unit is,

$$\frac{dC}{dt} = -(\frac{D}{\ell})(\frac{\partial C}{\partial x})_{x=0}$$

$$= \frac{C_o}{\ell} \sqrt{kD} \,[\mathrm{erf}\sqrt{kt} \;;\; + (e^{-kt}/\sqrt{kt\pi})] \tag{15}$$

Both, $k$ and $t$ are large and positive ($\mathrm{erf}\sqrt{kt} \sim \ell$; $e^{-kt} \sim 0$); therefore,

$$\frac{dC}{dt} \simeq \frac{C_o}{1} \sqrt{kD} \tag{16}$$

Here the rate is proportional to $1/\ell$ (compare Equation (9)). The rate of oxidation is directly proportional to the $O_2$-concentration or pressure and inversely proportional to $\ell$, as was the case for the previous example. (See Figures 6 and 7); $C_o$ is the oxygen concentration at $x=0$.

Polymer surfaces are often intentionally modified, usually by oxidation for the purpose of improving their adhesive properties.

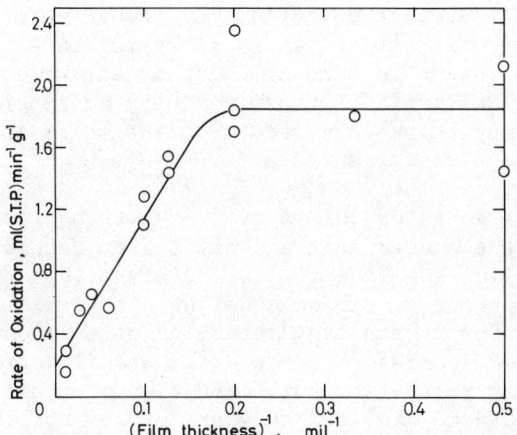

Figure 6. Variation of steady-state rate of polypropylene oxidation at 150°C with sample thickness.[17]

Oxidation which produces wettable, polar groups on the surface has been extensively applied to polyethylene.[19] Also surface grafting is carried out for the same purpose as well as treatment with alkali and acid, halogenation, sulfonation etc. Also crosslinking of polyethylene can be effected with fluorine.

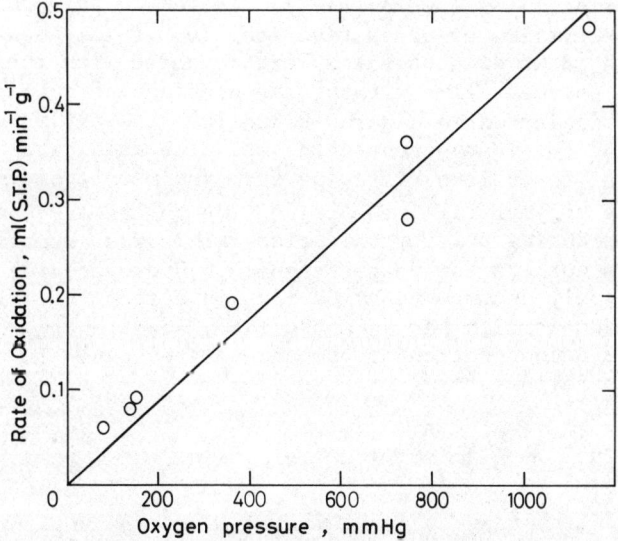

Figure 7. Variation of steady state rate of polypropylene oxidation at 110°C with oxygen pressure.[17]

Surface oxidation can also be caused by corona discharge or flame treatment. Surface oxidation may remove weak boundary layers and crosslinking will strengthen such layers. One treatment for achieving crosslinking is "crosslinking by activated species of inert gases" or "casing". Epitaxy can play an important role in obtaining a strong polymer surface from the melt; the substrate should have a high energy surface.

Even if the adhesive joints are very strong initially, they are exposed to the environment and may suffer deterioration. For instance, the simultaneous effect of moisture and stress can lead to failure. Apparently, silane coupling agents minimize this deterioration. For a detailed discussion of environmental effects and its mechanism see ref.[20] Here again stabilization against oxidation is similar to that of bulk oxidation i.e. similar antioxidants have to be used provided they are of low volatility.

<u>INTERFACIAL CATALYSIS</u> plays a very important part in surface (interfacial) degradation. Oxidation can be catalyzed by various metals. Catalysis can start at metal/polymer interfaces and metal ions can diffuse into the polymer. The catalytic action by copper or its oxides, for instance, during oxidative degradation of polyethylene and isotactic polypropylene, respectively, has been studied in detail. Some relevant results will be briefly discussed here.

Jellinek and co-workers[21] have studied the oxidative degradation of isotactic polypropylene catalyzed by well defined Cu or Cu-oxides (CuO and $CuO_{0.67}$). Microscope slides were covered by Cu or respective oxide films (200 Å thick). Isotactic polypropylene was coated on these films by dipcoating (ca. 10 μ thick; see Figure 8). The oxidation rates of such films were compared with the rates of oxidation of polymer films without the presence of catalysts. The technique is indicated in Figures 9 and 10. The oxide $CuO_{0.67}$ appeared to be the most potent catalyst. The main catalytic action was due to Cu-ions diffusing from the metal/polymer interface into the polymer film; the respective anions are carboxylate ions produced during polymer oxidation. This was actually shown by Allara and co-workers for polyethylene.[22] Hydroperoxide groups are catalytically decomposed by Cu-ions in a redox reaction. This catalyzed decomposition has an appreciably lower energy of activation than the decomposition of the uncatalyzed one. It can be formulated as follows,[21]

$$ROOH + Cu^{+2} \xrightarrow{k_{5a}} RO_2^{\cdot} + Cu^{+} + H^{+}$$

$$ROOH + Cu^{+} \xrightarrow{k_{5b}} RO_2^{\cdot} + Cu^{+2} + OH^{-}$$

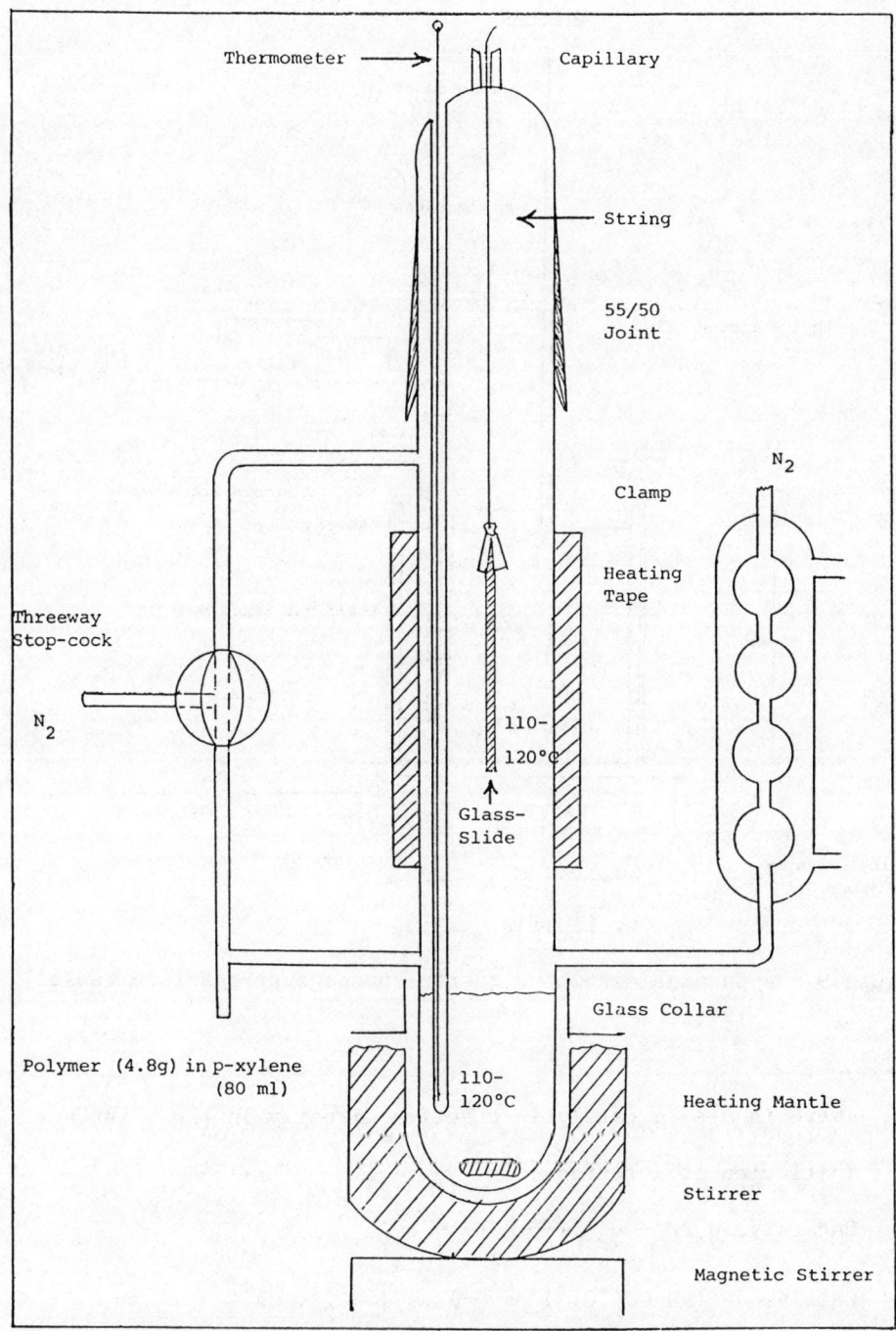

Figure 8. Dip-coating apparatus.[7]

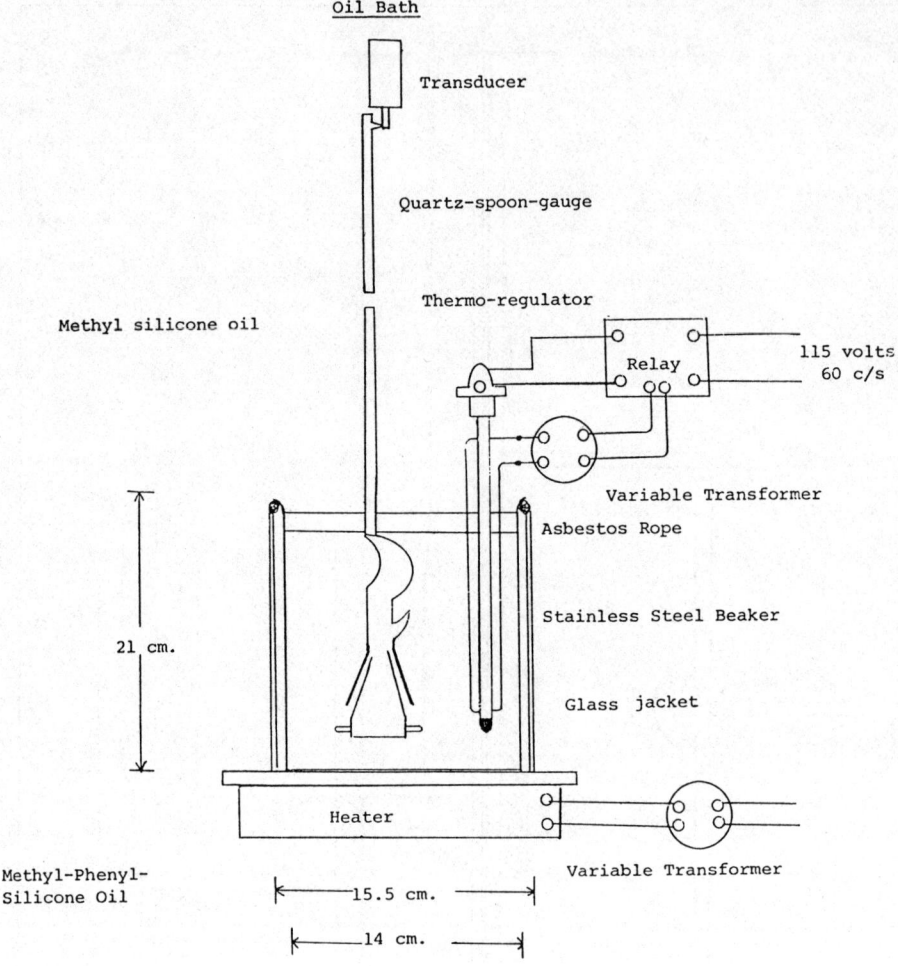

Figure 9. Apparatus assembly (quartz-spoon-gauge-reaction vessel).[7]

There is also a catalytic reaction directly in the interface.
Initiation at interface:[21]

Uncatalyzed $RH + O_2 \xrightarrow{k_i} R^\cdot + HO_2^\cdot$

Catalyzed $(RH\ Cu^{+2}O_2^-) \xrightarrow{k_i} R^\cdot + HO_2^\cdot + Cu^{+1}$

$Cu^{+1} \xrightarrow[\text{of air}]{\text{oxygen}} Cu^{+2}$

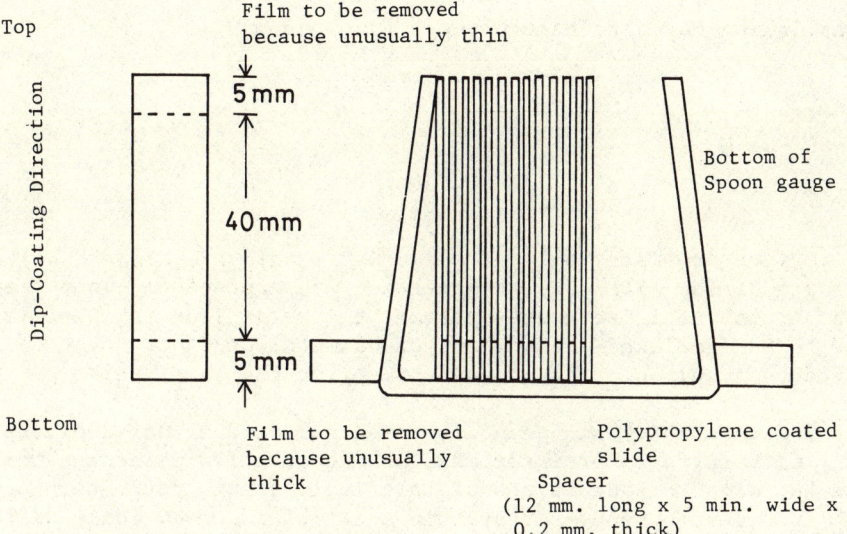

Figure 10. Reaction vessel and coated sides.[7]

An additional catalytic reaction takes place at the interface,[21]

$$CuO + RH \xrightarrow{k_{i,1}} R + CuOH$$

$$CuOH + RH \xrightarrow{k_{1,2}} R + Cu + H_2O$$

$$Cu \xrightarrow{air} CuO_x$$

The formation and decay of hydroperoxide were also studied in detail; this will be discussed in the next section. Many other metal catalyzed polymer oxidation reactions are known, but space does not permit to discuss them here.

It is of interest to note that Burkstrand[22a] using ESCA found that a polymer complex of polystyrene and oxygen with a Cu, Ni or Cr film, was vaporized onto the polymer surface and then treated by O-atoms (plasma treated). These complexes are probably formed by oxygen pulling off tertiary H-atoms from polystyrene and subsequent oxidation.

$$\begin{array}{c} H\ H\ H \\ | \ \ | \ \ | \\ \sim C - C - C \sim \\ | \ \ | \ \ | \\ H\ \phi\ H \end{array} \xrightarrow{O} \begin{array}{c} H\ H\ H \\ | \ \ | \ \ | \\ \sim C - C - C \sim \\ | \ \ | \ \ | \\ H\ \phi\ H \end{array}$$

Crosslinking may also take place,

$$\begin{array}{ccc} & \wr & \wr \\ & C & C \\ & | & | \\ -C- & O- & C- \\ & | & | \\ & C & C \\ & \wr & \wr \end{array}$$

Oxygen consisted only of a ca. 0.3 monolayer. The adhesive strength of the film/polymer interface was appreciably increased. In order to stabilize these systems, the metal ions in the film have to be complexed[22] or diffusion of metal ions has to be eliminated by a thin coat of an inert metal on the catalyst.[21]

The work which has been discussed so far does not deal explicitly with surfaces or interfaces of one or a few molecules thickness but with surface layers of relatively great depth (ca. 10000 A). In order to study surfaces in a stricter sense, quite different and only recently developed techniques and instrumental methods have to be used.[1]

It may be noted that chain scission determination is actually of an extremely high sensitivity not reached by many of the new instrumental methods. One rupture in every ten original chains can be determined by very simple means. Such scission represents an extremely small chemical reaction.

## INSTRUMENTAL METHODS[23]

A non-exhaustive condensed survey of these methods is given in Table 1. The techniques and instruments as such will not be discussed in any detail as several review articles[23] are available.

As pointed out above, Jellinek et al[21] studied the Cu and Cu-oxide catalyzed oxidation of isotactic polypropylene. One part of these studies was concerned with the formation and decomposition of hydroperoxide groups in absence and presence of catalyst. These reactions were followed by specular reflectance I.R. spectroscopy. A gold layer 320A thick was deposited on a microscope slide by vaporization. On this film, a $CuO_{0.67}$ film, for instance, was deposited (200 A thick). Finally, an isotactic polypropylene film (ca. 13μ thick) was dip-coated on this $CuO_{0.67}$ film (see Figure 8). Such slides were compared with slides without catalysts. The slide was located in a specially constructed thermostated I.R. cell which was filled with 670 torr of $O_2$. The oxidation temperatures were in the range of 90°C - 120°C.

An example of an I.R. reflectance spectrum is shown in Figure

Table I. Some Instrumental Surface Analysis Methods.

| Method | Abbreviation | Surface Depth | Remarks |
|---|---|---|---|
| I.R. Multiple Internal Reflection | IRS | ca. 5 μm | |
| Multiple Attenuated Reflection I.R. Spectrometry | MATR.-I.R. (ATR) | several hundred Å | |
| I.R. Reflexion Absorption (specular reflection) | R.A. | | |
| Surface Enhanced Raman Technique | | ca. 100 Å | |
| U.V. Photoelectron Spectroscopy | U.P.S. | a few tenths μm | |
| Soft X-ray Spectroscopy | | ca. 1 μm | |
| X-ray Photoelectron Spectroscopy | XPS or ESCA | ca. 50 Å | |
| Electron Induced Amber Electron Spectroscopy | AES | | Destructive |
| Secondary Ion Mass Spectroscopy | SIMS | | Destructive, if energetic |
| Ion Scattering Spectrometry | ISS | | Destructive, sensitive to isotopic masses |
| Rutherford Back-scattering | RBS | | $He^+$; 1.8 Mev destructive |
| Neutron and Ion Impact Radiation Scan | ITR | | Destructive |
| Inelastic Electron Tunneling Spectroscopy | IETS | | |
| Scanning Electron Microscopy | SEM | 10-70 nm | |

11. As anticipated, the -OOH concentration reaches a maximum value when the rates of formation and decay have become equal. The oxidation is most accelerated by $CuO_{0.67}$.

Such slides (without Au) were also investigated by ISS by Czanderna et al.[21] It was found that the $CuO_{0.67}$ and CuO substrate films were reduced and reoxidizied during degradation. (See Figures 12 and 13).

The chemical evidence found by Jellinek et al,[21] namely that the reaction mechanism is different in the presence and absence of reaction products (mainly water and carbon dioxide) was confirmed by ISS.[21] These reaction products could be continually removed from the reaction vessel (see Figures 9 and 10) by getters ($P_2O_5$;KOH).

Cu was found in the film by ISS agreeing with Allara's et al.[22] findings for the case of polyethylene.

The studies of Allara et al.[22] on the oxidation of polyethylene films employed some recent experimental methods. These authors

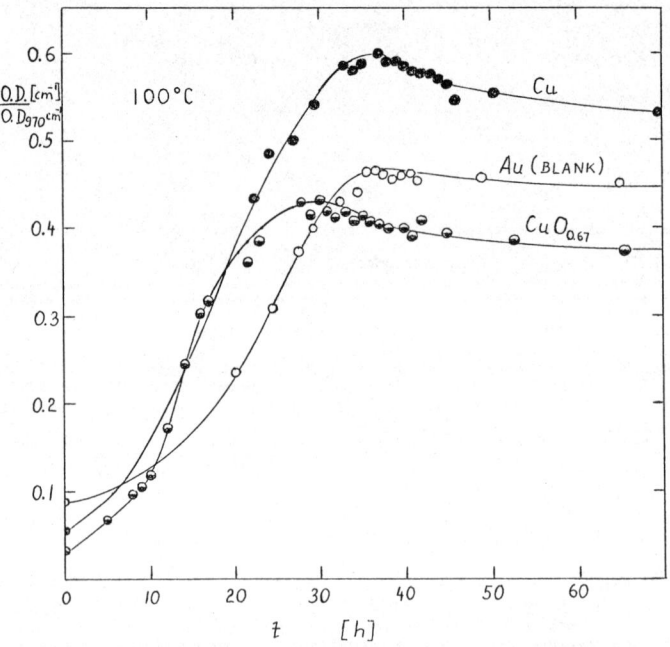

Figure 11. Hydrogen peroxide specular I.R. reflectance Spectra of isotactic polypropylene: $O_2$-pressure, 670 torr, 100°C. Standardized $OD/OD_{970}$ vs. time h. (H.H.G. Jellinek and I. Chodak, et.al., in press).

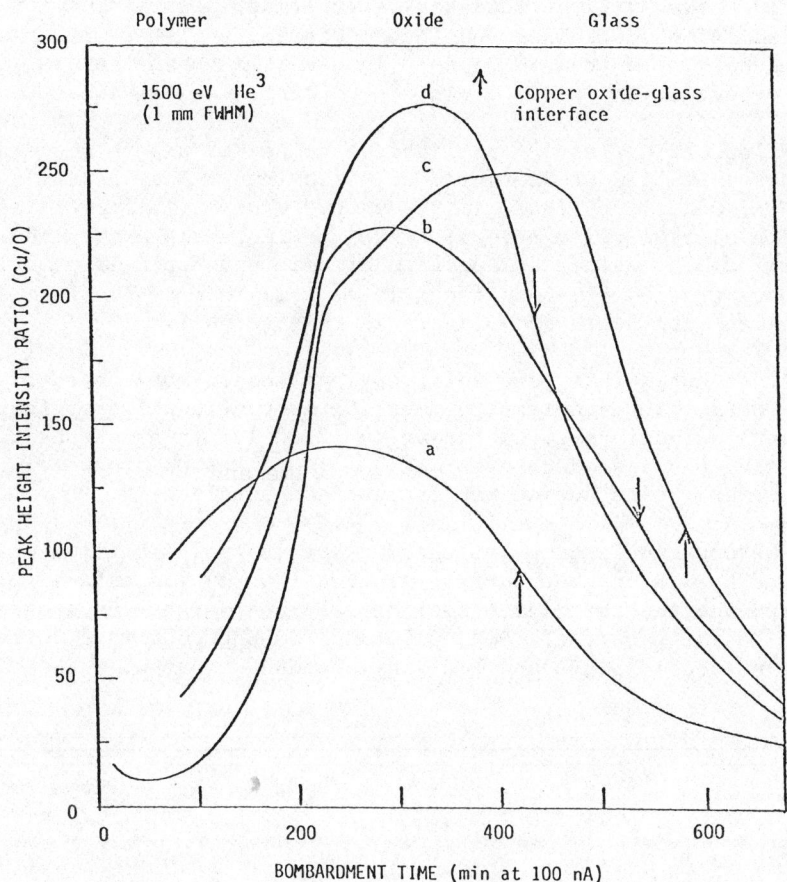

Figure 12. Plots of the Cu/O ratios obtained from isotactic polypropylene films by ISS, degraded in absence of reaction products: (a) non degraded; (b) degraded at 120°C for 80 min; (c) degraded at 120°C for 160 min.; (d) degraded at 100°C for 400 min.[21]

followed the formation of >C=O groups during oxidation by I.R. reflection methods (film thickness 0.15 mm). The depth of penetration of the I.R. radiation was 1.2 μm at 1710 cm$^{-1}$ and 1.4 μm at 1462 cm$^{-1}$. The greatest change in the spectra occurred in the 1700 cm$^{-1}$ to 1800 cm$^{-1}$ >C=O region. Both film surfaces (i.e. Cu/film/film/gas) were measured. The oxidation rate was higher at the metal/polymer interface than at the other surface. Carboxylate ions accumulate at the metal/polymer interface during oxidation and as indicated above.

Cu-ions were found to diffuse from the interface into the polymer film. The carboxylate ions serve as anions. This was directly shown by Rutherford back scattering, neutral and ion impact radiation scattering and laser emission spectroscopy.[22] Cu-ions migrate more than 1000A deep into a film during several hundred hours of oxidation at 40°C. The Cu-concentration is about 0.08 M.

Wiles and Carlsson[24] published a number of papers concerning the photooxidation of isotactic polypropylene. They determined by I.R. ATR, OOH groups (3400 cm$^{-1}$) and C=O groups (1715 cm$^{-1}$) as function of time (film thickness 22μ). Figures 14 and 15 show the normalized O.D. values as function of time and depth of penetration. The concentration of -OOH groups is at a maximum at the film surface. See also Reich and Stivala in this respect.[3]

It is known that pure polypropylene should not absorb radiation >2000A. Nevertheless, commercial isotactic polypropylene (Eastman) actually absorbs radiation >2900 A. This must be due to impurities and also due to formation of oxygenated groups along the polymer chains during manufacture and handling.[24]

Photochemical aging (weathering) of ABS polymer films was also followed by I.R.[18] The surface section (20 μm) was extensively oxidized whereas the middle section remained largely unchanged. On and near the surface, -OH, C=O, and -COOH groups were formed. 1,4 trans double bonds and 1,2 vinyl bonds decreased.

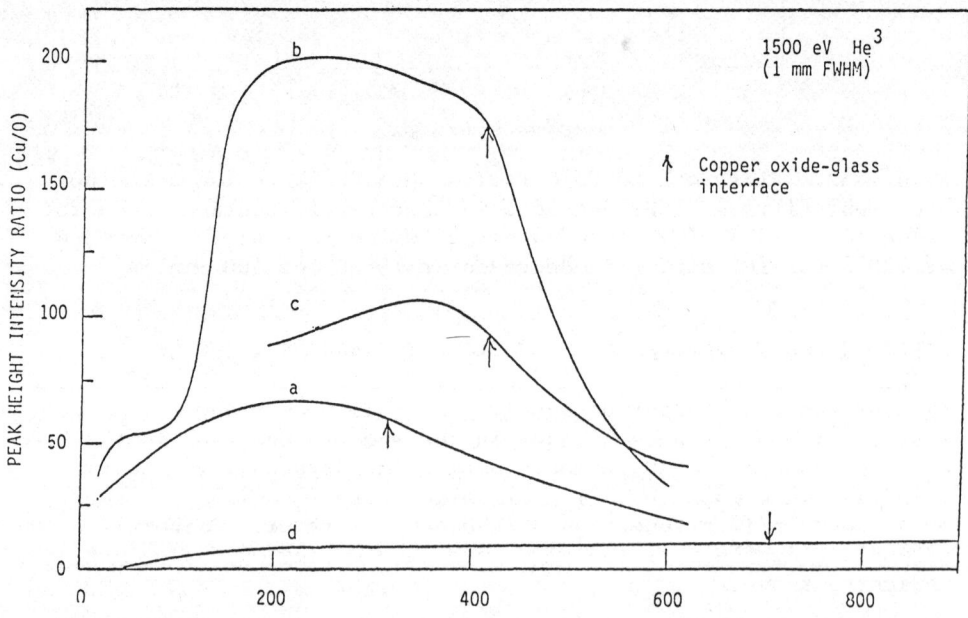

Figure 13. The same as Fig. 12 but in presence of reaction products (i.e. $H_2O$ and $Co_2$).[21]

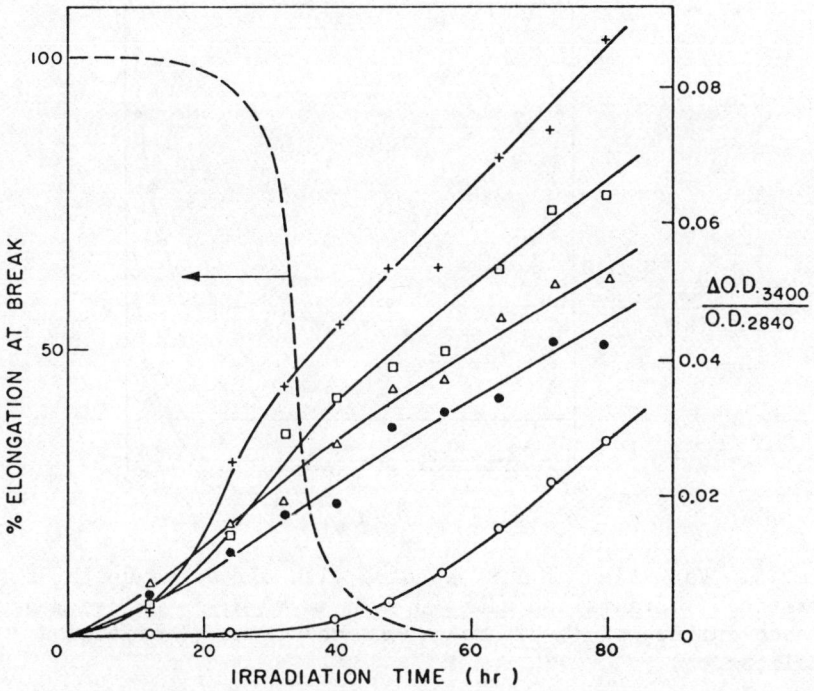

Figure 14. Hydroperoxide I.R. absorption as function of irradiation time. Transmission I.R. ○ ; ATR-depth values $d_e$: ● 0.87μ, △ 0.56μ, ☐ 0.15 ; + 0.12 μ.[24]

Chemical modification of polymer surfaces for obtaining desirable properties is often carried out, especially in adhesion work. High energy surfaces give better adhesion than low energy ones.[14a] This can be achieved, as pointed out above, by surface oxidation.[20] The type of groups formed and the extent of degradation can be followed with the recent instrumental techniques. In addition, surface grafting, etching, exposure to electrical discharge are often applied to the uppermost layers of polymer surfaces. Contact angles show drastic changes due to such treatments (polar groups are formed and the contact angle of polar liquids on such polymers decreases). ESCA is one of the most frequently used methods in this context[23] as it is non-destructive. ESCA has a very potent surface sensitivity. Signals come from the uppermost 50 A of the polymer surface. It is especially suitable for plasma formed polymers, as very thin films can be investigated. Quite often, as shown in this paper, the surface properties of polymers differ from those of the bulk polymer. Co-polymers (especially block co-polymers) are examples, and it is found that one of the blocks lies frequently and preferentially in the surface.

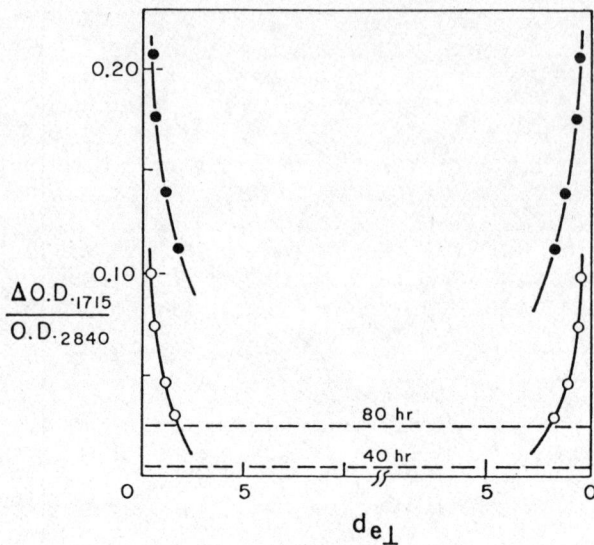

Figure 15. Variation of I.R. spectra with effective depth of I.R. penetration. Irradiation time: ○ 40 h, ● 80 h; $d_e$ values measured from each surface of △ 22μ film. Dashed lines transmission (normalization).[25]

A paper of interest as to the surface properties of polyethylene after treatment has been published by D. Briggs et al.[25] The surface of polyethylene in some cases containing an anti-oxidant was subjected to flame treatment for 1.2 s. The oxidation appeared quite mild and was similar to several treatments by chromic acid. It was confined to the uppermost polymer layer. This could previously not be detected until recently, but now ESCA can be applied. The depth of oxidation was between 40-90 A, much less than obtained by the acid treatment. Lap shear strength was increased by the flame treatment.

Examples studied by ESCA could be multiplied but space does not allow this.[26] (See respective presentations made in this symposium).

A few remarks may be made about ablation and burning.[27] Only polymers which do not char such as Teflon or polymethylmethacrylate will be considered i.e. polymers which depolymerize to monomer. (These are not used any more as ablatives, only the charring ones are taken. However, the reactions of the latter are not surface reactions in the proper sense of the word and cannot serve here as examples).

Ablative polymers are used for the purpose of cooling which is due to the fact that work has to be done at the surface. Endothermic processes take place (e.g. melting, vaporization, thermal decomposition, etc.). Some heat is dissipated by radiation from the very hot surfaces. Pyrolysis of an ablative polymer starts when its decomposition temperature is reached at the surface. Ablation may be different from bulk degradation as the surface is preferentially heated. One refers to 'surface pyrolysis' or "linear pyrolysis". The loss of material from the surface is termed "surface regression".

In this context, some experiments by the author and co-workers[28] with a novel apparatus may be mentioned. PMMA was surface pyrolyzed during burning in an air stream. The energy of activation as function of air flow rate was found to be that for vaporization of the monomer at low flow rates; this energy increased with increasing flow rate, eventually reaching the energy of activation for the chemical depolymerization reaction at high flow rates.

REFERENCES

1. K. L. Mittal, Pure Appl. Chem., 52, 1295 (1980).
2. H.H.G. Jellinek, Editor,"Aspects of Degradation and Stabilization of Polymers", Elsevier, New York, 1978.
3. L. Reich and S. S. Stivala, "Elements of Polymer Degradation", McGraw Hill, New York, 1971.
4. H.H.G. Jellinek, J. Polym. Sci. 14, 1249 (1976).
5. H. Kachi and H.H.G. Jellinek, J. Polym. Sci. 17, 2031 (1979).
6. D. L. Allara, J. Polym. Sci. 14, 93 (1976) and previous papers (see ref. 22).
7. H.H.G. Jellinek, H. Kachi, A. Czanderna and A. C. Miller, J. Polym. Sci. 17, 1493 (1979).
8. H.H.G. Jellinek, H. Kachi, S. Kittaka, M. Lee and R. Yokota, Colloid Polymer Sci. 256, 544 (1978). G. Frankenstein, M. Asce, T. Wuebben, A. M. Asce, H.H.G. Jellinek and R. Yokota, in "Proc. Symp. of Inland Waters for Navigation", Colorado State University, Aug. 10-12, 1976, American Society of Civil Engineers.
9. T. Iwaki and H.H.G. Jellinek, J. Colloid Interface Sci. 69, 17 (1979).
10. D. T. Clark and W. J. Feast, Editors, "Polymer Surfaces", J. Wiley and Sons, New York, 1978.

11. H.H.G. Jellinek, in "Proc. 2nd Internatl. Symp. on Degradation and Stabilization of Polymers", pp. 1-26, Dubrovnik, Oct. 4-6, 1978, see also S. S. Stivala and L. Reich, ibid. pp. 1-14.
12. H.H.G. Jellinek, J. Polym. Sci. 62, 281 (1963).
13. A. R. Shultz, in "Durability of Macromolecular Materials", R. K. Eby, Editor, ACS Symp. Series No. 95, pp. 29-43, American Chemical Society, Washington, D. C. 1979.
14. H.H.G. Jellinek, in "Encycl. of Polym. Sci. and Techn., Depolymerization", Vol. 4, pp. 740-793, J. Wiley & Sons, New York, 1966.
14a. K. L. Mittal, Polymer Eng. Sci. 17, 467 (1977); J. Vac. Sci. Technol. 13, 19 (1976).
15. H.H.G. Jellinek and S. N. Lipovac, Macromolecules 3, Part I, 231 (1970); ibid. Part II, 3, 237 (1970).
16. H.H.G. Jellinek and S. Igarashi, J. Phys. Chem. 74, 1409 (1970).
17. C. R. Boss and J.C.W. Chien, J. Polym. Sci. A-1, 4, 1543 (1966).
18. B. Ranby and J. F. Rabek, "Photodegradation, Photo-Oxidation and Photo-Stabilization of Polymers", J. Wiley & Sons, New York, 1975; also, B. Ranby, "Polymer Surfaces", D. T. Clark and W. J. Feast, Editors, Chap. 18, J. Wiley & Sons, New York, 1978, see also B. J. Bulkin, C. S. Chen and E. M. Pierce, Polym. Prepr. 22(1), 301 (1981).
19. H. A. Willis and V.J.I. Zichy in "Polymer Surfaces", D. T. Clark and W. J. Feast, Editors, Chapter 15, J. Wiley & Sons, New York, 1978. D. M. Brewis and D. Briggs, Polymer 22, 7 (1981).
20. H. Schonhorn, Chapter 10, see Ref. 19.
21. H.H.G. Jellinek, H. Kachi, A. Czanderna and A. C. Miller, J. Polymer Sci., Chem. Ed., 17, 1493 (1979); A. Czanderna, A. C. Miller, H.H.G. Jellinek and H. Kachi, J. Vac. Sci. Technol. 14, 227 (1977).
22. D. L. Allara et al., J. Colloid Interface Sci. 47, 697 (1974), Polym. Eng. Sci. 4, 12 (1974); J. Polym. Sci. Chem. Ed. 14, 93 (1976); in "Characterization of Metal and Polymer Surfaces". L. H. Lee, Editor, Vol. 2, p. 193, Academic Press, New York, 1977; J. Polym. Sci. Chem. Ed. 14, 1857 (1976); J. Catal. 45, 54 (1976); in "Stabilization and Degradation of Polymers", D. L. Allara and W. L. Hawkins, Editor, Adv. Chem. Series, No. 169, p. 293, American Chem. Soc., Washington, D. C. 1978.
22a. J. M. Burkstrand, Appl. Phys. Letters, 33, (5), 387 (1978).
23. A. Dilk, Anal. Chem. 53, 802A (1981); D. T. Clark and W. J. Feast, Editors, "Polymer Surfaces", J. Wiley & Sons, New York, 1978; see also D. W. Dwight and H. R. Thomas, Polym. Prepr. 22, (1) 302 (1981).
24. D. M. Wiles et al. Macromolecules 2, 597 (1969); ibid. 9, 695 (1976); ibid. 12, 1071 (1979).
25. D. Briggs, D. M. Brewis and M. B. Konieczko, J. Materials Sci. 14, 13441 (1979).

26. D. T. Clark, A. Dilks and H. R. Thomas, N. Grassie, Editor, Developments in Polymer Degradation-1", p. 87, Applied Science Publishers, London, 1977, see also H. J. Leary, Jr. and D. S. Campbell, Polymer Prepr. 22(1), 327 (1981).
27. E. C. Straus, in "Aspects of Degradation and Stabilization of Polymers", H.H.G. Jellinek, Editor, Chap. 11, Elsevier, New York, 1978; M. Ladacki, in "Chemistry in Space Research", R. F. Landel and A. Rembaum, Editors, p. 253, Elsevier, New York, 1972.
28. H. Kachi, H.H.G. Jellinek and M. Hall, J. Polym. Sci., Phys. and Phys. Chem. Ed., 19, 1131 (1981).

SURFACE OXIDATION REACTIONS OF UNSATURATED POLYMERS

B. Rånby, J.F. Rabek and J. Lucki

Department of Polymer Technology, The Royal
Institute of Technology, S-100 44  Stockholm,
Sweden

The photodegradation and photo-oxidation of unsaturated polymers initiated by UV irradiation has been studied for polydienes, polynorbornene and unsaturated model polyesters. These polymers are oxidized with various oxygen species, e.g. molecular oxygen (by addition to free radicals), singlet oxygen (by addition to double bonds, the "ene reaction"), atomic oxygen and ozone (by abstraction and addition reactions). Polydienes and polynorbornene are rapidly photo-oxidized with molecular and singlet oxygen, forming different oxygen-containing groups and chain degradation. The reactions are partly known, they start from the surface and proceed to deeper layers. Air pollutants (aromatic hydrocarbons, carbon particles, nitrous oxides) deposited on the polymer surfaces absorb UV quanta and are efficient initiators by energy transfer, e.g. in formation of singlet oxygen. The unsaturated polyesters absorb UV-light without sensitizer added, and they are rapidly photolyzed at the surface layer. Polyesters containing aromatic groups are selfstabilized by formation of ortho-hydroxyester endgroups, which are UV-absorbers, and semiquinone and quinone end-groups, which absorb both UV and short wavelength visible light. In this way the photolyzed surface layer prevents further penetration of the UV-radiation and protects the bulk of the polymer from degradation. This is important for commercial plastic materials, e.g. in glass-fiber reinforced unsaturated polyesters

Surface effects in the photodegradation and photooxidation of polymers are commonly observed. When exposed to sunlight or artificial UV-irradiation most polymer surfaces lose their physicochemical and mechanical properties. Ageing effects of these surfaces can be observed in the form of surface erosion, e.g. formation of cracks, reduced surface gloss, gradual chalking, change in colour, wetting properties and adhesion. Polymeric surfaces, e.g. in spacecrafts, automobiles and buildings in the form of plastics, composites and protective coatings are in particular exposed to sunlight. The most serious effect of increased ageing of polymer surfaces are caused by air pollutants, e.g. smog products. Various photochemical reactions including sensitized photooxidation and photodegradation may occur on exposure of polluted surfaces. Because of the complexity of smog chemistry[2] and the lack of detailed knowledge of the relevant elementary reactions, numerous assumptions and simplifications are made in the mechanistic interpretation of these reactions. Recent results in the laboratory study of accelerated ageing of polymers have made significant contributions to the understanding of the photooxidation reactions in outdoor exposure of various polymers.[3,4]

The most vulnerable polymers are those containing unsaturated bonds in their chain structure, e.g. polydienes, polynorbornene and unsaturated polyesters. Oxidation and photooxidative degradation of these polymers are highly complex. Several oxygen species, e.g. molecular oxygen, singlet oxygen, atomic oxygen and ozone, take part in these reactions[5-7], separately and/or in combination. All of these oxygen species are found in smog products. One aspect of our research program has been directed towards the study of molecular oxygen and singlet oxygen oxidation of polymers with unsaturation (isolated double bonds in the chain). Some of the main results obtained are discussed in this paper in terms of reaction mechanisms related to the photo-ageing in poluted atmosphere.

## Polydienes

Photooxidation of polydienes with molecular oxygen leads to the formation of allylic hydroperoxides according to the process:[6]

$$-CH_2-CH=CH-CH_2- \xrightarrow[-RH]{+R\cdot} -\overset{\cdot}{C}H-CH=CH-CH_2-$$

$$\xrightarrow[PH]{+{}^3O_2} -\underset{\underset{\text{OOH}}{|}}{C}H-CH=CH-CH_2- \qquad (1)$$

The light plays here a role in the following states:

1. Initiation stabe where free radicals R· are formed from the photolysis of internal and external impurities (some of them

may originate from the smog pollutants).

2. Energy transfer processes.

3. Photolysis of the hydroperoxide groups formed.

It is generally agreed that the degradation of macromolecules results from the beta-scission process of polymer oxy radicals which are formed from the photodecomposition of hydroperoxy groups.[8]

Singlet oxygen photooxidation also yields allylic hydroperoxides with shifted double bonds, according to the "ene process":[9,10]

$$-CH_2-CH=CH-CH_2- + {}^1O_2 \longrightarrow -CH=CH-CH-CH_2- \quad (2)$$
$$\qquad\qquad\qquad\qquad\qquad\qquad\qquad\qquad |$$
$$\qquad\qquad\qquad\qquad\qquad\qquad\qquad\qquad OOH$$

It is generally accepted that this reaction is due to singlet oxygen in the form $^1\Delta_g$ with an excess energy 22.5 kcal/mol which is a more long-lived form than the singlet oxygen $^1\Sigma_g$ with the excess energy 37.5 kcal/mol.

Singlet oxygen in both forms may be generated in photosensitized reactions in situ by energy transfer from excited air pollutants.[11]

In the case of cis-1,4-polyisoprene three different additions to double bonds are possible:

$$\qquad\qquad\qquad\qquad\qquad\qquad\qquad\qquad\qquad\qquad\qquad (3)$$

Evidence for these shifts of double bonds were found from the singlet oxidation of model compounds.[12,13]

The singlet oxygen photooxidation of polymers with pendant double bonds depends on the presence of methylene groups in close vicinity to the double bond. For example 1.2-polybutadiene[7] does not react with singlet oxygen whereas 1.2-poly(1,4-hexadiene)[10] is reactive:

$$-CH_2-CH- \;+\; {}^1O_2 \longrightarrow \text{non reactive} \qquad (4)$$
$$\begin{array}{c} | \\ CH \\ \| \\ CH_2 \end{array}$$

$$-CH_2-CH- \;+\; {}^1O_2 \longrightarrow -CH_2-CH- \qquad (5)$$

with side chains: $CH_2-CH=CH-CH_3$ on left, and $CH=CH-CH(OOH)-CH_3$ on right.

Other oxygen species are also reactive with double bonds. Detailed spectroscopical studies show that all oxygen species, i.e. singlet ($^1O_2$) and molecular oxygen ($^3O_2$) atomic oxygen (O) and ozone ($O_3$) react with cis-1,4-polybutadiene forming different oxygen-containing groups by different complex mechanisms.[7] During these oxidation processes chain scission accompanied by simultaneous crosslinking is always observed. The final result is a rapid decrease in all mechanical properties of this polymer. Several copolymers, e.g. ABS[14] and polyblends, e.g. with PVC or EVA[15], which contain polydiene segments as a component, are also vulnerable to photooxidation and degradation, due to the presence of unsaturated bonds.

## Polynorborne

This new commercially available elastomer has found some applications in the building and machine industries. Because polynorbornene molecules contain isolated double bonds, this polymer is susceptible to oxidation and photooxidation processes. The main reactions occur via free radical mechanism.[16] The asymmetric ESR single line spectra show the formation of two or even three different peroxy radicals located at the polynorbornene carbons 2 (or 5), 1 and 3 (or 4). These peroxy radicals are strongly resonance stabilized and long-lived even at temperatures up to 40°C. They abstract tertiary bonded hydrogen atoms in preference to secondary and primary bonded and form hydroperoxide groups. Decomposition of these hydroperoxide groups leads to the formation of polymer alkoxy radicals located at the polynorbornene carbons 2 (or 5), 1 and 3 (or4). Alfa-cleavage of these radicals produce ketone and/or aldehyde groups with opening of the pentane ring or formation of carbonyl groups on the ring according to the reactions:

Formation of these ketone and aldehyde groups was confirmed by IR and ESCA spectroscopy. Photooxidation of polynorborne is accompanied by chain scission and crosslinking reactions. In the solid state gel formation occurs up to the extent of 90%.

Polynorbornene is also susceptible to singlet oxygen[17] yielding hydroperoxides. The final result observed is chain scission and crosslinking.

## Unsaturated polyesters and their composites

The photooxidation of aliphatic unsaturated polyesters is accompanied by extensive chain scission, crosslinking and yellowing.[18] Coatings based on mixed unsaturated- aromatic polyesters are also susceptible to photooxidative degradation. On exposure to sunlight, there are gradual changes in chemical and physical properties, starting at the surface to give microcracks and chalking.

Unsaturated polyesters obtained by polycondensation of propylene glycol with maleic acid and ortho-phthalic anhydride (50:50) are photooxidized by molecular oxygen whereas they are completely non-reactive with singlet oxygen. The photolysis of unsaturated polyesters shows disappearance of double bonds and partial decomposition of ester carbonyl groups to CO and $CO_2$ molecules. This type of polyesters contains a distinct conjugated system of double bonds and carbonyl groups, which absorbs UV light and exists in resonance form:

$$\begin{matrix} \text{O} & \text{O} \\ \| & \| \\ -\text{C}-\text{CH}=\text{CH}-\text{C}- \end{matrix} \rightleftarrows \begin{matrix} \text{O} & \text{O} \\ | & | \\ -\text{C}-\text{CH}-\text{CH}-\text{C}- \end{matrix} \qquad (7)$$

Excitation of this system provides formation of intermediate biradicals in the resonance form:

$$\begin{matrix} \text{O} & \text{O} \\ \| & \| \\ -\text{C}-\text{CH}=\text{CH}-\text{C}- \end{matrix} \xrightarrow{h\nu} \begin{matrix} \cdot \text{O} & \text{O} \\ | & \| \\ -\text{C}-\text{CH}=\text{CH}-\text{C}- \\ \cdot \end{matrix} \rightleftarrows \begin{matrix} \text{O} & \text{O} \\ \| & \| \\ -\text{C}-\overset{\cdot}{\text{CH}}-\overset{\cdot}{\text{CH}}-\text{C}- \end{matrix} \rightleftarrows \begin{matrix} \text{O} & \text{O} \\ | & | \\ -\text{C}-\text{CH}-\text{CH}-\text{C}- \end{matrix} \qquad (8)$$

The spin components of the two excited electrons are neither "paired" nor "unpaired", they are independent and delocalized. The observed rapid crosslinking is interpreted as due to reactions of the intermediate radicals with a segment of the same or neighbouring macromolecules:

$$(9)$$

A primary photolytic chain scission occurs by the Norrish Type I reaction which gives formation of the CO and $CO_2$ observed:

## SURFACE OXIDATION REACTIONS OF UNSATURATED POLYMERS

$$\begin{array}{c}
\phantom{xxxxxxx}-O-\overset{O}{\underset{\|}{C}}-\underset{|}{CH}-\overset{\bullet}{CH}\; +\; CO\uparrow\; +\; \bullet O-\underset{|}{CH}-CH_2-O-\\
\phantom{xxxxxxxxxxxxxxxxxxx}CH_3\\[4pt]
\phantom{xxxxxx}-O-\overset{O}{\underset{\|}{C}}-\underset{|}{CH}-CH-\overset{O}{\underset{\|}{C}}\bullet\; +\; \bullet O-\underset{|}{CH}-CH_2-O-\\
\phantom{xxxxxxxxxxxxxxxxxxxxxxx}CH_3
\end{array}$$

$$-O-\overset{O}{\underset{\|}{C}}-\underset{|}{CH}-CH-\overset{O}{\underset{\|}{C}}-O-\underset{|}{CH}-CH_2-O-\qquad\qquad(10)$$

$$\begin{array}{c}
-O-\overset{O}{\underset{\|}{C}}-\underset{|}{CH}-CH-\overset{O}{\underset{\|}{C}}-O\bullet\; +\; \bullet \underset{|}{CH}-CH_2-O-\\
\phantom{xxxxxxxxxxxxxxxxxxxx}CH_3\\[4pt]
-O-\overset{O}{\underset{\|}{C}}-\underset{|}{CH}-\overset{\bullet}{CH}\; +\; CO_2\downarrow\; +\; \bullet \underset{|}{CH}-CH_2-O-\\
\phantom{xxxxxxxxxxxxxxxxxxxx}CH_3
\end{array}$$

The Norrish Type II intermolecular rearrangement yields terminal carboxylic groups:

$$-O-\overset{O}{\underset{\|}{C}}-CH=CH-\overset{O}{\underset{\|}{C}}-O-\underset{\underset{CH_3CH_2-O}{\diagdown}}{C}\overset{H\cdots O}{\diagup}\overset{}{\underset{\|}{C}}-CH=CH-\overset{O}{\underset{\|}{C}}-O-\;\xrightarrow{h\nu}\qquad(11)$$

$$\longrightarrow\; -O-\overset{O}{\underset{\|}{C}}-CH=CH-\overset{O}{\underset{\|}{C}}-O-\underset{\underset{CH_3}{\diagup}}{C}=CH_2\; +\; HO-\overset{O}{\underset{\|}{C}}-CH=CH-\overset{O}{\underset{\|}{C}}-O-$$

The photooxidation mechanism occurs via free radical mechanism with formation of polymer peroxy and alkoxy radicals and hydroperoxide and carbonyl groups. Formation of these groups was confirmed from IR spectra.

Of particular interest is the photooxidation of aromatic polyesters, studied by model compounds of poly(propylene-1,2-phthalates).[18] Chain scission gives radical end-groups (A), which by a normal sequence of $O_2$ addition, hydrogen abstraction, hydroperoxide decomposition and again hydrogen abstraction give orthohydroxyester end-groups (B), which absorb UV-light and act as stabilizers (C and D):

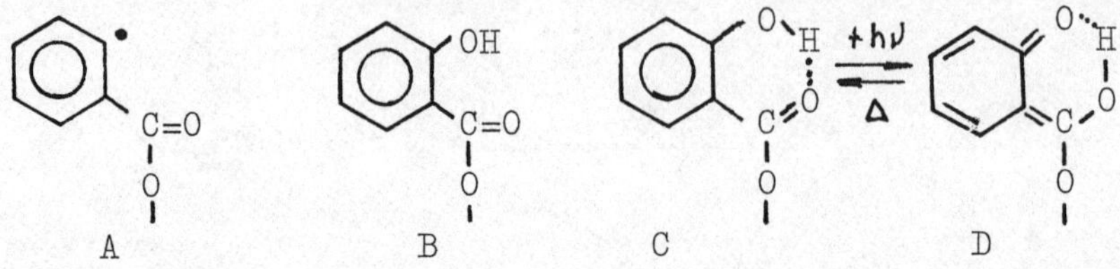

[Structures A, B, C, D shown with equilibrium C ⇌ D via +hν/Δ]  (12)

The same cyclic mechanism (C ⇌ D) is effective in commercial photostabilizers like phenyl esters of benzoic acid, which undergo Fries rearrangement, and ortho-hydroxy-benzophenones. Photodegraded surfaces of aromatic polyesters in this way become photostabilized to some extent, which limits further penetration of the photolysis.

Photolyzed aromatic polyester model compounds turn yellow.[18] From UV and IR absorption spectra, the formation of conjugated phenyl rings with ortho-substituted ester groups are indicated. The following mechanisms for formation of quinone and diquinone groups are proposed, starting from the alkoxy radical endgroup (A):

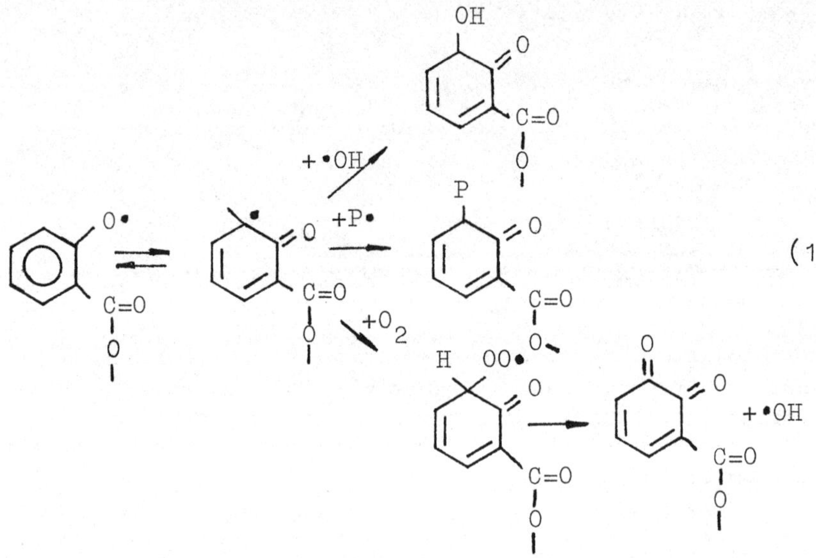

[Reaction scheme showing alkoxy radical progressing through +•OH, +P•, +O$_2$ pathways to form quinone and diquinone structures, producing +•OH]  (13)

Such quinone and diquinone groups absorb UV light with a tail into visible light, i.e. they form a light screen at the surface, which decreases further photolysis of deeper layers of the polyester.

Commercial polyesters containing both unsaturated and aromatic groups in the chains are crosslinked by styrene which reacts with some of the chain double bonds. It seems probable that commercial polyester curfaces, when photooxidized, develop the same stabilizing and light-absorbing groups as indicated for the aromatic model polyester studied in one project.[18]

## Other unsaturated polymeric systems

Several commercially produced polymers, e.g. polyolefins and poly(vinylchloride), may contain some small amounts of unsaturated bonds because of side-reactions and variations in the production and processing of the materials.

There is evidence that terminal vinyl groups in polyolefins may react with singlet oxygen:[19]

$$CH_2=CH-CH_2-CH_2- + {}^1O_2 \longrightarrow HOO-CH_2-CH=CH-CH_2- \qquad (14)$$

Similarily has been reported that polyene sequences formed during thermal and photochemical dehydrochlorination of poly(vinylchloride) are reactive with singlet oxygen.[20] In an oxidative atmosphere coloured conjugated sequences are bleached, due to attack of oxygen on long polyenes which reduces the length of conjugated chain sequences.

## CONCLUSIONS

Unsaturated polymers are photochemically degraded and oxidized mainly by radical mechanisms and addition of molecular oxygen ($^3O_2$). Singlet oxygen ($^1O_2$) oxidation is limited to addition to isolated double bonds with at least one vicinal $CH_2$ group (the "ene reaction"). Excited states and free radicals in the polymer structure are usually formed by energy or radical transfer mechanisms from internal impurities (modified groups) or external impurities (additives or pollutants). Several air pollutants are essential as sensitizers in photolysis of polymer surfaces which are attacked in primary reactions. The increasing photochemical air pollution on a global basis make commercial polymeric materials susceptible to ageing when exposed outdoors. The photolysis of several important rubber and plastic materials is reviewed from results obtained for model compounds. When photo-oxidized, aromatic polyesters form photo-stabilizing and lightabsorbing endgroups which decreases the rate of further reactions in deeper layers of the polymer.

## REFERENCES

1. B. Rånby, in "Polymer Surfaces" D.T. Clark and W.J. Feast, Editors p. 381, Wiley, Chichester, England, 1978.
2. "Photochemical Smog and Ozone Reactions", Adv. Chem. Ser. No. 113, American Chemical Society, Washington, D.D., 1972.
3. N.S. Allen, Editor, "Development in Polymer Photochemistry", Vol. 1 and 2, Applied Science Publ., London, 1980 and 1981.
4. N. Grassie, "Development in Polymer Degradation", Vol. 1-3, Applied Science Publ., London, 1980-1981.
5. B. Rånby and J.F. Rabek,"Photodegradation, Photooxidation and Photostabilization of Polymers", Wiley, London, 1975.
6. J.F. Rabek, J. Lucki and B. Rånby, Europ. Polym. J., 15, 1089 (1979).
7. J. Lucki, B. Rånby and J.F. Rabek, Europ. Polym. J., 15, 1101 (1979).
8. D. Lala and J.F. Rabek, Europ. Polym. J., 17, 7 (1981).
9. J.F. Rabek and B. Rånby, J. Polym. Sci., 14, 1463 (1976)
10. M.A. Golub, NASA Technical Memorandum No. 78604, June 1979.
11. J.F. Rabek and B. Rånby, Photochem. Photobiol., 28, 557 (1978)
12. H.C. Ng and J.E. Guillet, Photochem. Photobiol., 28, 571 (1978).
13. M.A. Golub, M.L. Rosenberg and R.V. Gemmer, Rubber Chem. Technol., 50, 704 (1977).
14. J.F. Rabek and B. Rånby, J. Appl. Polym. Sci., 23, 2481 (1979).
15. D. Lala, J.F. Rabek and B. Rånby, Polym. Degrad. Stabil., 3, 307 (1980-81).
16. S.K. Wu, J. Lucki, J.F. Rabek and B. Rånby, Polym. Photochem. 2, 73 (1982).
17. S.K. Wu, J. Lucki, J.F. Rabek and B. Rånby, Polym. Photochem. 2, 125 (1982).
18. J. Lucki, J.F. Rabek, B. Rånby and C. Ekström, Europ. Polym. J., 17, 919 (1981).
19. A.M. Trozzolo and F.H. Winslow, Macromolecules, 1, 98 (1968).
20. J.F. Rabek, B. Rånby, B. Östensson and P. Flodin, J. Appl. Polym. Sci., 24, 2407 (1979).

PHOTOOXIDATIVE DEGRADATION OF CLEAR ULTRAVIOLET ABSORBING ACRYLIC

COPOLYMER SURFACES

A. Gupta and R. H. Liang

Materials Research & Biotechnology Section, Jet
Propulsion Laboratory, California Institute of
Technology, Pasadena, California 91109

4800 Oak Grove Drive, Pasadena, California 91109

O. Vogl and W. Pradellok

Department of Polymer Science and Engineering,
University of Massachusetts, Amherst, MA 01003

A. L. Huston and G. W. Scott

Department of Chemistry, University of California,
Riverside, Riverside, California 92521

Photodegradation of copolymer of methyl methacrylate and 2[2´-hydroxy 5´vinyl-phenyl] 2H-benzotriazole has been investigated in order to determine the changes in the chemical composition of the surface of the copolymer on photooxidation. An electronic energy transfer mechanism has been postulated in order to interpret the observed photochemical changes in the polymer. Preliminary examination of the photophysical properties of the chromophore provides support for such a mechanism.

## INTRODUCTION

A potential route to weatherization of plastics is to coat them with a layer of transparent, ultraviolet absorbing polymer which is itself weatherable. A promising candidate is a copolymer of methylmethacrylate and a vinyl derivative of an ultraviolet absorber chromophore. If the concentration of the chromophore does not exceed 10-15 mole percent, the copolymer is likely to retain the mechanical and optical properties of PMMA for certain classes of ultraviolet chromophores. Hence, an investigation of the photostability of copolymers of this type was undertaken in order to determine the feasibility of development of UV absorbing transparent coatings.[1-2]

The outdoor photochemistry of these copolymers is expected to be initiated by energy absorption by the chromophores. In previous reports[3-6] we have shown that the lifetime of the reactive excited states is the critical parameter which controls the photoreactivity of the chromophores. From this point of view, presence of low lying excited singlet or triplet states is not a prerequisite for enhanced photostability. However, intersystem crossing processes generating triplet excited states should be inhibited for greater photostability since triplet states are relatively long lived. Flash spectroscopic studies and quantum yield measurements on 2-hydroxybenzophenone and its polymeric derivatives demonstrated that in nonprotic solvents, decay of transient absorbance occurs at a rate of $10^{11}$ sec$^{-1}$ triplet yield being negligible. However, in protic solvents, e.g., ethnanol, 2-hydroxybenzophenone has a small triplet yield and transient absorbance persists for 400 ps or more[6-8]. Photodegradation studies on a methyl methacrylate copolymer containing pendant 2-hydroxy benzophenone derivatives showed that a barely detectable but still significant triplet-triplet electronic energy transfer occurs in glassy films of this copolymer. Photo-sensitized hydrogen abstraction occurs at branch points where the chromophore is attached to the polymer backbone causing a gradual change in molecular weight distribution as aging progresses. Preliminary studies on 2(2´-hydroxy 5´-methylphenyl) benzotriazole (herein referred as I) derivatives demonstrated that ground state bleaching recovery is essentially complete after 225 ps (3 half lives) in EtOH solution at room temperature and 90 ps in a nonprotic solvent ($CH_2Cl_2$) following pulsed excitation at 335 nm[10]. These results indicated that there is no significant triplet production in this system in either solvent at room temperature. Triplet formation was inferred from phosphorescence observed at 90°K in this system by Kramer[9]. We observed fluorescence from I at room temperature in EtOH. However, the quantum yield is exceedingly low ($\sim 2 \times 10^{-5}$), and the decay time is found to be 17 ps, significantly shorter than the ground state bleaching recovery rate ($25 \pm 5$ ps)[10]. Radiationless transition in this and other systems is believed to be mediated by an intramolecular proton transfer process[10-16]. It has been suggested that the ground state is exclusively O-protonated, to which may be assigned the 302 nm band in the electronic absorption spectrum of I in fluid solutions, in analogy with the assignment of the lowest absorption band in methoxy and acetoxy derivatives of 2-hydroxy phenylbenzotriazole. However, the assignment of the 340 nm band requires the postulation of a ground state equilibrium between molecular populations in two quite substantial potential energy wells, since relative peak intensities in the absorption spectra at 302 and 340 nm change by only about 50% or less when the solvent is changed from methylcyclohexane to ethanol and finally to trifluoroethanol[15]. One possibility is that there is a tautomeric equilibrium between two O-protonated species, II and III as shown in Scheme 1.

This equilibrium may be perturbed to the left by protic solvents, with the assumption that the structure II is non-planar, due to the plane of the phenyl group being normal to the plane of the benzotriazole-like structure, III represents a planar structure which has a strong intramolecular hydrogen bond leading to extra stabilization of its first excited state through contribution from change resonance structures. Whether these schematic representations are accurate or not will be tested by performing detailed spectroscopic studies on the model system[11].

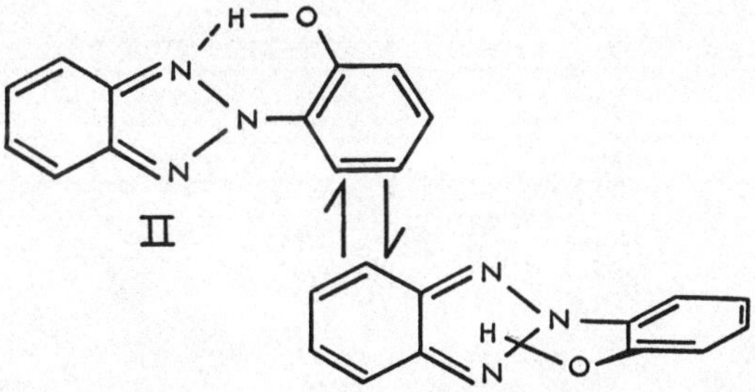

Scheme 1

A careful search for wavelength dependence of quantum yields of various photoprocesses at room temperature is also called for.

EXPERIMENTAL RESULTS AND DISCUSSION

Copolymer I was synthesized according to a synthetic scheme described previously by one of us[2,17]. This is given in Scheme 2.

A 15 mole percent copolymer was selected for evaluation. Electronic and IR spectra of the copolymer in fluid solution and in neat, thin films solvent cast from methylene chloride on pyrex plates were obtained. Films of the copolymer were then subjected to photodegradation by mounting free standing films on the surface of a "merry-go-round" type reactor[18], exposing them to filtered

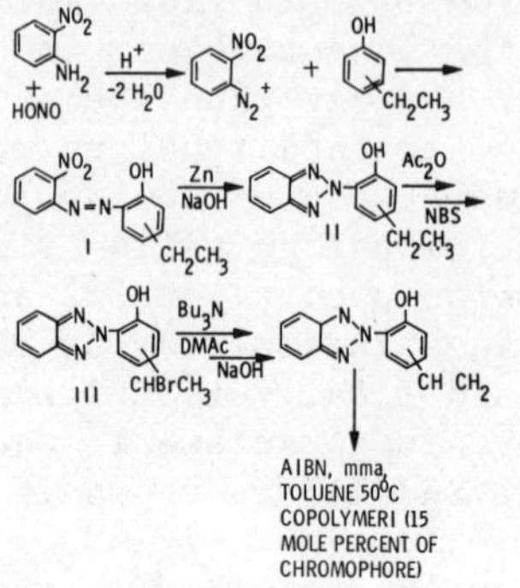

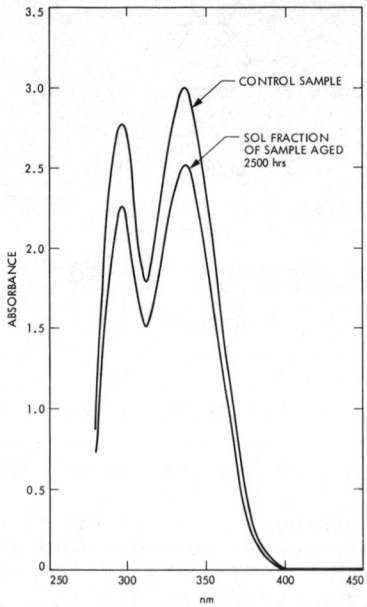

Scheme 2

(water and pyrex) ultraviolet flux from a medium pressure Hg lamp. Actinometry was performed using o-nitrobenzaldehyde as the actinometer[3]. Table I shows aging time, photon fluxes and the number of photons incident per chromophore unit. Absorption spectra, IR spectra (bulk and ATR) were recorded at these aging times. Figure 1 shows absorption spectra of identical concentrations of filtered solutions of control sample and a sample aged for 2500 hours.

Figure 1. Electronic absorption spectra of fluid solution of the solution of the copolymer solvent: $CH_2Cl_2$, 25°C.

Table I. Radiation Flux Deposited on Copolymer Films; Greater than 80% of All Radiation was Absorbed in the Given Wavelength Range.

| Aging Time (hr) | Photon Flux Einsteins/Sq. cm. (295-385 nm) | Photons Incident Per Chromophore Molecule |
|---|---|---|
| 750  | 1.10 | $8.00 \times 10^4$ |
| 1590 | 1.48 | $1.85 \times 10^5$ |
| 1770 | 2.45 | $2.55 \times 10^5$ |
| 2040 | 2.96 | $2.65 \times 10^5$ |
| 2170 | 3.12 | $2.70 \times 10^5$ |
| 2580 | 3.71 | $2.19 \times 10^5$ |
| 3700 | 5.33 | $4.00 \times 10^5$ |

These spectra show that the sol fraction in the aged sample remains chemically unchanged, at least as far as the chromophore is concerned. There is probably a slight increase in long wavelength absorption, but this difference in absorbance is barely outside the experimental error limits. However, the intensity of the absorption bands is decreased by approximately 20 ± 5% which could be a measure of the gel content. Figure 2 shows a plot of normalized absorbance of solutions of aged films relative to the control sample as a function of wavelength. No change is observed.

Molecular weight distribution of the copolymer were estimated from GPC measurements carried out on a high performance liquid chromotograph equipped with molecular exclusion type polystyrene columns. Slight gel formation was noted in samples aged for longer aging periods. Absorption spectra of films were obtained on aged and control samples. These data indicate that the absorption spectra in fluid solution remain unchanged, while in the films there is a slight increase in absorbance at 450-500 nm region for longer aging periods. Small changes observed in absorbance spectra of the aged films contained contributions from increases in back scattering, since scattering losses were observed in hemispherical transmission spectra. Gel fraction of the sample aged

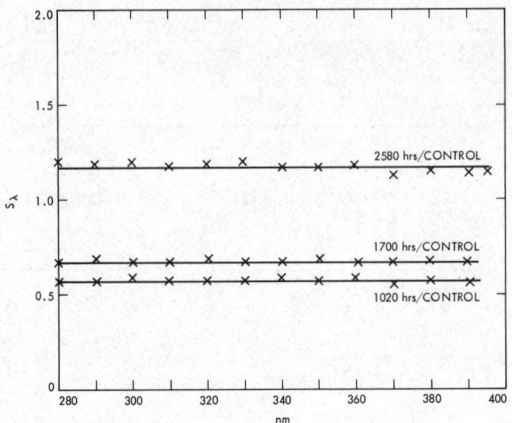

Figure 2. Ratio of absorbance of fluid solutions of aged samples to that of control samples as a function of wavelength, $CH_2Cl_2$, 25°C, $S_\lambda = a_\lambda$ sample/$A_\lambda$ control.

for 3500 hours was estimated to be 35 percent or less.

Figure 3 shows transmission IR spectra on aged and control samples; computerized subtraction shows no significant change in peak areas in any of the samples after the spectra are normalized for film thickness. Figure 4 shows relfectance IR spectra on the same films. A systematic increase in absorbance in the 3200 $cm^{-1}$ region is noted, indicating that hydroxyl groups are being formed

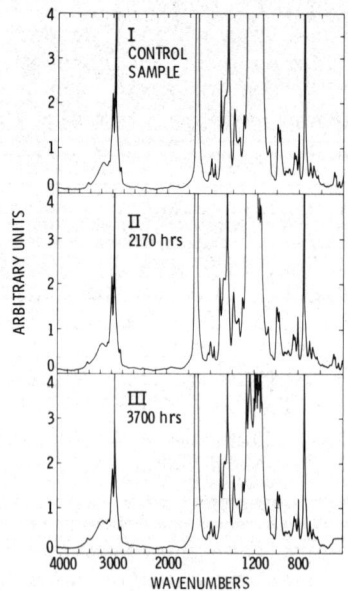

Figure 3. Transmission IR spectra of copolymer films.

# PHOTOOXIDATIVE DEGRADATION

at the surface due to photooxidation. The IR peaks due to the chromophore apparently remain unchanged, although some changes in C-H peak amplitudes may be noted. Contact angles were measured for a set of reference liquids and surface energy was estimated for both surfaces (exposed to the light and the back surface) at above-mentioned aging times. Table II shows these data.

Table II. Surface Tension ($\gamma_c$) of Samples Measured as a Function of Aging Time.

| Aging Time (hr) | Exposed Surface | Back Surface |
| --- | --- | --- |
| 0 | 38-39 | 36-42 |
| 750 | 40-44 | 20-24 |
| 1000 | 45 | 29-30 |
| 1770 | 55 | 24-39 |
| 2040 | 59 | 27-38 |
| 2170 | 50 | 25-32 |

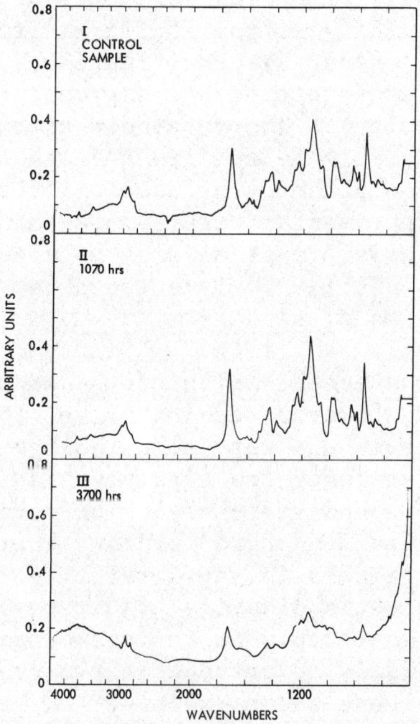

Figure 4. Relfectance (ATR) Ir Spectra of Copolymer Films.

The surface energy of the other surface does not change significantly. Table III gives molecular weights and molecular weight distributions of aged samples.

Table III. Molecular Weight Distribution Data on Aged and Control Samples.

| Sample | $\overline{M}_n$ | $\overline{M}_w$ | $\overline{M}_n/\overline{M}_w$ |
|---|---|---|---|
| Control #1 | 93,000 | 188,000 | 2.03 |
| 750 hrs. | 98,000 | 202,000 | 2.05 |
| 1700 hrs. | 75,000 | 273,000 | 3.62 |
| 2170 hrs. | 75,000 | 273,000 | 3.62 |
| | | | |
| Control #2 | 64,000 | 120,000 | 1.90 |
| 3700 hours[1] | 69,000 | 171,000 | 2.49 |

[1] Gel formation noted; approximate estimate of gel formation is 20% in the sample aged for 2170 hours and ~ 30% in the sample aged for 37000 hours.

Absorption recovery kinetics were obtained using a single third harmonic pulse ($\Delta t \simeq 10$ ps) at 355 nm from a modelocked $Nd^{3+}$: silicate glass laser system (Figure 5). Approximately 90% of the UV pulse was used to bleach the ground state absorption of the sample (0.3 mJ/pulse). The remainder of this pulse (probe pulse) followed a variable delay line and was used to monitor changes in sample absorption as a function of time following excitation. Complete experimental details are being published elsewhere[10]. Figure 6 shows a typical plot obtained on the copolymer at room temperature. Table IV gives the decay lifetimes measured on the copolymer and model systems.

Synthesis and characterization of the copolymer followed published procedures, however, the nature of the ground and excited state equilibria are not yet established. The lifetime of the excited singlet state of the copolymer is presumably less than or equal to 25 ps. Ground state bleaching recovery data on the model compounds and its deuterated hydroxy analog demonstrate that the proton transfer process is rapid relative to the observed decay kinetics. The extraodinarily high rate of the proton transfer process suggests that an incipient preformed hydrogen bond O--H···N may already be present in the ground state regardless of the solvent environment.

These data indicate that the photooxidation is occurring only

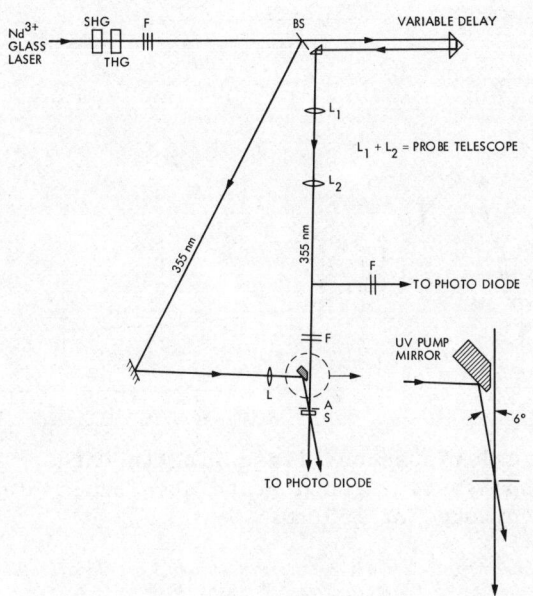

Figure 5. Schematics of the Picosecond Flash Kinetic Set Up for Measurement of Ground State Bleaching Recovery Rates.

Table IV. Ground State Bleaching Recovery Lifetimes of the Co-polymer and Model Chromophores.

| Sample | Solvent/Temp. | Ground State Bleaching Recovery Lifetime |
|---|---|---|
| [2(2´-hydroxy 5´-vinyl-phenyl) benzotriazole co [mma] | $CH_2Cl_2$m 25°C | 25 ± 5 ps |
| 2(2´-hydroxy 5-methyl-phenyl) benzotriazole | MCH, 25°C | 33 ± 5 ps |
| --do-- | $CH_2Cl_2$, 25°C | 29 ± 5 ps |
| --do-- | EtOH, 25°C | 74 ± 6 ps |
| 2(2´-deutrooxy 5-methyl-phenyl) benzotriazole | MCH, 25°C | 22 ± 3 ps |

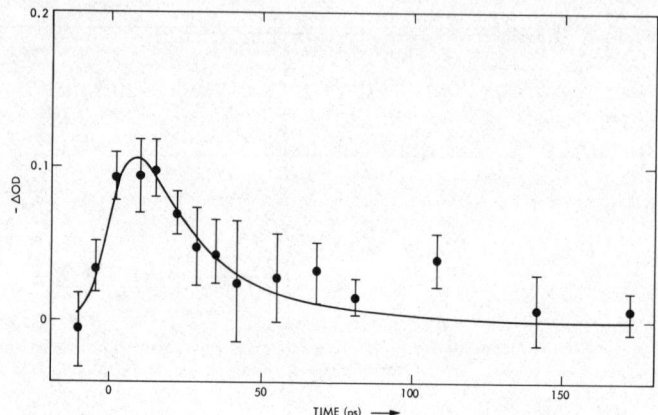

Figure 6. Typical picosecond flash kinetic data on fluid solutions of the copolymer. Ground state absorbance bleaching recovery is being monitored at 355 nm.

at the exposed surface. It is not possible to rule out complete absorption of light by hydroperoxy groups at the surface monolayer however, it can be estimated that the total photon flux directly absorbed by the hydroperoxy groups is so small that it cannot by itself account for the observed photooxidation. Further kinetic studies of the change in surface energy occurs over a relatively short period (first 800 hours). The rate of photooxidation then slows down as the potential sites of photooxidation (hydroperoxy groups and tertiary hydrogens) situated at the surface are consumed. Hence the sensitization mechanism must be presumed to operate at the surface.

It may be expected that photooxidation at the surface of these copolymers may be inhibited by applying an inert surface layer. This was achieved by blending the copolymer with pure PMMA and solution casting the blend. When solutions cast from methylene chloride, the blend forms an outer layer of pure PMMA. The rate of photooxidation is reduced to virtually zero in these blends, as determined from preliminary experiments.

## ACKNOWLEDGEMENTS

This paper represents one phase of research carried out at the Jet Propulsion Laboratory, California Institute of Technology, sponsored by the National Aeronautics and Space Administration, Contract NAS7-100. Work at JPL and University of California, Riverside was supported by the RAD task of the Solar Thermal Power Systems project sponsored by the Department of Energy. Work at the University of Massachusetts was supported by the FSA project, Environmental Isolation Task, also sponsored by the Department of Energy.

## REFERENCES

1. A. Gupta, R. H. Liang, J. Moacanin, D. Kliger, R. Goldbeck, J. Horwitz, and V.M. Miskowski, Europ. Polym. J., 17, 484 (1981).
2. S. Yoshida and O. Vogl, ACS Polymer Preprints, 22(1), 201 (1980).
3. A. Gupta, A. Yavourian, S. Di Stefano, C. D. Merritt and G. W. Scott, Macromolecules, 13, 821 (1980).
4. A. Gupta, D. Kliger and G. W. Scott, in "Proceedings of the International Symp. on Photodegradation, Photooxidation and Photostabilization of Organic Coatings", ACS Symposium Series No. 151, p. 27, Amer. Chem. Soc., Washington, D. C., 1981.
5. A. Gupta, R. H. Liang, J. Moacanin, R. Goldbeck and D. Kliger, Macromolecules, 13, 212 (1980).
6. C. D. Merritt, G. W. Scott, A. Gupta and A. Yavrouian, Chem. Phys. Lett., 68, 169 (1980).
7. S. Y. Hon, W. M. Heterington, G. M. Korenowski, K. B. Eisenthal, Chem. Phys. Lett., 68, 282 (1980).
8. A. A. Lamola and P. J. Sharp, J. Phys. Chem., 70, 2634 (1966).
9. T. Werner, G. Wossner and H. E. A. Kramer, in "Proceedings of the International Symp. on Photodegradation, Photooxidation and Photostabilization of Organic Coatings", ACS Symposium Series No. 151, Amer. Chem. Soc., Washington, D. C., 1981.
10. A. J. Huston, G. W. Scott and A. Gupta, to be published.
11. H. Shizuka, K. Matsui and I. Tanaka, J. Phys. Chem., 81, 2243 (1977).
12. T. Werner, J. Phys. Chem., 83, 320 (1979).
13. W. Klopffer, Advances in Photochem., 10, 311 (1977).
14. H. J. Heller and H. R. Blattman, Pure Appl. Chem., 36, 141 (1974).
15. J. A. Otterstedt, J. Chem. Phys., 58, 5716 (1978).
16. A. Huston, G. W. Scott and A. Gupta, Unpublished results.
17. G. Yoshida and O. Vogl, Makromol. Chem., in press.
18. F. G. Moses, R. S. H. Lin and B. M. Monroe, Mol. Photochem., 1, 245 (1969).
19. R. W. Anderson, Jr., D. E. Damschem, G. W. Scott and J. D. Tally, J. Chem. Phys., 71, 1134 (1979) and references therein.
20. W. Pradellok, A. Gupta and O. Vogl, J. Polymer Sci., Polymer Chem. Ed., in press.

# INTERACTIONS BETWEEN SEVERAL RADIATION SOURCES AND CERTAIN POLYMER SURFACES; REFLECTANCE - TRANSMITTANCE CHARACTERISTICS

J. R. Hallman  
Lakeland Community College  
Mentor, OH 44060

C. M. Sliepcevich  
University of Oklahoma  
Norman, OK 73069

J. R. Welker  
Applied Technology Corporation  
Norman, OK 73069

During ignition testing of several polymeric materials, it was found that the surface reflectance/absorptance characteristics had a definite influence on the time of ignition. Since several of the polymers were either transparent or translucent, the monochromatic surface reflectance/absorptance had to be adjusted to account for energy transmittance. As a result, reflectance and transmittance measurements were made over the monochromatic wavelengths of 0.3 - 10.0 microns. Average values of reflectivity and transmissivity were calculated by the equation:

$$(f)_{av} = \frac{\int f_\lambda e_\lambda \, d\lambda}{\int e_\lambda \, d\lambda}$$

where $(f)_{av}$ is either the average reflectivity, $r_{av}$, or average transmissivity $\tau_{av}$, where $f_\lambda$ is either the monochromatic reflectance $r_\lambda$ or the monochromatic transmittance $\tau_\lambda$. $\lambda$ is the wavelength and $e_\lambda$ is the monochromatic emissive power of the heat source. Average values of reflectivity and transmissivity were calculated for several heat sources; radiant panel, carbon arcs (2900°K, 3900°, and 6000°K), solar radiation, benzene flame, and blackbody radiation. Plots of average reflectivity and transmissivity are illustrated, and charts of transmittance vs. wave length, and reflectance vs. wave length are included.

## INTRODUCTION

Polymeric materials are being used in ever-increasing amounts in many varied applications which has caused, and is causing, much concern to those involved in fire prevention and control. Numerous methods have been developed to investigate ignition phenomena, burning rates, radiant heat transfer and other associated combustion characteristics. In certain of these studies, it was found that ignition times were dependent upon both the type and amount of the energy source, and the changing surface characteristics of the material exposed. During testing, very diversified results were obtained in the determination of the ignition times of polymers and rubbers by Hallman, et. al.[1-3].

Investigations of surface characteristics of opaque polymers have indicated that energy absorption at the surface is a function of the polymer composition, and the surface changes occurring during heating. These studies and results have been previously reported for several polymers and rubbers.[4-6]

However, tests on transparent materials gave different results, and are being presented in this paper for use by those who have need of specific polymer reflectance and transmittance data. The information is presented as average reflectivity and transmissivity for certain transparent/translucent polymers when subjected to certain sources of radiation, including solar, benzene flame, radiant panel, blackbody (tungsten lamp), and three types of carbon arcs. These heat sources are being used in various ignition testing systems and devices in fire prevention studies.

Opaque material surfaces, when exposed to radiant energy sources, reflect and absorb energy according to Kirchhoff's law:

$$\alpha_\lambda + r_\lambda = 1 \qquad (1)$$

Where $\alpha_\lambda$ is the absorptance factor and $r_\lambda$ is the reflectance factor.

Where the material is partially transparent, the energy is distributed according to a modified Kirchhoff's law[7].

$$\alpha_\lambda + r_\lambda + \tau_\lambda = 1 \qquad (2)$$

Where $\alpha_\lambda$ is the absorptance, $r_\lambda$ is the reflectance and $\tau_\lambda$ is the transmittance.

In order to calculate the average reflectivity and transmissivity of a material for a particular radiant source, it is necessary to use the spectral emmissive properties of the source[3]. The blackbody radiation distribution was calculated using the relationship developed by Planck for the spectral distribution of emissive

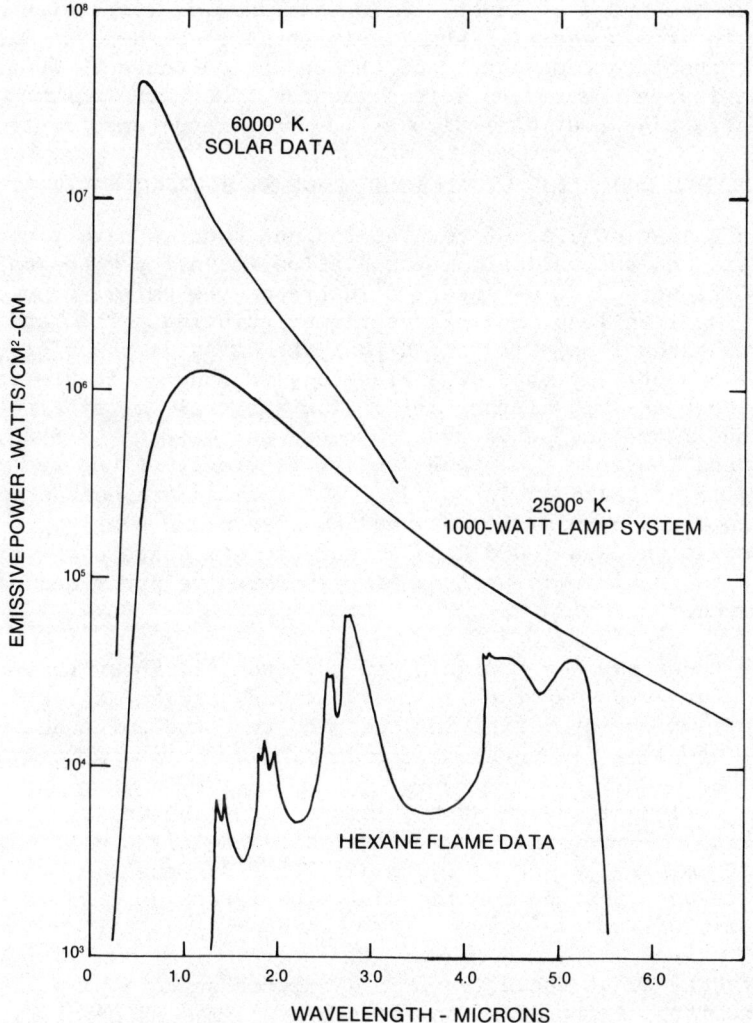

Figure 1. Monochromatic emissive power of several heat sources.

power[8]:

$$e(\lambda) = \frac{2\pi C_1 d\lambda}{\lambda^5 (\exp^{C_2/\lambda T} - 1) R^2} \quad (3)$$

where $e(\lambda)$ is the spectral energy at the wavelength $\lambda$, $C_1 = hC_o^2$ and $C_2 = hC_o/k$ (h = Planck's constant, k is the Boltzmann constant, and $C_o$ is the speed of light), R is the index of refraction correction for the speed of light in air (here assumed to be 1.0), and T is the absolute temperature of the radiation source. Where relative emissive power data were presented, they were assumed to be referred to blackbody radiation at the specified temperature.

## THERMAL RADIATION STUDIES AND POLYMER SURFACE ACTIVITY

Other past studies of several radiant sources have yielded information on the variation of radiation intensity with monochromatic wavelength[9-12]. Figure 1 illustrates the emissive powers of solar*, tungsten lamp, and hexane flame radiation. The manufacturer of the tungsten lamp** indicated that the emissive power of the lamps is essentially that of a blackbody at 2500°K, it provides a maximum peak at 1.15 microns[11,12]. The spectral emissive power of the hexane flame and other hydrocarbons was measured by Ryan, Penzias and Tourin[10]. As indicated by Figure 1, hexane has primary emissive power peaks at 2.7 μ and 4.3 μ , which are characteristic of water vapor and carbon dioxide, and smaller peaks near 2 μ. The solar radiation data (6000°K solar surface temperature) have been included to enable a comparison of both emissive power and wavelength span.[9]

The tungsten lamp radiation is not true blackbody radiation; however, for purposes of this study, it was assumed to behave as a graybody which, by definition, assumes that the emittance is constant at all wavelengths.

Figure 1 indicates that the lamp emissive power is greatest in the band 0.7 - 2.5 microns while the hexane flame will have the strongest emissive power in the bands 2.5 - 3.2 and 4.2 - 5.4 μ. This difference must be considered in the interpretation of incident radiation.

Reflectance of radiation from a polymer surface as well as the transmittance of radiation through a polymer material is a function of the wavelength of the radiation and the polymer material itself.

Figure 2 illustrates the variation of monochromatic reflectance with wavelength for Plexiglas , while Figures 3 and 4

---

*The solar radiation curve was drawn as an approximation from the solar monochromatic emission data and the use of the measured solar temperature of 6000°K in the Stefan-Boltzman radiation equation.

**GE quartz infrared lamp, types 2M/T3/1CL/HT-230-250V and 1000T3/CL/HT-230-250V.

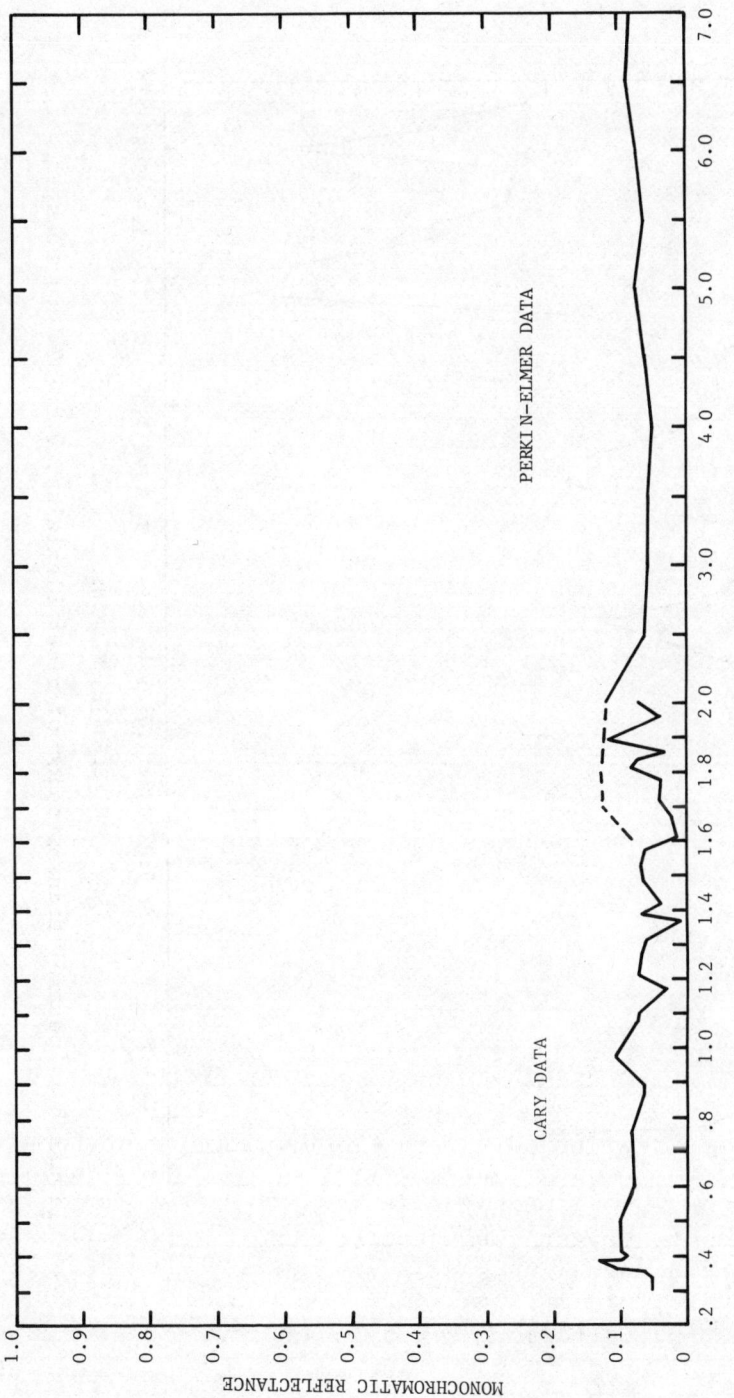

Figure 2. Sepctral reflectance of clear Plexiglas, 0.3 - 2.0 microns.

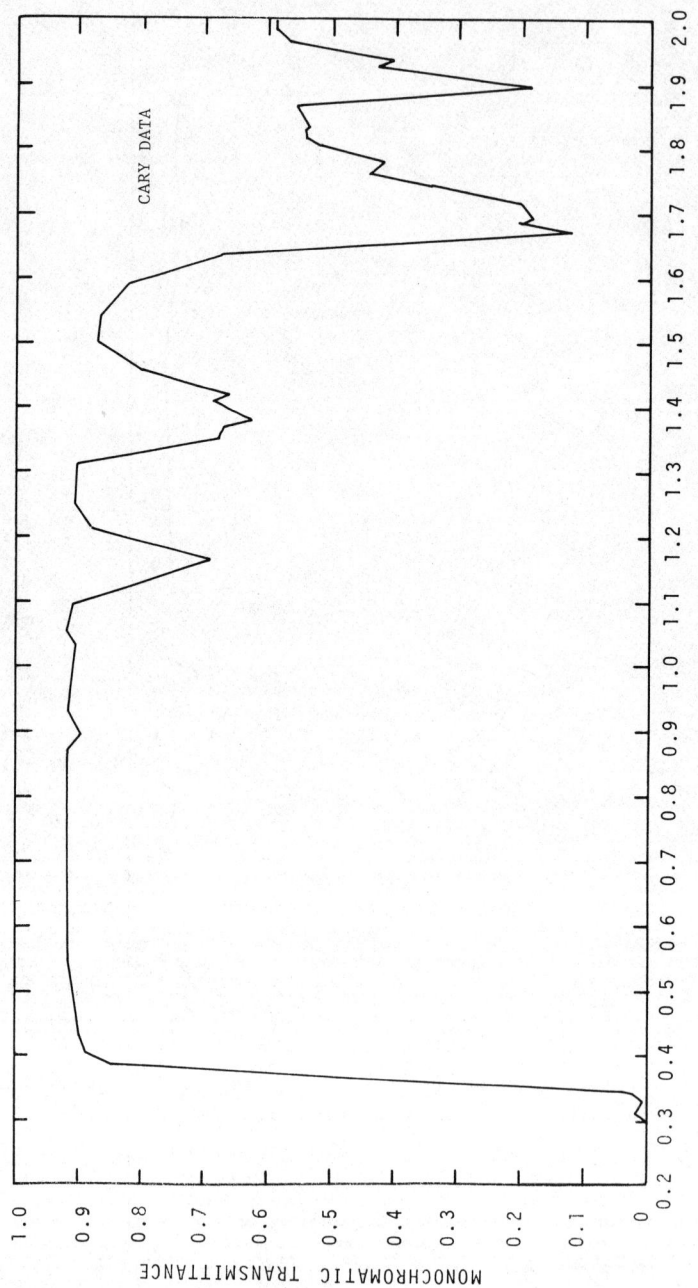

Figure 3. Spectral transmittance of clear Plexiglas, 0.3 - 2.0 microns.

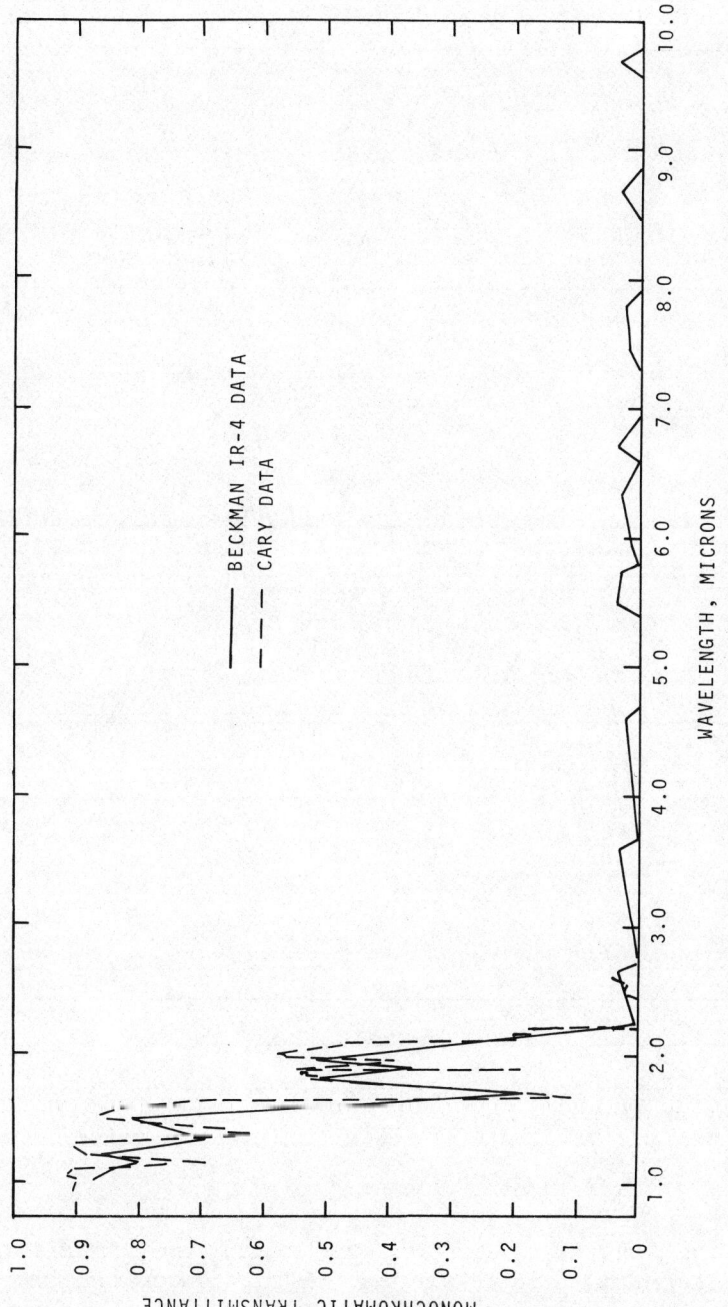

Figure 4. Spectral transmittance of clear Plexiglas, 1.0 – 10.0 microns.

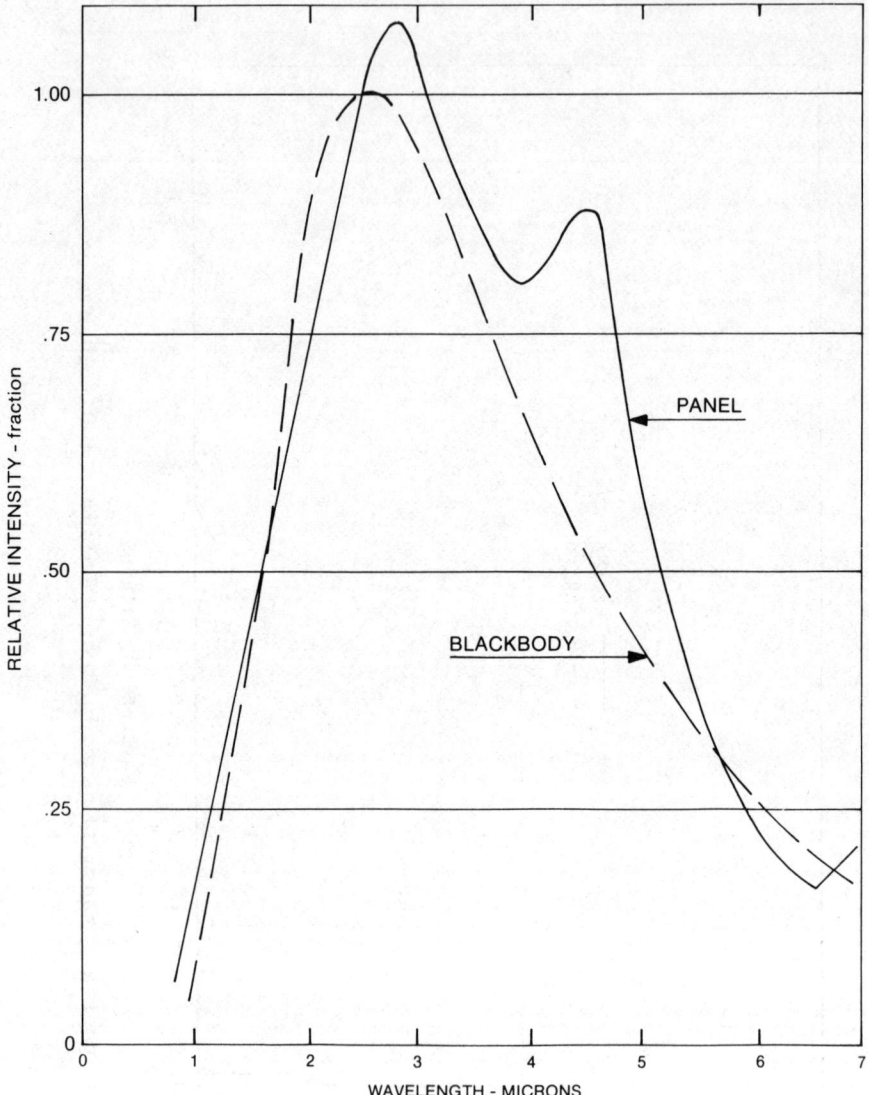

Figure 5. Relative intensity of a radiant panel compared to a blackbody at 1143°K.

indicate the particular variation of monochromatic transmittance with wavelength, also for Plexiglas. As is illustrated, there is a wide variation in both the reflectance and transmittance curves which must also be considered in the interpretation of average values over the designated wavelengths.

Details of the measurement procedures can be found in Reference 1.

The average reflectivity of a polymeric material for radiation from the radiant sources is obtained by the relationship

$$r_{av} = \frac{\int r_\lambda e_\lambda d\lambda}{\int e_\lambda d\lambda} \tag{4}$$

where $r_{av}$ is the average reflectivity of the target material and radiation source, $r_\lambda$ is the monochromatic reflectance of the target, $\lambda$ is the wavelength, and the $e_\lambda$ is the monochromatic emissive power of the heat source. Details of the method are given in Reference 4. The average transmissivity can be obtained from the relationship

$$\tau_{av} = \frac{\int \tau_\lambda e_\lambda d\lambda}{\int e_\lambda d\lambda} \tag{5}$$

where $\tau_{av}$ is the average transmissivity of the polymer, $\tau_\lambda$ is the monochromatic transmittance of the material, $\lambda$ is the wavelength and $e_\lambda$ is the monochromatic emissive power of the heat source.

Since the data obtained during the investigation were given in discrete wavelengths, both Equations (4) and (5) were converted to a trapezoidal rule simulation for use in a standard computer program, as illustrated by Equation (6).

$$r_{av} = \frac{\Sigma r_\lambda e_\lambda \Delta\lambda}{\Sigma e_\lambda \Delta\lambda} \tag{6}$$

The average transmissivity, $\tau_\lambda$, was calculated using a similar equation.

It has been found, by the authors, that monochromatic emissive power relationships for different radiation sources are not easily obtained in the literature. For that reason, the various plots of intensity vs. wavelength are included for general information as well as provide a data source for those requiring emissive power information for different radiation systems. Further, Figures 13-30 indicate the relationship of spectral reflectance and spectral transmittance for several transparent or translucent polymers.

The gas-fired radiant panel and its radiation characteristics have been studied by several groups to determine its spectral intensity[13-15]. The gas-fired radiant panel used to determine the flammability and/or flame spread of materials, including polymers, is run at a fixed panel temperature as specified by ASTM -E162[16]. Figure 5 illustrates the relative intensity of the radiant panel compared to a blackbody source at 1143°K.

The actual monochromatic emissive power for the panel was calculated by multiplying the blackbody emissive power by the relative intensity. The relative intensity is the ratio of the intensity at any wavelength to the maximum intensity for a blackbody at the temperature considered. The wavelength at which the maximum blackbody intensity occurs can be determined from Wein's displacement law. The radiant panel and the carbon arcs to be discussed below are not perfect graybodies and their measured temperature were not always the same as the blackbody temperature to which the investigators compared them. In this paper, the data presented in the energy plots are those used by the investigator referenced, even though the result indicates relative intensities greater than blackbody values. Similarly, emitted energy is used if the original data were presented in that form. Although resulting in some confusion, it was desired to use the actual data in a form as nearly as possible to that given in the original references.

Table IA tabulates the average reflectivity for serveral transparent/translucent polymers for solar, benzene flame and the radiant panel radiation.

Table IB tabulates the average transmissivity for these same polymers and radiation sources. A comparison of average reflectivity and transmissivity values between the gas-fired radiant panel and a 1143°K blackbody radiator indicates a general difference of less than five percent in either values for a given polymer[6]. A comparison indicates that the radiant panel emission spectrum is nearly the same as a blackbody source and that for most studies it can be considered to be a blackbody.

Figures 6-8 indicate the average reflectivity and average transmissivity for the polymers for blackbody conditions over the temperature range of 1,000°K to 6,000°K by 1000 degree increments. As is illustrated, the average values tend to become constant as the blackbody temperature increases but the specific values are dependent upon the particular polymer. These data are included as a reference comparison to the various heat sources as well as for information on average reflectivity and transmissivity when using near a blackbody radiator such as tungsten lamps.

Carbon arc spectral emission characteristics have been studied by several investigators who have developed three levels of operating temperatures: low intensity 2900°K, standard 3900°K, and high intensity 5500 - 6000°K. Most of the monochromatic arc emission data did not extend beyond a wavelength of 3.0 microns. It was assumed that the emission values for the arcs were the same as the reference blackbody emissions for wavelengths 3.0 microns to 7.0 microns. A comparison of the reference blackbody with the arc plots shows little variation from 2.5 microns to 3.0 microns.

Table IA. Variation of Average Reflectivity of Polymers with Solar, Benzene Flame, and the Radiant Panel as Radiation Sources.

| MATERIAL | 1143°K PANEL | SOLAR | BENZENE FLAME |
|---|---|---|---|
| LEXAN | .167 | .555 | .112 |
| KEL-F | ----------DATE NOT RELIABLE---------- | | |
| POLYSTYRENE | .102 | .067 | .098 |
| CELLULOSE ACETATE BUTERATE | .076 | .054 | .061 |
| PLEXIGLAS | .080 | .061 | .046 |
| POLYPROPYLENE | .100 | .344 | .066 |
| POLY (VINYL CHLORIDE) | .084 | .078 | .071 |

Table IB. Variation of Average Transmissivity of Polymers with Solar, Benzene Flame, and the Radiant Panel as Radiation Sources.

| MATERIAL | 1143°K PANEL | SOLAR | BENZENE FLAME |
|---|---|---|---|
| LEXAN | .026 | .072 | .024 |
| KEL-F | .153 | .047 | .124 |
| POLYSTYRENE | .218 | .838 | .125 |
| CELLULOSE ACETATE BUTERATE | .150 | .826 | .064 |
| PLEXIGLAS | .149 | .842 | .054 |
| POLYPROPYLENE | .050 | .036 | .078 |
| POLY (VINYL CHLORIDE) | .183 | .770 | .082 |

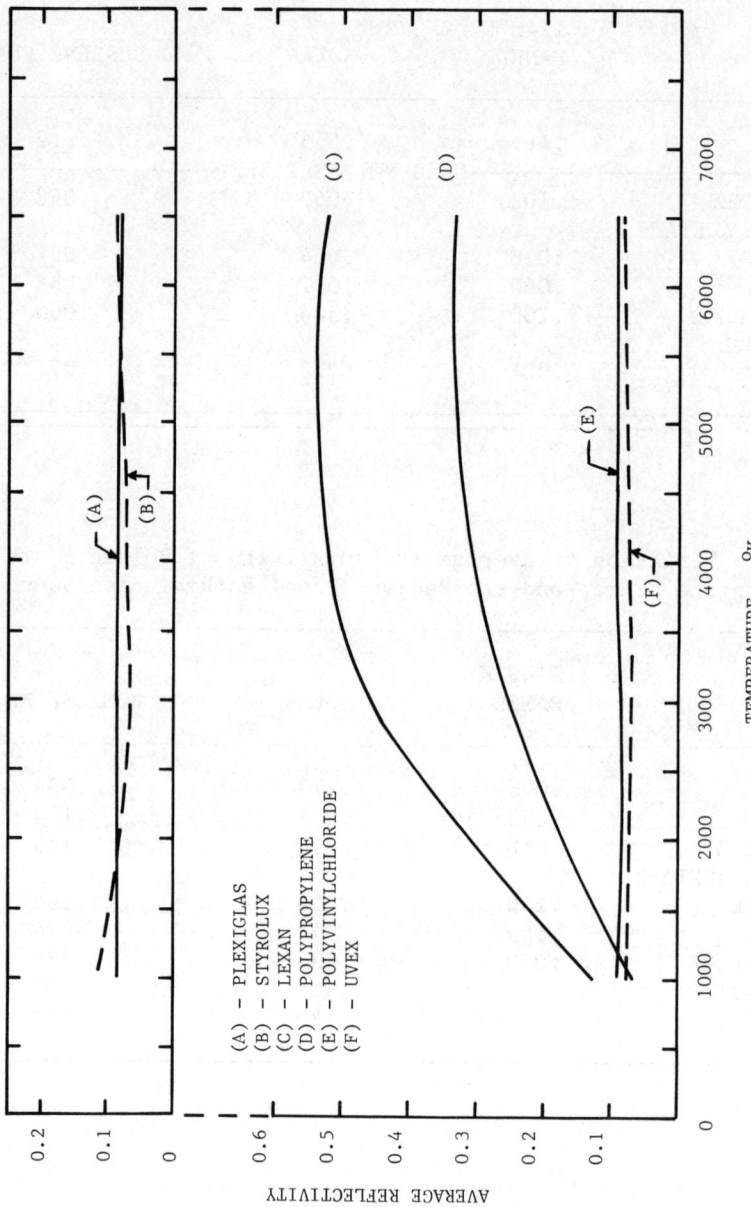

Figure 6. Variation of average reflectivity of polymers with blackbody temperatures.

# REFLECTANCE-TRANSMITTANCE CHARACTERISTICS

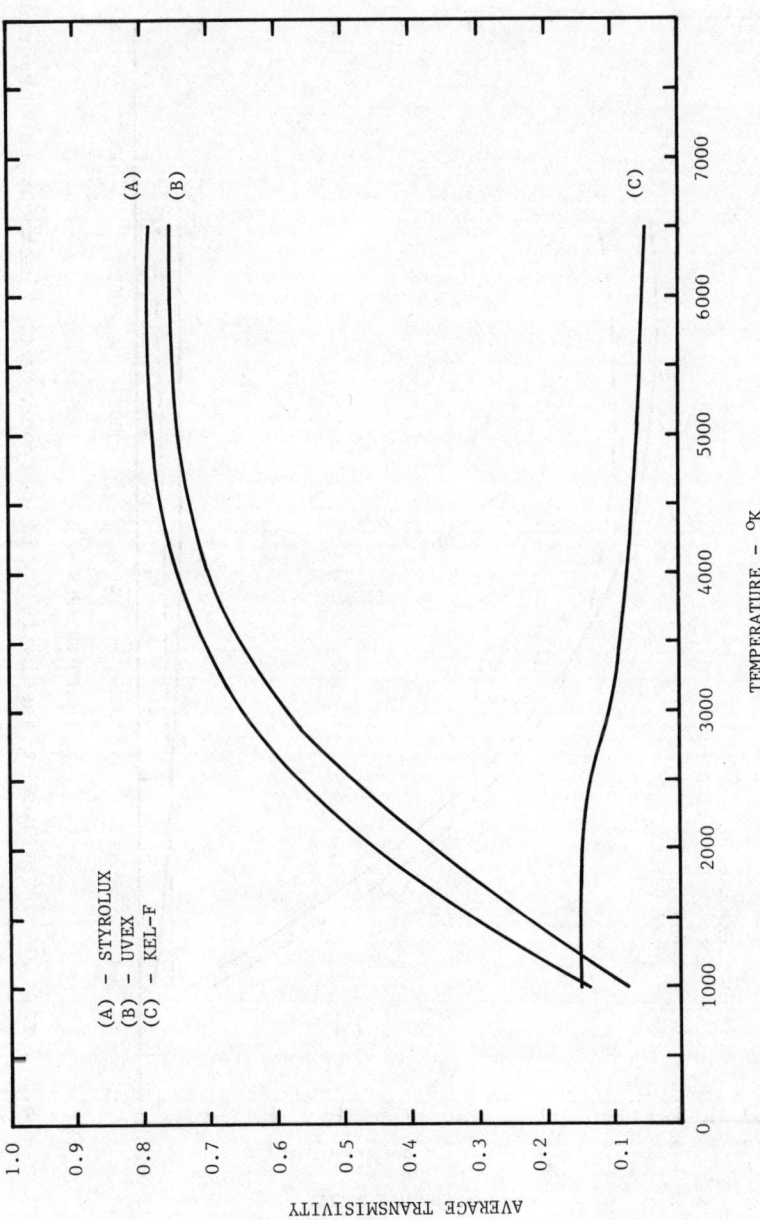

Figure 7. Variation of average transmissivity of polymers with blackbody temperatures.

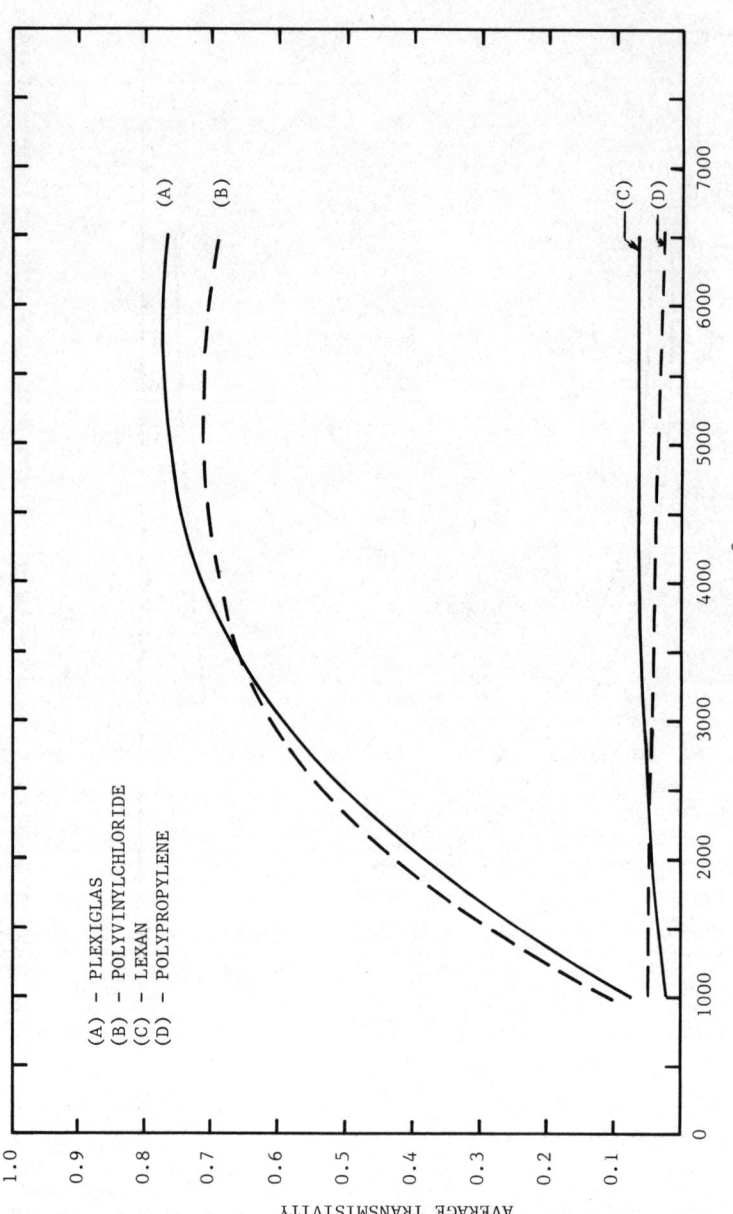

Figure 8. Variation of average transmissivity of polymers with blackbody temperatures.

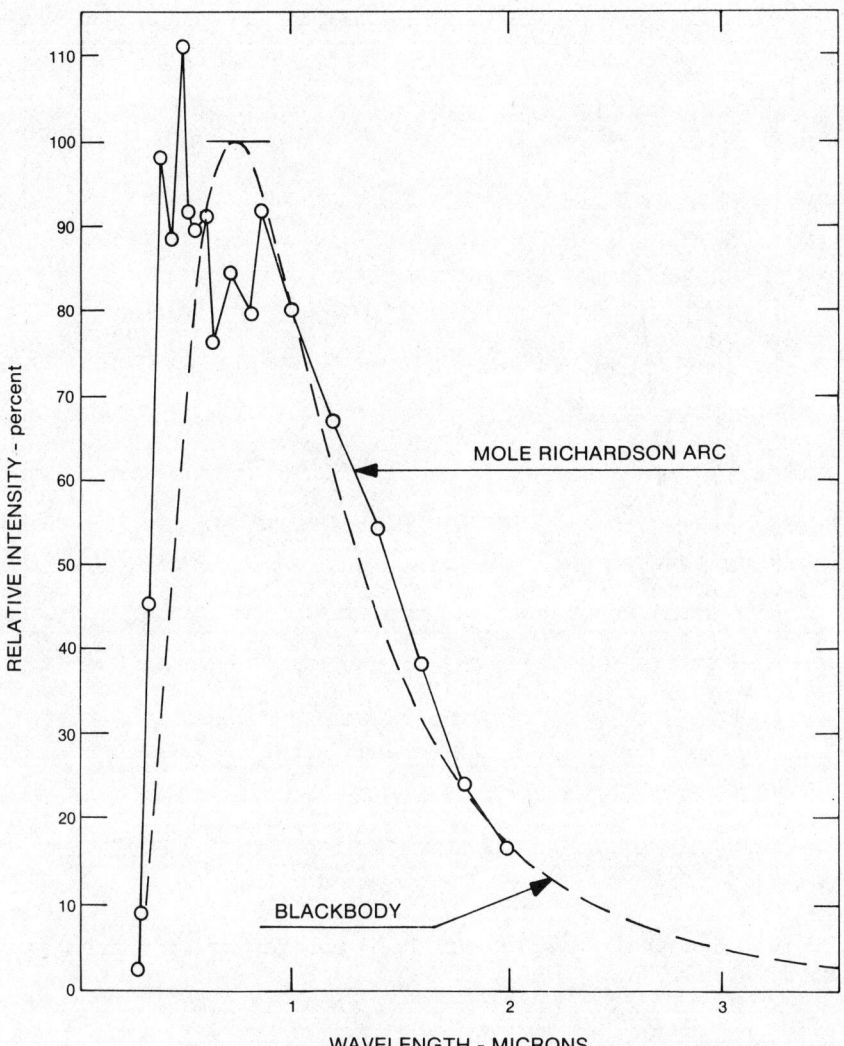

Figure 9. Relative intensity of the Mole Richardson arc compared to a blackbody at 4000° K.

The low intensity carbon arc, 2900°K, was studied by Magdeburg and Schley[17] using two different types of electrode materials:

(1) Anode R.W. II, 6.35 mm Diameter (cored), cathode Norris D., 7 mm Diameter.

(2) Anode Norris H., 6 mm Diameter (cored), cathode Norris D., 7 mm Diameter.

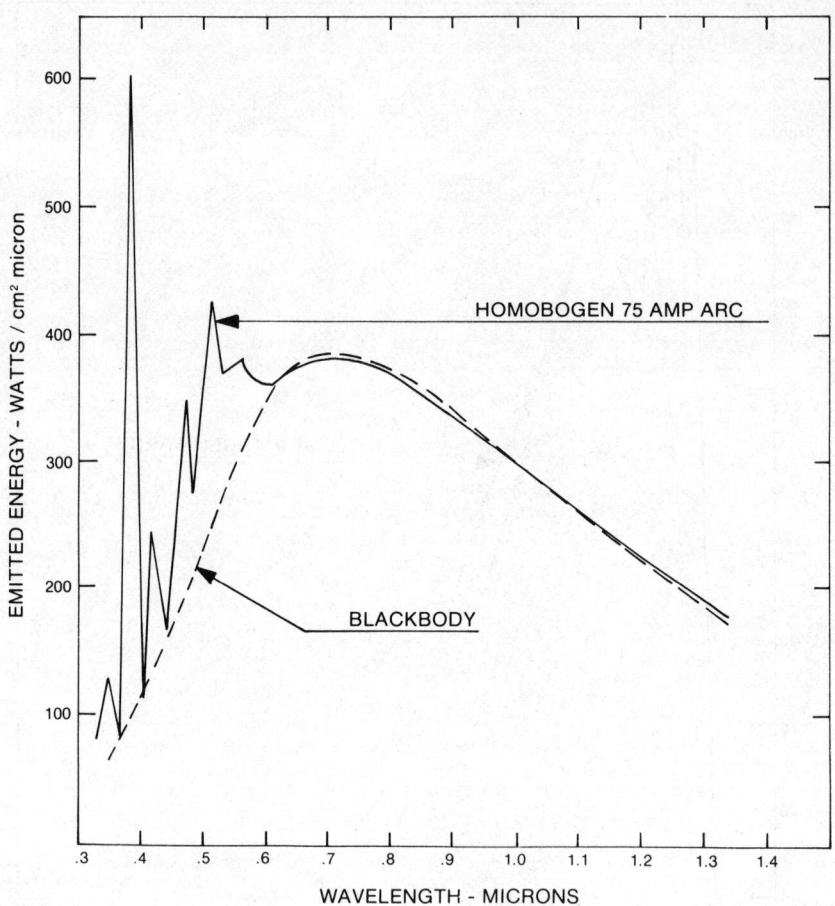

Figure 10. Spectral emission of the Homobogen arc compared to a blackbody at 3900°K.

The spectral data are listed in the paper; they were used to calculate the average reflectivity and average transmissivity according to Equations (4) and (5). Average reflectivity and transmissivity data for the polymers using the 2900°K arc sources are listed in Tables IIA and IIB. A comparison of the 2900°K arc data with the 3000°K blackbody data of Figures 6-8 shows variations of 5 - 20% difference between values for a particular polymer. These same variations were found when comparing like temperatures during absorptance studies[6].

The standard or normal carbon arc is most generally accepted to have an operating temperature of 3800° - 4000°K[18-22].

Table IIA. Variation of Average Reflectivity of Polymers with Several Carbon Arc Radiation Sources.

| MATERIAL | MORRIS H | RW 11 | 3800°K NULL-LOZIER | 3900°K HOMOBOGEN | 4000°K MOLE RICHARDSON | 5800-6060°K BUTLER | 5800-6060°K BECKBOGEN |
|---|---|---|---|---|---|---|---|
| LEXAN | .513 | .518 | .513 | .544 | .522 | .551 | .558 |
| KEL-F | ---------- | ---------- | ---------- DATA UNCERTAIN | ---------- | ---------- | ---------- | ---------- |
| POLYSTYRENE | .065 | .068 | .050 | .045 | .070 | .077 | .079 |
| CELLULOSE ACETATE BUTERATE | .091 | .087 | .068 | .076 | .073 | .077 | .081 |
| PLEXIGLAS | .030 | .068 | .076 | .078 | .095 | .075 | .078 |
| POLY PROPYLENE | .312 | .303 | .313 | .325 | .313 | .345 | .360 |
| POLY (VINYL CHLORIDE) | .099 | .098 | .093 | .090 | .089 | .095 | .093 |

Table IIB. Variation of Average Transmissivity of Polymers with Several Carbon Arc Radiation Sources.

| MATERIAL | 2900°K MORRIS H | RW 11 | 3800°K NULL-LOZIER | 3900°K HOMOBOGEN | 4000°K MOLE RICHARDSON | 5800° BUTLER | – 6060°K BECKBOGEN |
|---|---|---|---|---|---|---|---|
| LEXAN | .062 | .062 | .065 | .068 | .065 | .069 | .072 |
| KEL-F | .076 | .078 | .066 | .053 | .070 | .058 | .039 |
| POLYSTYRENE | .759 | .753 | .786 | .843 | .773 | .825 | .847 |
| CELLULOSE ACETATE BUTERATE | .697 | .696 | .733 | .783 | .726 | .799 | .824 |
| PLEXIGLAS | .731 | .735 | .745 | .806 | .713 | .834 | .856 |
| POLY PROPYLENE | .038 | .040 | .038 | .037 | .039 | .036 | .031 |
| POLY (VINYL CHLORIDE) | .688 | .692 | .705 | .750 | .692 | .763 | .754 |

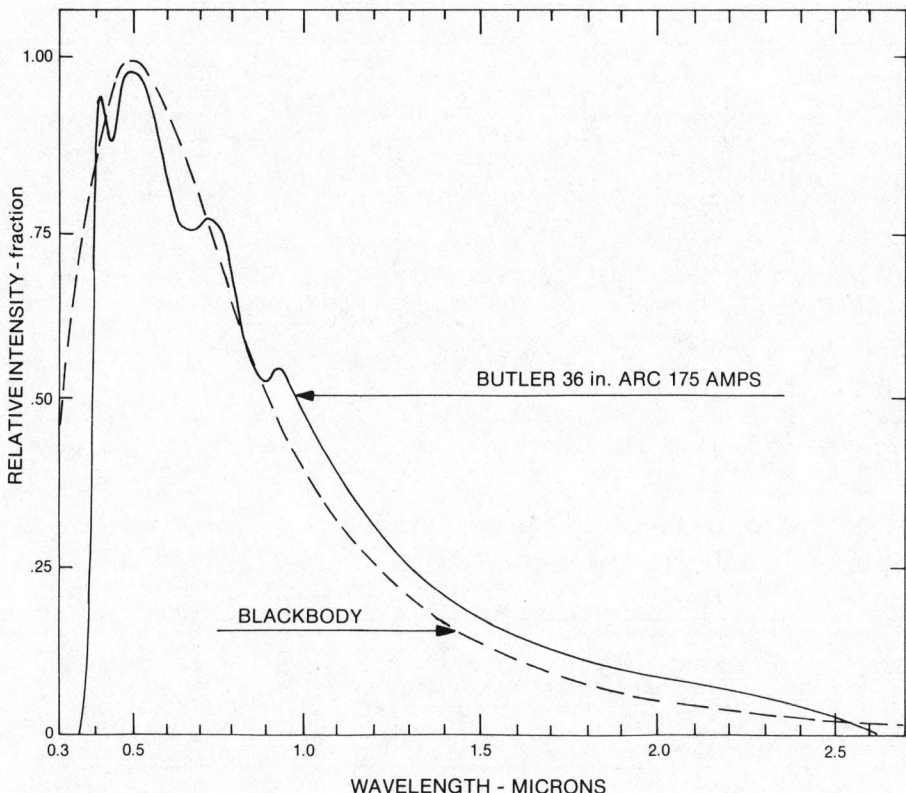

Figure 11. Relative intensity of the Butler arc compared to a blackbody at 5800°K.

Figure 9 presents the spectrum of the 4000°K Mole Richardson arc radiation while Figure 10 illustrates the spectrum for the 3900°K Homobogen arc system. The 3800°K spectra data of the Null-Lozier arc[18] are given in the referenced paper; no plots are presented. The Homobogen arc spectral distribution is given in intensity units, and the Mole Richardson is given in relative intensity.* Since the Mole Richardson arc data are given at relative intensity compared to 4000°K blackbody, the values for the emissive power were calculated using Equation (3) and the relative intensity.

---

*Relative intensity refers to the peak output of a blackbody at the reference temperature for both the blackbody curve and the carbon arc source.

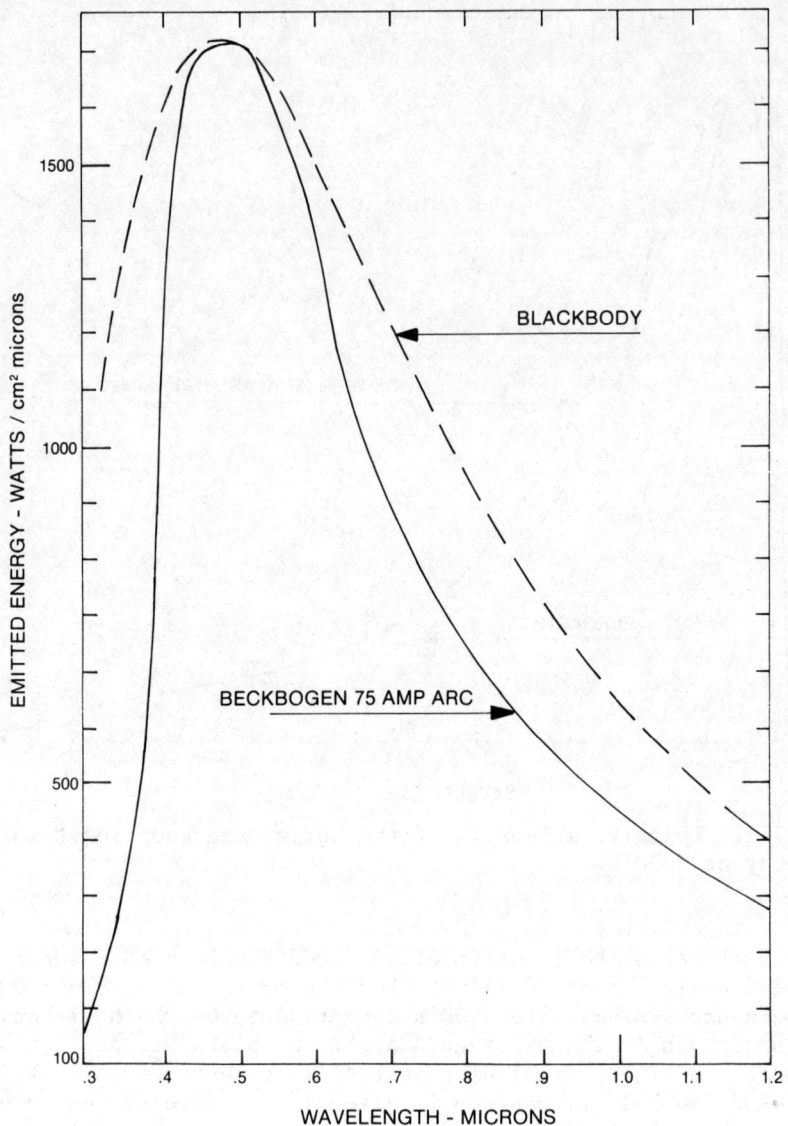

Figure 12. Spectral emission of the Beckbogen arc compared to a blackbody at 6060°K.

The average reflectivity and transmissivity were then calculated using those emissive powers.

Tables IIA and IIB list the average reflectivity and average transmissivity for the materials exposed to each type of arc. A comparison of these Table II values with those of the 4000°K blackbody of Figures 6-8 show variations of 5 - 15% difference. The same variation was found for absorptance values[6].

The high intensity carbon arc system 5500°K-6000°K was studied and reported by at least two groups, Butler, et. al.[23] and Finkelnburg[20,21]. Figure 11 illustrates the Butler 36 inch arc emission spectrum, while Figure 12 represents the Beckbogen arc spectrum. The Butler data are for a 5800°K blackbody while the Beckbogen arc data are for a 6060°K blackbody. Tables IIA and IIB list the average reflectivity and transmissivity for the various polymers for the latter two arc sources. Again a comparison of the arc data with 6000°K blackbody shows variations of 5 - 15% difference. The absorptance study comparisons show lesser deviations.[6]

## CONCLUSIONS

During the determination of ignition characteristics of several polymers[2], it was found that the reflectance-transmittance characteristics of the polymer surface caused different times of ignition. Examination of the test results indicated that materials that were transparent or translucent yielded different ignition times in comparison to the same material when opaque. When the reflectance/transmittance values were applied to the heat equation, ignition times for polymers were equated for flames and 2500° tungsten lamps. In addition, other radiation sources were investigated and the average transmissivity and reflectivity values obtained were formed to permit correlation with the previous ignition tests. Discussion of results can be found in the work of Hallman[1].

The application of transmittance and reflectance values to radiant heat transfer has aided in solving some types of heating problems. The data in this paper are given to supplement that already published for absorptance values for various polymer and rubber materials[4,6]. The data should be useful to those engaged in fire research, materials testing, and other studies of physical characteristics of materials as well as the development and application of energy balances in radiation systems.

## REFERENCES

1. J. R. Hallman, "Ignition Characteristics of Plastics and Rubber", Unpublished Ph.D. Dissertation, University of Oklahoma, 1971.
2. J. R. Hallman, J. R. Welker and C. M. Sliepcevich, SPE J., 28 No. 9, 43 (1972).
3. J. R. Hallman, J. R. Welker and C. M. Sliepcevich, Polym. Plast. Technol. Eng., 6(1), (1976).
4. J. R. Hallman, J. R. Welker and C. M. Sliepcevich, Polymer Eng. Sci., 14, 717 (1974).
5. Idem in "Characterization of Metal and Polymer Surfaces", Vol. II, L. H. Lee, Editor, pp.429-446, Academic Press, Inc., New York, 1977.

6. J. R. Hallman, C. M. Sliepcevich and J. R. Welker, J. Fire Flammability, 9, 353 (1978).
7. R. Birkebak, "Measurement Techniques in Thermal Radiation", Paper Presented at a Short Course, given at the University of Tennessee Space Institute, Tullahoma, Tenn., May 1967.
8. R. Siegal and J. R. Howell, "The Blackbody, Electromagnetic Theory and Material Properties", Thermal Radiation Heat Transfer, Vol. I (NASA SP. 164), NASA Office of Technical Utilization, 1968, pp. 1-37.
9. "Solar Function at Sea Level", Smithsonian Miscellaneous Collection, 74 (1), Smithson. Inst. Astrophy. Obs. Ann V. 105 (1932).
10. L. R. Ryan, G. J. Penzias and R. H. Tourin, "An Atlas of Infrared Spectra of Flames, Part I. Infrared Spectra of Hydrocarbon Flames in the 1-5μ Region", Scientific Report No. 1, Contract AF19 (604)-6106. Air Force Cambridge Research Laboratories, Bedford, Mass., July 1961.
11. *B. T. Lee and N. H. Alvares, Paper Presented at the 1967 Spring Meeting, Western States Section, the Combustion Institute, La Jolla, California, April 1967.
12. T. J. Love, "Radiative Heat Transfer", Merrill Publishing Co., Columbus, Ohio, 1968.
13. D. L. Simms and J. Miller, "Some Radiation Characteristics of a Gas-Fired Panel", Fire Research Note No. 217/1955, Research Programs Objective B2/2 (P), Fire Research Station, Boreham Wood, Hertfordshire, England, 1955. Used by permission.
14. D. L. Simms and J. E. Coiley, British Appl. Phys., 14, 292 (1963).
15. J. J. Comeford, Combustion and Flame, 18, 125 (1972).
16. ASTM Standards, Part 18, Thermal and Cryogenic Insulating Materials, Building Joint Sealants: Fire Tests, etc., American Society for Testing and Materials, Philadelphia, PA, 1976.
17. H. Mageburg and U. Schley, Z. fuer Angew. Phys., 20 (5), 465 (1966).
18. U. R. Null and W. W. Lozier, J. Optical Soc. Amer., 52, 1156 (1962).
19. Private Communication, A. F. Robertson, Center for Fire Research, National Bureau of Standards, Washington, D. C., July 13, 1976.
20. W. Finkelnburg, "Hockstrom Kohlebogen, Physik and Technik einer Hoch Temperature Bogenentlodung", pp. 90-99, Springer-Verlag, Berlin, 1948.
21. W. Finkelnburg, "The High Temperature Carbon Arc", Office of the Military Government for Germany (U.S.), FIAT Final Report No. 1052, pp. 52-59, 176, 177 (1947).
22. Private Communication, P. L. Hinkley, Building Research Establishment, Fire Research Station, Boreham Wood, Hertfordshire, England, August 13, 1976.

---

\*   Material Available, Vanderbilt University Library.

23. *C. P. Butler, Paper Presented at the Conference on Solar Energy: The Scientific Basis, The University of Arizona, Tucson, Arizona, October 31 and November 1, 1955.

APPENDIX

| | |
|---|---|
| LEXAN | – ZELUX, Westlake Plastic, polycarbonate, sample 1/8" thick, polished sides. |
| KEL-F (translucent) | – Saunders Corporation of Los Angeles, Type 5441 #S310 sample 1/8" thick, polished sides. |
| POLYSTYRENE | – STYROLUX, Westlake Plastics, sample 1/8" thick, polished sides. |
| CELLULOSE ACETATE BUTERATE | – UVEX, Eastman Kodak, sample 1/8" sheet stock. |
| PLEXIGLAS | – "C", unshrunk, Rohm and Haas, sample 1/8" sheet stock. |
| POLYPROPYLENE (translucent) | – Siberling Rubber Co., natural, sample 1/8" thick, polished sides. |
| POLY (VINYL CHLORIDE) | – Union Carbide, rigid, sample 1/8" sheet stock. |

---

\* Material Available, Vanderbilt University Library, Nashville, Tennessee.

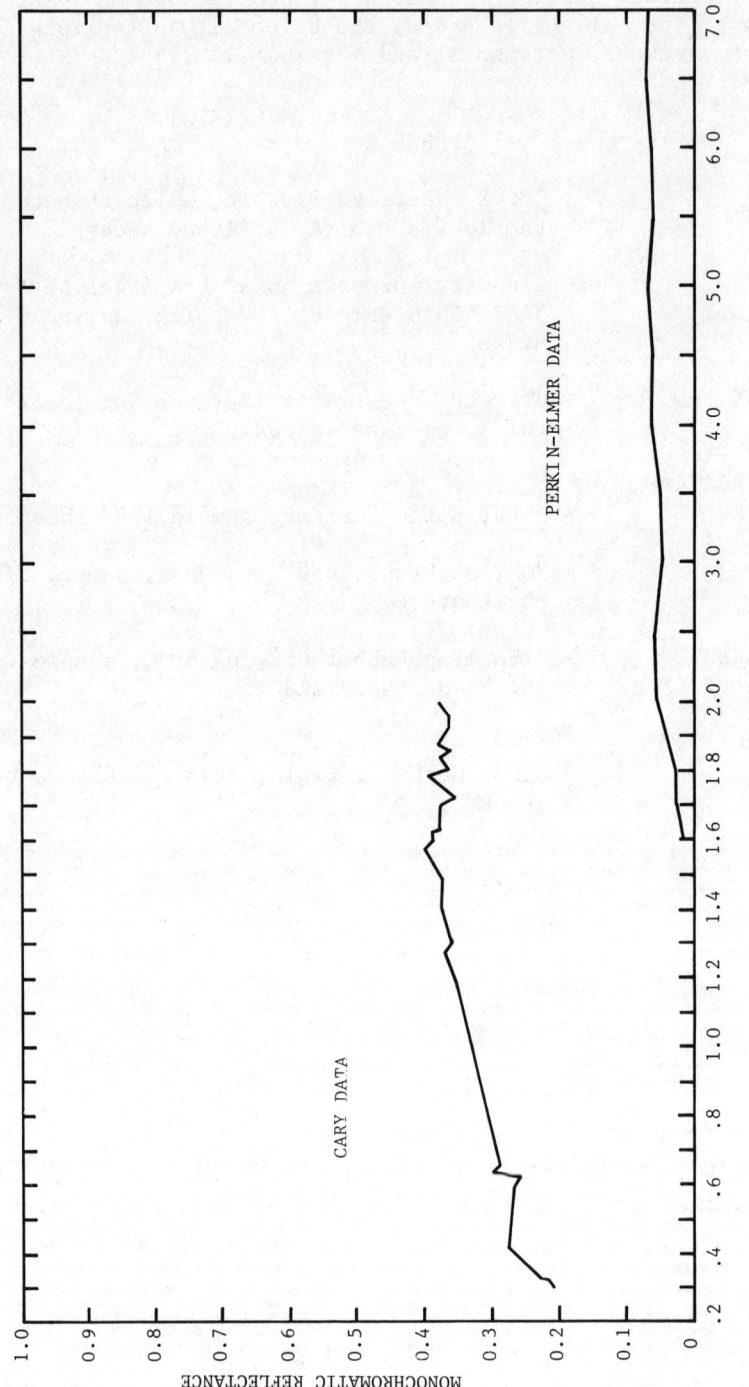

Figure 13. Spectral reflectance of KEL-F, 0.3 – 7.0 microns.

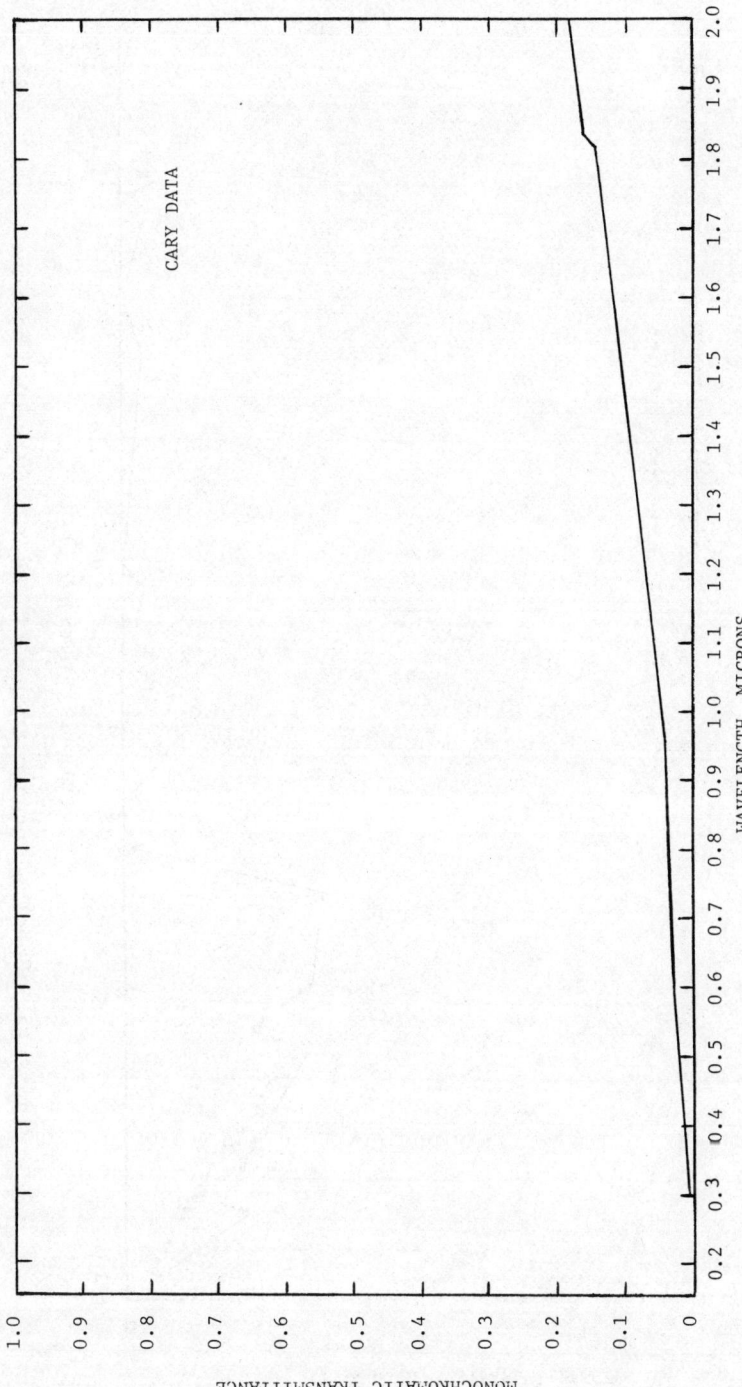

Figure 14. Spectral transmission of KEL-F, 0.3 - 2.0 microns.

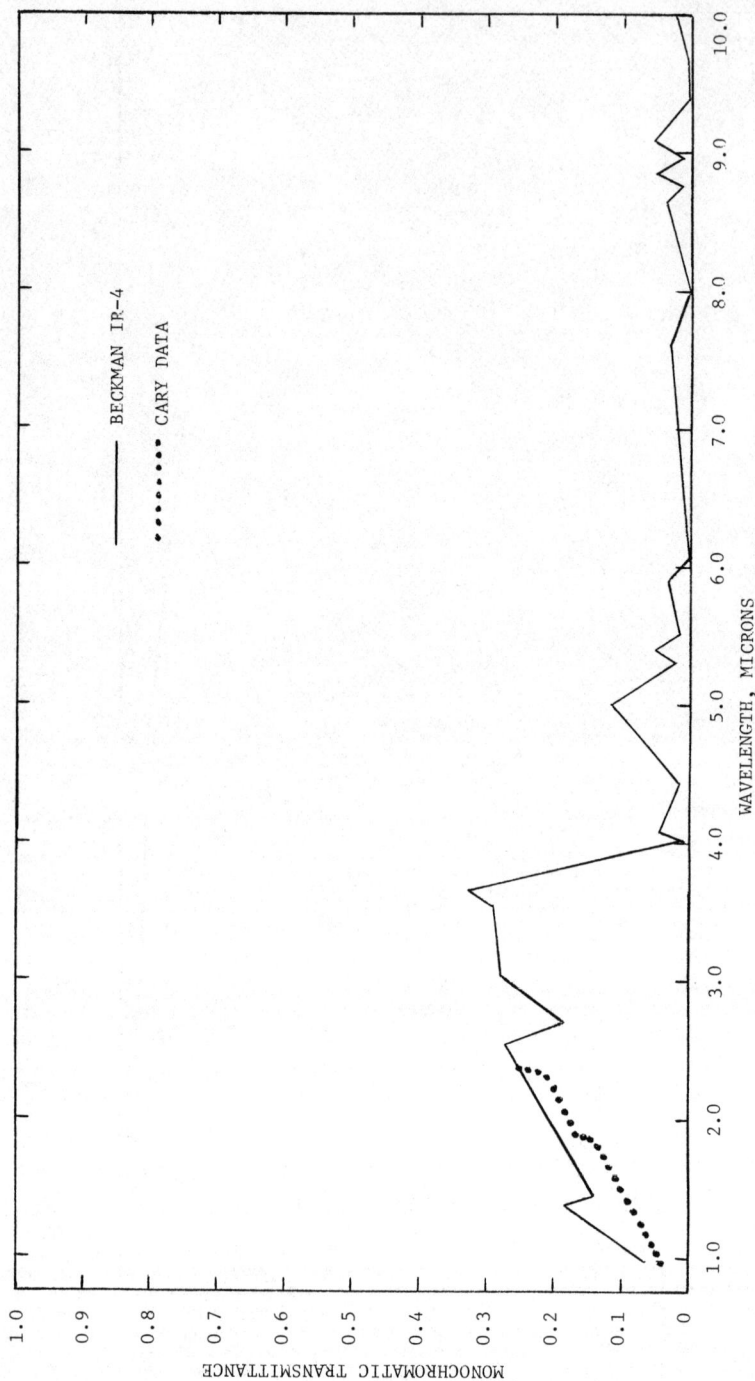

Figure 15. Spectral transmittance of KEL-F, 1.0 – 10.0 microns.

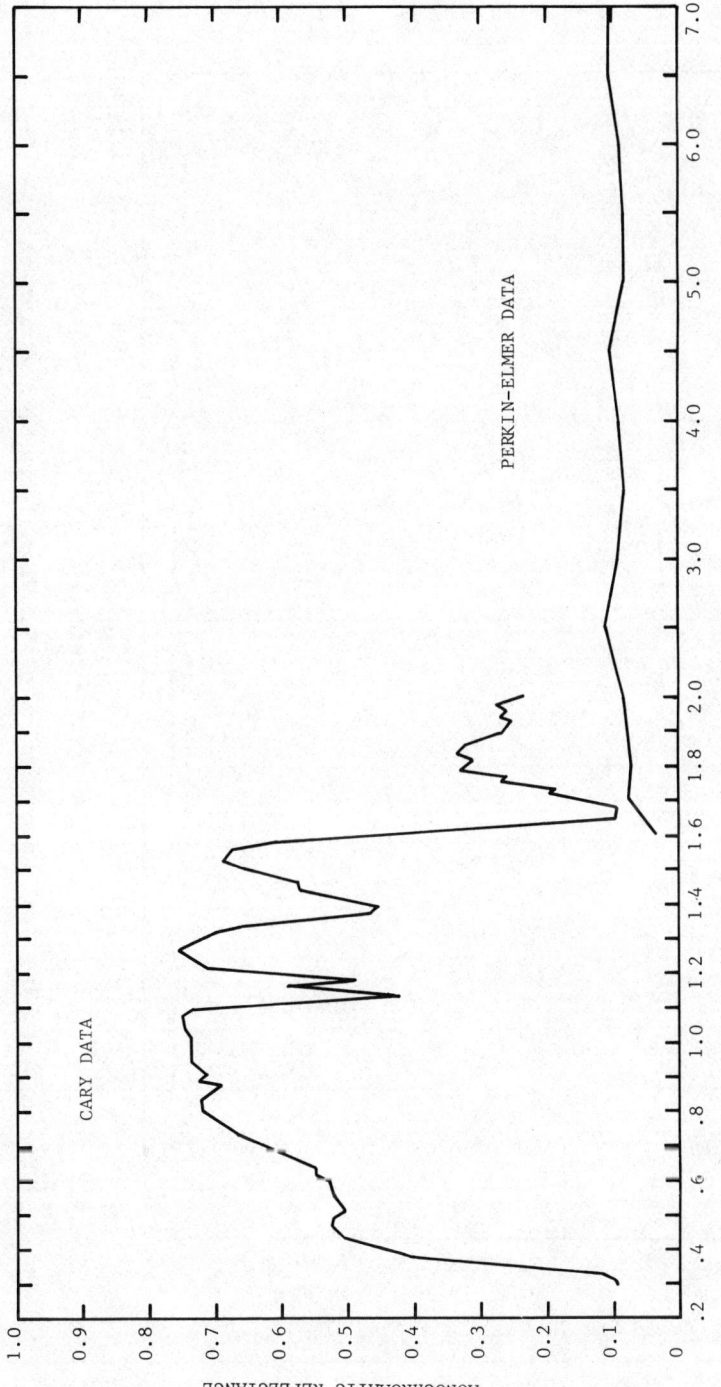

Figure 16. Spectral reflectance of polished LEXAN, 0.3 – 7.0 microns.

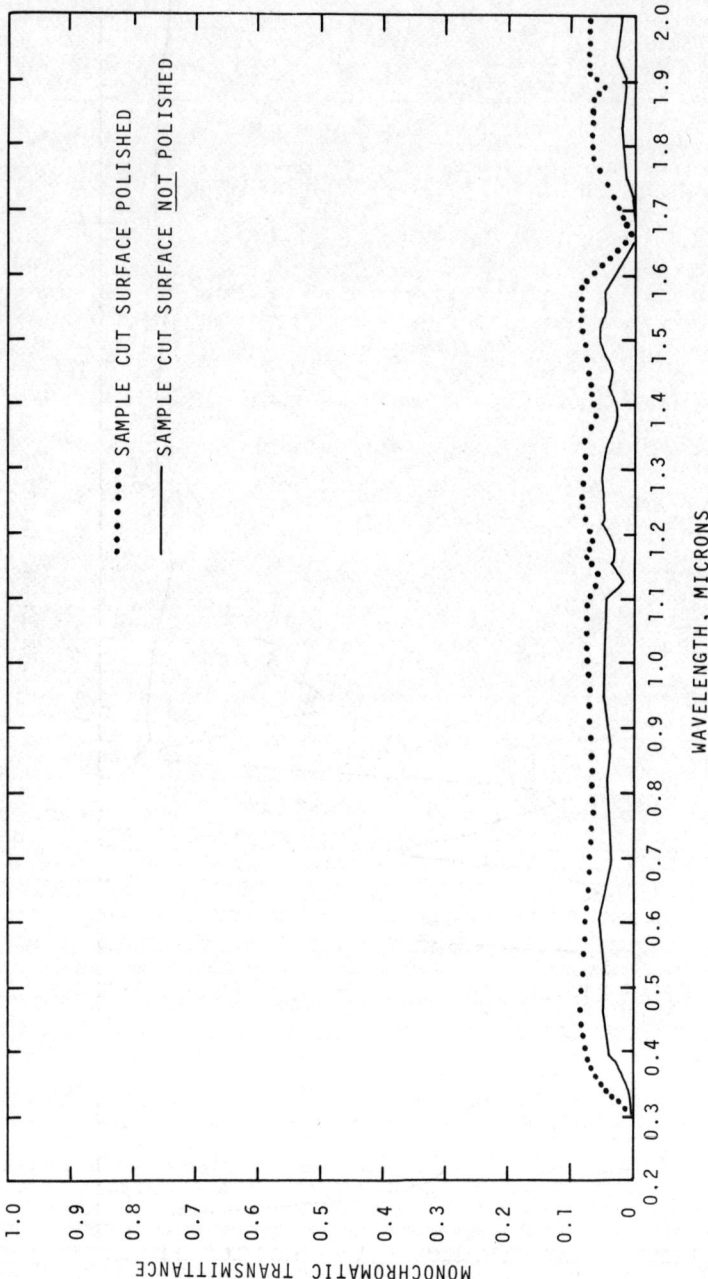

Figure 17. Spectral transmission of LEXAN with surface variations.

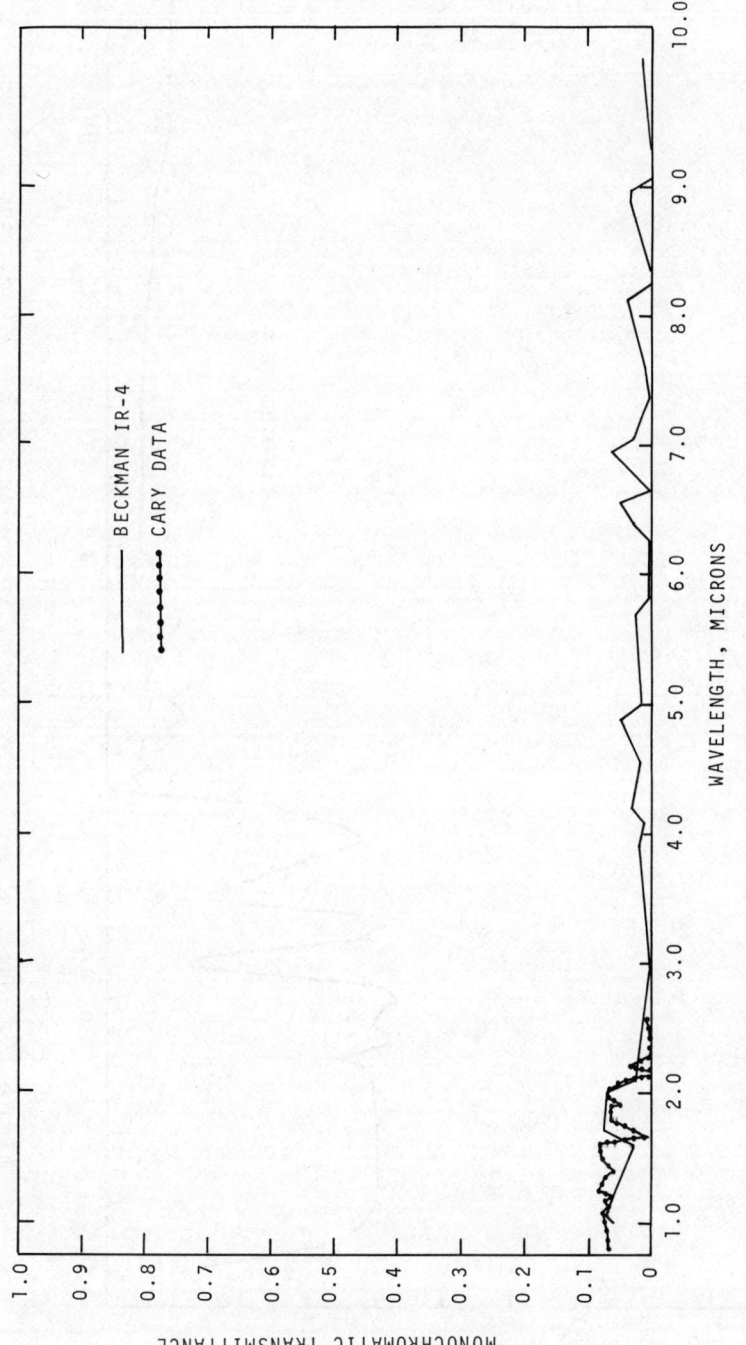

Figure 18. Spectral transmission of LEXAN, polished surfaces.

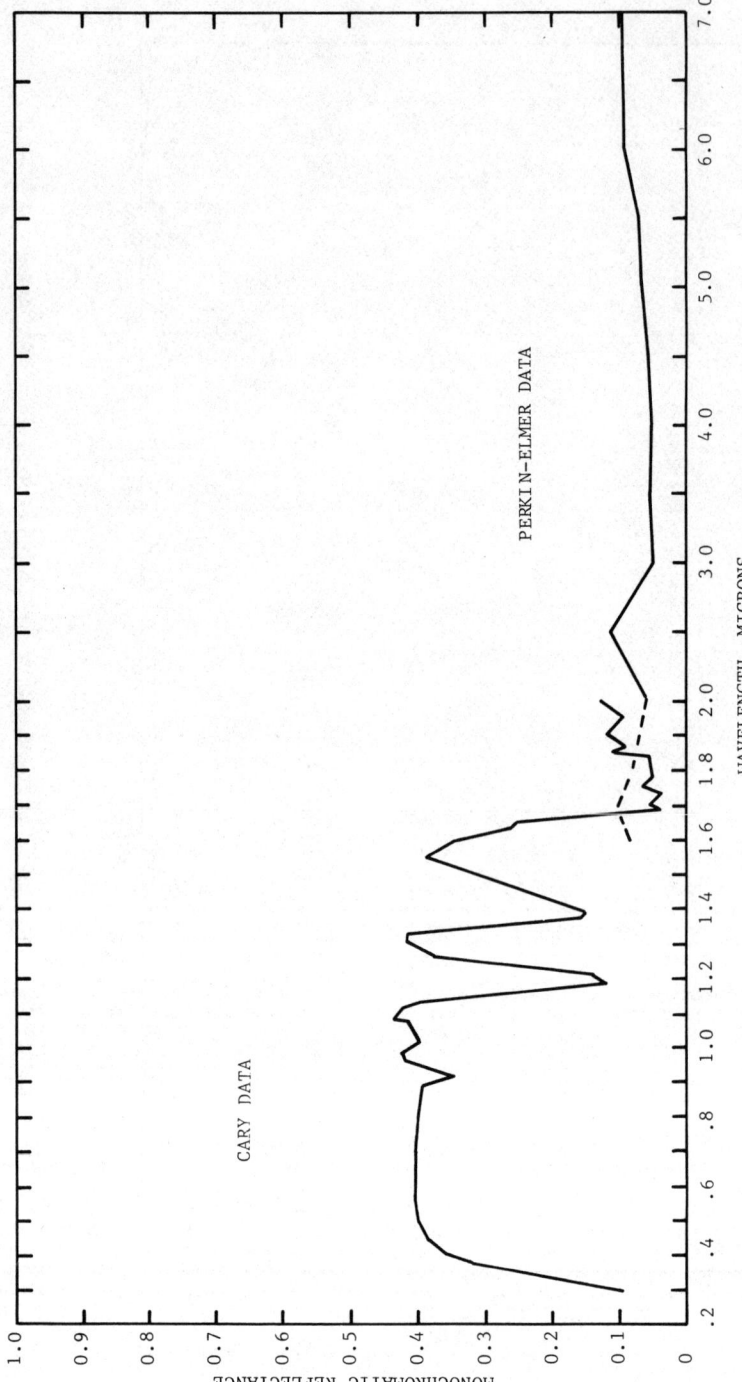

Figure 19. Spectral reflectance of polypropylene, 0.3 - 7.0 microns.

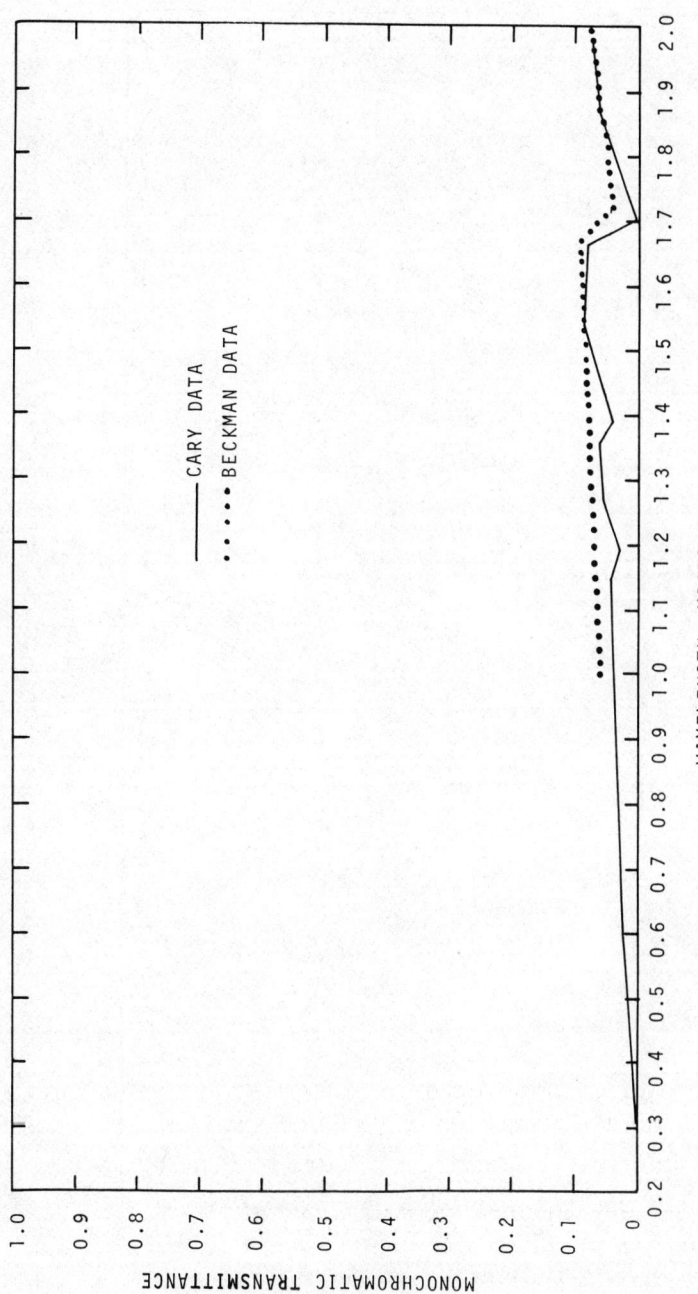

Figure 20. Spectral transmittance of polypropylene, 0.3 - 2.0 microns.

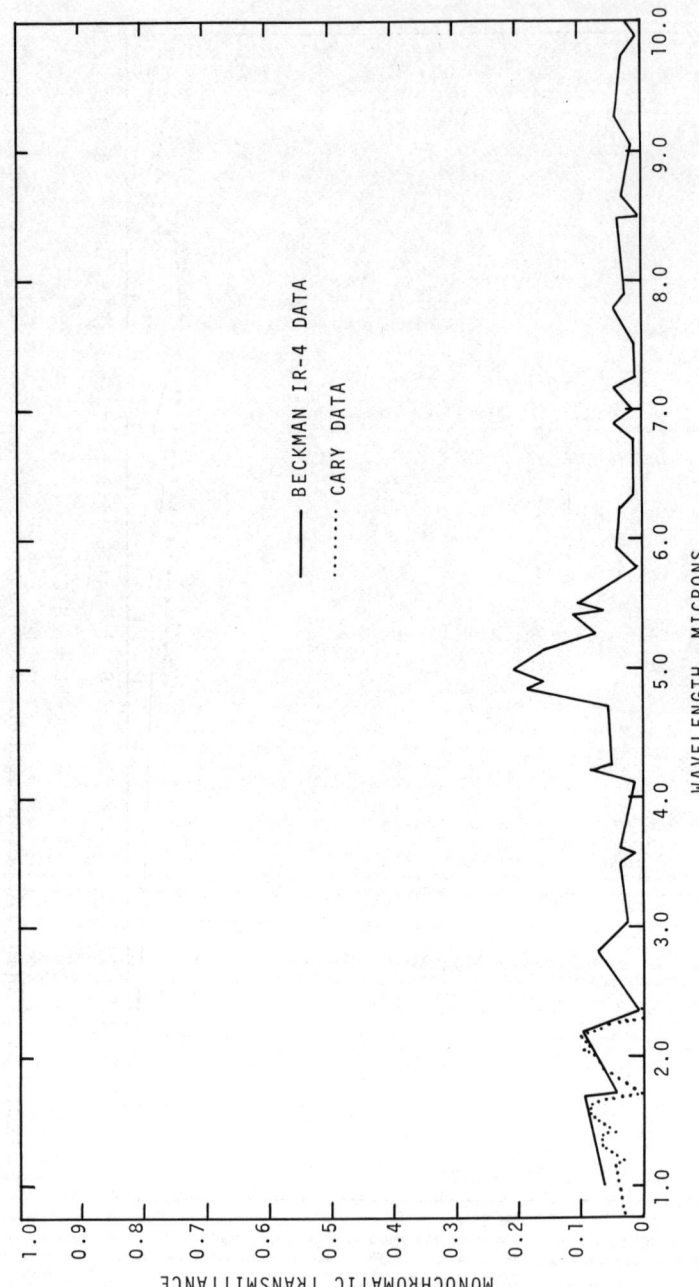

Figure 21. Spectral transmittance of polypropylene, 1.0 – 10.0 microns.

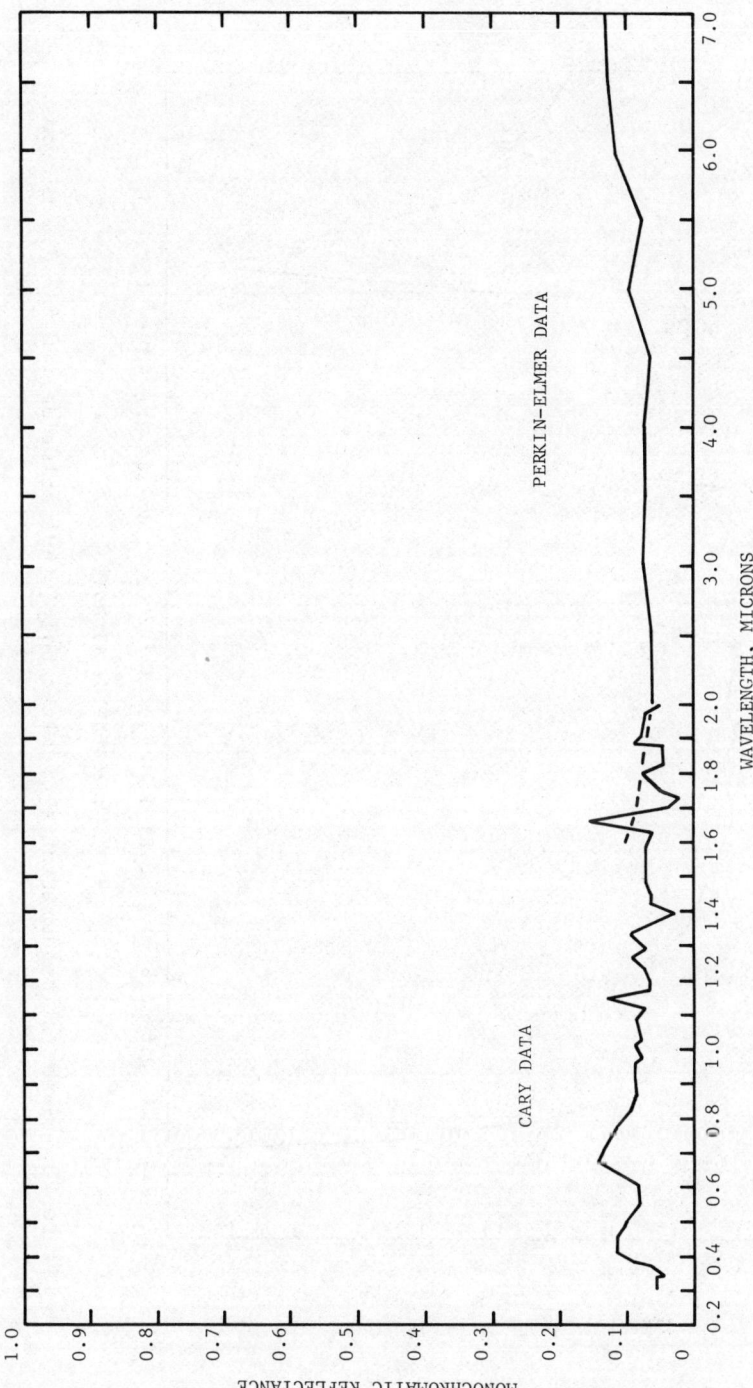

Figure 22. Spectral reflectance of clear poly(vinyl chloride), 0.3 - 7.0 microns.

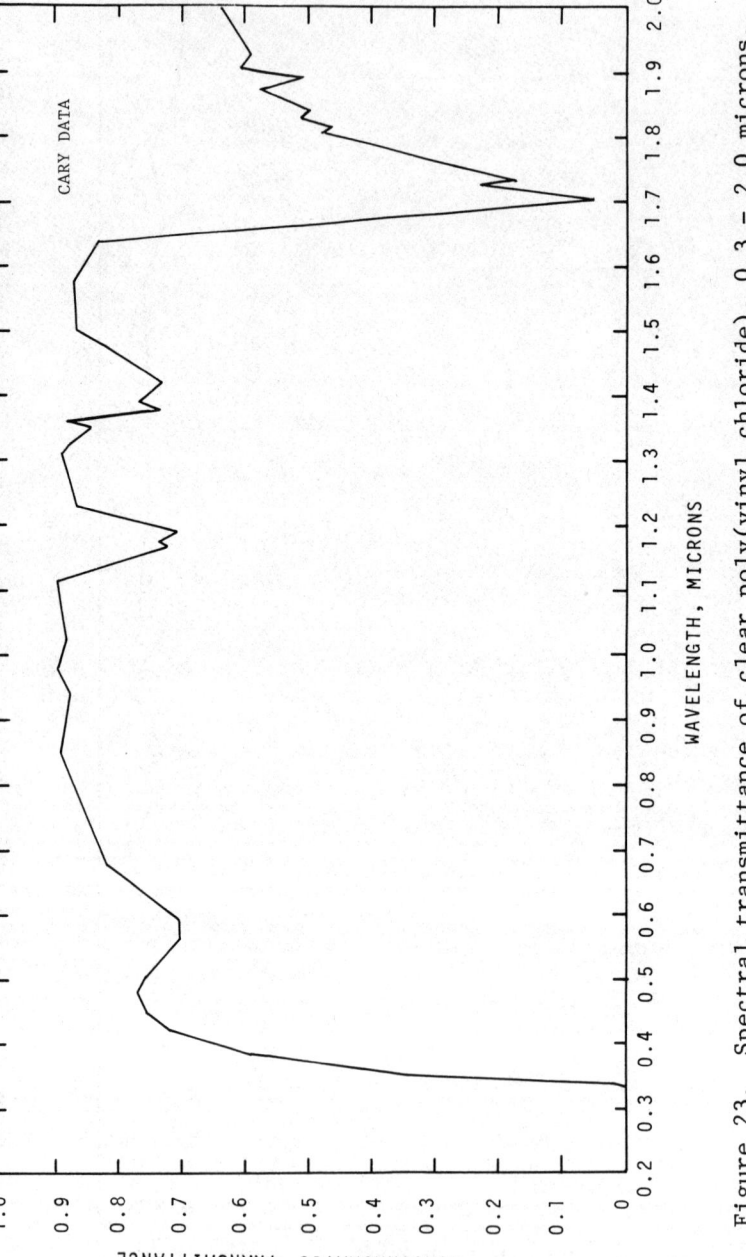

Figure 23. Spectral transmittance of clear poly(vinyl chloride), 0.3 - 2.0 microns.

# REFLECTANCE-TRANSMITTANCE CHARACTERISTICS

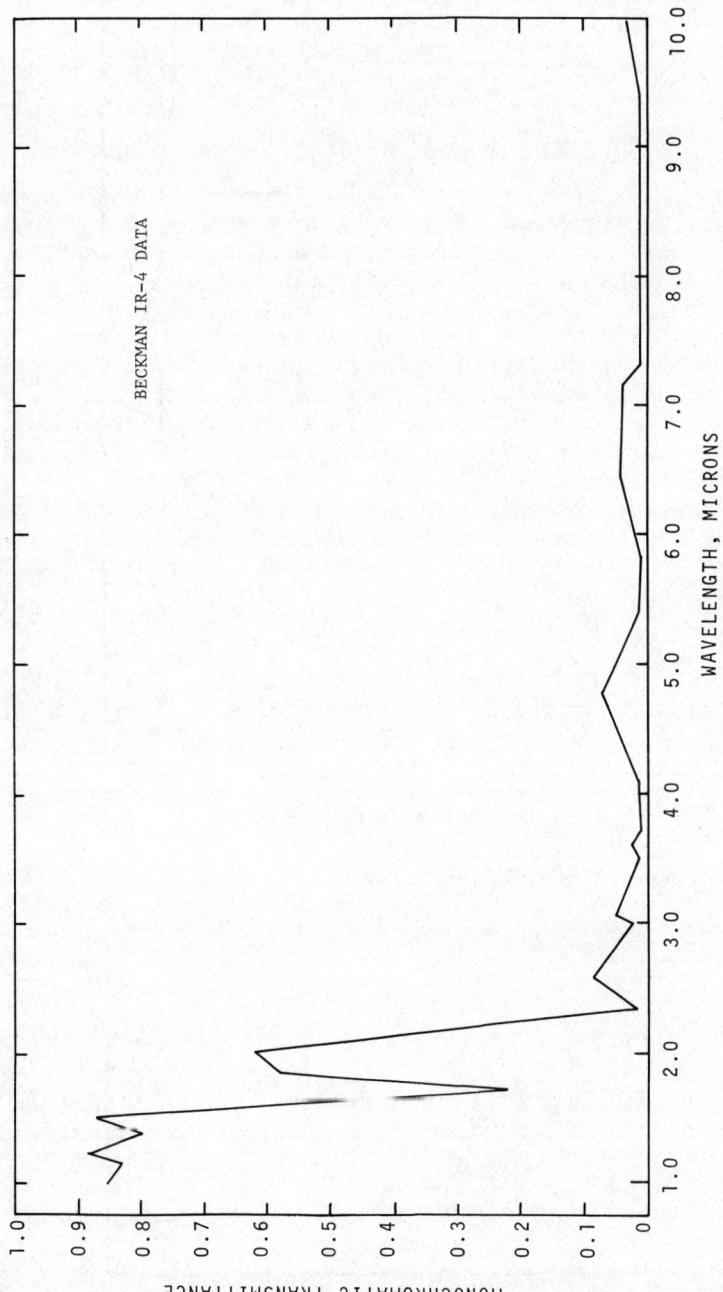

Figure 24. Spectral transmittance of poly(vinyl chloride), 1.0 – 10.0 microns.

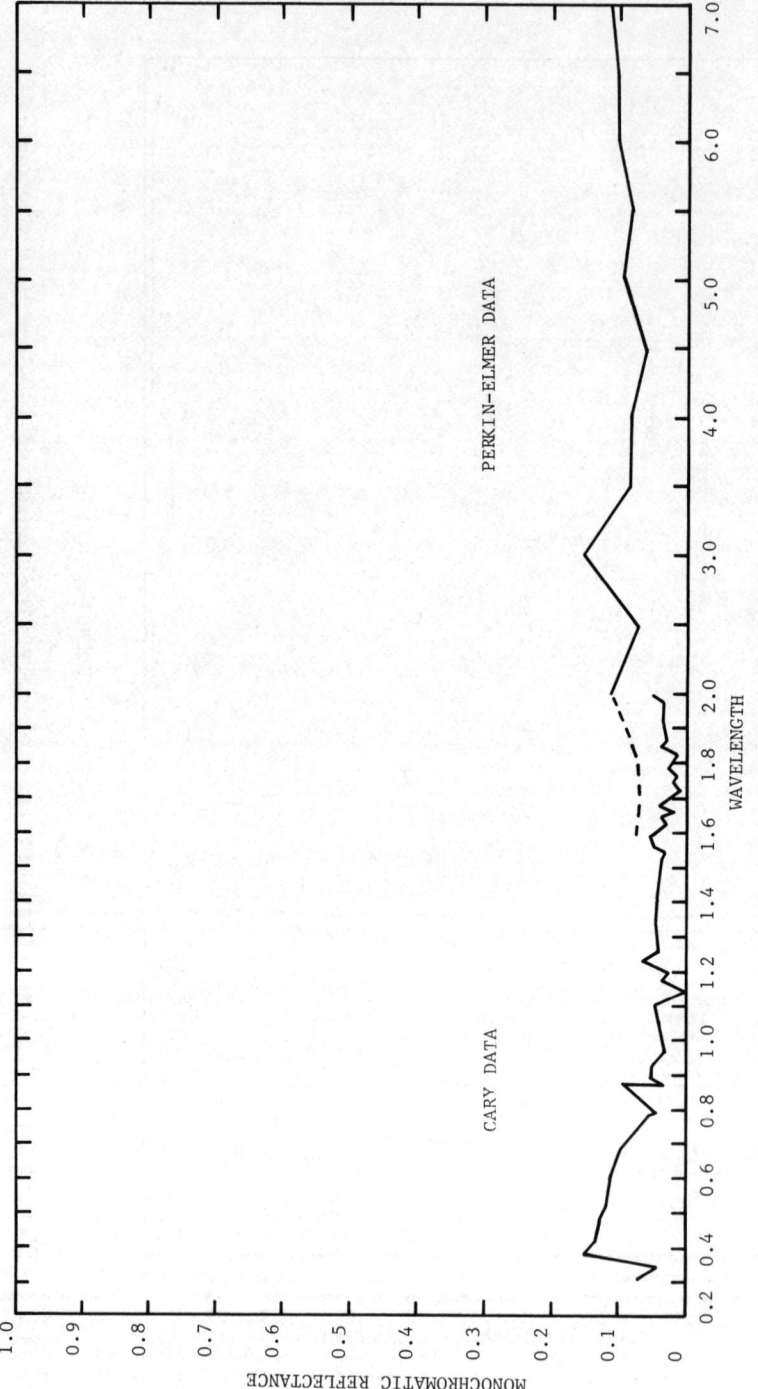

Figure 25. Spectral reflectance of STYROLUX, 0.3 - 7.0 microns.

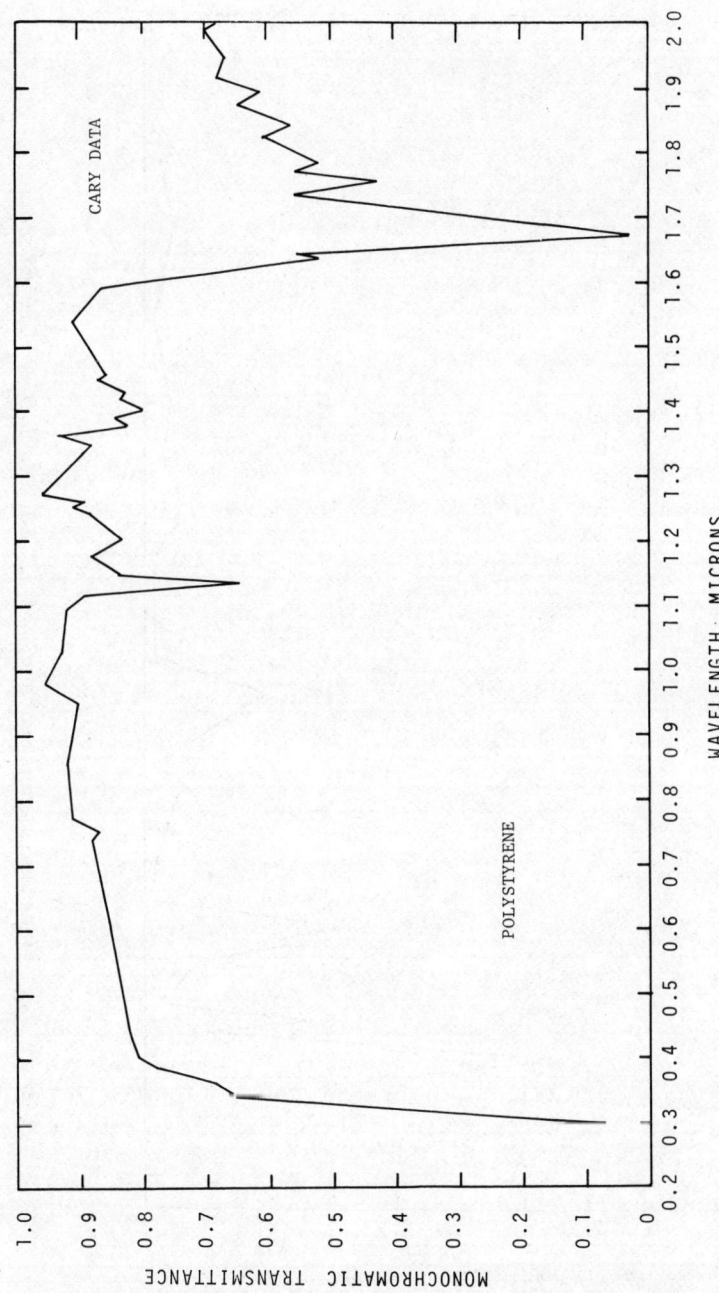

Figure 26. Spectral transmittance of STYROLUX, 0.3 – 2.0 microns.

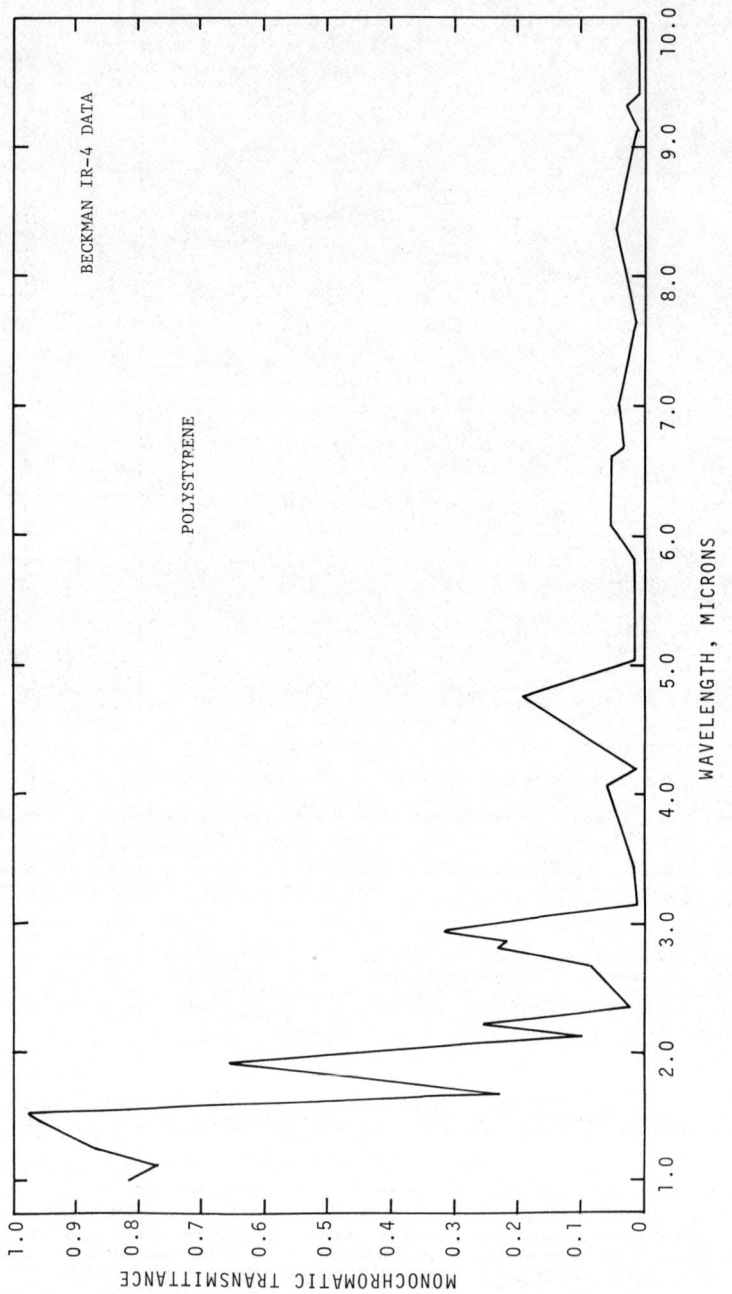

Figure 27. Spectral transmittance of STYROLUX, 1.0 – 10.0 microns.

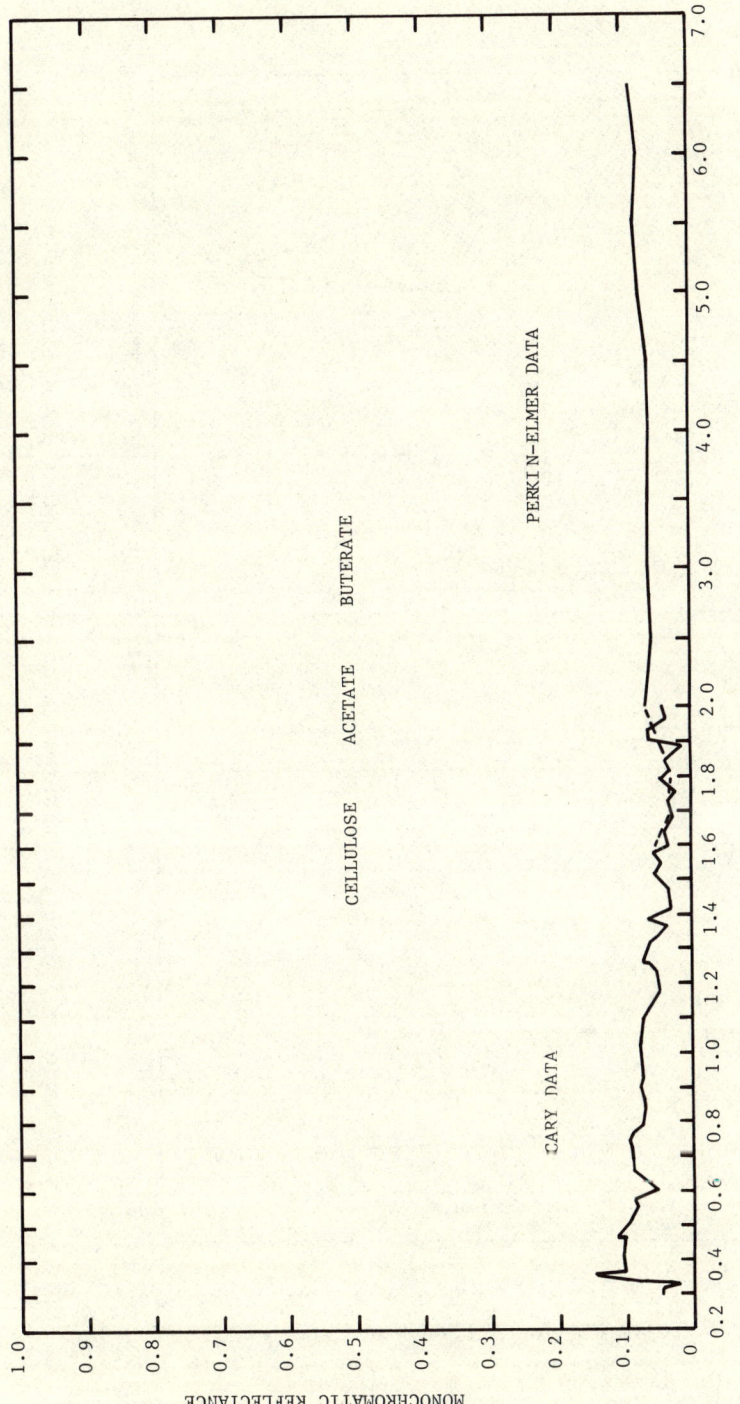

Figure 28. Spectral reflectance of UVEX, 0.3 - 7.0 microns.

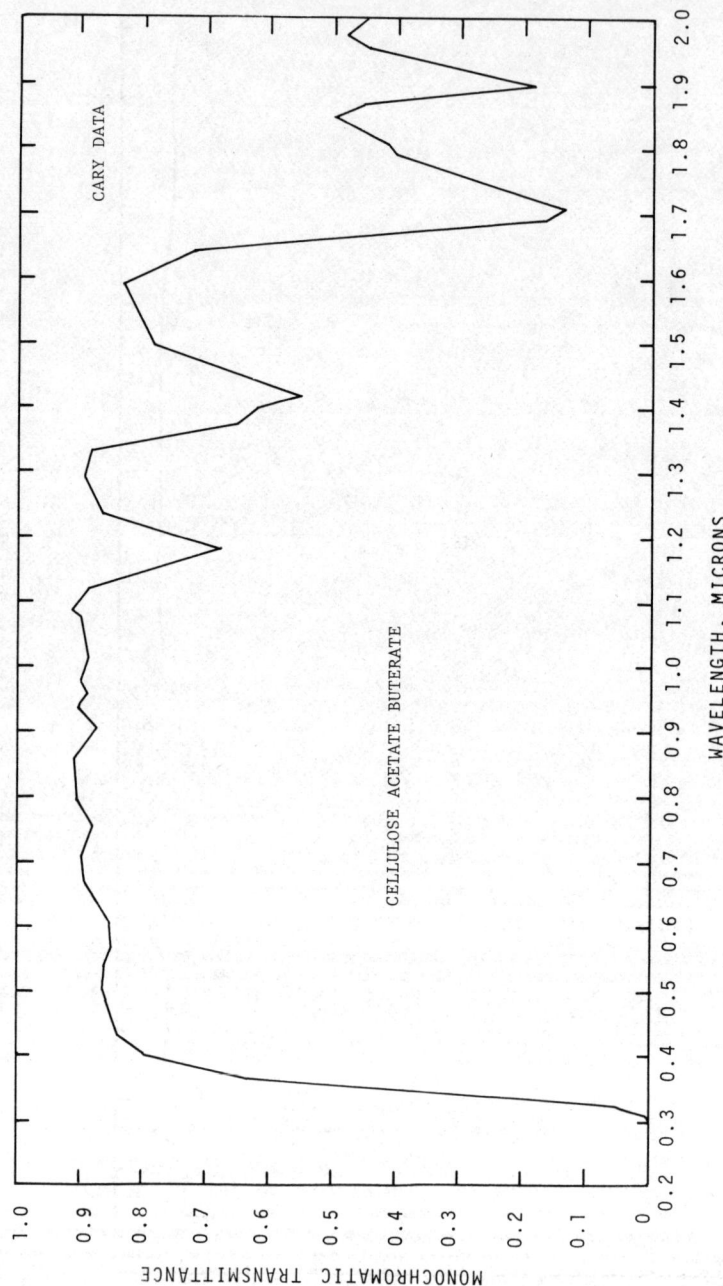

Figure 29. Spectral transmittance of UVEX, 0.3 – 2.0 microns.

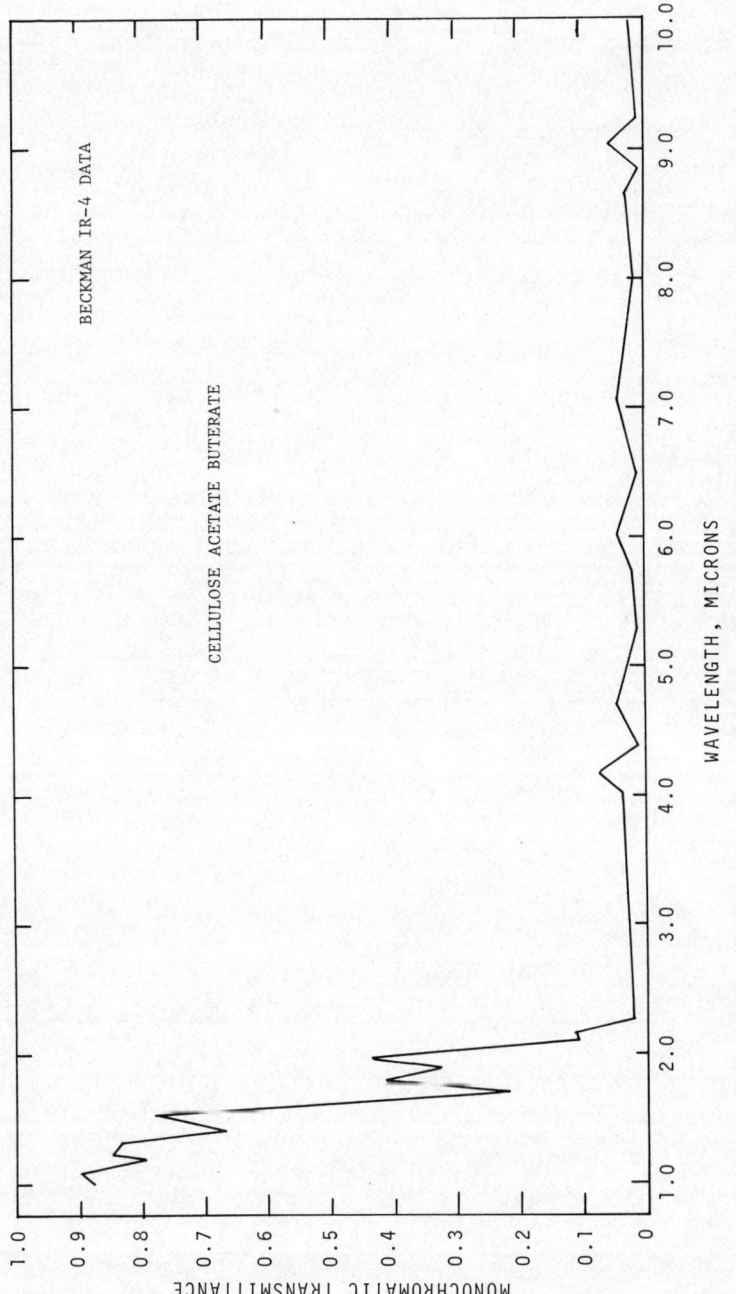

Figure 30. Spectral transmittance of UVEX, 1.0 – 10.0 microns.

# CHARACTERIZATION OF DENSE SULFONATED POLYSULFONE MEMBRANES[a]

M. A. Dinno[b], Y. Kang[c], D. R. Lloyd[d], J. E. McGrath and J. P. Wightman

Chemistry Department
Polymer Materials and Interfaces Laboratory
Virginia Polytechnic Institute & State University
Blacksburg, Virginia 24061

Sulfonated polysulfone has been shown to possess desirable physical properties for use as a reverse osmosis desalination membrane. An extensive study of dense membranes made from sulfonated polysulfone having 0.16, 0.34, 0.42, 0.53, 0.68, and 1.0 degree of sulfonation is described. The effect of the degree of the sulfonation and the counter ion, namely Na and K, on the properties of the membrane were examined. The nature of the water within the membrane was found by infrared spectroscopy to be less hydrogen-bonded when compared to the bulk water. The stoichiometry of the surface of the membrane was different by ESCA analysis from the bulk of the material. Electrical properties of the membranes were investigated by measuring the conductance of the membrane. The electrolyte/ /membrane/electrolyte system showed ohmic behavior. The membrane conductance under applied transient current was greatly dependent on the nature of the cation, whereas the nature of the anion had no effect. The activation energy for ion transport had minimal dependence on the membrane thickness, but depended on the nature of the electrolyte system. The effective size of the current-carrying channels seemed to increase with higher degrees of sulfonation, probably due to a swelling process.

## INTRODUCTION

Surface analysis of reverse osmosis membranes has been a rather neglected area within membrane science, although the interfacial and surface properties of membranes critically influence the extent of separation in reverse osmosis[1]. Especially in an asymmetric reverse osmosis membrane such as the one shown in Figure 1, it is the ultra-thin dense layer at the top that is considered responsible for the solute-water separation. The transition from the top dense layer to a much thicker porous region is clearly seen in Figure 1.

In this study, membrane properties were measured and correlated with various surface characteristics of dense sulfonated polysulfone (SPSF) membranes. Specifically, the analysis of dense SPSF membranes by electron spectroscopy for chemical analysis

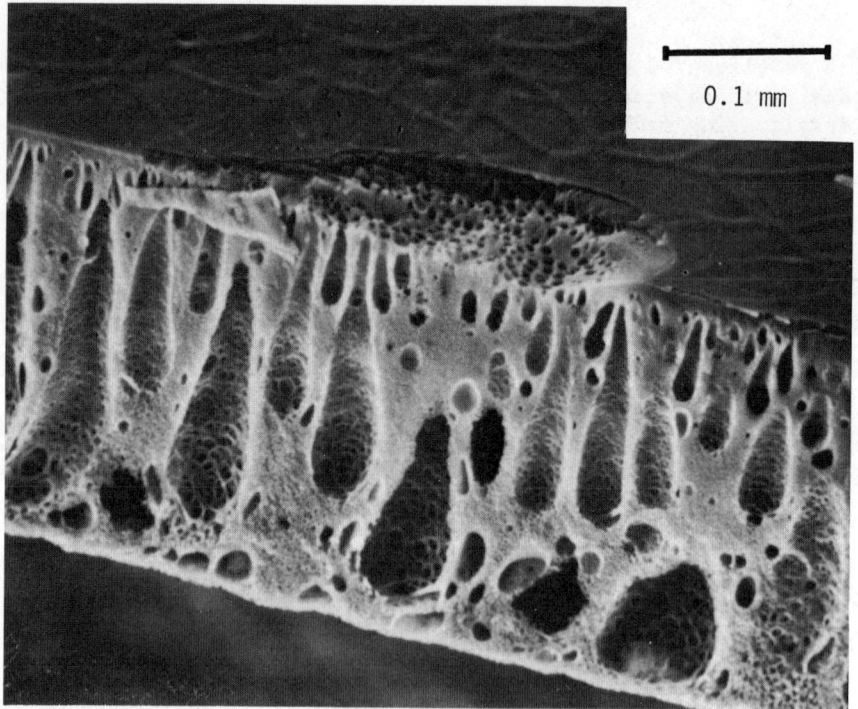

Figure 1. SEM photomicrograph of asymmetric SPSF-Na(0.42) membrane.

(ESCA) and the measurement of electrical properties of dense SPSF membranes in various electrolyte systems are reported here. Other studies of dense SPSF membranes from our laboratory, including contact angles, water uptake, ion exchange capacities, and preliminary ESCA analysis, were reported earlier.[2]

Surface chemical composition is an important property of a membrane and it can be markedly different from the bulk composition.[3] ESCA is a technique which can generally probe the top 5 nm of a solid, and variations in the surface chemical composition with depth can be obtained by systematic changes in the photoelectron takeoff angle[4]. The angle-resolved ESCA technique results in enhanced surface analysis[4-6]. Studies of polymer surfaces with ESCA are abundant in the literature[7-9], but the analysis of desalination membranes with ESCA has not yet been reported.

The electrical and electrokinetic properties of polymeric membranes[10-12] have received attention for some time, although studies on asymmetric desalination membranes have been reported only recently[13-17]. Demisch and Pusch[16] determined the electrical and electroosmotic coefficients of asymmetric cellulose acetate membranes as a function of NaCl concentration. The mobility of ions, which could effect salt separation in asymmetric cellulose acetate membranes was investigated by Kinjo and Sato[17] and the possible influence of bound water on ionic mobility within the membrane was suggested. These authors also investigated the electrical properties of dense cellulose acetate membranes reporting a drastic reduction, three orders of magnitude in the mobilities of $Na^+$ and $Cl^-$ ions in the dense membrane compared to their mobilities in the aqueous solution. In general, the properties reported in the literature are limited to intrinsic membrane potentials and electrical resistances.

## EXPERIMENTAL

### Materials

Sulfonated polysulfones of degrees of sulfonation (D.S.) 0.16, 0.34, 0.42, 0.53, 0.68, 1.00 were synthesized as the sodium and potassium salts. Dense membranes were prepared by casting from a solution of DMF or DMSO; the details of dense membrane preparation are given elsewhere[2]. Typically 0.599 M (3.5%) $NaCl_{(aq)}$ solutions were prepared. Additional electrolytes studied were solutions of $NaClO_3$, KCl, $MgCl_2$ and choline chloride. Glass-distilled water was used as the solvent.

## ESCA

ESCA analysis of dense SPSF membranes was done using a duPont 650 electron spectrometer with a magnesium anode (1253.6 eV) as the x-ray source. Binding energies were calibrated by taking the background carbon 1s photopeak at 284.6 eV. Quantitative data were obtained by correcting the areas under the photoelectron peaks, using published photoelectron cross-sections[18].

## Electrical Properties

The transport of ions across dense SPSF membranes was induced by a square pulse of transient current. Using the apparatus shown in Figure 2, a current source supplied currents between 0.5 and 100µA via 3-layer agar electrodes. The solution in both lucite chambers was circulated to and from the reservoir at the top by bubbling through a 95% $O_2$ + 5% $CO_2$ gas mixture. The change in potential across the membrane due to the current pulse was measured by calomel electrodes placed in the reservoirs and was plotted out on a strip chart recorder. This change in potential is not to be confused with the intrinsic membrane potential that may exist due to surface charge distribution.

## RESULTS AND DISCUSSION

### ESCA of Dense Membranes

The surface chemical compositions of membranes were determined using ESCA analysis and the results are tabulated in Tables I and II for the SPSF-Na and SPSF-K membranes, respectively. Atomic fractions of C, S, O, and Na or K for both the top and bottom sides of the membranes are listed in addition to the values of the Na/S and the K/S atomic ratios. Here, the top and bottom sides refer to the side of the dense membrane interfaced with air and glass plate during the casting process, respectively. The precision of the values of the atomic ratios is about 20%.

Neither the top nor the bottom sides of the membranes showed any regular dependence of chemical composition on the degree of sulfonation. Based on the ESCA analyses, the degree of sulfonation at the membrane surface does not reflect the bulk

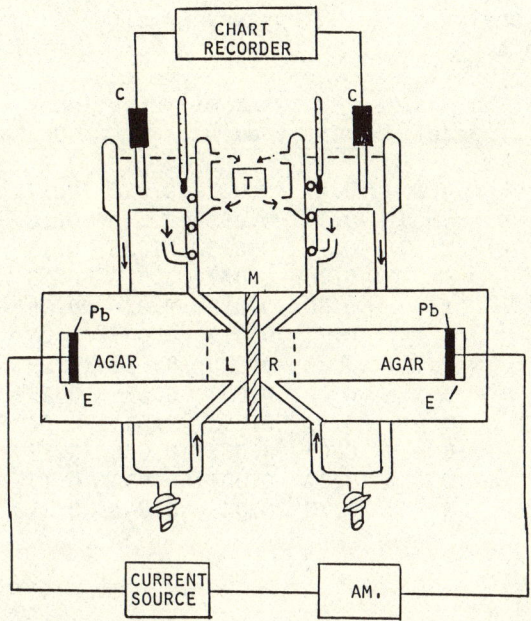

Figure 2. Diagram of apparatus for electrical property study. AM - microammeter; C - calomel electrodes; E - current sending electrodes; M - membrane; T - temperature control system.

degree of sulfonation. In fact, the atomic ratios of sulfur to counter ions, as determined experimentally by ESCA, differed significantly from the theoretical values calculated from the degree of sulfonation. Wide-scan ESCA spectra were obtained on SPSF-Na (0.42)* membranes, and only Si was found as a contaminant. The small Si photoelectron peak generated by the top surfaces of SPSF-Na membranes was significantly reduced by argon etching (≈2nm etched), which is indicative of minimal surface contamination.

## Angle-Resolved ESCA Analysis

Information regarding the surface chemical compositions for SPSF-Na and SPSF-K membranes compared to their bulk compositions

*Abbreviation meaning the sodium salt of sulfonated polysulfone having a degree of sulfonation of 0.42.

Table I.  ESCA Analysis of the Surface Chemical Composition of SPSF-Na Membranes.

| Membrane | Side | Atomic Fraction | | | | | Theor. |
| --- | --- | --- | --- | --- | --- | --- | --- |
| | | C | Na | S | O | S/Na | S/Na |
| SPSF-Na(0.16) | Top(T) | 0.77 | 0.027 | 0.027 | 0.17 | 1.0 | 7.2 |
| | Bottom(B) | 0.75 | 0.033 | 0.030 | 0.18 | 0.91 | |
| (0.34) | T | 0.78 | 0.013 | 0.026 | 0.18 | 2.0 | 3.9 |
| | B | 0.76 | 0.037 | 0.026 | 0.18 | 0.70 | |
| (0.42) | T | 0.80 | 0.010 | 0.026 | 0.16 | 2.6 | 3.4 |
| | B | 0.72 | 0.012 | 0.043 | 0.23 | 3.6 | |
| (0.53) | T | 0.76 | 0.018 | 0.033 | 0.19 | 1.4 | 2.9 |
| | B | 0.70 | 0.037 | 0.035 | 0.23 | 0.95 | |
| (0.68) | T | 0.70 | 0.053 | 0.051 | 0.20 | 0.96 | 2.5 |
| | B | 0.65 | 0.067 | 0.056 | 0.23 | 0.84 | |
| (1.0) | T | 0.82 | 0.007 | 0.030 | 0.15 | 4.3 | 2.0 |
| | B | 0.72 | 0.028 | 0.041 | 0.21 | 1.5 | |

Table II.  ESCA Analysis of the Surface Chemical Composition of SPSF-K Membranes.

| Membrane | Side | Atomic Fraction | | | | | Theor. |
| --- | --- | --- | --- | --- | --- | --- | --- |
| | | C | K | S | O | S/K | S/K |
| SPSF-K(0.12) | Top(T) | 0.80 | 0.013 | 0.022 | 0.18 | 1.7 | 9.3 |
| | Bottom(B) | 0.72 | 0.029 | 0.035 | 0.21 | 1.2 | |
| (0.32) | T | 0.77 | 0.031 | 0.028 | 0.17 | 0.90 | 4.1 |
| | B | 0.73 | 0.042 | 0.031 | 0.20 | 0.74 | |
| (0.48) | T | 0.68 | 0.015 | 0.022 | 0.29 | 1.5 | 3.1 |
| | B | 0.89 | 0.018 | 0.009 | 0.094 | 0.5 | |
| (0.72) | T | 0.76 | 0.019 | 0.028 | 0.19 | 1.5 | 2.4 |
| | B | 0.69 | 0.042 | 0.046 | 0.22 | 1.1 | |
| (0.87) | T | 0.81 | 0.013 | 0.039 | 0.40 | 3.0 | 2.1 |
| | B | 0.85 | 0.019 | 0.023 | 0.11 | 1.2 | |

was obtained from angle-resolved measurements, based on the total areas under the sulfur and sodium (or potassium) photoelectron peaks. The results of a 90° normal mode analysis are compared with the 30° surface enhanced analysis in Table III. Both SPSF-Na (0.42) and SPSF-Na (1.0) membranes showed decreases in S/Na and C/Na ratios on going from the 90° to the 30° analysis. This indicates a higher concentration of sodium on the membrane surface compared to the bulk. In contrast, the concentration of potassium on the surface was less than the bulk concentration in SPSF-K membranes, as evidenced by the increased values of the S/K and C/K ratios. The enhanced concentration of counter ions on the SPSF-Na membrane surface may explain the greater hydrophilicity of SPSF-Na membranes compared to SPSF-K membranes, as determined by their greater water uptake and by the lower contact angles of water reported previously[2].

As a follow-up study of membrane hydrophilicity, the structure of water sorbed by SPSF-Na (1.0) membranes was analyzed by near infrared spectroscopy. The values of $\lambda_{max}$ that correspond to the O-H stretching ($\nu_3$) and bending ($\nu_2$) absorption bands at various temperatures are listed in Table IV. Within experimental error, $\lambda_{max}$ did not show any temperature dependence. However, comparison of $\lambda_{max}$ for the membrane sorbed water with that of water, also listed in Table IV, shows that the $\lambda_{max}$ of water in the membrane is shifted to a shorter wavelength. This shift is attributed to weaker hydrogen bonding of water in the SPSF-Na (1.0) membrane compared to that in bulk water. A similar observation was made by Luck et al.[19] with asymmetric cellulose acetate membranes.

Table III. Comparison of Angle-Resolved ESCA of SPSF-Na and SPSF-K Membranes.

|      | SPSF-Na(0.42) | | SPSF-Na(1.0) | | SPSF-K(0.48) | | SPSF-K(0.87) | |
|------|------|------|------|------|------|------|------|------|
|      | 90°  | 30°  | 90°  | 30°  | 90°  | 30°  | 90°  | 30°  |
| S/X* | 2.3  | 1.7  | 4.7  | 0.65 | 1.4  | 2.8  | 3.0  | 6.6  |
| C/O  | 3.7  | 4.5  | 4.0  | 3.9  | 3.8  | 4.9  | 5.8  | 5.5  |
| C/S  | 20.  | 21.  | 11.  | 16.  | 28.  | 35.  | 21.  | 25   |
| C/X* | 46.  | 36.  | 53.  | 10.  | 38.  | 97.  | 62.  | 160. |
| O/S  | 5.5  | 4.8  | 2.8  | 4.0  | 7.3  | 7.2  | 3.6  | 4.5  |

*X: Na or K

Table IV. IR Absorption Due to Bound Water in SPSF-Na(1.0) Membrane at 100% Relative Humidity.

| Temperature(°C) | $\lambda_{max}$ (Å) |
|---|---|
| 22.0 | 19,190 |
|  | 19,190 |
| 20.0 | 19,190 |
|  | 19,190 |
| 18.0 | 19,190 |
|  | 19,180 |
| 16.0 | 19,180 |
|  | 19,180 |
| 14.0 | 19,190 |
|  | 19,200 |
| 12.0 | 19,190 |
|  | 19,210 |
| 10.0 | 19,200 |
|  | 19,210 |
| 22.0 | 19,190 |
|  | 19,200 |
| Avg. | 19,190 ± 10 |
| Pure Water | 19,310 ± 20 |

## Electrical Properties of Dense Membranes

Using square current pulses, the I-V (current-potential) characteristics of the electrolyte/membrane/electrolyte system were studied. The I-V plots followed a typical ohmic behavior for low currents. However, when currents larger than 40 μA were used, a slight deviation from ohmic linearity was observed. Such a behavior is expected and will be discussed below.

The effect of the size of the cation on the membrane ionic conductance was studied. The electrolyte systems chosen were sodium chloride and choline chloride. Figure 3 includes potential vs current results obtained for the four combinational systems namely choline chloride/membrane/choline chloride, NaCl/membrane/choline chloride, choline chloride/membrane/NaCl, and NaCl/membrane/NaCl. It is important to note that the first quadrant shows the effect of current passage from the left-hand chamber (see Figure 2) through the membrane to the right-hand chamber (positive direction). The third quadrant represents

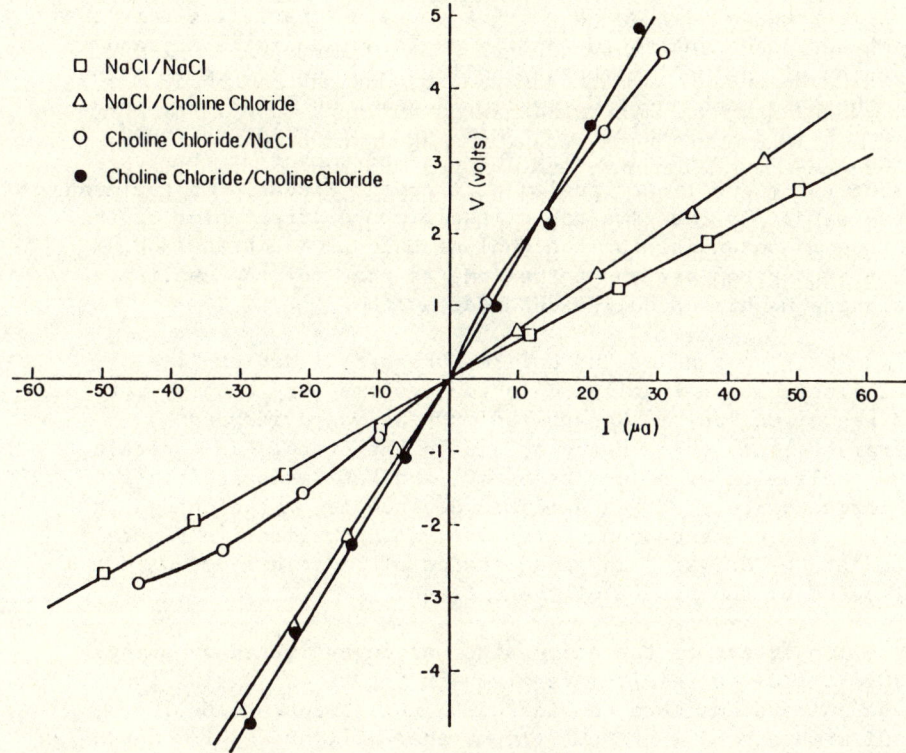

Figure 3. Membrane potential difference in volts as a function of applied current in µA for NaCl and choline chloride systems: SPSF-Na (0.42) membrane.

results obtained by reversing the direction of current flow (negative direction). A first-quadrant comparison of the NaCl/membrane/NaCl system with the choline chloride/membrane/choline chloride system indicates a 3-fold increase in the membrane resistance for the latter system. This increase reflects the greater resistance for transport of the choline ion across the membrane. Migration can be pictured as occurring through current-carrying channels within the membrane.

For the NaCl/membrane/choline chloride system, the positive direction of applied current induces migration of $Na^+$ ions across the membrane from left to right and of $Cl^-$ ions from right to left. If the ionic migration due to the applied field is pictured to be through current carrying channels within the membrane, then one could readily explain the similarity and overlap of the NaCl/membrane/choline chloride system with that of the NaCl/membrane/NaCl system as shown in Figure 3. This model will also predict the ohmic behavior at low currents, since the ionic conductivity for a fixed number of charge carriers in the linear region will depend linearly on the mean free path. By the same token, the first-quadrant plot of the change in potential for the choline chloride/membrane/sodium chloride system was about the same as that for the choline chloride/membrane/choline chloride system.

The responses of the NaCl/membrane/NaCl and choline chloride/membrane/choline chloride systems were both symmetrical to the direction of the applied current, as evidenced by straight lines through the origin in the graph. In contrast, the choline chloride/membrane/NaCl and NaCl/membrane/choline chloride systems showed a marked dependence of the change in potential on the current direction, and the lines in Figure 3 bend at the origin. The conductance of these systems were thus affected by the nature of the cations.

The effect of the anion size was investigated by using $NaClO_3$, and the results are given in Figure 4. Again, four types of systems were considered: $NaClO_3$/membrane/$NaClO_3$, NaCl/membrane/$NaClO_3$, $NaClO_3$/membrane/NaCl, and NaCl/membrane/NaCl. As is clearly shown in Figure 4, the electrical behavior is indifferent to the change in the size of the anion. In conclusion, the cations alone play a major role in the ion transport process through the current-carrying channel, although the mechanism for this transport process requires further investigation.

## Model of the Current-Carrying Channel

This investigation was performed by analyzing the shapes of voltage responses to a series of transient 1 min current pulses of 50μA in the NaCl/membrane/choline chloride system. Traces are shown in Figure 5 of the voltage response to three sequential current pulses applied from the left chamber to the right chamber (positive current), followed by another three pulses in the reverse direction (negative current.) The initial current pulse in the positive direction showed an instantaneous spike in the potential which then decreased slowly. With the second and third successive pulses, the magnitude of the spikes and the height of the plateaus decreased. Negative current also induced an 'exponential-like' shape, but the height of the

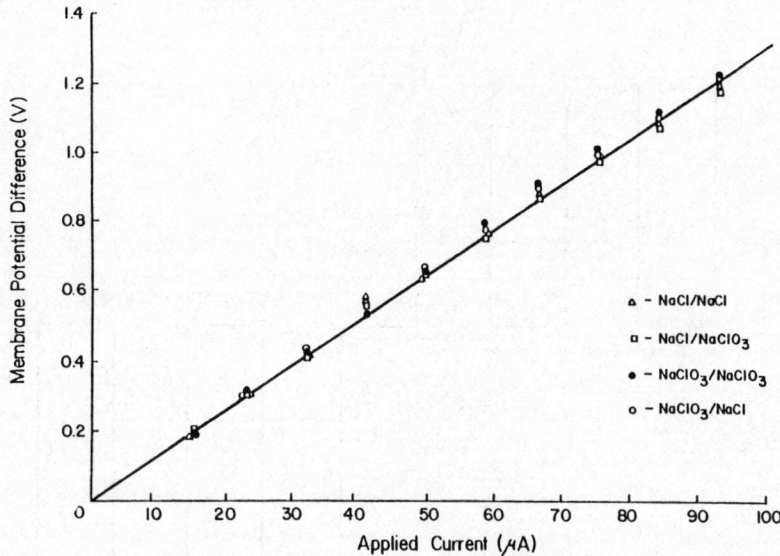

Figure 4. Membrane potential difference for NaCl and NaClO$_3$ systems: SPSF-Na(0.42) membrane.

plateau increased with excessive pulses. It should be noted here that the positive current induced the migration of Na$^+$ ions from the left chamber across the membrane to the right chamber. Choline$^+$ ions passed in the opposite direction when a negative current was applied.

Based on the shapes of the voltage response curves, the following deductions were made: (1) when positive current passed, the initially high change in potential indicated high resistance; the decrease in the potential reflected a reduction in the resistance. Therefore, as Na$^+$ and Cl$^-$ ions passed through the membrane, the resistance decreased. (2) When negative current was passed, continuous application of the current forced the migration of choline$^+$ ions and Cl$^-$ions through the membrane and increased its resistance. Combining these two deductions with the presence of Cl$^-$ions in both systems, one simple interpretation would be that the transport of Na$^+$ ions opens up the current-carrying channel whereas the choline$^+$ ions block the channel. The results obtained for the MgCl$_2$(aq) system were similar to the others except for one interesting

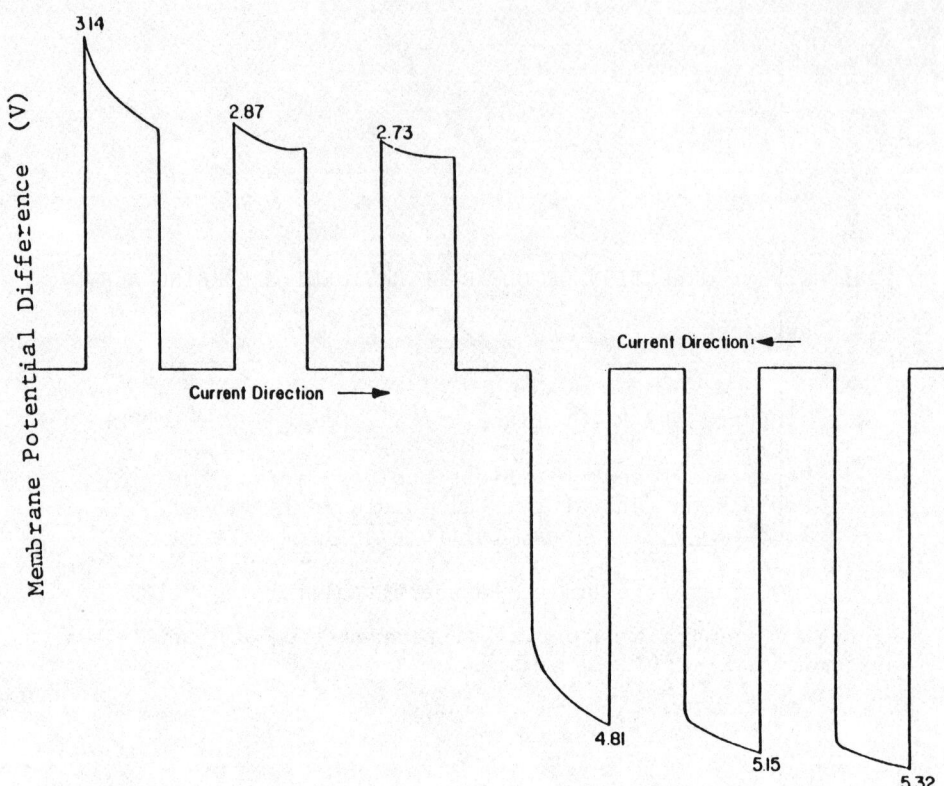

Figure 5. Overvoltage response to transient current by SPSF-Na(0.42). System: NaCl/membrane/choline chloride. Current: 50µA.

aspect. The change in potential set up in response to the applied current was not reversible for the $MgCl_2$/membrane/$MgCl_2$ system. The initial change in potential with 0.5µA was about the same as the analogous change in potential for the NaCl(aq) system. At currents greater than 10µA, however, the change in potential increased to values similar to those obtained in the choline chloride system, and they remained high even if a 0.5µA current was subsequently applied.

The conductances of several systems at three sequential currents: 0.5µA, 40µA, and 0.5µA are listed in Table V. The study was performed on one membrane but with different electrolyte systems. The NaCl/membrane/NaCl system was repeated periodically to assure the reproducibility. For all the systems studied except for the $MgCl_2$ system discussed above, the value of the conductance remained relatively constant for the three sequential current pulses. Furthermore, the conductance for choline

Table V. Conductance of SPSF-Na(0.42) Membrane in Various Electrolyte Systems.

| System | 0.5μA | Applied Current 50(20)μA | 0.5μA |
|---|---|---|---|
| NaCl/NaCl | 15.4* | 18.0 | 16.7 |
| MgCl$_2$/MgCl$_2$ | 11.0 | (3.5) | (3.9) |
| NaCl/MgCl$_2$ | (6.6) | (5.9/3.9) | (5.7) |
| NaCl/NaCl | 13.7 | 15.8 | |
| KCl/KCl | 19.2 | 19.2 | 19.2 |
| NaCl/KCl | 19.2 | 19.2 | 19.2 |
| NaCl/NaCl | 15.4 | 17.1 | 15.4 |
| ChCl/ChCl** | (3.5) | (3.3) | (3.5) |
| NaCl/ChCl | (5.6/4.6) | (5.6/3.4) | |
| NaCl/NaCl | 11.0 | 14.5 | 12.8 |
| LiCl/LiCl | 12.8 | 15.0 | 13.3 |
| NaCl/LiCl | 13.2 | 15.8 | 13.3 |

*Units: μmho cm$^{-2}$
**ChCl: Choline Chloride [(CH$_3$)$_3$NCH$_2$CH$_2$OH]$^+$Cl$^-$

chloride/membrane/choline chloride system was about a factor of five smaller than the conductance for NaCl/membrane/NaCl system. An explanation for this phenomenon follows and provides the basis for the construction of a channel model. Consider a channel with a certain radius, as shown in Figure 6. The sizes of hydrated Na$^+$ ions and the choline$^+$ ions are smaller and larger, respectively, relative to the channel radius. Thus a higher resistance may be expected for the choline ions. The actual physical characteristics of these channels may not be altered by high currents. However, new factors seem to develop, namely (i) clogging of the channels with large enough ions that could barely go through and (ii) the formation of accumulation and depletion layers at the interface, which will alter its energy barrier and therefore manifest what could be the rate limiting step.

The results obtained with the Mg$^{++}$ system support the presence of the first factor. At low currents, the hydrated Mg$^{++}$ ions will migrate through the membrane in a manner similar to that of the Na$^+$ ions. However, with an increase in current, a large number of Mg$^{++}$ ions would push into the membrane thus causing a large increase in collisions and scattering, which would result obviously in shortening the mean free path to near zero (clogging). This simple model is consistent with the experimental results. While it calls for further verification,

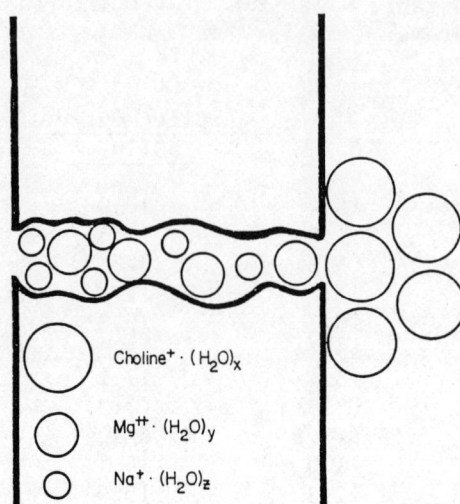

Figure 6. Model for current-carrying channel of SPSF-Na(0.42) membrane.

techniques developed in keeping with the proposed model could be useful in determining the channel sizes of a variety of polymer membranes.

### Electrical Properties of Membranes of Different Degrees of Sulfonation

The effect of the degree of sulfonation on the electrical resistances of membranes was also investigated. The conductances for the various systems studied are listed in Table VI. As can be seen, polysulfone showed the lowest conductance, which increased with increasing degree of sulfonation. In fact, the conductance of the SPSF-Na (1.0) membrane was identical to the conductance of the bulk solution. Hence it can be concluded that increased sulfonation leads to swelling of the membrane in water, which results in larger channels and therefore significantly decreases the selectivity of the membranes. This swelling may explain the poor performance in desalination systems of polysulfone with a high degree of sulfonation, although a certain degree of sulfonation is essential to impart hydrophilicity and produce the necessary flux.

Table VI. Conductance of Various Membranes.

| Membrane | Conductance ($\mu$mho cm$^{-2}$) |
|---|---|
| Polysulfone | 1.65 |
| SPSF-Na(0.42) | 21.4 |
| SPSF-Na(1.0) | (59600) |
| Solution | (59600) |

Applied Current: 0.5µA (50µA)
System: NaCl/Membrane/NaCl

In conclusion, the conductance studies made it possible to obtain information concerning the channel size. In addition, by simply measuring the membrane resistance with the present apparatus, the membrane performance may be predicted as will be discussed below.

## Activation Energies for Ion Transport

The activation energies for ion transport were calculated[20,21] on the basis of an Arrhenius-type equation,

$$\underline{R} = A \exp(E_a^{\ddagger}/RT)$$

where $\underline{R}$ is the membrane resistance, $E_a^{\ddagger}$ is the activation energy, R is the gas constant, T is the Kelvin temperature, and A is a factor that represents the hopping frequency of the ions across the energy barrier. The natural log of membrane resistance was plotted against 1/T to obtain the value of $E_a^{\ddagger}$.

The effect of variations in membrane thickness and differences in electrolytes on the magnitude of the $E_a^{\ddagger}$ were investigated. The thicknesses of the cast SPSF-Na(0.42) membrane prior to drying were 8, 12, and 16 mil; the thicknesses of the dried membranes were 0.57, 0.98, and 1.4 mil, respectively. The activation energies for these three membranes are tabulated in Table VII. These values for the activation energies compare well with the activation energies of desalination of SPSF membranes (8 to 10 kcal/mole) reported by Vinnikovz and Tanny[22]. It is interesting to note that the value of $E_a^{\ddagger}$ remained relatively constant for all three thicknesses. Since the I-V plots of these membranes, especially at low currents exhibited a pure ohmic behavior, it was not surprising to find that $\underline{R}$ increased as a function of thickness. However, the fact that the activation energy $E_a^{\ddagger}$ was found to be independent of the thickness indicates

Table VII. Activation Energies ($E_a^{\ddagger}$) of SPSF-Na(0.42) Membranes for 0.5 μA in NaCl(aq) Systems.

| Membrane Thickness | | $E_a^{\ddagger}$ (kcal/mole) |
|---|---|---|
| Cast (mil) | Dry (mil) | |
| 8 | 0.57 | 8.7 ± 0.8 |
| 12 | 0.98 | 9.2 ± 0.1 |
| 16 | 1.4 | 9.4 ± 0.1 |

that the rate limiting step is caused by a mechanism other than that of ionic scattering and collisions within the membrane channels as discussed above. It appears that the main contribution to the energy barrier is the initial interaction of ions with the membrane at the membrane/solution interface. Obviously, this initial step is present in any membrane/electrolyte system; thus, the secondary step of ion transport within the membrane either contributes much less to the process or has the same $E_a^{\ddagger}$ value regardless of the membrane thickness. However, the difference would be manifested as parallel curves (same slope but different intercept A) in the Arrhenius plots.

The effect of the nature of the electrolytes on the activation energy is seen in Table VIII. The resistances were considerably higher for the $MgCl_2$(aq) and choline chloride(aq) systems as compared to NaCl(aq). The values of $E_a^{\ddagger}$ were about the same for the $MgCl_2$(aq) and NaCl(aq) systems and much lower for the choline chloride system. This rather surprising result may be interpreted as follows: for the NaCl(aq) and $MgCl_2$(aq) systems, the average value of $E_a^{\ddagger}$ for ion transport across the SPSF-Na(0.42) membrane was around 9 kcal/mole, independent of the membrane thickness and the nature of the electrolyte solution. Comparison of this $E_a^{\ddagger}$ value to the value of 13 kcal/mole for the

Table VIII. Effect of Electrolytes on $E_a^{\ddagger}$ for 8 mil SPSF-Na(0.42) Membrane.

| System | $E_a^{\ddagger}$ (kcal/mole) |
|---|---|
| NaCl/NaCl | 9.2 |
| $MgCl_2$/$MgCl_2$ | 8.6 |
| ChCl/ChCl* | 2.4 |
| 3.5% NaCl solution | 3.3 |

*ChCl : Choline Chloride

disruption of bound water indicates that the ion transport process may involve the breakage of bound water. In addition, similarities in the energy barriers for the NaCl and $MgCl_2$ systems can be interpreted as arising from similar transport mechanisms.

The notably lower activation energies for the choline system can be understood as follows: The relatively large choline$^+$ ions may not pass through the membrane channel, thus exhibiting a very small temperature dependence (if any). Therefore, the energy barrier for the choline$^+$ ion transport can be regarded as infinite. The actual conductance process must be carried out by either $Cl^-$ ion or hydroxyl ion resulting from the breaking of free water. This speculation is supported indirectly by noting that the activation energy for disturbance of free water by electrolytes is 3.3 kcal/mole.

### Electrical Properties and Membrane Performance

Electrical properties were measured for various membranes after they were used in reverse osmosis desalination to study the possible correlation between electrical properties and membrane performance. Membranes were air-dried upon removal from the reverse osmosis cells and placed in the system used for measuring the electrical properties described above. The results of this study are given in Table IX. The tabulated values for the conductances(G) have been corrected for the 0.599M solution resistances of 16.6, 22.8, and 11.4 ohm cm$^{-2}$ for NaCl, choline chloride, and $MgCl_2$, respectively.

Table IX. Membrane Performance and Electrical Properties.

| Membrane | Rate(g/min) | Rejection (%) | G($\mu$mho cm$^{-2}$) | | |
|---|---|---|---|---|---|
| | | | NaCl/NaCl | ChCl/ChCl | $MgCl_2$/$MgCl_2$ |
| PS | 6.7x10$^{-6}$ | ---- | 0.53 | 0.89 | 0.70 |
| SPSF-Na(0.34) | 7.4x10$^{-3}$ | 96 | 3.0x10$^4$ | 6.0x10$^2$ | 1.6x10$^3$ |
| SPSF-Na(0.53) | 8.0x20$^{-3}$ | 92 | 2.4x10$^5$ | 4.3x10$^3$ | 1.5x10$^4$ |
| SPSF-Na(0.68) | 5.4x10$^{-2}$ | 86 | 1.0x10$^6$ | 1.2x10$^5$ | 6.4x10$^4$ |
| SPSF-Na(1.0) | 1.4x10$^{-1}$ | 66 | -- | -- | -- |

*ChCl: Choline Chloride

In regard to membrane performance, a well-established trend was noted between rate and the degree of sulfonation. The polysulfone membrane gave practically zero rate. However, membrane rate improved significantly with increasing degrees of sulfonation; this increase in rate reflecting the increase in the degree of sulfonation. Similarly, a decrease in the salt rejection was observed for the greater degrees of polysulfonation. The membrane conductance was small for the polysulfone membrane and there was minimal dependence of the conductance on the particular electrolyte system. Thus, it could be concluded that the size of the ion transport channel in the polysulfone membrane is small indeed.

For the sulfonated membranes, however, the membrane conductances were significantly greater and showed practically no resistance at the highest degree of sulfonation. In addition to the mutually higher membrane conductances, other differences between the three electrolyte systems were minimal at the highest degree of sulfonation. For example, with SPSF-Na(0.34), the conductances for the NaCl, choline chloride, and $MgCl_2$ electrolyte systems varied by an order of magnitude. However, with SPSF-Na(1.0), there was no significant difference in conductance between the three electrolyte systems.

The results listed in Table IX seem to be consistent with the model developed in the previous sections. The sulfonation of polysulfone introduces channels similar to those found in various kinds of ion exchange materials. Increasing the degree of sulfonation, in an attempt to improve the hydrophilicity of the membrane/electrolyte interface, caused an increase in the number of these channels and/or their diameter and thereby increasing ionic conductance and decreasing salt rejection. However, while the number of these channels or their size may be very important in determining the ionic current density through the membrane, the rate limiting step appears to be at the membrane/electrolyte interface as presented earlier, where the activation energy for ionic conductance was found to be independent of the thickness of the membrane.

The following conclusions can be made from this study: (1) membrane conductance must be greater than 1 $\mu ohm \cdot cm^{-2}$ for the membrane to be suitable as a reverse osmosis membrane; (2) ion transport channels should have critical diameters that differentiate between electrolytes to give different electrical conductances; these conductance measurements provide a quick determination of the suitability of a membrane for reverse osmosis desalination.

## SUMMARY

Sulfonated polysulfone (SPSF) membranes were relatively free of surface contamination. Si, the only observed contaminant, was readily removed by argon etching. The surface chemical compositions of SPSF membranes were quite different from their bulk chemical composition. The surface concentrations of counter ions were greater for SPSF-Na membranes compared to SPSF-K membranes, resulting in greater surface hydrophilicity for SPSF-Na membranes. Membrane conductance was affected greatly by the nature of the cations in the electrolyte and the degree of membrane sulfonation. Measured activation energies showed minimal dependence on membrane thickness but they were highly dependent on the nature of the electrolyte system indicating that the rate limiting step is the membrane/electrolyte interface. Conductance measurements proved to be convenient for estimation of the sizes of current-carrying channels and prediction of the membrane performances in reverse osmosis desalination.

## ACKNOWLEDGEMENTS

The authors thank Andy Mollick for the design of the IR cell holder and Beth Dillinger for her technical assistance. The assistance of Dr. James S. Jen in running the argon-etching ESCA experiments is very much appreciated. Partial financial support for this work, including a graduate research assistantship for one of us (YK), was provided by OWRT Grant Nos. 14-34-0001-9404 and 14-34-0001-1488. Additional financial support was provided under ONR Contract Nos. N0001479C0971 and N0001481K0663.

## REFERENCES

[a] Based on the Ph.D. Dissertation of Yoonok Kang.
[b] Present address: Physics Department, University of Mississippi, University, MS 38677.
[c] Present address: Industrial Lab., Bldg. No. 34, Kodak Park, Rochester, NY 14650.
[d] Present address: Chemical Engineering Department, University of Texas, Austin, TX 78712.

1. S. Sourirajan, Editor, "Reverse Osmosis and Synthetic Membranes: Theory-Technology-Engineering" Chapter 1, National Research Council of Canada Publications, Ottawa, Ontario, Canada, 1977.

2. D. R. Lloyd, L. E. Gerlowski, C. D. Sunderland, J. P. Wightman, J. E. McGrath, M. Iqbal and Y. Kang in "Synthetic Membranes." A. F. Turbak, Editor, Vol. I, ACS Symposium Series No. 153, pp. 327-350. American Chemical Society, Washington, D.C., 1981.
3. D. W. Dwight, J. E. McGrath and J. P. Wightman, J. Appl. Polym. Sci: Appl. Polym. Symp., $\underline{34}$, 35 (1978).
4. C. S. Fadley, R. J. Baird, W. Siekhaus, T. Novakov and S. A. L. Bergstrom, J. Electron Spectrosc. Relat. Phenom., $\underline{4}$, 93 (1974).
5. D. T. Clark, Advances in Polym. Sci., $\underline{24}$, 125 (1977).
6. D. T. Clark, D. B. Adams, A. Dilks, J. Peeling and H. R. Thomas, J. Electron Spectrosc. Relat. Phenom., $\underline{8}$, 51 (1976).
7. D. T. Clark in "Progress in Theoretical Organic Chemistry", I. G. Csizmadia, Editor, Vol. 2, Elsevier, Amsterdam, 1976.
8. B. D. Ratner, P. K. Weathersby, A. S. Hoffman, M. A. Kelly, and L. H. Sharpen, J. Appl. Polym. Sci., $\underline{22}$, 643 (1978).
9. H. R. Thomas and J. J. O'Malley, Macromolecules, $\underline{12}$, 323 (1979).
10. K. S. Spiegler, Desalination, $\underline{9}$, 367 (1971).
11. F. L. Ramp, Desalination, $\underline{16}$, 321 (1975).
12. C. A. Kumins and A. London, J. Polym. Sci., $\underline{46}$, 395, (1960).
13. S. Horigome and Y. Taniguchi, J. Appl. Polym. Sci., $\underline{21}$, 343 (1977).
14. S. G. Wong and J.C.T. Kwak, Desalination, $\underline{15}$, 213 (1974).
15. K. W. Choi and D. N. Bennion, Ind. Eng. Chem., Fundam., $\underline{14}$, 296 (1975).
16. H. U. Demisch and W. Pusch, J. Colloid Interface Sci., $\underline{76}$, 445 (1980).
17. N. Kinjo and M. Sato, Desalination, $\underline{27}$, 71 (1978).
18. J. H. Scofield, J. Electron Spectrosc. Relat. Phenom., $\underline{8}$, 129 (1976).
19. W. A. P. Luck, D. Schioberg and U. Siemann, J. Chem. Soc., Faraday Trans. 2, $\underline{76}$, 136 (1980).
20. A. Singh, Ph.D. Dissertation. Virginia Polytechnic Institute and State University, Blacksburg, Virginia, 1980.
21. H. Eyring and E. M. Eyring, "Modern Chemical Kinetics", p. 51, Reinhold Publishing Corp., New York, 1963.
22. N. Vinnikova and G. B. Tanny, in "Synthetic Membranes," A. F. Turbak, Editor, Vol. I, ACS Symposium Series No. 153, pp. 351-365, American Chemical Society, Washington, D.C., 1981.

# SURFACE PHENOMENA IN LATEX FILMS VULCANIZATION

O. Shepelev and M. Shepelev

Nir-Lat

Kibbutz Nir-Oz, P.M. Negev 85122, Israel

In the sulphur vulcanization of latex films, the formation of cross-links proceeds only on the surface of the latex globules. The rate of reaction of vulcanization is determined by content of rubber in the surface layer of globules. The rate of reaction of vulcanization is zero order with respect to sulphur. Therefore it is possible, by regulating the amount of vulcanizing agents in the latex, to define the quantity of rubber taking part in the vulcanization reaction, and to calculate the average thickness of the surface reaction zone. The peculiar mechanical characteristics of these films are explained by their unusual network structure.

## INTRODUCTION

Recently a number of works have been published in which the peculiarities of the structure of vulcanized latex films are described in detail. By means of a number of different methods, it was determined that these films have a microheterogeneous structure which represents the combination of a highly vulcanized polymer on the surface layer of latex globules and an unvulcanized polymer inside the globules. The relation between the topology of the vulcanizing network of latex films and their properties has been described[1-4]. The present work is devoted to an investigation of the kinetics of vulcanization process of latex films.

## MATERIALS AND METHODS OF INVESTIGATION

We investigated the vulcanization kinetics of films which were obtained from styrene-butadiene latexes. The basic characteristics of latexes are shown in Table I.

Table I. Basic Properties of Latexes Used in This Work.

| Type of latex | Total solids, % | Average globules diameter, Å | Styrene-butadiene ratio |
|---|---|---|---|
| Polysar IY | 43.5 | 1,800 | 45 : 55 |
| Polysar 646 | 47.1 | 2,700 | 45 : 55 |
| Polysar 725 | 66.1 | 5,000 | 34 : 66 |

A vulcanizing group of variable composition, pph, which is shown in Table II, was introduced in latexes to conduct sulphur vulcanization of films. Films were obtained by air drying of blends on a polyethylene surface for 48 hours at room temperature. Vulcanization of the films was carried out in an air oven at 100, 120 and 140°C.

During the investigation we tested, by weight method, the maximum swelling of vulcanized films in xylene, solfraction content, content of free sulphur[5] and mechanical properties of the films. Calculations of the vulcanization network parameter were carried out according to the procedure described earlier[6,7*].

---

*) Estimations of the network parameter by Flory-Rehner equation had formal nature in view of non-statistical distribution of cross-links.

Table II. Composition of the Vulcanizing Groups for Latex Blends.

| Material | Number of latex blend | | | | | | | | |
|---|---|---|---|---|---|---|---|---|---|
| | 1 | 2 | 3 | 4 | 5 | 6 | 7 | 8 | 9 |
| Sulphur | 0.100 | 0.500 | 1.00 | 1.50 | 2.00 | 2.50 | 3.00 | 5.00 | 10.0 |
| Zinc dimethyldithiocarbamate | 0.033 | 0.165 | 0.33 | 0.50 | 0.67 | 0.83 | 1.00 | 1.65 | 3.3 |
| Zinc mercaptobenzothiazole | 0.033 | 0.165 | 0.33 | 0.50 | 0.67 | 0.83 | 1.00 | 1.65 | 3.3 |
| Zinc Oxide | 0.100 | 0.500 | 1.00 | 1.50 | 2.00 | 2.50 | 3.00 | 5.00 | 10.0 |

For the sake of comparison, films prepared from solutions of blends 4-9 in xylene by drying for 72 hours at room temperature were tested.

## EXPERIMENTAL DATA AND DISCUSSION

Figures 1 and 2 show the degree of polymer crosslinking and free sulphur content as a function of vulcanization time for the films prepared from latex Polysar IY. In the case where the initial sulphur content in the films was 2.5 pph or more (blends 6-9), the rate of addition of sulphur to rubber does not change during the vulcanization process, (i.e., it can be described by a straight line).

At the same time, the curve of the change of free sulphur content during vulcanization of latex films obtained from blends 1-5 is rather accurately described by a straight line in semilogarithm coordinates (Figure 3). Thus, depending on the sulphur content in the blend, vulcanization can proceed with 1-st order kinetics with respect to sulphur - as occurs for blends 1-5 - if the sulphur content is 2 pph or less; or with zero order if the sulphur content is 2.5 pph or more.

As long as sulphur content in the blends is considerably less than that of rubber (for example, in the blend 6 sulphur : rubber ratio is 1:40), independence of the reaction rate on the sulphur concentration can be explained only by concluding that the vulcanization reaction affects only a small part of the rubber - the surface layers of latex globules - while the main part of rubber, the inner volume of latex globules, remains unvulcanized.

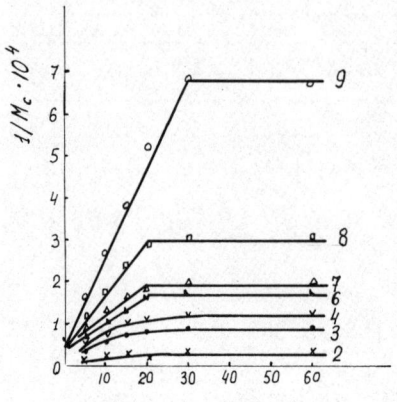

Figure 1. Dependence of $1/M_c$ on the duration of vulcanization at 120°C for films made from latex Polysar IY. Numbers on the curves correspond to the numbers of blends.

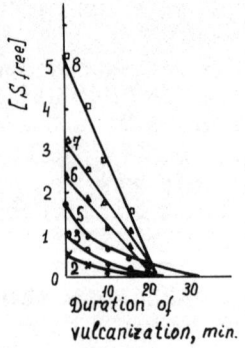

Figure 2. Dependence of free sulphur content, [S free], in the films made from latex Polysar IY on the duration of vulcanization at 120°C. Numbers on the curves correspond to the numbers of blends.

The process of latex films vulcanization can be presented in the following way: particles of sulphur in the film are surrounded by latex globules and, until the moment of vulcanization, no marked diffusion of sulphur into rubber occurs. This is confirmed by the same size of sulphur particles in the dispersion and dried film.[8] As the vulcanization process takes place at a temperature higher than the temperature of sulphur melting (114°C), the sulphur

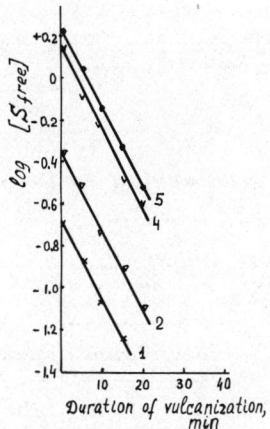

Figure 3. Dependence of the logarithm of free sulphur content, log [S free], in the films made from latex Polysar IY, on the duration of vulcanization at 120°C. Numbers on the curves correspond to the numbers of blends.

particles are already melted at the beginning of the process and the interaction between sulphur and the surface layer of the globules begins. As soon as the quantity of sulphur in the reaction zone decreases, because of its addition to rubber, the new portion of sulphur comes into the reaction zone as it melts in the interglobular space. This process continues until all the melted sulphur from the interglobular space is spent. The amount of sulphur in the reaction zone is controlled, apparently, by the solubility of sulphur in rubber at the temperature of vulcanization. If the concentration of sulphur exceeds its top solubility in the surface layers of globules, the reaction proceeds according to zero order; if the quantity of sulphur is less than this limit, the reaction goes according to 1-st order. Thus, knowing the solubility of sulphur in rubber and the top quantity of sulphur at which the vulcanization reaction goes according to 1-st order, it is possible to estimate the thickness of the layer, $\Delta$, in which the vulcanization takes place. Assuming the following conditions
- solubility of sulphur in rubber at 120°C is 9.8 pph;
- content of sulphur at which vulcanization proceeds according to the 1-st order is between 2 and 2.5 pph;
  $\Delta$ was calculated to be 75Å.

Naturally, the maximum quantity of sulphur at which vulcanization proceeds according to zero order depends on the size of latex particles.

Assuming the value of Δ to be constant for styrene-butadiene latexes, it is possible to calculate the minimum quantity of sulphur necessary for zero order vulcanization reaction for latexes of different globules size. Results are presented in Table III along with experimentally determined values.

Table III. Dependence of the Minimum Sulphur Quantity, S Min, at Which the Reaction Proceeds According to Zero Order, on the Globules Size.

| Type of Latex | Average diameter of globules, Å | S Min, pph | |
|---|---|---|---|
| | | Calculated | Experimentally determined |
| Polysar IY | 1,800 | - | between 2.0 - 2.5 |
| Polysar 646 | 2,700 | > 1.4 | between 1.0 - 1.5 |
| Polysar 725 | 5,000 | > 0.8 | between 0.5 - 1.0 |

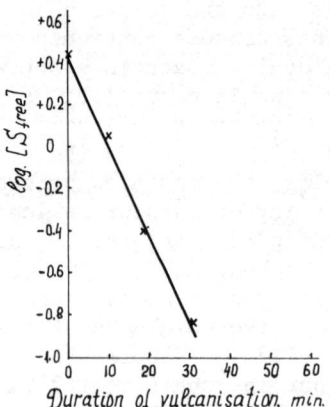

Figure 4. Dependence of the logarithm of free sulphur content, log [S free], in the films obtained from the solution in xylene of blend analogous to blend 7, on the duration of vulcanization at 140°C.

# SURFACE PHENOMENA IN LATEX FILMS VULCANIZATION

The good agreement between the calculated and experimentally determined values clearly shows that our assumptions are justified. The kinetics of vulcanization of films, obtained from the solution in xylene of a blend analogous in composition to blend 7, was also investigated. From the data presented in Figure 4, one can see that the vulcanization of these films is described by 1-st order kinetics (the curve of dependence of free sulphur content on vulcanization duration becomes straight in semi-logarithmic coordinates).

In Figure 5, dependence of stress at 100% elongation and tensile strength of films on the degree of polymer cross-linking is shown. One can see that there exists a correlation between the degree of rubber vulcanization and mechanical properties of the films. The latter is simply determined by the $1/M_c$ value, i.e. the results of testing of films, obtained from all blends 1 - 9, are described by this curve. This fact obviously confirms that the process of vulcanization, independent of its duration or the content of the vulcanizing group, proceeds according to one mechanism, i.e., on the surface of the latex globules.

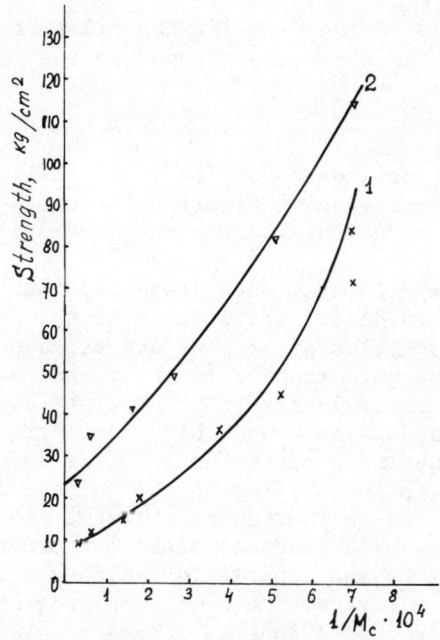

Figure 5. Dependence of stress at 100% elongation - curve 1, and tensile strength - curve 2, on the degree of polymer cross-linking in the films obtained from latex Polysar IY.

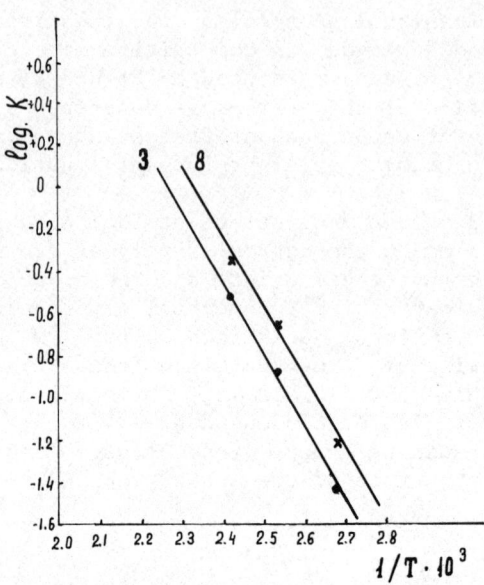

Figure 6. Dependence of log K on 1/T for films from latex Polysar IY. The numbers on the curves correspond to the numbers of blends.

In Figure 6 the dependence of log K on 1/T is shown. On the basis of this dependence, the activation energy, $E_a$, of the vulcanization process was calculated to be 16 kcal/mol.

From all the experimental data presented thus far it is possible to imagine in detail the process of vulcanization in these latex films. The particles of sulphur are arranged in the interglobular space and do not dissolve in the rubber until the vulcanization begins. At the vulcanization temperature, melting of the sulphur particles takes place, and they proceed to dissolve into surface layers of the latex particles. If the amount of sulphur is less or equal to the solubility of the sulphur in globules surface layers, all the sulphur comes into the reaction zone and the vulcanization reaction proceeds with 1-st order kinetics with respect to sulphur. If the quantity of sulphur exceeds its solubility in the surface layers, there will be progressive transport of sulphur into the reaction zone as it reacts with the rubber. In this case, the reaction rate remains constant in time until all the sulphur is exhausted. Under these circumstances the reaction proceeds according to zero order kinetics.

The results of mechanical tests, presented in Figure 5, show that the tensile strength of vulcanized latex films surpasses considerably that of analogous vulcanized rubbers (30-35 kg/cm$^2$). This fact, already determined repeatedly for other types of rubber[2], can be explained in terms of the peculiarity of network of latex films distribution.

## CONCLUSIONS

It is shown that the localization of the vulcanization process on the surface of latex globules suggests that this process takes place according to zero order kinetics, if sulphur content exceeds it solubility in the globules surface layers.

The approximate thickness of globules layers, in which the vulcanization takes place, can be calculated.

## REFERENCES

1. V. V. Chernaya, S. A. Shteinberg, Yu. V. Grubman and M. I. Shepelev, Dokl. Akad. Nauk SSSR, 197, 112 (1971).
2. V. V. Chernaya, S. A. Shteinberg and M. I. Shepelev, Plast. Kaut., 18, 209 (1971).
3. M. I. Shepelev, in "Sovremennie dostizhenia v oblasti fizikokhimii lateksov" (Modern Achievements in the Physical Chemistry of Latexes, the Structure of Vulcanizates and the Technology of Latex Processing), p. 90, Moscow, 1971.
4. L. Zlatkevich and M. Shepelev, J. Polym. Sci., 16, 427 (1978).
5. British Standard, 903, Part B6 (3), 1958.
6. P. J. Flory and J. Rehner, J. Chem. Phys., 11, 521 (1943).
7. P. J. Flory, "Principles of Polymer Chemistry", Cornell University Press, Ithaca, N. Y., 1953.
8. S. A. Shteinberg, "Some Peculiarities of Vulcanization Process and Structure of Latex Films", Ph.D. Thesis, Yaroslavl (USSR), 1975.

# FACTORS INFLUENCING LATEX AUTOHESION

M. Gur-Aryeh and M. Shepelev

Nir-Lat, Latex Products

Kibbutz Nir-Oz, P.M. Negev 85122, Israel

Processes taking place during polymer adhesion in the system: latex film (substrate) - latex (adhesive) are considered. It is shown that the main factor promoting the complete autohesion of substrate and adhesive layers is the complete and significant contact between the globules surfaces. Coagulation of adhesive on the substrate surface, which prevents realization of full contact between globules, can be suppressed by treating the substrate surface with non-ionic surface-active agents. An increase of contact area is achieved by means of substrate vulcanization, which raises the surface relief.

## INTRODUCTION

Latex technology provides the possibility to manufacture products of complicated configuration - products with internal cavities, chambers, tubes, etc. A sketch of one such product is shown in Figure 1.

In order to manufacture such products, it is necessary to apply two or more consecutive latex layers to the mold. In places where an internal cavity must be obtained, the corresponding pattern is applied between the latex layers. The material for this pattern consists of a special composition which prevents the adhesion of latex layers. In all other places the maximum autohesion between latex layers must be achieved.

In previous studies devoted to the problem of autohesion of latex films, only those films which were obtained by drying were considered[1-3]. Conclusions of these investigations are not relevant to latex films obtained by ionic deposition.

## MATERIALS AND METHODS OF INVESTIGATION

In this study, the degree of autohesion between two layers deposited on the surface of a glass mold was investigated. The Latex blend consisted of (in weight parts): rubber of natural centrifugated latex - 100; sulphur - 2; zinc dimethyldithiocarbamate - 1; zinc mercaptobenzothiazole - 1; 2,2'-methylenebis (6-tert-butyl-4-methylphenol) - 0.5.

For ionic deposition of latex on the surface of a glass mold, the coagulant consisting of the following materials was used (in weight parts): Calcium chloride - 25, Clay - 25, water - 50.

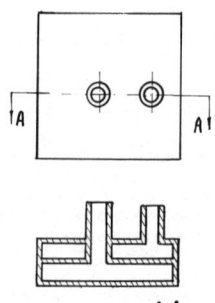

Figure 1. Sketch of rubber good with inner chambers.

The glass molds were first dipped in the coagulant. The first layer of latex was deposited for 7 min. The films obtained were dried at 70°C for various times. The molds with the first layer of latex were then again dipped into latex for 7 min. The second layer was dried for 180 min. at 70°C. The completed assembly was taken and vulcanized for 60 min. at 120°C.

In addition, the influence of the vulcanization period of the first latex layer on the degree of autohesion was investigated. In this case, the first layer was dried at 70°C for 180 min. and vulcanized for various intervals at 120°C. The second layer of latex was deposited only after vulcanization of the first layer.

The effect of the treatment of the surface of the first latex layer with different materials on the degree of autohesion was studied. For this purpose, we washed the surface of the substrate with a small amount of one of the following materials: potassium oleate solution (ionic surface-active agent) or alkylarylpolyether alcohol solution (non-ionic surface-active agent).

The degree of autohesion was determined by separating the latex layers (each sample being 20 mm wide) on the tensile testing machine. The speed of the upper grip movement was 200 mm per min.

After gold was applied to the surface of the latex films, photographs of these films were made with the help of a scanning electron microscope.

## EXPERIMENTAL RESULTS AND DISCUSSION

With the application of the second layer, after drying the surface of the first layer, we found that the autohesion value was not dependent on the duration of drying and that the autohesion value was very low (about 0.2 kg/cm). The reasons for the low autohesion might be as follows. During the first latex layer formation and drying, the diffusion of $Ca^{++}$ ions proceeds from the coagulant layer through the layer of latex thus creating a layer of coagulating salt on the surface of the dried latex film. When the mold is dipped in the latex the second time, $Ca^{++}$ ions present on the substrate surface cause the coagulation of the latex around the mold. This coagulation process prevents the creation of close contact between latex globules and the first layer; i.e., because of coagulation, the latex globules lose their mobility before they are able to fill in the hollows and pores of the substrate surface and thus a close contact between the substrate and adhesive layers is not formed.

Therefore, we tested methods of preventing or slowing down the coagulation process by treating the first layer with solutions

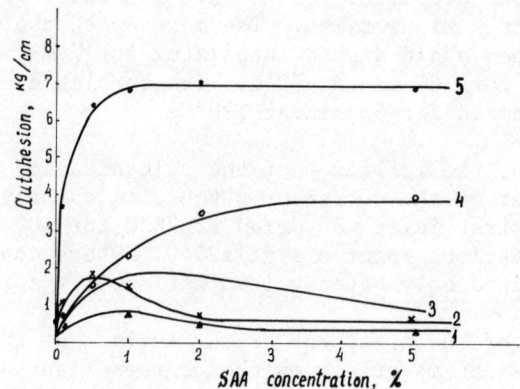

Figure 2. Influence of surface-active agent (SAA) concentration, and duration of substrate drying on films autohesion.
1 - ionic SAA, drying for 30 and 60 min.;
2 - non-ionic SAA, drying for 30 min.;
3 - ionic SAA, drying for 180 min.;
4 - non-ionic SAA, drying for 60 min.;
5 - non-ionic SAA, drying for 180 min.

of ionic and non-ionic surface-active agents of different concentrations. The results of this investigation are shown in Figure 2. As shown, the treatment of the first layer with surface-active agents considerably raises the degree of autohesion between the layers, especially in the case of non-ionic agent. An increase in drying time raises the autohesion value of those films in which the first layer was treated with a surface-active agent.

These results can be explained as follows. The substrate surface is defined as film relief consisting of half-spheres of globules, hollows and pores. In the non-vulcanized latex film the relief is smoothed out by surface tension energy. During the drying of films, at 70°C, a partial vulcanization of the polymer takes place. The vulcanization of latex films proceeds on the surface of latex globules[4], and the hardening of the globule surface prevents the smoothing out of the relief. Increases in the drying time will, in turn, increase the film surface.

However, it is impossible to realize the advantages of this increase of surface without the treatment of the first layer. When dipping the mold into latex, the $Ca^{++}$ ions on the surface of the first latex layer cause coagulation of the latex before contact is

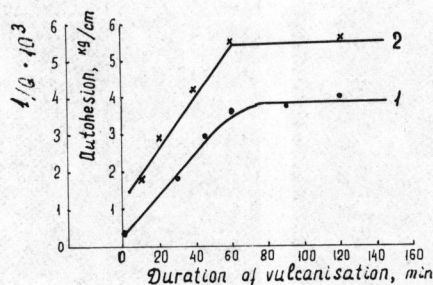

Figure 3. Influence of substrate vulcanization duration at 120°C on autohesion (1) and on the inverse of film swelling in xylene $1/Q$ (2). Q - maximum swelling of the film sample in xylene.

achieved between the latex globules and the surface of the first layer. The treatment of the first layer with a surface-active agent increases the degree of autohesion. Maximum autohesion is achieved when a non-ionic surface-active agent is used. The ionic surface-active agent, together with $Ca^{++}$ ions, form a non-soluble salt which slows down the coagulation of the latex due to decreased concentration of $Ca^{++}$ ions. But this salt also prevents contact between latex globules and the film because the insoluble Ca salt of the surface-active agent stays only on the film surface. The increase in ionic surface-active agent concentration to more than 1% will decrease autohesion, thus the problem of contact, because of the formation of calcium salt, becomes more important than the slowing of coagulation.

The non-ionic surface-active agent does not form water-soluble salts. An increase in the agent's concentration does not prevent the contact between the substrate (the first latex layer) and the adhesive (the second latex layer) layers. If the first latex layer is dried for 30 min., autohesion is decreased if the non-ionic surface-active agent concentration is raised above 0.5%. The increase in the drying time of the latex layer enlarges the surface to such a degree that there is no autohesion value reduction even if the non-ionic surface-active agent concentration is raised to 5%.

Considering the importance of the substrate surface and the possibility to achieve contact with the surface for complete autohesion between the substrate and the adhesive layers, we investigated the possibility to raise autohesion through vulcanization of

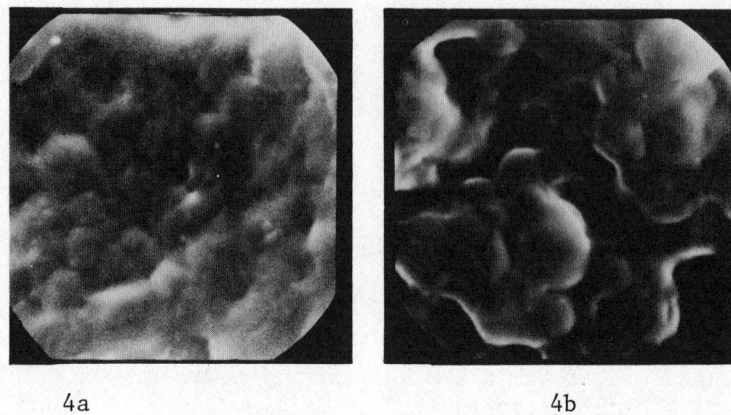

Figure 4. Electron micrographs of substrate surface (x 20,000).
4a - non-vulcanized film (drying 180 min.)
4b - vulcanized film (drying 180 min., vulcanization 60 min.)

the first layer before application of the second latex layer.

In Figure 3, the influence of the vulcanization period of the first latex layer on the autohesion value is shown. The data in the same figure indicate that vulcanization during the first 60 min. results in maximum amount of globule surface cross-links and that no further increase of cross-linking takes place when the vulcanization period is increased to 120 min. This is based on the film swelling in xylene data. The degree of relief of the film can change only until cross-linking of the globule surface takes place (during the first 60 min. of vulcanization). Once the process of vulcanization is completed, the area of the substrate, apparently, does not change.

The increase in the surface area of the latex film, as a result of vulcanization, was determined directly by means of electron micrographs which are shown in Figure 4. The comparatively smooth surface of non-vulcanized film, which was dried for 180 min. at 70°C, (Figure 4a) becomes rough with clearly visible pores, hollows and caverns after vulcanization for 60 min. at 120°C (Figure 4b). It is clear that the surface of the substrate increases after vulcanization.

In the case of vulcanized first layer, it was expected that treatment of the first latex layer with a surface-active agent would provide a high degree of autohesion. The results of our investigation are shown in Figure 5. The treatment of the first

# FACTORS INFLUENCING LATEX AUTOHESION

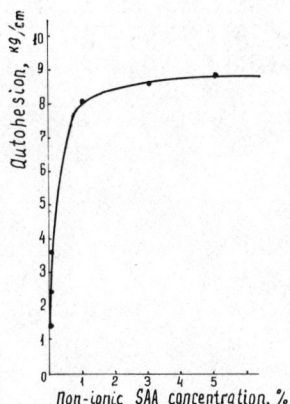

Figure 5. Influence of non-ionic surface-active agent concentration, used for treatment of the substrate, on the autohesion of latex layers. Substrate was vulcanized for 60 min.

layer with 1% water solution of non-ionic surface-active agent promotes complete adhesion of layers, as the failure was cohesive in nature. A further increase in non-ionic surface-active agent concentration does not change the degree of autohesion.

## CONCLUSIONS

1. The possibility of achieving complete autohesion between two layers of latex films, obtained by means of coagulant dipping method, is demonstrated.

2. We have shown that the main factors providing autohesion between two latex films are the strongly developed surface of the substrate and the absence of coagulation of the adhesive layer on the substrate surface.

## REFERENCES

1. S.S. Voyutzkii and B. V. Starkh, "Fiziko-khimia processov obrazovania plyonok iz dispersii visokopolimerov"(Physical Chemistry of Films Formation Processes from Dispersions of High Polymers), Moscow, 1954.
2. S. S. Voyutzkii, "Autogezia i adgezia visikopolimerov" (Autohesion and Adhesion of Highpolymers), Moscow, 1960.

3. V. A. Kosareva, P. I. Rakhlin and B. A. Perlin, "Latexi" (Latices), Voronez, p. 37, 1975.
4. V. V. Chernaya, S. A. Shteinberg, Yu. V. Grubman and M. I. Shepelev, Dokl. Akad. Nauk SSSR, <u>197</u>, 112 (1971).

# Part IV
# Tribology and Triboelectrification

TRIBOLOGY OF POLYMERS: STATE OF AN ART


B. J. Briscoe

Department of Chemical Engineering & Chemical Technology
Imperial College of Science & Technology
Imperial College, London SW7, UK


Tribology is about the friction, lubrication and wear of contacting solids in relative motion. The paper reviews these three topics in very general terms. The emphasis is upon the inherent subjectivity which has arisen when very complex processes are simplified in order to render them amenable to description in qualitative and quantitative terms. Tribology has been described as an art; in many respects it is, although we strive to transform the total subject into a science.

## INTRODUCTION

### The Semantics

The term "tribology" was derived by combining the Greek "tribos", to rub, with the Greek suffix "ology". The literal meaning is thus "the science of rubbing". The word was coined about thirty-four years ago and has been in common usage since the early part of the last decade. A number of authors have sought to give a concise and complete description of the current sense of the term.[1-4] There is general agreement that the subject covers all aspects of the interactions between solid bodies in relative motion; the word replaces the phrase "the friction, lubrication, and wear of solids". These three subjects are unusual in the extent of their multidisciplinary nature; practitioners range from the theoretical physicist through to the practising field engineer. Amongst the contributing well-established disciplines are material and polymer science, metallurgy, chemistry, mechanical and chemical engineering, and applied mathematics. Each major discipline offers a special contribution to the science and practice of tribology, although because of the importance of the subject commercially, the practical aspects have moved ahead more rapidly than the science. Indeed, the term "tribology" generally conveys the impression of a technological discipline closely allied to engineering practice. For this reason tribology might have been better named tribonics[5] to underline this sentiment. The Russians in fact use this word. On the other hand, certain authors have concluded that the subject is neither a science nor a technology but an art[4,6] which would, of course, further undermine the significance of the suffix "ology". The understanding of many important parts of the subject is sufficiently diffuse and incomplete to prevent an explicit description in scientific terms and hence the notion of an art form has developed.† There are, however, many facets to the subject which form perfectly reasonable and self-contained units of science or engineering. Today, the spectrum of the quality of knowledge is at least as broad as the range of discipline involved.

So far no mention has been made of the materials which are the subject of this review; polymers, and in particular, organic polymers. There is an important reason for this belated introduction; in many aspects of tribology the principle descriptive ideas are generally applicable to all solids. Organic polymers do, however, represent a good example of the general case.

---

† Dr. Gert Solomon strongly advocated this approach about ten years ago. He likened the level of our understanding of certain aspects of tribology to that which exists in religious pursuits.

### General Outline and Scope of the Present Review

The present review will outline the tribology of polymers under three separate headings; friction, wear, and lubrication. The emphasis will be upon presenting a brief account of the basic mechanisms of those features of the subject that have emerged as separate and identifiable processes. We shall see that many of these subdivisions are quite artificial, but they are a necessary expediency to render a range of very complex phenomena amenable to description. The art in tribology is in a large measure in the nature of these subdivisions and inherent simplifications.

The first topic to be reviewed is the dry friction of nominally clean polymer surfaces.

## POLYMER FRICTION

### Introduction: the Two-Term Model of Friction

The basis of the main subdivision in the mechanisms of polymer friction lies in the early work on metallic friction. Many authors contributed to formulation of the two-term non-interacting dissipation process model of friction. The monographs by Bowden and Tabor[7,8] were particularly influential in establishing its credibility. Augustin Coulomb was perhaps the scientist who introduced the general idea. The model has also been widely applied to polymeric systems[6,9] and the companion review on the "Tribology of Rubber-like Material" by Moore[10] adopts this model.

The essence of the two-term model of friction is shown in Figure 1. The frictional work is considered to be dissipated in two separate regions in the interfacial region. The interfacial zone is thought of as being very narrow and corresponds to high rates of energy dissipation. This class of processes correspond to the dissipation mechanisms usually grouped under the heading of "adhesion models" of friction. The other process involves deformation within a larger volume of material and hence lower rates of energy dissipation. The ploughing and deformation mechanisms are within this category. The distinction of two non-interacting processes is artificial and is no more than an expediency which greatly simplifies a complex range of processes. In practice, the application of the approach is very subjective. This subjectivity is compounded by the additional need to define the physical limits of the deformation zones. Before discussing each dissipation process further it should be mentioned that, while the two-term model has its many advocates, several authors[6,11] have expressed concern about its validity and urged care and restraint in its application.

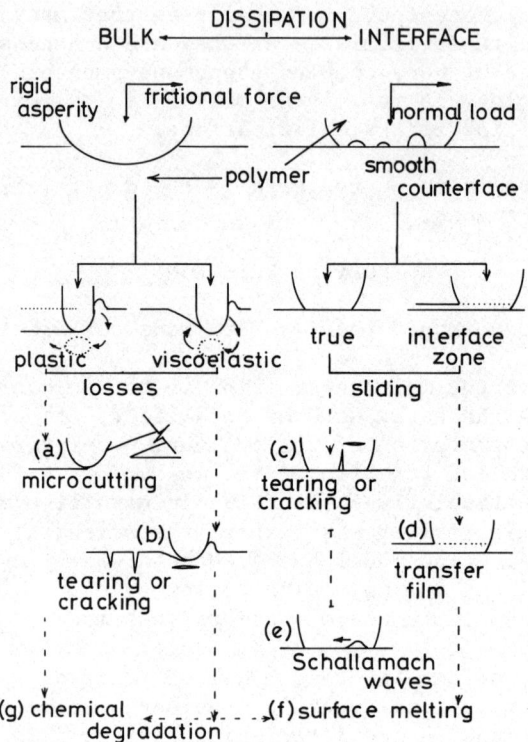

Figure 1. A schematic outline of the salient features of the two-term model of friction and wear. The frictional work is supposed to be dissipated by two separate non-interacting processes: the interfacial and deformation components. a) plastic grooving, leading to microcutting; b) viscoelastic grooving causing fatigue cracking and tearing with subsurface heating and damage; c) true interfacial sliding: high, effective rates of surface strain and heating. Potential for extensive chemical degradation; d) interface zone shear: rupture within the polymer and transfer wear; e) a subgroup of true interfacial sliding; the propagation of Schallamach waves.

## Large Volume Deformation Friction

Deformation or ploughing friction envisages energy dissipation <u>within</u> the polymer in volumes which are comparable with overall contact or asperity dimensions. The polymer is subjected to cyclic stresses at rates of strain comparable with the contact frequency. The contact frequency is defined as the displacement velocity divided by the contact length. The nature and magnitude of the energy dissipation is a function of the response of the polymer to deformation. Figure 1 shows a very simplified description of the principle dissipation processes assuming that the polymer exhibits reversible viscoelastic deformations at low strains followed by either plastic flow or fracture at large strains. A similar scheme has been presented elsewhere by the present author.[6,9]

The problem of the nascent viscoelastic groove has been widely investigated both experimentally and theoretically. Tractions generated by adhesive forces may be suppressed by either efficient lubrication or rolling and a relatively simple qualitative description of the process may be produced. Much of the fundamental work has been carried out with elastomers and Moore reviews these aspects in his paper.[10] Other reviews are also available.[6,9,12,13,14] Similar studies have been carried out with thermoplastics.[15,16] The major conclusion of these studies is the close connection between the rolling and lubricated friction and the viscoelastic properties of the polymer.[7,8,9,17,18] There has been numerous theoretical treatments of the problem and while the dissipation process is conceptually a simple one, a detailed treatment is laborious. The essence is shown in Figure 1. Work is fed into the polymer at the front of the asperity and the polymer is deformed. The polymer is viscoelastic and does not recover this strain instantaneously, particularly at the rear of the asperity. In terms of the stress distribution, a net restraining force is developed. The frictional work is dissipated beneath the contact in a series of cyclic deformations. The extent of this loss is a function of the total deformation and dissipative character of the polymer. The major commercial application of this type of friction is in the optimisation of the lubricated traction of automobile tyres on wet roadways.[19] Similar considerations apply in the proper design of synthetic shoe soles and the rubber tips of walking sticks.[20]

If the polymer is deformed beyond its elastic limit additional frictional work is dissipated in plastic deformation or the propagation of viscoelastic cracks. Tabor[21] has presented several analyses of the plastic cases which are appropriate for ductile polymers and the importance of tearing processes may be seen in early work carried out by Schallamach.[22] Bethune[23] has also presented some clear examples of the sorts of cracking damage seen in glassy polymers. These forms of irreversible deformation are more important in the context of wear which is the subject of a subsequent discussion.

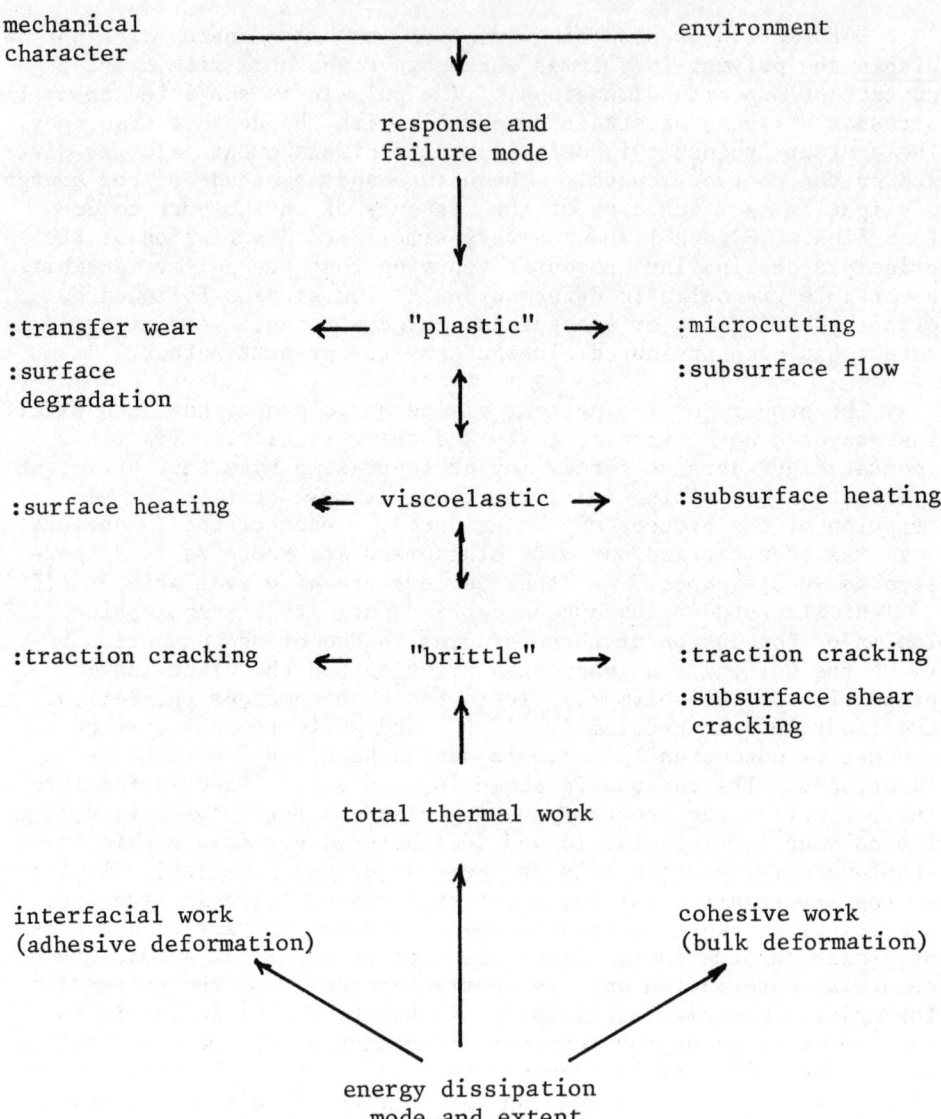

Figure 2. The interactions between energy dissipation mode and polymer damage processes which reveal the basic mechanisms of wear.

## Interfacial Friction: Observation of the Cold Welding of Contacting Solids

These models were initiated by Desaguliers.[1] The frictional force $F_a$, is thought of as arising from the shear of adhesive junctions at the areas of intimate solid-solid contact. Thus

$$F_a = \tau A \quad (1)$$

where $\tau$ is an interface shear strength and A the real area of contact. The central problems are the determination of $\tau$ and A as functions of the contact variables. The evaluation of the real contact area remains an elusive task and clearly the estimation of $\tau$ requires that the real contact area be known. Very little can be said about the quantitative estimation of the real contact area. Some success has been found in predicting the value theoretically and this provides the most common route to the evaluation of $\tau$. There are also a number of systems such as smooth elastometric contacts where the value may be estimated directly (see later).

### The Interfacial Shear Strength

The interfacial shear strength, $\tau$, was equated with the bulk shear strength for metallic systems by Bowden and Tabor.[8] A similar approximation was made by King and Tabor[25] and found to be quite a reasonable approximation for many organic polymers. More recently, a closer examination of this connection between bulk rheology and friction has been undertaken.[9,24] It appears that during this mechanism of friction the frictional energy is dissipated within a very narrow region adjacent to the interface. The thickness of this regime is not known with any certainty for any polymer but we believe that it is generally of the order of 50 to 200 nanometer in extent. This estimate arises from experiments on thin polymeric film deposited upon smooth rigid counterfaces. In these experiments the frictional force is largely independent of the thickness in this range of film thickness. The magnitude of the interfacial shear strength, $\tau$, is also very similar to the value found when a monolithic specimen of the same polymer is slid over a clean smooth counterface providing that the contact pressures generated in the two cases are of similar magnitude[26-28]; the comparison is, of course, made at similar temperatures and sliding velocities. This type of comparison illustrates the two routes whereby $\tau$ may be estimated. In the film studies, a thin layer of polymer is deposited upon a glass plate and a glass hemisphere is loaded against the film.[29] Assuming that contact deformation is wholly elastic and that the film does not greatly distort the form of the classical Hertzian contact the contact area may be calculated. If it is further assumed that during sliding this is the area sheared and that no direct glass-glass contact may be calculated. Using this method the functionality of interfacial shear strength with contact pressure, temperature, sliding velocity, and contact time have been examined for many polymers. The interfacial shear strength, $\tau$, varies with[9,26,29] the mean contact pressure, P, to a good approximation as

$$\tau = \tau_o + P \text{ ; constant temperature, sliding velocity} \quad (2)$$
$$\text{and contact time}$$

Recalling Equation (1) and noting that the coefficient of friction, μ, is defined as frictional force per unit load Equation (2) becomes

$$\mu = \frac{\tau_o}{P} + \alpha \quad (3)$$

If we assume that for heavily loaded polymeric contacts the average contact pressure is close to the hardness of the polymer, $P_o$, we may compute this coefficient of friction and compare this with the direct measurement.[31] The agreement between the value predicted from thin film studies[27,28] and direct measurements on monolithic specimens is very good. It is also worth noting that the <u>bulk</u> shear yielding of polymers follows a relationship similar to Equation (2). The pressure coefficients are often very similar in magnitude but the unconstrained values for thin films are usually about a factor of ten less than those measured in the bulk.[24,32]

The interfacial shear strength has also been investigated as a function of temperature, sliding velocity and contact time.[26,33,34] The contact time variable characterises the time during which an element of the surface layer is subject to the normal load. Many organic films exhibit viscoelastic retardation of response during compression and hence the increase in interface shear stress with contact pressure (Equation 2)) is not fully realised. Polymethylmethacrylate (PMMA) contacts, for example, may often exhibit a decrease in friction with increasing sliding velocity and this effect has been tentatively ascribed to a combination of this retardation in the effective contact pressure and a reduction in contact area produced by the effective stiffening of the polymer with loading frequency.[35] This type of analysis assumes that isothermal conditions prevail which is a reasonable assumption in the PMMA case described. When adiabatic heating occurs the interfacial shear strength will generally decrease. This influence of temperature upon the interfacial shear strength of many polymers has been studied in detail[26,33,36] and it seems as if this property changes with temperature in a manner which might have been anticipated from observing the temperature dependence of bulk yield. When making this connection, it is necessary to bear in mind that the interface of the polymer will be subject to rather high and transient pressures and be deforming at high strains at very high rates of strain. Unfortunately bulk deformation data are not yet available under these conditions and a measure of subjectivity is required during the extrapolation of bulk behaviour.

The continuum approach introduced above specifies nothing about the nature of the friction process other than that adhesive forces provide the interaction needed to enable interface shear processes, which seem to be similar to bulk shear processes, to dissipate the frictional work. The friction process is simply a case II adhesive failure albeit an unusual one. Several important questions remain particularly concerning the extent and position of both the zone of relative displacement and energy dissipation. Two extreme cases have been considered; true interfacial sliding with energy dissipation in adjacent regions and a zone of interfacial shear deformation where both relative displacement and dissipation are localised.[9,36] This distinction is shown in Figure 1(c),(d). Rubbers often exhibit apparently true interfacial sliding although in many cases interfacial dislocations or Schallamach waves propagate through the interface[38-40], and the frictional work is dissipated as the net work required to peel and reform the interface as the wave passes. This process establishes a clear link between viscoelastic loss in the adhesion process and the frictional work. The empirical relation between loss and friction is central in rubber friction and was established by Grosh.[41] Glassy polymers and cross-linked resins may also dissipate the frictional work during true interfacial sliding since often no transfer of polymers is detected when they are slid against clean and smooth counterfaces.[42] Some semicrystalline polymers in contrast transfer quite coherent films of polymers under similar conditions and by inference we suppose that the shear work is dissipated within a zone of finite thickness in the polymer. PTFE and polythenes behave in this manner near room temperature. The magnitude of the interfacial shear strength and its functionality with contact variables does not seem to sense this difference in the friction process. The nature of the sliding process is discussed in the context of transfer wear in the next section.

Friction of Polymers: the State of the Art

The deformation mechanisms of friction are quite well understood and rely upon a combination of contact mechanics and material science. Our predictive capacity is good to a first approximation although many details remain to be resolved for both plastic and viscoelastic deformations.[9,11,37] Many improvements could be achieved by developing a better understanding of complex geometries produced during deformation, in particular the role played by surface tractions. The analysis of surface tractions *per se* presents more intractable problems, not least the difficulty associated with defining quantitatively the area of real contact. Multiple asperity contacts[43] coupled with the uncertain effects of junction growth[21,44] and contraction[45] during the imposed shear have not proved amenable to analysis, save in rather special cases such as heavily loaded contacts[26], certain fibre contacts[46], and smooth elastomers.[39] The general features of some of the dissipation processes have been identified although up to the present only a rather pictorial understanding has been developed.

# POLYMER WEAR

## Classifications of Polymer Wear Processes

There seem to have been two types of classification of polymer wear: phenomenological and mechanistic. Lancaster[47], and Lancaster and Evans[48] have concentrated upon a scheme which specifies such processes as abrasion, transfer, erosion, fatigue, and cavitation in keeping with observation of worn parts and wear debris. Tabor and the present author[49,50] have attempted to extend the two-term friction model by including a spectrum of material response to speculate upon the damage produced by the frictional work. The scheme is shown in Figures 1 and 2. The friction work is thought of as entering the polymer in two zones. One is the narrow zone adjacent to the interface and corresponds to the adhesion component of friction. The other zone is much more extensive and corresponds to the deformation or ploughing component of friction. The distinction is clearly artificial but it has the virtue of grouping the polymer's response to deformation into high strain and high strain rate and comparatively low strain and strain rate categories respectively. The classification also distinguishes regions of high and low energy dissipation densities and as such introduces the relative importance of thermal heating. The adhesion induced wear mechanisms, for example, include thermally induced chemical erosion of the polymer surface. The various wear processes will now be reviewed within this context.

## Deformation Wear Processes

The two major damage processes involve plastic grooving including microcutting, and the propagation of quasi-brittle cracks. In some circumstances very substantial subsurface heating may be involved with consequent melting.[49] In the two main types of material response it is generally not clear whether the material is removed by a unit deformation or by multiple deformation fatigue processes. The examination of the worn surface or the wear debris does not generally provide an equivocal judgment. It is convenient to divide polymers into thermoplastics and cross-linked materials when reviewing deformation wear.

### Thermoplastics

If a rigid asperity, modelled as a cone of slope $\theta$, is pulled over the surface of a ductile polymer, such as low density polythene, a permanent groove is produced if the normal load, W, is sufficient to generate a contact stress in excess of the flow stress of the polymer. In reality the cone will have a finite radius of curvature at its tip. Assuming a very simple deformation model[49] we may show that the wear per unit sliding distance, Z, is related to the important variables by

$$Z \approx K' \left\{\frac{W \tan \theta}{H}\right\}^{\frac{2}{3}} \quad (4)$$

where H is a time dependent hardness. The details of the damage process are contained in the parameter K which expresses the probability of generating a wear particle per unit sliding distance. The parameter contains the material toughness as well as the stress intensification provided by the asperity. For a multiple asperity contact Equation (4) becomes

$$Z = K \frac{W \tan \theta}{H} \quad (5)$$

The K term now contains the role played by repeated deformation on the same element of surface area. The $\theta$ is some mean value of slope. Kraghelskii[52] was amongst the first to develop this simple picture. Probably the best test of this model for organic polymers has been carried out by Hollander and Lancaster.[53] They obtained a good correlation between single pass abrasive wear and the mean slope of the asperities. The linear dependence upon load is also quite well established.[47] The fundamental basis for the scaling factor, K, remains largely unresolved. At an unsophisticated level three factors are important in governing its magnitude. First, the efficiency of the cutting process will be a marked function of $\theta$. Lancaster and Warren and Eiss[54] have identified the importance of some critical asperity angle above which cutting will cease. Halliday[55] identified this transition in metals. Second, the intrinsic toughness of the polymer itself is obviously important. For fixed toughness values Ratner[56] and others, particularly Lancaster[47,49], have established a very good correlation between abrasive wear resistance and the work to rupture of the same polymer as measured in conventional tensile deformation tests. This correlation also contains the sensitivity of the polymer to the cutting angle of the asperity. It is also surprising the correlation[47] is good in view of the disparity in deformation conditions at asperities and in simple tensile failures. The third concern is over the fatigue nature of the process; one can readily envisage that repeated plastic deformations will produce filaments and lips of material that may be removed by subsequent contacts.

The practical cases of abrasion are somewhat more complex as the surface topography of the counterface is continuously being modified by either polishing by the polymer or by the inclusion of wear debris into valleys between the surface asperities.[48,49] There are, in addition, the influences of surface heating and the strain induced modification of the polymer's rupture properties. There are also important effects to be considered in lubricated systems where the environment may facilitate stress cracking or

or crazing on one hand, or plasticisation on the other (see later).
In principle the use of environment provides a useful tool for arbitration between primarily ductile and primarily brittle failures.
In systems which are prone to stress cracking during microcutting[47,48]
crazing solvents would be expected to enhance the wear process.
There are indications that this is the case for polymethylmethacrylate.
Similar conclusions may be drawn from the examination of the geometry
of wear debris.[49]

### Elastomers

Moore has reviewed elastomers in a companion paper[10] and only
a few general points will be emphasised here. Many pioneering
fundamental studies were carried out by Schallamach.[57-59] He worked
with both model asperities and abrasive papers as well as automobile
tyres. These works established the importance of tearing produced
by either traction with rather obtuse asperities and geometric engagement with more acute asperities. The importance of tear resistance or the work to rupture is quite evident. The studies on
abrasive papers revealed the intrinsic fatigue nature of wear
process in the patterns produced in the worn surface or in correlation between wear rate and fatigue life.[60] Unidirection abrasion
produces characteristic patterns called "abrasion patterns" and
higher overall wear rates than where the direction of abrasion is
changed periodically. A number of authors have also shown that
fatigue inhibitors can suppress abrasive wear[61] and Gent[62] has recently demonstrated that quite high free radical concentrations are
produced in abrasive wear of rubbers.

### Deformation Wears: the State of the Art

The general features of deformation wear processes are well
recognised. Equally well recognised are the limitations regarding the
detailed understanding of many features of the process. For elastomers we have a good practical connection between abrasive wear and
fatigue life although we cannot at present predict the size and shape
of the wear debris.

We also have the Ratner-Lancaster correlation between abrasion
and toughness; yet to be resolved are the reasons why this correlation is found to be so good in practice.

### Interfacial Wear Processes

When many thermoplastics slide over rigid, clean, smooth substrates the interfacial shear strength of the adhesive junction is
observed to be greater than that of the polymer.[37,42] The contact
ruptures within the polymer and a layer of polymer is transferred
to the counterface. Polyethylene and PTFE at room temperature are

notable examples. A number of glassy polymers such as polystyrene and polyethylmethacrylate as well as highly crosslinked materials do not appear to transfer polymer to the counterface under similar conditions. Those polymers which form transferred layers are susceptible to what is termed transfer wear.

## Transfer Wear

The transfer wear process is envisaged as follows:[6,42,29,50,63-69] A transferred layer of polymer is formed upon the counterface. Subsequent contacts of the polymer against this film causes it to be displaced from the contact zone and a further layer of film is deposited. The continual removal and replenishment of transfer film provides a rather efficient micro-machinery process and often the resulting wear rate is surprisingly large. This is the case when a pin of PTFE is slid over a flat disc in a circular path; typically about 10 nm of polymer is removed from the pin during each pass over the wear path. The transfer film thickness is thought to be of this order in these experiments.[49]

The overall view of the transfer wear process may be divided into its basic elements. These are

(i) the initial adhesion between the polymer and the counterface
(ii) the locus or zone of shear
(iii) the resulting thickness of the transferred layer, its coherence and morphology
(iv) the process of film detachment
(v) the final displacement of loose film from contact zone.

Very little of a quantitative nature can be said about these individual steps. The initial adhesion will be largely due to van der Waals forces with perhaps a coulombic force contribution. The *a priori* prediction of the locus of failure has not proved feasible. Crosslinked systems do not appear to form transfer layers nor do thermoplastics below their glass transition temperatures. In this context the transferred material is thought of as being chemically undergraded or "whole" polymer. A certain amount of chain scission is inevitable[70] but it is not of the same order as that produced when, say, automobile tyres lay down their characteristic transferred layers on roadways with the accompanying blue smoke. The glass transition temperature in this context also has a special meaning: it is unlikely to correspond exactly to bulk values because of the hydrostatic stresses and deformation rates with the contact. Several authors have examined the effects of frictional heating during sliding and it is clear that local surface thermal softening often occurs when the ambient temperature is well below any thermally induced transition region. This may produce a gross surface melting. Tanaka has illustrated this point very elegantly and has shown a good correlation between the melt rheology of the polymer and the level of frictional work.[71] It is not clear if such substantial thermal effects play a role at modest sliding speeds with semicrystalline polymers such as PTFE and polythenes. In these cases the transfer process appears to have the character of a cold drawing process with perhaps a measure of adiabatic shear heating. The factors which then control the thickness of the transfer layer in what might be described as "cold transfer" are less obvious. In the case of gross surface melting, the melt rheology and thermal properties of the polymer are clearly important. It has been suggested that under conditions of "cold transfer" it is the morphology or geometry of the polymer which controls the transfer film thickness. Pooley and Tabor[42], for example, were able to show that transfer films of PTFE on glass were sometimes only a few molecular chain diameters in thickness, perhaps 10 nm. These works also noted what they termed the role of molecular topography upon the extent and form of the transferred layer. Smooth molecules such as high density polythene and PTFE will form very thin and highly oriented layers once sliding motion has developed. This process occurs at relatively low sliding speeds and room temperature and corresponds to low values of the frictional force (coefficient of friction ca. 0.08). There are indications that at higher speeds and lower temperatures where molecular relaxation is inhibited that the frictional work increases and the transferred layer increases in thickness and at the same time loses much of its orientation. This type of transfer was termed "lumpy" transfer and is characteristic of rough molecular profile polymers such as low density polythene. A similar form of transfer is noted on the inception of sliding motion with PTFE and may also be produced by combined linear motion with a spin about the load axis.[72] These experiments serve to emphasise the importance of orientation during sliding when transfer occurs and presumably this is an important factor in controlling the film thickness. The position is less clear on the matters of film coherence and the location of interface rupture. Why some polymers transfer not at all, for example, polymethylmethacrylate, or only to a minimal extent like ultra high molecular weight polythenes in contrast to the ready transfer of PTFE is largely unresolved.

The understanding of the behaviour of these transferred layers during multiple contacts is even less refined. Multiple contacts may gradually produce a coherent layer and under gentle stresses produce an equilibrium film thickness. Transferring polymers also seem to be reluctant to transfer onto themselves indefinitely. Mechanisms of film removal are equally speculative although microscopic evidence suggests that tractive stresses may "pull" the film away from the counterface. Delamination[73], as suggested for certain forms of metallic wear is often evident in micrographs of transferred layers. Transfer film formation is an important feature of the performance of many dry bearings. In these systems it is usual however to incorporate relatively hard filler particles into the polymers.

## Filled Polymers

The majority of tribological applications of polymers involve the inclusion of relatively hard second phases[74]; certain civil engineering expansion joints and the well-known use of ultra high molecular weight polythene in prosthetic joints are notable exceptions. In many cases the filler is present for economic or cosmetic reasons, but there are many cases where the filler is chosen primarily because it conveys a substantial improvement in tribological performance. It is not uncommon to find that an appropriate filler may reduce the rate of wear by three orders of magnitude. The benefits may arise indirectly through enhanced thermal conductivity and creep resistance or through a subtle modification of the wear mechanism. It is the latter that are of particular interest in the context of interfacial wear phenomenon.

When a rigid filler is incorporated into a film transferring polymer, the rate of film transfer and removal is greatly reduced in many cases. This is very evident with PTFE and its composites. It seems that major benefits are produced in the low friction, thin film transferring polymers like PTFE and the linear polythenes. The role played by the filler particles is uncertain. A number of authors advocate the picture of rigid filler particles protruding from the worn surface of the composite with perhaps a film of transferred polymer acting as a lubricant.[75,76] The rate of wear is then strongly controlled by the intrinsic wear resistance of the filler. The evidence is mainly from microscopic examination of worn composites where proud and worn fillers may be readily observed. There is also supporting evidence from studies on the environmental sensitivity of wearing composites where the data suggest that the total wear rate is controlled by the rate of wear of the filler.[77] In addition to this simple picture two other effects have been cited as of potential importance.

First, the filler may retard reorientation at the composite interface and thereby suppress the rate of transfer film deposition. Second, the filler may produce local stress intensifications within the transferred layer and hence, by some uncertain means, produce a more strongly attached transferred film.[77,78] These two effects would be potentially capable of retarding the overall rate of transfer wear and there is a body of indirect evidence which indicates that they are likely to play a significant role. For example, there are certain synergistic effects when mixed oxide fillers are incorporated into PTFE and high density polythene and the composites slid on ferrous metals.[78,79] A number of authors have speculated that mild chemical reactions may occur at the interface between the transferred film and the metal and that the oxides act as catalysts and sites for local intense heating.[78,79] These chemically induced bonds would be stronger than van der Waals forces and hence the film would be more securely attached to the counterface. Surface chemical analysis seems to confirm that this mild chemical bonding may be produced.[80,81] Excessive chemical activity such as that sometimes produced by sliding filled composites over clean metals in vacuum results in higher wear[81]; chemical erosion has begun to replace transfer wear as the dominant mechanism.

### Chemical Wear Mechanisms

An element of chemical degradation most certainly occurs in all forms of wear. This may range from the rather mild chain scission which occurs in abrasion and transfer wear to the gross cleavage which occurs with certain Russian bearing formulations based upon aromatic polyamides.[82,83] The probability of mild chemical degradation occurring in transferred wear has already been mentioned. The formation of free radical species during the abrasion of rubbers has also been demonstrated. There are a few general reviews[83-85] of the role of chemical degradation during wear and a number of authors have used surface temperature calculations in conjunction with thermal degradation properties to predict trends during chemical wear.[86] This type of model seems particularly suitable for describing the wear of brake materials such as filled epoxy resins at high temperatures.

### Interfacial Wear Processes: State of the Art

The important features of interfacial wear processes continue to evolve. Transfer wear processes are common but it is still not yet clear which factors control its occurrence and its general characteristics when it is observed. The extent of chemical degradation and its contribution to wear is also only vaguely understood.

## LUBRICATION OF POLYMERS

### General Features

If the contacting bodies are separated by only a few molecular layers of fluid, the attractive forces are almost entirely attenuated and adhesive junctions are not formed. The frictional work is now dissipated through a combination of viscous shear within the fluid and bulk deformation within the contacting bodies. For a rigid sphere loaded on the surface of a polymer, the magnitude of lubricated sliding friction is often found to be close to the value found for the rolling friction of the dry contact providing that adhesion is suppressed between the counterfaces.[8,38,39] This type of fluid lubrication is termed elastohydrodynamic lubrication or sometimes hydrodynamic lubrication. These forms of lubrication have been studied extensively both theoretically and experimentally although the major interest has been in metallic systems. There have been, however, many significant contributions made in polymeric contacts particularly with highly deformable rubbers and to a lesser extent with glassy polymers. Fluid lubrication is important in the action of wind-screen wiper blades[87], automobile tyres[10], and seals[88]; the companion review[10] on "Elastomer Tribology" reviews some of these applications. The important features of these lubrication studies involve the description and measurement of the shape and dimensions of the fluid film, the traction during sliding and the squeeze film behaviour under normal loading. Rubbers have the experimental attraction that the deformation of the contact region is greatly exaggerated in comparison with metal systems and this fact makes them amenable for use in defining contact deformation. Optical interference methods have been widely adopted[89-92] to measure film thickness and pressure measurements have been made[93] as a function of position throughout various contact geometries. Mean or average film thickness measurements have also been carried out using conducting rubber.[94] There have also been many studies of the traction in fluid lubricated contacts with both elastomers[95,96,97] and polymeric fibres.[98,99,100] These studies have also been complemented by many theoretical or analytical studies of rheology of the contact.[101-103] There have also been many studies of more rigid systems which provide useful guidance. However in the case of, metallic systems prodigiously large increases in fluid viscosity are conferred upon the entrapped fluid induced by the very large contact stresses.[104] The fluids in elastomeric contacts are in comparison isoviscous in nature.

As a general conclusion we may say that within this regime of lubrication theory and experiment are in good agreement and the properties of the surfaces are rather unimportant. A measure of wetting may be required and surface roughness and contact geometry are known to affect the efficiency of this lubrication mechanism. Similar conclusions may be made for other polymers although it seems that elastohydrodynamic lubrication is most effective for elastomers. However, with all polymers complications arise when the fluid film thickness becomes comparable with the molecular dimensions of the lubricant. This is the regime of boundary lubrication.

## Boundary Lubrication of Polymers

Many of the special features of the boundary lubrication of polymers are introduced in a companion paper[87] and will not be repeated in detail. It is now evident that the fluid should reduce the adhesion between the contacting solids and for certain special contacts this effect can be quantified. The example cited is for an elastically deforming contact formed between orthogonal monofilaments of polyethylene terephthalate. In these contacts the fluid's main role is to reduce the autoadhesion through the modification of the interfacial free energies.[105,106] An extension is the role of electrical repulsive forces in electrolyte solutions; the electrical double layer forces are apparently capable of effectively increasing the contact area and hence reducing the contact pressure at the asperities. Roberts and Tabor identified similar effects in elastomeric contacts.[90]

In the cases of boundary lubrication described so far, it is not always necessary to invoke a change in the interfacial shear strength to account for the reduction in the frictional work. The value of $\tau$ remains unchanged and it is the contact area which is reduc There are, however, cases where true boundary lubrication is observed; that is the interfacial shear strength is modified by the adsorption of suitable species. The lubrication of polyethylene terephthalate by perfluorodecalin is such an example; apparently the fluid actually ingresses into the interface. A more illustrative example has been given by Senior and West.[10] They sulphonated polyethylene and noted a substantial increase in the frictional force. Amines were found to be particularly effective in lubricating these surfaces and restored the frictional work to values comparable, and often below, that of the original surface. The strong interaction between the sulphonate group on the polymer surface and the lubricant's basic functional group is sufficient to produce a coherent lubricant layer. A similar but more precise approach has been adopted by Butterfield et al[108] using the heat of adsorption as a measure of boundary lubrication efficiency.

A different type of boundary lubrication is observed in the lubrication of nylon in aqueous media. The surface of the nylon is plasticised and the shear strength, $\tau$, of the interfaces decreases. The bulk of the polymer is largely unaffected and the contact area remains unchanged; the friction therefore decreases. Surface plasticisations of this kind are certainly quite common and a recent example is that of silicone fluid and polyphenylene oxide.[109] Except for cases of extreme plasticisation of the polymer, it would appear that fluid lubrication will always reduce the frictional force - not an obvious truism.

Whilst lubrication invariably decreases the frictional work, it is by no means the case that the wear will be reduced. Indeed the wear may increase. Many fluids are capable of introducing stress cracking and crazing in polymers and the wear may therefore increase because the polymer becomes more susceptible to mechanical damage.[111-112] Lancaster and Evans[48] have examined these effects for a number of systems. The rate of wear is found to be somewhat greater for those combinations where the solubility parameters of the fluid and the polymer are comparable in magnitude. This also corresponds to the correlations for the maximum reduction in strength due to solvent-induced crazing and cracking. It is not obvious whether a similar condition might not lead to a maximum degree of plasticisation in other systems. In these cases the wear rate may then be at a minimum on smooth counterfaces although there is no direct structural evidence for this effect. On rough surfaces one might support that the abrasion resistance would be reduced. The fluids may, of course, have the general virtue of reducing thermal heating and in any event some hydrodynamic lift from the fluid would be anticipated. Thus although the material may be weakened, the wear rate is still reduced because of improved thermal dissipation and reduced friction.

Apart from the effects of matrix weakening discussed, the fluid may also undermine the formation of strongly attached transferred layers on the counterface. In this sense they have the converse effect to that tentatively suggested for fillers earlier. Aqueous surfactant solutions are very effective at removing transferred PTFE layers from glass surfaces.[42]

While many polymers are not generally lubricated effectively by external means as far as wear suppression is concerned, the friction and wear have been successfully modified by the use of internal lubricants. It is common practice to incorporate free surface active additives into polymers during processing to convey suitable changes to the surface of the product. The inclusion[113,114] of amides into low density polythene is the classical example. The $C_{18}$ amides are probably molecularly dispersed in the melt but phase separate upon cooling and eventually migrate to free surfaces. The layers are most effective at reducing friction and adhesion; other additives may be used to dissipate electrostatic charges. In these systems it seems that stored elastic strains around the precipitate provide the driving force for molecular diffusive transport.[115] The amide/polythene system also has a self-regulating mechanism to maintain a uniform film thickness; if the amide is washed or worn away, more diffuses out to repair the lesion. The nature of the interface layer in this system is not fully characterised although the layer is thought to be crystalline.[116] Whether the surface zone is a plasticised polymer layer overlaid with more additives or a more abrupt concentration gradient is unknown. The former model is, of course, close to a plasticised externally lubricated surface. Apart from the deliberate inclusion of external lubricants for service use there are undoubtedly cases where contaminants or additives included for other purposes fulfil this role. The waxy bloom on rubbers and free plasticisers on PVC are examples. There are also a group of additives which act as

internal lubricants by forming pockets of lubricants which are
broken open as the specimen is worn away.  An example is the inclusion of relatively high molecular weight silicone fluids into
polyacetals and polystyrene.[118]  Molecular transport through the
matrix is rather slow and the polymer must wear to expose the droplets of oil.  Hydrocarbon oils are also used to lubricate nylons
and polythenes.[118]  The concentrations are quite high compared
with the 5% or so used in amide/polythene systems and a combination
of molecular and capillary flow may occur in these cases.

### Lubrication of Polymers: State of the Art

Lubrication of polymers provides a mixed blessing; in some
cases lubrication is relatively efficient, in others the cure is
worse than the disease.  Thick film hydrodynamic lubrication with
elastomers is widely practised and indeed it proves difficult to
effectively undermine this process in automobile and human traction
on wet roads and pavements.  Some polymers are, however, greatly
weakened by the contact with lubricants and while the friction may
decrease, the wear increases.  In addition, hydrostatic stresses
may be generated within cracks thereby opening them and also the
adhesion of transferred layers to counterfaces will usually be reduced.[48]  The detrimental effects of lubricants is most evident
in boundary lubrication where the lubricant film thickness is comparable with the molecular dimensions of the fluid.  Certain lubricants which mildly plasticise the surface layers or migrate from
within the polymer matrix have many attractive features, not least
of them their self regulating character.

## CONCLUSIONS

The trinity of polymer tribology is very much in its infancy;
indeed organic polymers are relatively recent acquisitions and new
polymeric species continue to appear at regular intervals.  The
broad features of the three parts continue to evolve and many reviews are now available.[6,9,13,41,47,51,85]  The understanding of
a number of areas is quite sophisticated.  This is the case for
deformation friction and the elastohydrodynamic lubrication of
viscoelastic contacts.  Other parts are, by contrast, only understood in rather casual terms in spite of the fact that a large body
of experimental data is now available.  The abrasive wear process
and the interfacial frictional mechanisms are of this sort.  Other
areas have been only very sparsely treated and little is understood
in quantitative terms.  The processes of chemical and transfer
wear are arguably in this category.  There are also certain important areas such as boundary lubrication which have been almost
completely neglected.  The introduction of new engineering polymers
and composites has served to highlight these difficulties.  Matrials such as the poly(ether ketones), poly(ether ether ketones),
poly(sulphones), poly(phenylene sulphide) and their alloys appear

to have great potential but our rather inadequate understanding of polymer tribology means that much of their evaluation must be empirical. Similarly the use of polymers in special applications presents problems. The use of ultra high molecular weight polythene in human joint replacements is a good example.[119,120]

There have however been many promising recent developments in related scientific fields such as fatigue[12] and several additions to the analytical techniques available to the polymer tribologist. Surface analytical methods, developed initially for metals, such as ESCA[122,123], Auger analysis[81,125] and SIMS reveal important changes in the chemistry of sliding contacts and the presence and extent of transferred layers. Buckley has recently reviewed these and other techniques.[125] There have also been great improvements in the description of surface topography.[12] Electron microscopy and diffraction of surface layers, wear debris, and surfaces still have great potential value in elucidating wear as friction processes[127,128]. Russian workers have also found value in mass spectrometry and X-ray diffraction studies.[85] The recent development of neutron activation analysis for monitoring the extent of wear is an extremely valuable tool for the accurate measurements of small amounts of wear[129]. Ellipsometric studies of transfer films also appear to be attractive.[130]

The scientist and practitioner of the tribology of polymers is thus often faced with a paucity of firm precedents and analytical tools. This fact coupled with the intrinsic complexity of the subject provides a poor footing for mechanistic inquiries and predicting a given system's behaviour. These exercises require a large measure of subjectivity based upon the materials response to deformation in what are believed to be the principle types of dissipation and damage processes. This stylised approach is often frustrated because of a lack of basic knowledge of the processes involved. For example, when does abrasive wear become transfer wear as the surface roughness is reduced? Technically this point may be important as at least two polymers show a minimum rate of wear in the roughness regime where this transition is thought to take place.[131,132] The treatment of such problems requires the sort of skill which tribologists now take for granted but prompted Dr. Gert Solomon to invoke the notion of art in the subject of tribology.

## ACKNOWLEDGEMENTS

The author is grateful to Mary Briscoe and Nordica Low for their assistance in the preparation of this article.

## REFERENCES

1.  D. Dowson, "History of Tribology", Longmans, London, 1979.
2.  D. Summers-Smith, "An Introduction to Tribology in Industry", The Machinery Publishing Company, London, 1969.
3.  H. Czichos, "Tribology: A Systems Approach to the Science and Technology of Friction, Lubrication, and Wear", Elsevier, Amsterdam, 1978.
4.  D. F. Moore, "Principles and Applications of Tribology", Pergamon Press, 1975.
5.  D. Tabor, private communication.
6.  B. J. Briscoe and D. Tabor, in "Polymer Surfaces", D. Clark and J. Feast, Editors, Ch. 1, J. Wiley, London, 1978.
7.  F. P. Bowden and D. Tabor, "Friction and Lubrication of Solids, I", Oxford Press, Oxford, England, 1950.
8.  F. P. Bowden and D. Tabor, "Friction and Lubrication of Solids, II", Oxford Press, Oxford, England, 1964.
9.  B. J. Briscoe, in "Friction and Traction", D. Dowson, M. Godet, C. M. Taylor and D. Berthe, Editors, IPC Press, Guildford, England, 1981.
10. D. F. Moore, this proceedings.
11. K. L. Johnson, in "Friction and Traction", D. Dowson, M. Godet, C. M. Taylor and D. Berthe, Editors, IPC Press, Guildford, England, 1981.
12. D. Moore, "The Friction and Lubrication of Elastomers", Pergamon Press, 1972.
13. D. Tabor, in "Advances in Polymer Friction and Wear", L. H. Lee, Editor, Vol. 5A, Plenum Press, New York, 1974.
14. J. A. Greenwood, H. Minshall and D. Tabor, Proc. Roy. Soc., A259, 480 (1961).
15. K. C. Ludema and D. Tabor, Wear, 9, 329 (1966).
16. D. G. Flom and A. Bueche, J. Appl. Phys., 30, 1725 (1959).
17. D. Tabor, Proc. Roy. Soc., A229, 198 (1955).
18. W. O. Yandell, Wear, 17, 229 (1971).
19. G. F. Morton, Plastics Rubber International, 6, 256 (1981).
20. A. Kennaway, Bulletin Prosthetics Research, 130 (1977).
21. D. Tabor, in "Surface and Colloid Science", E. Matijevic, Editor, Vol. 5, John Wiley, New York, 1972.
22. A. Schallamach, J. Polymer Sci., 9(5), 385 (1952).
23. B. Bethune, J. Material Sci., 11, 199 (1976).
24. B. J. Briscoe and A. C. Smith, Polymer, 22, 158 (1981).
25. S. King and D. Tabor, Proc. Physical Society, BLXVI, 729 (1953).
26. B. J. Briscoe and A. C. Smith, "Review on "Deformation Behaviour of Materials" III, No. 3, 151 (1980).
27. J. Amuzu, B. J. Briscoe and D. Tabor, Trans. Amer. Soc. Lubrication Engineers, 20(40), 354 (1977).
28. J. Amuzu, B. J. Briscoe and M. Chaudri, J. Appl. Phys. D, 9, 133 (1976).

29. B. J. Briscoe, B. Scruton and R. F. Willis, Proc. Roy. Soc., A333, 99 (1973).
30. L. C. Towle, J. Appl. Phys., 42, 2368 (1971).
31. D. Tabor, "Hardness of Metals", Oxford Press, Oxford, England, 1951.
32. B. J. Briscoe and D. Tabor, Wear, 34, 29 (1975).
33. B. J. Briscoe and D. Tabor, J. Adhesion, 9, 145 (1978).
34. B. J. Briscoe and A. C. Smith, J. Phys. D. Applied Physics, 15, 579 (1982).
35. B. J. Briscoe and A. C. Smith, Trans. Amer. Soc. Lubrication Engineers, 23(3), 232 (1980).
36. B. J. Briscoe and A. C. Smith, in "Friction and Traction", D. Dowson, M. Godet, C. M. Taylor and D. Berthe, Editors, IPC Press, Guildford, England, 1981.
37. B. J. Briscoe, in "Adhesion IV", K. W. Allen, Editor, Applied Science Publishers, 1981.
38. A. Schallamach, Wear, 17, 301 (1971).
39. G. A. D. Briggs and B. J. Briscoe, Phil. Mag., A38, 387 (1978).
40. A. D. Roberts and A. R. Thomas, Wear, 33, 45 (1975).
41. K. A. Grosh, Proc. Roy. Soc., A274, 21 (1963).
42. C. M. Pooley and D. Tabor, Proc. Roy. Soc., A329, 251 (1972).
43. J. Archard, Nature, 172, 918 (1951).
44. J. J. Courtney-Pratt and E. Eisner, Proc. Roy. Soc., A238, 529 (1957).
45. A. Savkoor and G.A.D. Briggs, Proc. Roy. Soc., A356, 103 (1977).
46. B. J. Briscoe and S. K. Kremnitzer, J. Phys. D. Applied Physics, 12, 505 (1979).
47. J. K. Lancaster, in "Polymer Science", A. D. Jenkins, Editor, Ch. 14, North Holland Publishing Company, Amsterdam, 1972.
48. D. Evans and J. K. Lancaster, in "Material Science and Technology: Wear", D. Scott, Editor, Vol. 13, Academic Press, New York, 1979.
49. B. J. Briscoe and D. Tabor, in "Fundamentals of Tribology", N. P. Suh and N. Saka, Editors, MIT Press, Cambridge, Mass., 1980.
50. B. J. Briscoe and D. Tabor, British Polymer J., 10, 74 (1978).
51. B. J. Briscoe, Tribology International, p. 231, August 1981.
52. I. V. Kraghelskii, "Friction and Wear", Butterworths, London, 1965.
53. D. E. Hollander and J. K. Lancaster, Wear, 25, 155 (1973).
54. J. H. Warren and N. S. Eiss, in "Wear of Materials, 1977", W. A. Glaeser, K. C. Ludema and S. K. Rhee, Editors, p. 494, American Society of Mechanical Engineers, New York, 1977.
55. J. S. Halliday, Proc. Inst. Mech. Eng., 169, 777 (1955).
56. S. N. Ratner, Il. Farberova, O. V. Radyukevich and E. G. Lure, Soviet Plastics, 7, 37 (1964).
57. A. Schallamach, J. Polymer Sci., 9, 385 (1952).
58. A. Schallamach, Wear, 1, 384 (1958).
59. A. Schallamach, in "The Chemistry and Physics of Rubber-like Substances", L. Bateman, Editor, Ch. 13, MacLaren & Sons, London, 1965.

60. D. H. Champ, E. Southern and A. G. Thomas, in "Advances in Polymer Friction and Wear", L. H. Lee, Editor, Vol. 5A, p. 133, Plenum Press, New York, 1974.
61. G. I. Brodskii and M. M. Reznikovskii, in "Abrasion of Rubbers", D. I. James, Editor, M. E. Jolley, Translator, MacLaren & Sons, London, 1967.
62. A. N. Gent and C. T. R. Pulford, Wear, 49, 135 (1978).
63. K. Tanaka and T. Miyata, Wear, 41, 383 (1977).
64. D. R. Wheeler, NASA Technical Report 1728, 1980.
65. V. K. Jain and S. Bahadur, in "Wear of Non-Metallic Materials", p. 487, Mechanical Engineering Publications Ltd., London, 1978.
66. S. K. Rhee and K. L. Ludema, in "Wear of Non-Metallic Materials", p. 11, Mechanical Engineering Publications Ltd., London, 1978.
67. W. A. Brainard and D. H. Buckley, Wear, 26, 75 (1975).
68. H. Czichos, in "Wear of Non-Metallic Materials", p. 285, Mechanical Engineering Publications Ltd., London, 1978.
69. V. A. Belyi, I. V. Kragelskii, V. G. Savkin and A.I.Sviridyonok, in "Wear of Materials, 1977", W. A. Glaeser, K. C. Ludema and S. K. Rhee, Editors, p. 532, American Society of Mechanical Engineers, New York, 1977.
70. B. C. Arkles and M. J. Schireson, Wear, 39, 177 (1976).
71. K. Tanaka and Y. Uchiyama, in "Advances in Polymer Friction and Wear", L. H. Lee, Editor, Vol. 5A, p.499, Plenum Press, New York, 1974.
72. B. J. Briscoe and T. A. Stolarski, Nature, 281, 206 (1979).
73. N. P. Suh, Wear, 44,1 (1977).
74. Tribology International, special issue on "Dry Bearings", 6, 213 (1973).
75. B. Arkles, S. Gerakaris and R. Goodhue, in "Advances in Polymer Friction and Wear", L. H. Lee, Editor, Vol. 5A, p. 663, Plenum Press, New York, 1974.
76. K. Tanaka, in "Wear of Materials, 1977", p. 510, American Society of Mechanical Engineers, New York, 1977.
77. B. J. Briscoe, M. D. Steward and A. J. Groszek, Wear, 42, 99 (1977).
78. B. J. Briscoe, A. Pogosian and D. Tabor, Wear, 27, 19 (1974).
79. G. C. Pratt, Trans. J. Plastics Inst., 32, 255 (1964).
80. G. Pocock and P. Cadman, Wear, 37, 129 (1976).
81. S. V. Pepper, J. Applied Phys, 45(7), 2949 (1976).
82. V. V. Korshak, S-S. Gribova, A. YuS. Paulova, S. Nekrasov, Yu L. Avetisyan, P. N. Gribkova and I. K. Serov, Polymer Science USSR, 22,568 (1981).
83. M. O. W. Richardson, in "Advances in Polymer Friction and Wear", L. H. Lee, Editor, Vol.5B, p.89, Plenum Press, New York, 1974.
84. G. M. Bartenev and V.V. Lavrentev, "Friction and Wear of Polymers", Tribology Series 6, L.H. Lee and K.C. Ludema, Editors, Elsevier, New York, 1981.
85. V.A.Belyi, A.I.Sviridyonok, M.I.Petrokovetso and V.G. Savkin, "Friction and Wear in Polymer-Based Materials", Pergamon Press, New York, 1981.

86. T. Liu and S.K. Rhee, in "Wear of Materials, 1977", W.A. Glaeser, K. C. Ludema and S. K. Rhee, Editors, American Society of Mechanical Engineers, New York, 1977.
87. A. D. Roberts, Engineering Materials and Design, 55, January 1969.
88. C.R. McClune and D. Tabor, Tribology International, 219, August 1978.
89. A.D. Roberts and P.D. Swales, Brit. J. Appl. Phys. (J. Phys. D.) Series 2, 2, 1317 (1969).
90. A.D. Roberts and D. Tabor, Proc. Roy. Soc., A325, 323 (1971).
91. M. T. Kirk, Nature, 194, 965 (1962).
92. J.F. Archard and M.T. Kirk, Inst. Mech. Engrs. Lubrication and Wear Convention, London, 1963, Paper 15, 181.
93. G.R. Higginson, Int. J. Mech. Sci., 4, 204 (1962).
94. P. D. Swales, D. Dowson and J. L. Latham, Inst. Mech. Engrs., Elastohydrodynamic Lubrication Symposium 1972, London, Paper C4/72, p. 22.
95. A. D. Roberts and K. L. Johnson, Wear, 27, 225 (1974).
96. D. F. Moore, Wear, 26, 413 (1973).
97. B. D. Gujrati and K. C. Ludema in "Advances in Polymer Friction and Wear", L. H. Lee, Editor, Plenum Press, New York, 1974.
98. W. W. Hansen and D. Tabor, Textile Research J., 27, 300 (1957).
99. C. Schlatter, R.A. Olney and B.N. Baer, Textile Research J., 29 200 (1959).
100. "Surface Characteristics of Fibres and Textiles (Parts I and II)," M. J. Schick, Editor, Marcel Dekker, New York, 1977, Paper No. 1, Part I.
101. K. P. Baglin and J.F. Archard, Inst.Mech. Engrs., Second Symposium on Elastohydrodynamic Lubrication 1972, London, Paper C3/72.
102. B. J. Hamrock and D. Dowson, Trans. ASME, 100, 236 (1978).
103. J. A. Greenwood, J. Phys. D. Appl. Phys. 3, 1970 (1977).
104. W. Hirst and A.J.Moore, Proc.Roy.Soc., London, A365, 537 (1979).
105. M. Adams, B. J. Briscoe and S.K. Kremnitzer, this proceedings volume.
106. M. Adams, B. J. Briscoe and S.K. Kremnitzer, in "Proc. 34th Int. Conf. Societe de Chimie Physique, Microscopic Aspects of Adhesion and Lubrication", Elsevier, 1982.
107. M. Senior and G. West, Wear, 18, 311 (1971).
108. R. Butterfield, D. Falmer, and E.M. Scurr, Wear, 18, 243 (1971).
109. W. Sketchler, J.K. Lancaster and T. Quinn, Trans. Amer. Soc. Lubrication Engineers (to be published 1982).
110. E. H. Andrews, in "The Physics of Glassy Polymers", R. W. Haward, Editor, Applied Science Pub. Ltd., London, 1973.
111. B. J. Briscoe, S.M. Richardson and D.J. Walsh, in "Spec. Periodical Report, Macromolecules", J. Kennedy, Editor, Ch. 5, Royal Society of Chemistry, London, 1982.

112. D. Evans, in "Wear of Materials, 1977", W. A. Glaeser, K.C. Ludema and S. K. Rhee, Editors, p. 47, Amer. Soc. of Mech. Engrs., New York, 1977.
113. S. C. Cohen and D. Tabor, Proc. Roy. Soc., A291, 186 (1966).
114. B. J. Briscoe, V. Mustafaev and D. Tabor, Wear, 19, 399 (1972).
115. S. H. Nah and A. Thomas, J. Polymer Sci., Polymer Phys. Ed., 18, 511 (1980).
116. D. Allen, B.J. Briscoe and D. Tabor, Wear, 25, 393 (1973).
117. M. P. L. Hill, P. L. Millard and M. J. Owen, in "Advances in Polymer Friction and Wear", L. H. Lee, Editor, p. 469, Plenum Press, New York, 1974.
118. A. H. Ozhannakmezoq and M. W. Pascoe, in "The Wear of Non-Metallic Materials", D. Dowson, M. Godet and C.M. Taylor, Editors, p.60, Mechanical Engineering Publications, London,1976.
119. D. Dowson, in "Polymer Surfaces", D. Clark and J. Feast, Editors, J. Wiley, London, 1978.
120. K. J. Brown, J. R. Atkinson and D. Dowson, Trans. ASME J. Lubrication Technology, 104, 17 (1982).
121. R. W. Hertzberg and J. A. Manson, "Fatigue of Engineering Plastics", Academic Press, 1980.
122. D. Clark, in "Polymer Surfaces", D. Clark and W. J. Feast, Editors, J. Wiley, London, 1978.
123. D. Clark, this proceedings volume.
124. T. Lui, S.K. Rhee and K. L. Lawson, "Wear of Materials, 1979", K. C. Ludema, W. A. Glaeser and S. K. Rhee, Editors, p. 595, ASME, New York, 1979.
125. D. H. Buckley, "Surface Effects in Adhesion, Friction, Wear and Lubrication", Elsevier, New York, 1981.
126. T. R. Thomas and M. King, "Surface Topography in Engineering", BHRA Fluid Engineering, Bedford, England, 1977.
127. W. Bonfield, B. C. Edwards and A. J. Markham, Wear, 37, 383 (1976).
128. L. Engel, H. Klingele and H. Schaper, "An Atlas of Polymer Range", Wolfe Science Books, Munich, Germany, 1981.
129. N. S. Eiss, J. H. Warren and S. D. Doolittle, Wear, 38, 125 (1976).
130. J. E. Kosicki, W. Zwierzycki and M. Sielicki, 3rd International Tribology Congress, EUROTRIB '81, Vol. IV, p.198,Warsaw, 1981.
131. T. K. Mustafaev, D. Dowson, B.J. Gillis, in "The Wear of Non-Metallic Materials", D. Dowson, M. Godet and C. M. Taylor, Editors, p. 65, Mechanical Engineering Publications, London, 1976.
132, D. Dowson, J. M. Challen, K. Holms, and J. R. Atkinson, in "The Wear of Non-Metallic Materials", D. Dowson, M. Godet and C. M. Taylor, Editors, Mechanical Engineering Publications, London, 1976.

TRIBOLOGICAL PROPERTIES OF RUBBERLIKE MATERIALS

D. F. Moore

Department of Mechanical Engineering
University College
Dublin, Ireland

Friction is regarded as a fundamental energy-dissipative mechanism[1,2] as a result of which surface degradation and wear occur. In the case of rubberlike materials, such a mechanism exhibits distinct visco-elastic properties, and it is common to distinguish between adhesional and hysteresis friction. It is demonstrated that such a distinction offers a convenient means of visualizing the important aspects of the entire frictional mechanism, but it need not be the only interpretation of experimental results. This viewpoint is reinforced by the difficulties encountered in continuing the distinction between micro-hysteresis and macro-adhesion on fine textured surfaces. A broad analogy between the mechanisms of friction and wear is developed[3]. Thus, the adhesion and hysteresis contributions to friction on rough surfaces produce abrasive and fatigue wear respectively, whereas Schallamach waves of detachment on smooth surfaces appear to produce wear by rolled fragments. Under lubricated conditions, a simple theory of elasto-hydrodynamic separation is described for rubberlike materials sliding and slipping on an underlying macrotexture[4]. Here, it is demonstrated that macro-elastohydrodynamic and micro-elastohydrodynamic interaction effects occur in proportion to the scale of surface roughness. On fine surfaces, stiffening of the rubberlike material causes it to behave like a rigid body[5].

## INTRODUCTION

Tribology may be defined as the friction, lubrication and wear of engineering surfaces in relative motion. The objective is to understand in basic terms the mechanisms involved when two surfaces interact during sliding, and to subsequently apply this knowledge to the optimization of industrial processes. In the case of elastomeric or rubberlike materials, we assume that they slide on a rough, rigid base surface with or without interfacial lubrication.

Typical engineering applications include the friction and wear performance of pneumatic tyres, the mechanism of walking with rubber-soled shoes on a wet pavement, the wiping performance of automobile windscreen wipers, the slip-grip action of flexible belts on rough pulleys, the leakage of O-rings and gaskets, the dynamic performance of flexible rotary seals, the action of solid lubricant linings in bearings and in reciprocating machinery, and the performance of rubber shock-mounts and vibration dampers.

## ADHESION AND HYSTERESIS

The friction force generated between sliding bodies consists of both adhesion and deformation terms as is well-known, thus:

$$F = F_{adh} + F_{def} \qquad (1)$$

The adhesion term is a surface effect, and may be regarded as occurring to a depth in either surface which does not exceed molecular dimensions (i.e., Angstrom units), whereas the deformation term can be classified as a bulk phenomenon having its ultimate effect at the sliding interface. In the case of metal-on-metal contact, the formation of localized junctions due to pressure welding and their subsequent rupture during the action of sliding accounts for the adhesional mechanism—whereas the ploughing of harder asperities through the matrix of the softer metal constitutes the deformation term[6]. For rubberlike materials, the adhesion term is best described as a thermally-activated, molecular-kinetic stick-slip mechanism which occurs according as the flexible surface "flows" over the macro-texture of the base surface. Several theories have been developed to explain the nature of this phenomenon[7,8]. Such a flowing action produces a bulk deformation effect within the rubberlike material, and this contributes the hysteresis component of friction[9,10].

We can view adhesion as a _resulting_ (macroscopic) or as a _causative_ (molecular) mechanism. Considering the macroscopic viewpoint, we can then write for the total adhesional force, $F_{adh}$:

$$F_{adh} = \sum_{i=1}^{M} F_i = \sum_{i=1}^{M} A_i s_i \qquad (2)$$

where $A_i$ is the local macroscopic area of contact between an elastomer and its rough substrate (for rubberlike materials), and $s_i$ the local effective shear strength of the interface. On a molecular level, we can write for each location i:

$$F_i = n_i j_i \qquad (3)$$

where $n_i$ is the number of molecular junctions or bonds between the elastomer and base at location i, and $j_i$ is the effective local junction strength of each bond as a result of the molecular-kinetic stick-slip action. From Equations (2) and (3), it follows that:

$$s_i = \frac{n_i}{A_i} j_i \qquad (4)$$

so that the local shear strength of the macroscopic model is a function of the molecular bond strength $j_i$.

A detailed treatment of the various theories of adhesion is beyond the scope of this paper. However, Table I below summarizes the various contributions by type, author and year[8] up to about 1970. During the last decade or so, variations and refinements of these theories have appeared, and it is interesting to note that the basic molecular-kinetic stick-slip explanation of the adhesional mechanism remains intact.

Table I. Summary of Adhesion Theories, 1928 - 1970.

| Classification: | Type: | Author and Year: |
|---|---|---|
| Molecular-kinetic Theories | Rate Theories | Bartenev (1954) |
| | | Schallamach (1963) |
| | | Lavrent'ev and Ostreiko (1968) |
| | Mixed Theory | Bulgin et al. (1962) |
| Mechanical Model Theories | "Gedankenmodell" Theories | Prandtl (1928) |
| | | Rieger (1967) |
| | Phenomenological Theory | Savkoor (1965) |
| Other Theories | Simple Theory | Ludema and Tabor (1966) |
| | Unified Theory | Kummer (1966) |

The generation of hysteresis friction forces between a sliding elastomer and a rigid base surface requires that the latter exhibits distinct asperities on a macroscopic scale. Considering the interaction of a single symmetrical asperity with the loaded rubberlike material, it is seen at once that the pressure distribution in the region of contact is (a) symmetrical when no relative motion exists at the interface, and (b) unsymmetrical when such relative sliding commences at finite speeds. The physical reasoning for the asymmetry effect is that the rubberlike material exhibits an inertia effect, so that it tends to accumulate in front of an asperity and shows a delayed recovery from the indentation effect behind each asperity. The resolution of pressure distribution into vertical and horizontal components shows at once that whereas the vertical components combine to equal the incremental load $\delta W_i$ for both cases of static and sliding contact, the horizontal components are equal and opposed only for the symmetrical distribution. For the unsymmetrical case corresponding to sliding, there is a net horizontal unbalanced pressure force which is opposite in direction to that of sliding, and this contributes the hysteresis or bulk deformation mechanism, $F_{hyst}$.

This simple description of the hysteresis effect on a single asperity of a surface indicates how a friction force evolves from pressure asymmetry, but it makes no mention of energy loss which occurs in the process. In point of fact, we have explained the physical nature of hysteresis by concentrating more on the effect rather than the cause of the phenomenon in the interests of simplicity. Thus, for the sliding elastomer, work is done within the forward zone of contact in deforming and deflecting the rubberlike material, wereas within the rearward contact zone almost all of this work is returned. The difference between the work expended and that returned is the energy loss which gives rise to frictional forces.

The separation of the frictional mechanism for rubberlike materials into adhesional and hysteresis (deformation) components in accordance with Equation (1) offers a particularly graphic and therefore physically acceptable understanding of the phenomenon. However, we are not at all bound to adhere to this convention in interpreting frictional data. The fundamental and common element in both the adhesional and hysteresis mechanisms is energy loss associated with relative motion at the sliding interface. This energy is accompanied by either macroscopic or by microscopic surface deterioration as a consequence of the energy transfer mechanism. If we divide the energy loss by some appropriate distance dimension, we define automatically a friction force. This concept is especially useful if the energy loss takes place at or in the vicinity of a sliding interface, and the firction force so defined is taken to occur at the interface. In certain cases, however, internal friction losses may occur within the bulk of a system subjected to cyclic stress variation and far removed from any possible sliding interface. In this situation, it becomes meaningless to speak of a friction force. The best example which comes to mind is the rolling behavior of a pneumatic tyre on a rigid pavement surface[11]. Here, the continual flexing of sidewalls and carcass dissipates energy as heat within the tyre, and we speak of internal friction (also called hysteresis——but <u>not</u> hysteresis friction). If braking action is now externally applied to the tyre, additional energy losses are caused near the tread/road boundary, and these can be conveniently expressed as the adhesional and the hysteretic components of an interfacial friction force. We must emphasize that energy loss is the fundamental factor in both the internal and external friction mechanisms, and the concept of a frictional force is a convenient means of expressing such energy loss only in the vicinity of a sliding interface.

The distinction between adhesion and hysteresis becomes less obvious as the scale of surface texture diminishes. In this case, the very definition of hysteresis friction as that arising from energy dissipation in the bulk of a material implies necessarily a diminution for progressively finer scales of texture. At the same time, adhesional friction although theoretically acting at a surface operates within a thin surface layer due to the make-stretch-break-relax sequence of molecular stick-slip action. It is clear that on finer surfaces there is an overlap in the visualization of what we might term macro-adhesion and micro-hysteresis effects, and the distinction expressed by Equation (1) may not be as real between these two components as was first imagined. Hence the need for establishing that friction is primarily an energy dissipative stick-slip event, irrespective of whether such energy dissipation may be expressed in terms of adhesion and hysteresis force components.

## FRICTION AND WEAR

Wear is most conveniently defined as the extent of surface deterioration following exposure to friction. Thus, in general terms, the greater the causative frictional force the greater the resultant wear occurring——but a direct proportionality between the two is neither to be expected nor in accord with experimental observation. Not enough is known at present about the exact nature of wear[12], and it appears to be an extremely difficult if not impossible task of predicting the magnitude of wear effects in a given sliding environment.

Despite the present early stage of comprehension, a direct relationship between the volume of abraded material $\Delta V$ and the coefficient of sliding friction f has been shown to exist[4] for rubberlike materials. One of the more common indicators of wear is the energy index of abrasion, $K_E$, which is defined as follows:

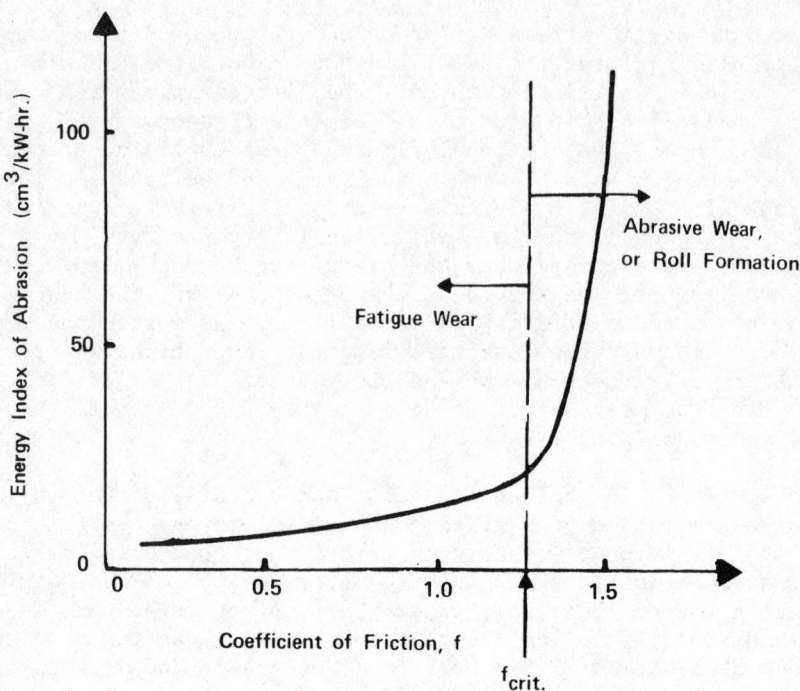

Figure 1. Wear as a function of frictional coefficient for different rubbers[7,12].

$$K_E = \frac{A'}{f} = \frac{\Delta V}{f\,W\,L} \tag{5}$$

where $A'$ is the abrasion factor for the elastomer, W the normal load and L the sliding length. The dependence of $K_E$ on the coefficient of sliding friction f is more complex than the inverse relationship in Equation (5) implies. In fact, Figure 1 shows the highly nonlinear relationship between $K_E$ and f for rubberlike materials—here, it is at once apparent that the interrelationship is a function of the severity of the wear mechanism, this in turn being dependent on the type of surface. Thus, on rounded smooth textures fatigue wear suggests a relatively mild mechanism, whereas on either sharp angular or perfectly smooth surfaces the severe forms of wear (due to abrasive tearing and wear by roll formation respectively) predominate.

There are, of course, three distinctive wear mechanisms[12] for rubberlike materials:

1. <u>Abrasive Wear.</u>  A sharp texture in the base surface causes abrasion and localized tearing of the sliding elastomer.  In

such cases, micro-cutting and longitudinal scratches are observed on the abraded rubber.

2. <u>Fatigue Wear</u>. If the base surface exhibits blunt or rounded rather than sharp asperities, the surface of the rubberlike material undergoes cyclic deformation, and failure eventually occurs as a result of fatigue.

3. <u>Roll Formation</u>. On smooth surfaces, a third mechanism of wear specific to highly elastic and soft materials has been observed to cause the formation of rolled shreds at the sliding interface, and eventual tearing of the rolled fragment.

In many cases, combinations of these mechanisms occur, and there is a multiplicity of physical forms which wear assumes. Thus, following the bond formation, stretch and break mechanism responsible for adhesion, the effects of bond rupture may appear as shredding, tearing, buckling, plucking, rolling into a whorl, etc., accompanied by local irreversible changes in elastomeric structure and properties. Here the roles of surface texture, sliding speed and normal load are of crucial importance in determining the approximate form of the predominating wear mechanism——but the whole must remain an extremely complex and to a large extent unpredictable phenomenon. The important point is that wear of necessity accompanies friction, being a consequence of the latter, so that "zero wear" models as proposed by some authors are a contradiction in terms. It may be that in some cases the amount of wear debris is too minute and infinitesimal to detect even with the most sophisticated experimental equipment, but this should not be taken as evidence that wear, in fact, has not occurred. As a general rule, greater wear effects occur on higher friction surfaces.

The most recent phenomenon peculiar to smooth surfaces has been given the name "waves of detachment" due to the pioneering work of Schallamach[13]. These waves of detachment have been observed to cross the contact patch between a soft rubberlike material and a hard smooth substrate, and detailed features of the mechanism have appeared in the literature. There can be no doubt that such waves arise from surface buckling of the soft elastomer as a consequence of frictional stress. Indeed, it is also certain that the energy involved in peeling rubber away from the contact area during the formation of waves of detachment accounts for a large part of frictional energy dissipation and forms part of the total buckling mechanism. The appearance or non-appearance of waves of detachment in a given sliding situation is, of course, merely a manifestation on a macroscopic level of the fundamental and underlying molecular-kinetic, stick-slip adhesional phenomenon that applies in all situations where at least one of the sliding bodies is a rubberlike material. Variations in the interpretation of this fundamental mechanism must appear from time to time according as

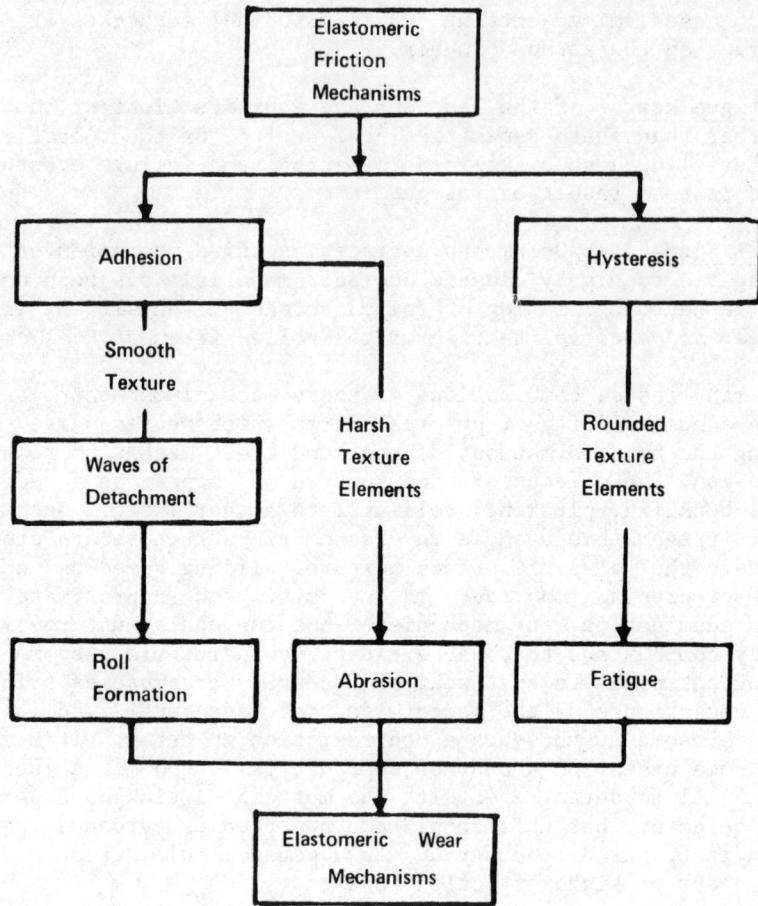

Figure 2. Interrelationship of friction and wear mechanisms in elastomers[3].

experimental techniques are refined and our understanding of the molecular nature of matter evolves, but the very essence of all frictional phenomena according to our present knowledge remains an energy-dissipative bonding, stretch and rupture cycle of events in molecular dimensions.

There appears to be a direct correlation between the three principal forms of elastomeric wear listed earlier and the fundamental and causative frictional mechanisms of adhesion, hysteresis and waves of detachment as shown in Figure 2, thus:

ABRASIVE WEAR ——————————— ADHESIONAL FRICTION

FATIGUE WEAR ———————————— HYSTERESIS FRICTION

ROLL FORMATION ————————————— WAVES OF DETACHMENT

Perhaps also friction and wear are related in some distinct manner to the increase of entropy principle of the Second Law of Thermodynamics. While such a relationship may yield no further physical insight into the nature of wear itself, it would surely provide a worthwhile unifying influence by now including wear and surface deterioration as part of the natural progression from order to randomness in the universe.

## ELASTOHYDRODYNAMICS

The term "elastohydrodynamic lubrication" applies to a particular form of lubrication between engineering surfaces which deform elastically under load. The film thickness between the sliding surfaces is far from sufficient to fill the void spaces in the texture, and the lubricant generally clings to one or both of the sliding members due to surface tension effects. Although the phenomenon of elastohydrodynamic lubrication was first discovered as a result of the need for understanding the lubrication of gears and roller bearings, recently there has been an increased interest in the lubrication of soft, flexible surfaces made from elastomeric or rubberlike materials. The deformations occurring with rubberlike materials are, of course, substantially greater than in the case of metallic surfaces, and they occur much more readily in the former case. Thus, the elastohydrodynamic phenomenon is of far greater significance in the case of elastomeric surfaces. This is also evident from the fact that whereas in the case of intermetallic contact substantial local deformation of the contact areas occurs only at fluid pressures which are sufficiently high to alter the viscosity of liquid lubricants, such pressures are never reached in elastohydrodynamic contact between a rubberlike material and its mating surface.

If we now consider the interaction of a flexible surface and a single asperity of the rigid substrate under relative sliding conditions and in the presence of a lubricant, we use the term macro-elastohydrodynamics[4,7]. Such asperities have no upper limit in size, but it is conjectured that the smallest size or dimension is of the order 10 to 100 microns[5]. Distinctly smaller asperities than these also interact in a lubricating/elastic environment and give rise to what we prefer to call micro-elastohydrodynamic effects. If we consider the draping configuration of an elastomer about a single asperity, it is clear that in the absence of tangential motion the normal load would effectively squeeze out the lubricant at the asperity tip. The effect of a tangential velocity of sliding, however, is to create a hydrodynamic pressure wedge in

the converging part of the lubricant film, and this is resisted by
the elasticity of the rubberlike material——so that a state of vertical elastohydrodynamic equilibrium is attained. According as the
speed of sliding is increased, so also the intensity of the hydrodynamic pressure effect rises until at a critical speed it is
sufficient to push the lubricant effectively over the asperity tip,
thereby creating the macro-elastohydrodynamic condition.

We can counteract this, of course, in a number of ways. Most
important of all, we can adjust the angularity or the sharpness of
the surface texture. It appears that for extremely sharp or pointed asperities, there can never exist a trace of lubricant[11], so
that dry contact prevails at the peaks. According as the angularity
degenerates into roundness, there is a greater danger of the hydrodynamic pressure wedge effect overcoming the resisting elastic
pressure——and a micro-roughness must be incorporated at asperity
peaks to break up the lubricating film and establish contact. The
need for a micro-roughness has been well documented in previous
publications by the author[4,7,11], and the criterion for establishing its required magnitude is given as follows:

$$\varepsilon_{mr} \geqslant h_c^* \qquad (6)$$

where $\varepsilon_{mr}$ is the micro-roughness amplitude and $h_c^*$ the critical value
of elastohydrodynamic film thickness at asperity peaks for a given
set of operating conditions.

Care must be exercised, however, that the degree of micro-
roughness does not cause sever abrasion losses. Figure 3 shows that
in the case of road surfaces, there is a design texture range between 10 and 70 microns to preserve high values of wet friction
with slipping and sliding tyres, while still keeping abrasion within
reasonable limits.

One effect of crucial importance in the elastohydrodynamic
contact problem is the scale of surface texture[5] in relation to the
elasticity or flexibility of the rubberlike material. Thus, in the
case of tyre tread elements slipping and sliding on the macro-
texture of a typical asphaltic, pebbled or concrete surface, the
chain length of rubber molecules is sufficiently small to permit a
draping effect about surface asperities. Hence the phenomenon of
viscous hydroplaning[4,11] occurs as manifested by Equation (6) above.
The draping effect is in turn modified by an effective stiffening
of the rubberlike material according as the frequency of indentation $\omega$ is increased, where:

$$\omega = \frac{V}{\lambda} \qquad (7)$$

Here, V denotes sliding speed and $\lambda$ the mean wavelength of the
surface macro-texture. It is seen clearly from this equation that

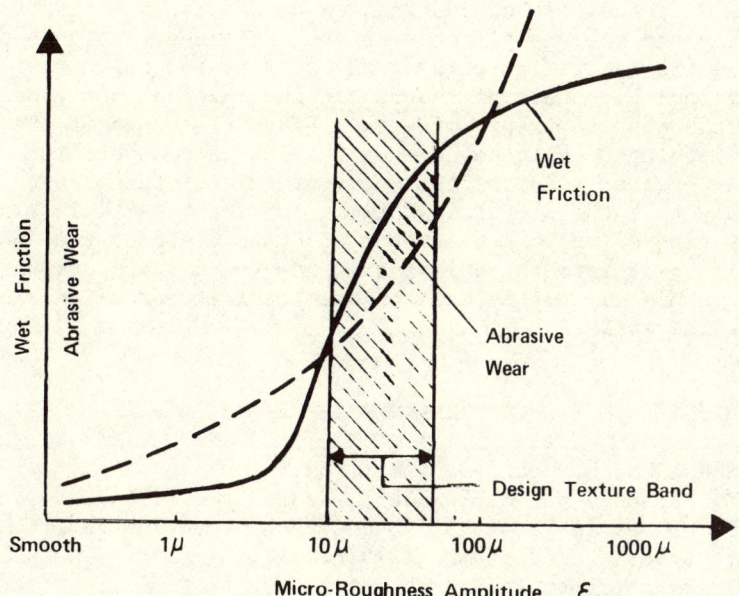

Figure 3. Effect of micro-roughness on tyre abrasion and wet friction[11].

for a given sliding speed V, the frequency of loading and unloading of the rubberlike surface is inversely proportional to the scale of surface texture. Thus, considering the micro-texture itself which generally occupies the peak of each macro-texture element, $\lambda$ is reduced by a factor as great as 100, and the surface molecules now exhibit a stiffening effect which is so pronounced that any draping tendency whatever can be neglected. It is believed that squeeze-film theory offers a simple and convincing explanation of the effects of micro-elastohydrodynamic interaction in a sliding situation[5]. In this case, it would appear that the draping pattern which is so pronounced for macro-asperities at reasonable speeds of engineering interest, is strongly influenced by both the stiffening and asymmetry effects[14] according as the rate of loading of asperities increases. For small values of micro-roughness, it may well be that the asymmetry effect disappears altogether, whereas the effect of stiffening now precludes any indentation whatever. Thus, the elasticity of the rubber surface is of no importance whatever at asperity peaks, and the material behaves like a rigid body.

A final word must be said about the lubricity of rubberlike materials in a lubricating environment. The molecular chain structure of such materials is not greatly different in magnitude or form from that of the contacting lubricating oils, so that the physical boundary between the two is not clearly defined. This contrasts with the case of lubrication between metallic surfaces. For elastomeric materials, there exists a diffuse zone perhaps 50 or 100 Angstrom units in depth at the surface boundary whose properties are neither those of the rubberlike material nor those of the lubricant but a mixture of each. Since the scale of the finest polishing techniques is such that surface asperities are not enormously greater than the thickness of this diffuse zone, it is most likely that the properties of the lubricant participating in the generation of the micro-elastohydrodynamic effect are substantially different from the bulk lubricant ——and therefore unpredictable. This may partly explain the lubricity or slipperiness effect at the surface.

## REFERENCES

1. A. Schallamach, Wear, 6, 375 (1963).
2. G. M. Bartenev and A. I. Elkin, Wear, 8, 8 (1965).
3. D. F. Moore, in "Proc. 3rd Leeds/Lyon Symposium on Tribology", Univ. Leeds, Paper VI (ii), 1976.
4. D. F. Moore, Inter. J. Mech. Sci., 9, 797 (1967).
5. D. F. Moore, in "Proc. 4th Leeds/Lyon Symposium on Tribology", Institut National des Sciences Appliquées de Lyon, Lyon, Paper XII (i), 1977.
6. F. P. Bowden and D. Tabor, "The Friction and Lubrication of Solids", Clarendon Press, Oxford, 1964.
7. D. F. Moore, "The Friction and Lubrication of Elastomers", Pergamon Press, Oxford, 1972.
8. D. F. Moore and W. Geyer, Wear, 22, 113 (1972).
9. R. R. Hegmon, Rubber Chem. & Technol., 42, 1122 (1969).
10. D. F. Moore and W. Geyer, Wear, 30, 1 (1974).
11. D. F. Moore, "The Friction of Pneumatic Tyres", Elsevier Scientific Publishing Co., Amsterdam, 1975.
12. D. I. James, "Abrasion of Rubber", McLaren & Sons Ltd., London, 1967.
13. A. Schallamach, "The Physics of Tire Traction", p. 167, Plenum Press, New York, 1974.
14. D. F. Moore, Wear, 21, 179 (1972).

A SURVEY OF THE ADHESION, FRICTION AND LUBRICATION OF POLY-
ETHYLENE TEREPHTHALATE MONOFILAMENTS

        M.J. Adams        Unilever Research, Port Sunlight
                              Laboratory, Merseyside L63 3JW

        B.J. Briscoe       Dept. of Chemical Engineering &
                              Chemical Technology, Imperial College
                              London SW7 2BY

        S.L. Kremnitzer   Physics & Chemistry of Solids
                              Cavendish Laboratory
                              Madingley Road
                              Cambridge, CB3 OHE

    The paper describes recent and continuing studies
of the autoadhesion and friction between orthogonal
polyethylene-terephthalate (PET) monofilaments. The
main theme is an estimation of the area of contact
which is treated as though it is generated between single
asperities. The adhesion and its attenuation in various
fluids and for different surface roughnesses is measured
and the results are analysed using recent theoretical
treatments developed for other systems. Friction data
are interpreted using the 'adhesion model' which is
based on interface shear neglecting junction size
variations during sliding. In the examples discussed,
a comparison has been made of experimental data and
theory. Generally the two compare well and emphasise
the importance of such phenomena as surface topography,
wetting, electrical double layer forces, adsorption
and ploughing forces during sliding.

## INTRODUCTION

It has proved to be expedient to regard polymer friction as arising from energy dissipation in two separate and non-interacting processes. Bowden and Tabor [1] adopted this approach with metallic systems and termed the two processes the ploughing and adhesive components. More recently Briscoe and Tabor [2] reviewed the value of this approach in polymeric systems. The ploughing component of the frictional work is thought of as being dissipated in a relatively large volume adjacent to the contacts. The losses arising from the adhesion component are attributed to the shearing of adhesive junctions and are confined to a much smaller region close to the interface. For polymer-polymer contacts ploughing losses are often negligible and thus the friction can be analysed using the adhesion model of friction. There are still a number of fundamental questions which have yet to be resolved in the application of this model. These include, for example, the mechanisms of non-hydrodynamic lubrication in the presence of liquids, the non-Hertzian normal load dependence of friction for a point contact, a theoretical description of intermittent motion and the relationship between surface roughness and friction. The present paper addresses a number of these problems.

Some of the limitations for achieving progress in such areas arises from problems associated with the interpretation of data obtained with the various experimental systems used. They generally involve a sphere on flat configuration which is a prerequisite for fundamental friction studies since the contact mechanics are in principle well-defined. The inherent disadvantages of large specimens are the difficulties in ensuring single asperity contact, the use of applied normal loads which are considerably greater than the separation forces of the contact due to surface interactions and the tendency of large contacts to be subject to a dwell time effect. Furthermore, some form of force detection device is required to monitor the friction and this introduces additional complications since the frictional processes will be influenced by the dynamic response of the equipment. Many of these difficulties can be overcome by studying the behaviour of monofilaments.

A procedure, developed by Pascoe and Tabor [3] is based on the study of two fibres in an orthogonal configuration. The bending of one of the fibres is used as a means of applying a normal load and its deflection, produced by movement of the other fibre, is a measure of the frictional force. Since the curvature of the fibres is so acute the contact area will be very small and hence contact is likely to be between single asperities [4]. The size of the contact can be determined from the autoadhesive force which is obtained by bending measurements during pull-off. Thus the system allows the measurement of friction at micro-contacts with a well-defined force monitoring system. The applied normal loads can be made comparable to the magnitude of the surface forces.

The work reported here is concerned with applying this approach to PET fibres and it consists of a broad survey of a number of different aspects. In particular, the effects of surface roughness, liquid environments and solid lubricants have been studied together with some preliminary work on intermittent motion. A fairly comprehensive picture of the effects of liquids has emerged which can be described in terms of wetting, electrical double layer forces, boundary lubrication and plasticisation. A more detailed account of the wetting effects are presented elsewhere [5]. For solid boundary lubricants a ploughing contribution which increases with film thickness has been identified. Some preliminary data on surface roughness effects have been obtained. Examination of stick-slip behaviour has provided a useful basis for our current studies in this area [6] which is closely related to the statistical nature of the frictional process [7]. In addition, the work has led to a more general description of the normal load dependence of friction for polymers in point contact [8].

In many cases the autoadhesive load is comparable with the normal load for these systems. Surrounding the contact with fluid is found to be capable of inducing two effects. The general result is that the normal load is attenuated by an amount that is calculable using established procedures. In some cases the fluid may actually ingress into the contact zone and act as a boundary lubricant. The change of the frictional force in these circumstances, however, cannot be calculated. Interesting effects have been observed in electrolyte solutions where double layer forces are believed to be capable of generating an effective tensile load at the periphery of the contact region

## MATERIALS

PET fibres were obtained from Imperial Chemical Industries Limited, Harrogate and Courtaulds Limited, Coventry. The fibre types and corresponding titania contents (percentage by weight of $TiO_2$) are given in Table I. They were circular in cross-section with radii in the range 7-12 μm. For the contact angle measurements, PET in the form of "Mylar" film (trade name of E I du Pont de Namours & Co) was used. The PET was cleaned by Soxhlet extraction with petroleum ether (60/80) followed by immersion in cold dichloroethylene for 10 mins and then hot distilled water for 1 hour.

Table I. The Table Shows the Types, Supplier and % $TiO_2$ of the PET Fibres Used and Includes the Corresponding Fibre Radii and Measured Pull-off Forces.

| Fibre | Supplier | % $TiO_2$ | Fibre Radius μm | Pull-off Force μN |
|---|---|---|---|---|
| 622 | ICI | 1.5 | 7.2 | 0.85 ± 0.1 |
| 122 | ICI | 0.05 | 7.3 | 0.7 ± 0.05 |
| 80/25 | Courtaulds | 0.05 | 8.6 | 1.51 ± 0.2 |
| 124 | ICI | 0.05 | 9.3 | 1.85 ± 0.3 |
| JM562 | ICI | 0.5 | 11.4 | 2.23 ± 0.2 |
| GS280 | ICI | 0.05 | 11.5 | 2.90 ± 0.5 |

Surface chemically pure water was prepared using the method described by Taylor and Mingins [9]. n-Heptane (Analar) was obtained from Hopkin and Williams and passed through an alumina column followed by fractional distillation. Perfluorodecalin (trade name "Flutec") was obtained from Imperial Smelting, Rio Tinto Zinc and used without further purification. n-Decanol (Analar) from British Drug House was purified using an alumina column. n-Octylamine (Puriss) was obtained from Fluka and filtered under nitrogen before use to remove any carbonates. Sodium dodecyl sulphate (specially purified for biochemical applications) from BDH was further purified using the method of McGhee[10] which involved recrystallisation from aqueous ethanol followed by protracted Soxhlet extraction with petroleum ether (40/60).

## EXPERIMENTAL

Contact angles on "Mylar" films were measured using the sessile drop technique [11]. Surface tensions were measured by the null-buoyancy Wilhelmy plate method using a roughened glass plate showing zero contact angle to water. The plate was supported from an electronic microbalance and the values are quoted to ± 0.1 $mNm^{-1}$.

The measurement of friction was carried out using two pieces of equipment to obtain a range of loads between $10^{-8}$ and $10^{-4}$ N. The lower load range (up to approx $10^{-6}$ N) was covered using a cantilever apparatus similar to that described by Pascoe and Tabor [3]. It is based on the application of loads by pressing one monofilament against another through a cantilever arrangement. This equipment could also be used for measuring the pull-off force between fibres. The principles are shown schematically in

Figure 1. Fibre 1 is incarcerated under slight tension at both ends in a cell which can be used for submerging the fibre in a liquid. The cell is mounted on the motorised travelling stage of a microscope and adjusted to ensure that the fibre is level and co-axial with the direction of movement. Fibre 2 is mounted on a second holder at one end only and is brought into orthogonal contact with the first fibre (Figure 1b, i). The applied load is varied by adjusting the bending of the partially free fibre (Figure 1b, ii). The force is calculated from the vertical deflection $\Delta h$;

$$\Delta h = W\ell^3/3E_3 I \qquad (1)$$

where I is the second moment of inertia of the cross-section of the fibre with radius r which is given by $\pi r^4/4$; $E_3$ is the longitudinal elastic modulus of the filament. The length from the point of contact is $\ell$ and ranges from 0.5 to 5 mm. Movement of the microscope stage causes Fibre 2 to bend because of friction at the contact. The frictional force F is computed from the lateral deflection $\Delta f$;

$$\Delta f = F\ell^3/3E_3 I \qquad (2)$$

The effective length $\ell$ increases with deflection and the appropriate correction was made.

For measurement of the friction at higher loads a hanging fibre apparatus was constructed. A schematic diagram is shown in Figure 2. A long monofilament (Fibre 2, approx 20 cm) is suspended from a microbalance under a mass M. A second filament (Fibre 1) is mounted horizontally on a frame constructed of spring steel to allow the tension to be controlled with the aid of strain gauges. This filament is brought into contact with the freely hanging monofilament which is displaced by a known amount. The horizontal assembly is then moved up and down along the axis of the hanging filament and the resulting frictional force is recorded by an electronic microbalance. For an upward movement, the recorded weight is $Mg-F_u$ where $F_u$ is the frictional force in this direction and when the motion is downward, the recorded weight is $Mg + F_D$ where $F_D$ is the frictional force when the fibre moves down. To a good approximation the true frictional force is obtained from the mean of $F_u$ and $F_D$. The normal load W* may be increased by increasing M (typical range, 50-500 mg), decreasing the length (between 15 and 25 cm) of the hanging filament L or increasing its deflection D. The angles of deflection were necessarily small to reduce the effect of change in angle, and hence load, during motion. In these experiments, the maximum angle of deflection was $5^\circ$ and at this maximum angle, the variation during motion was $\pm 0.04^\circ$. Details of the analysis

have been published elsewhere by Briscoe and Kremnitzer[4].
The applied normal load $W^*$ is given by:-

$$W^* = Mg \tan \theta \qquad (3)$$

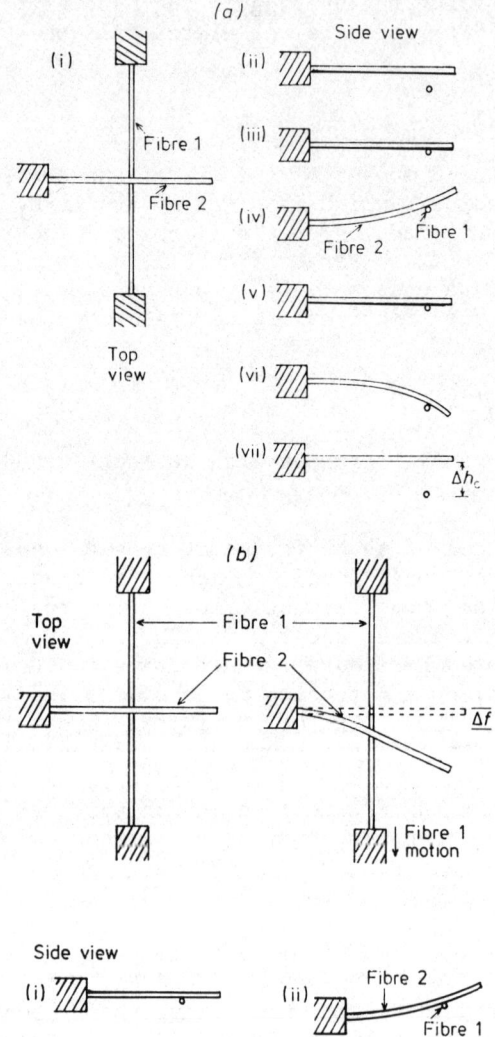

Figure 1. Schematic diagram of the 'cantilever apparatus' for (a) adhesion and (b) friction measurements. Details are given in the text.

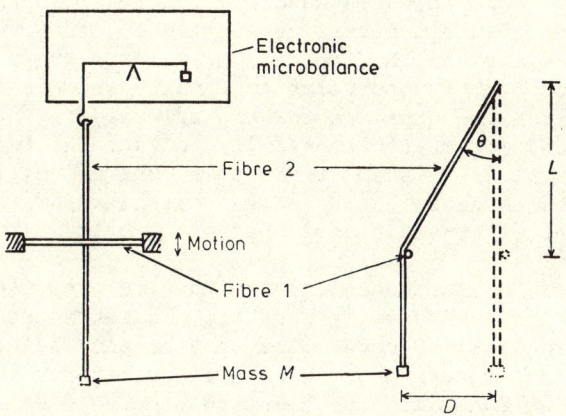

Figure 2. Schematic diagram of the 'hanging fibre' friction apparatus. Details are given in the text.

The adhesion between fibres was measured using the cantilever equipment. Two monofilaments are brought into contact (Figure 1a, v) and the critical deflection $\Delta h_c$ required to separate them (Figure 1a, vi and vii) is measured. $\Delta h_c$ is plotted as a function of $\ell^3$ and the critical separation force $W_c$ obtained from the gradient (Equation 1).

## RESULTS AND DISCUSSION

### Adhesion of Dry Fibres

Adhesion measurements of dry monofilaments were made on six pairs of each of the fibre types. The results are given in Table I; the reproducibility was ± 5% for a single pair of fibres. Figure 3 shows a plot $\Delta h_c$ against $\ell^3$ for a pair of type 80/25 fibres. The adhesive force is clearly a constant for the pair (Equation 1). The pull-off force was also observed to be independent of the rates of separation of the monofilaments which indicates that the viscoelastic work is not appreciable [12]. We shall adopt this assumption throughout the remainder of the paper. In essence, the contact mechanics are assumed to be governed by a time invariant elastic response.

We may envisage that the contact between two rough orthogonal fibres will be of two kinds; a multiple asperity contact or between single asperites. Smooth fibres can be treated as a special case of single asperity contact where the mutual radius of curvature of the asperities is equal to the radii of the fibres. The direct means of diagnosis is not available. However, friction measurements on the fibres over a large range of normal forces were found to fit the relation $F = KW^n$ where $n = 0.75 \pm 0.1$ which is the value expected for a single asperity contact between polymers; for multiple asperity contact the index n approaches unity [13]. It seems reasonable, therefore, to treat the adhesion and later friction as that arising from the single asperity contact of curved elastic bodies.

The analysis of the adhesion data thus resolves to that of two curved elastic bodies in contact. The strength of the adhesive junction will be determined by a balance between the surface attractive forces and the bulk elastic forces opposing deformation. Two theories have been proposed to describe such a contact. The first is due to Derjaguin, Muller and Toporov [14,15] (DMT theory) which was developed for hard materials ($E > 10^{-9}$ Nm$^{-2}$). On the assumption that the deformation is Hertzian and that separation occurs when the contact area is reduced to zero, they derived the following expression for the force of detachment

$$W_c = 2\pi R \Delta \gamma \qquad (4)$$

where R is the mutual radius of curvature and $\Delta \gamma$ is the thermodynamic work of adhesion. An alternative approach was proposed by Johnson, Kendall and Roberts [16] (JKR theory), which in contrast to the DMT theory neglects interfacial forces outside the contact region. The theory leads to a non-Hertzian contact geometry with the formation of a 'neck' and a separation force given by

$$W_c = \frac{3}{2} \pi R \Delta \gamma \qquad (5)$$

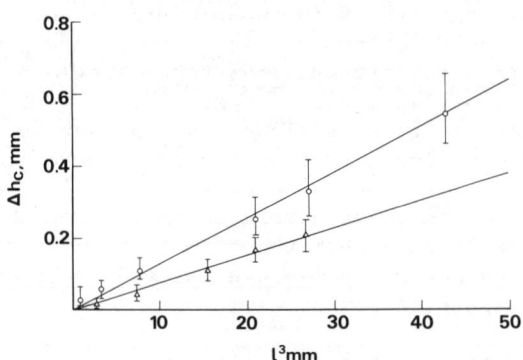

Figure 3. The adhesion of crossed fibres showing the critical displacement as a function of $\ell^3$ for dry fibres (O) and immersed in water ($\Delta$).

This relation was confirmed experimentally using rubber spheres. Israelachvili, Perez and Tandon [17], however, measured the pull-off force and contact geometry between crossed cylindrical surfaces of molecularly smooth mica coated with materials of known surface free energy and found that DMT theory gave the correct value of $W_c$ while the contact geometry was consistent with the JKR theory. The results confirm Tabor's [18] suggestion that the interfacial forces outside the contact zone cannot be neglected when the neck height is comparable to the equilibrium separation distance between the surfaces (approx 0.3 nm). Such a situation prevails for high moduli material (e.g. mica) or for small radii of curvatures. In the case of glassy polymers, Tabor [18] calculated that the pull-off force would approach the DMT value for radii of curvature of the order of 1 μm. This problem has been examined recently by Muller, Yushchenko and Derjaguin [19]. Their analysis suggests that the JKR theory may sometimes be more appropriate particularly for dry contacts between PET fibres. At present it does not appear to be possible to identify which of the two approaches is more reasonable or indeed if some hybrid is more suitable. In any event the choice does not lead to large errors and we will use the DMT equation for our analysis

The thermodynamic work of adhesion is given by

$$\Delta\gamma = \gamma_s^1 + \gamma_s^2 - \gamma_{ss}^{12} \tag{6}$$

where $\gamma_s^1$ and $\gamma_s^2$ are the surface free energies of the two bodies and $\gamma_{ss}^{12}$ is the interfacial free energy. For like bodies $\gamma_{ss}^{12}$ is zero and $\Delta\gamma$ reduces to $2\gamma_s$. El-Shimi and Goddard [20] have calculated $\gamma_s$ to be 49.5 mNm$^{-1}$ for PET from contact angle measurements using the harmonic-mean equation [21,22].

The mutual radii of curvature may now be calculated from Equation (4). The results are plotted as a function of the macroscopic fibre radius in Figure 4. It shows that R is strongly related to the gross fibre radius and influenced little by the $TiO_2$ content of the fibres. In the SEM the fibres appeared smooth with a relatively small number density of visible asperities.

SEM examination also showed that for the same weight percentage of $TiO_2$, the smaller fibres appeared rougher than the larger ones which may account for the radial dependence of the adhesion. If there is a greater concentration of particles on the surface than in the bulk, then the larger surface area to volume ratio in the smaller fibres may result in an apparent increase in $TiO_2$ content at the surfaces. Further analysis of this type of system is being continued at present.

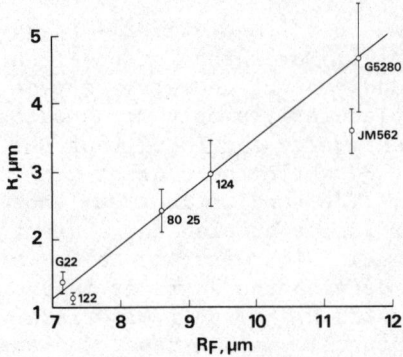

Figure 4. The asperity radius R calculated from the pull-off force using Equation (4) and plotted as a function of fibre radius $R_F$

## Adhesion in Water

Figure 3 also shows the results for adhesion measurements in water. The measured value of Wc for this fibre is reduced from 1.6 μN in the dry state to 0.9 μN in water.

In a liquid environment the pull-off force $W_c^L$ becomes (cf Gent & Schultz [23])

$$W_c^L = 2\pi R \Delta \gamma^L \quad (7)$$

with $\Delta \gamma^L = \gamma_{SL}^1 + \gamma_{SL}^2 - \gamma_{SS}^{12}$

where $\Delta \gamma^L$ is the thermodynamic work of adhesion in a liquid environment and $\gamma_{SL}$ is the solid-liquid interfacial surface tension; for like solids $\Delta \gamma^L$ reduces to $2\gamma_{SL}$. A number of empirical or semi-empirical expressions have been proposed for the calculation of $\gamma_{SL}$; for example details of the various approaches have been discussed by Mittal [24,25]. They have included the geometric mean equation proposed by Good [26] and the harmonic mean equation due to Wu [21,22,27]. More recently, Fowkes [28] has proposed that dipole contributions are negligibly small and that only dispersion forces and electron donor-acceptor interactions (which include hydrogen bonding) should be considered. On the basis of Wu's measurements on incompatible binary polymer liquids and melts we have chosen to use the harmonic mean equation which is given as follows:-

$$\gamma_{SL} = \gamma_S + \gamma_L - \frac{4\gamma_S^d \gamma_L^d}{\gamma_S^d + \gamma_L^d} - \frac{4\gamma_S^P \gamma_L^P}{\gamma_S^P + \gamma_L^P} \qquad (8)$$

where $\gamma_S = \gamma_S^d + \gamma_S^P$

and $\gamma_L = \gamma_L^d + \gamma_L^P$

where $\gamma^d$ and $\gamma^P$ are the dispersion and polar components of the surface tension. Combination with Young's equation gives

$$\gamma_L^d \cos\theta = -\gamma_L + \frac{4\gamma_S^d \gamma_L^d}{\gamma_S^d + \gamma_L^d} + \frac{4\gamma_S^P \gamma_S^P}{\gamma_S^P + \gamma_L^P} \qquad (9)$$

where $\theta$ is the contact angle. From measurements of contact angles on PET using water and methylene iodide, El-Shimi and Goddard [20] calculated that $\gamma_S^d$ and $\gamma_S^P$ are 36.5 and 13.0 mNm$^{-1}$ respectively using this equation. They also found that for paraffin wax $\gamma_S^d = \gamma_S = 25.5$ mNm$^{-1}$. Substitution of this value into Equation (9) and the contact angle of water on paraffin wax gives $\gamma_L^d$ for water as 21.7 mNm$^{-1}$ and hence $\gamma_L^P$ as 50.5 mNm$^{-1}$. Using these results to calculate the interfacial tension between water and PET gives a value of 25.9 mNm$^{-1}$ for $\gamma_{SL}$.

The pull-off force of 1.6 μN in the dry state is equivalent to an asperity radius of 2.6 μm (Equation 4). For $\gamma_{SL}$ = 25.9 mNm$^{-1}$ and R = 2.6 μm, Equation 7 predicts a pull-off force of 0.84 μN which is very close to the experimental value of 0.9 μN.

The calculation of the separation force in water has assumed that only van der Waals and polar forces contribute to the adhesion. In water there is a possibility of electrical double layer forces reducing the adhesion. No attempt was made to determine the isoelectric point. It would appear from the close agreement between the measured and calculated values that in fact these forces are small for the PET in water.

## Static Friction of Dry Fibres

Friction measurements with the cantilever equipment gave the rather remarkable result that the frictional force tended to zero when the applied <u>negative</u> normal load tended to $W_C$. A typical set of data is shown in Figure 5. Presumably the torsional flexibility of the fibres produces a 'rolling' failure under tangential stress which corresponds to a normal peeling failure. Thus the frictional force can be thought of as arising from a combination of an applied load $W^*$ and an "adhesive" load $W_c$. As a first approximation this total load, W, will be assumed to be the arithmetical sum, $W^* + W_c$. This total load will be used to compute the contact area using Hertzian analysis where the contact area A is given by [29]

$$A = \pi (KWR)^{2/3} \tag{10}$$

where $K = \frac{3}{2} (\frac{1-\nu^2}{E1})$

where $\nu$ and $E_1$ are Poisson's ratio and the tranverse elastic modulus respectively. The yield stress for PET (for oriented sheets, values in the range 5 to 35 x $10^7$ $Nm^{-2}$ have been measured [30]) was not exceeded in any of the contact pressures reported here. The numerical values of the terms in Equation (10) are discussed later.

We will neglect junction growth [1] and junction contraction [31] We believe that these effects are comparatively small and in any event are not readily evaluated with accuracy. The frictional work may be regarded as arising from two sources; ploughing or grooving and interface shear of adhesive junctions [1]. The possibility of ploughing losses under the present experimental conditions can be neglected [32,33] so that the frictional work may be attributed to the fracture energy of junctions formed by the adhesive forces between asperites in contact [1,2]. The interfacial shear strength of the junctions is given by

$$\tau = F/A \tag{11}$$

where F is the frictional force and A the area of contact. Empirical studies on thin polymeric films have shown that to a good approximation $\tau$ is given by [2,32,34]

$$\tau = \tau_o + \alpha P \tag{12}$$

where P is the contact pressure P (= W/A) and $\tau_o$ and $\alpha$ are constants depending on the film material.

Combinations of Equations (10, 11 and 12) leads to the following expression for the frictional force

$$F = \tau_o \pi(KWR)^{2/3} + \alpha W \tag{13}$$

The numerical values of R have already been obtained. $\tau_o$ and $\alpha$ values are available for similar materials [32,35]. We have used $\alpha$ = 0.36 and $\tau_o$ = 6 x $10^6$ $Nm^{-2}$. The measured longitudinal modulus $E_3$ in bending for these fibres was found to be ca. $10^{10}$ Nm$^{-2}$. From Ward [30], this corresponds to a draw ratio of approx 3. At the highest draw ratio of approx 8, $E_3/E_1 \simeq 27$ and we have taken one half of this value which corresponds to $E_1 = 0.74 \times 10^9$ Nm$^{-2}$. The Poisson's ratio is taken as 0.4. Calculated values of the frictional force at zero applied load i.e. $W = W_C$, are shown in Figure 6 along with experimentally determined values; the two are in good agreement.

As has already been indicated, SEM examination showed that the smaller fibres, which are associated with the smaller asperities appeared rougher than the larger fibres. The data can be interpreted as a reduction in friction with increasing surface roughness. Previous measurements to correlate the two factors for fibres have met with varying degrees of success. For example Howell et al [13] reported that fibre friction, measured using a capstan device, diminishes appreciably if the surface of the cylinder or the fibre is roughened; other workers suggest that the correlation is not at all clear and depends on the characterisation of the roughness [36,37]. Our data suggest that introducing titania roughens the surface, reduces the contact area and hence the friction. The surface concentration of titania is controlled by at least two factors; the bulk concentration and the radii of the fibres. We see no simple means at present of predicting the change in effective topography for a given fibre/filler system.

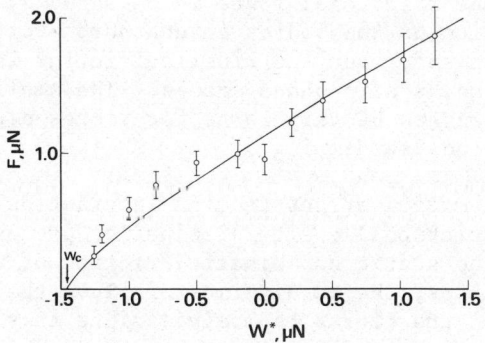

Figure 5. Static frictional force (cantilever equipment) for dry PET fibres as a function of applied load W*. The theoretical curve was calculated using Equation 13.

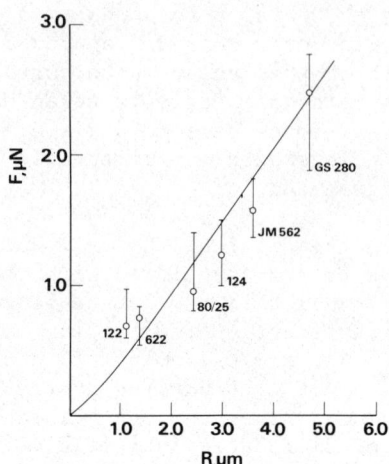

Figure 6. Frictional force (cantilever apparatus) at zero applied load as a function of asperity radius (from Figure 4). The theoretical curve was calculated from Equation 13.

## Dynamic Friction

A characteristic feature of unlubricated fibres is their propensity to exhibit intermittent motion in sliding. The response of most polymers is of this type because the static friction is generally greater than the kinetic friction and often the dynamic friction decreases with increasing sliding velocity. At a point of stick, the frictional force is balanced by that arising from the deformation of the bodies in contact. Further sliding breaks the contact and, when the kinetic friction is less than the static friction, a slip phase ensues. The amplitude of the intermittent motion can be very large for fibres because of their high flexibility and low damping.

Intermittent motion arises from an interaction between the machine and the frictional energy dissipation process [38,39]. The latter includes the static and kinetic friction of the contact. In the case of fibres, the machine must include the force detection system and the fibres themselves since they have appreciable degrees of elastic freedom. The response of the machine is governed by such factors as the natural frequency, inertia, damping and sliding velocity. It is obvious that the transient behaviour of the slip phases will be strongly dependent on such factors but at present it is not clear why the static friction should be so affected. The classical example is the effect of sliding velocity; at low velocities the intermittent motion is exhibited in its most gross form as stick-slip. At high velocities relaxation oscillation occurs and the static frictional value decreases [40].

In this work we have observed that distinct changes in the amplitude and duration of the stick-slip, measured on the hanging fibre equipment, occur as the tension in the horizontal fibre is varied. Figure 7 shows typical traces at high and low tensions. Figure 8 shows the static friction at zero applied load as a function of the tension, T; the data are non-linear.

We have not analysed this system variable in detail but it is notable that the static friction scales quite accurately as the reciprocal of the square root of the tension. The data of Figure 8 have been plotted in this form to demonstrate the relationship. The fundamental resonance frequency $\nu$ of a string under tension is given by:-

$$\nu = \frac{1}{2\ell} \sqrt{\frac{T}{\rho}} \qquad (14)$$

where $\rho$ is the linear density. It would appear therefore that increasing the natural frequency has an analogous effect to increasing the sliding velocity. Additional experimental work is required to develop this connection further.

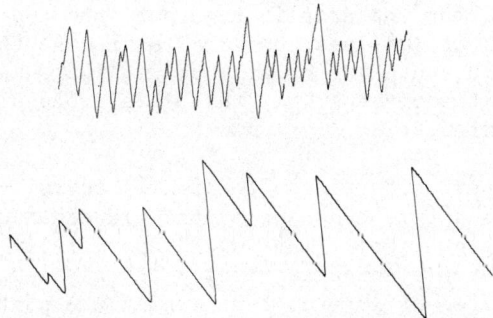

Figure 7. Typical traces from the hanging fibre equipment. The upper trace corresponds to the horizontal fibre at high tension and the lower to zero tension.

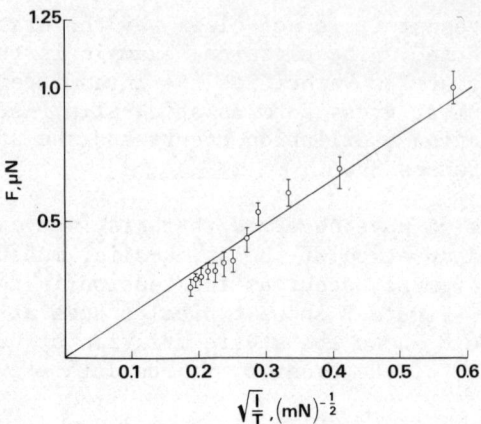

Figure 8. The static frictional force measured with the hanging fibre equipment at zero applied normal load as a function of fibre tension.

The foregoing remarks have referred to the static frictional force but the important parameter is the frictional work per unit sliding distance and we believe that this should be virtually independent of the dynamics of the system. Other factors which need to be considered are the conformity of the fibres and the machine response. It is arguable that the conformity of the horizontal fibre is greater at lower tension and the contact area, and hence the frictional force, would be greater. In general, however, for the data in Figure 8, the applied normal force is a factor of $10^4$ less than the tension so that the relative change in contact area will be small. Unless more subtle effects operate it seems unlikely that this process is the source of the observed trends.

In these experiments a complication is introduced because of the relatively long time constant of the microbalance which supports the hanging fibre. This would appreciably attenuate the frictional force at high tensions which also corresponds to the high frequencies of intermittent motion. A part of the observed response may originate in this machine effect.

The configuration of the cantilever equipment is nominally the same as the hanging fibre equipment but the stroking motion is inherently different from the bowing action of the hanging fibre. It was found that the frictional force was insensitive to changes in the tension of the fibre incarcerated at both ends. Since there is no transducer system for detecting forces, i.e. the fibre in the machine, it makes an ideal apparatus for studying intermittent motion and further studies are being currently pursued.

### Friction in aqueous sodium dodecyl sulphate solutions

The friction between large rubber contacts and glass in the presence of aqueous sodium dodecyl sulphate (SDS) solutions has been studied extensively by Roberts and Tabor [41]. Many interesting results have emerged particularly the potential lubricating action of electrical double layers. In this work we have investigated whether similar behaviour is prevalent in the microcontacts occurring between PET fibres. Measurements were made for a range of applied normal loads using the cantilever equipment. The friction was determined in water and in 3.4 and 8.5 mM SDS with and without 0.1 M NaCl. The results are shown in Figure 9. In water the friction is less than that of the dry fibres (not shown) and further decreases in friction are seen in 3.4 and 8.5 mM SDS respectively. The addition of 0.1 M NaCl results in the friction at both SDS concentrations being very similar to that at 3.4 mM SDS with no salt added. A cursory inspection of these trends indicates that electrical double layer forces are certainly of importance. We now attempt to analyse the data in more detail assuming that double layer forces are operating. The lubricating action of pure water is considered first.

We first recall that the attenuation of the autoadhesion in water may be wholly attributed to the change in the work of adhesion. In terms of the previous analysis of the frictional force the adhesive component of the load $W_C$ is reduced. For the present it will be assumed that the specific interfacial shear work is unaltered; i.e. the water does not ingress into the contact zone but merely reduces the contact area for a given applied load. We have found that protacted immersion of PET fibres in water has little effect on the bending modulus and we conclude that the mechanical properties of PET fibres are reasonably insensitive to water. The friction in water can therefore be calculated using Equation (13) with the values of K, $\alpha$ and $\tau_0$ used for the dry fibres. The asperity radius R was calculated to be 2.9 μm from the measured pull-off force ($W_C$ = 1.8 μN) for the fibres in the dry state. We assume that this mean quantity is also unchanged. The results of these calculations are shown in Figure 9 (top curve). The calculated curve is in good agreement with the measured values.

The sodium dodecyl sulphate and salt solution data are now

considered. An indication that repulsive electrical forces are operating may be judged from the fact that on simple wetting grounds the salt dosed solution should provide marginally better lubrication than the single component surfactant solution (see Table II). The presence of salt ions, however, is presumably suppressing the electrical double layer forces induced by surfactant ion adsorption.

Table II. Surface Tensions of SDS and SDS/NaCl Solutions and Corresponding Advancing Contact Angles on Mylar Film.

| Conc. of SDS mM | Conc. of NaCl mM | $\gamma_L$ mNm$^{-1}$ | $\theta$ deg |
|---|---|---|---|
| 0 | 0 | 72.2 | 73 |
| 3.4 | 0 | 53.2 | 60 |
| 8.5 | 0 | 39.2 | 52 |
| 3.4 | 100 | 33.6 | 46 |
| 8.5 | 100 | 33.5 | 47 |

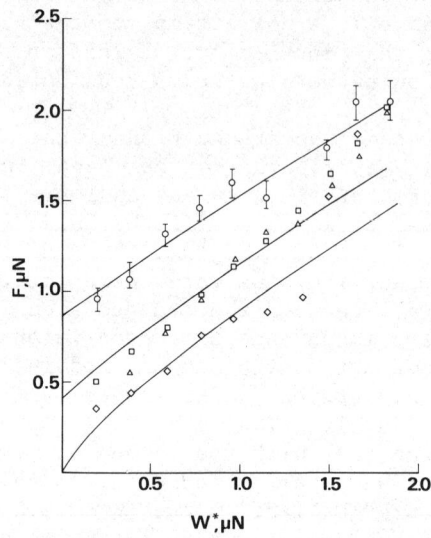

Figure 9. The friction force (cantilever apparatus) as a function of applied normal load in water (O), 3.4 mM aq SDS (Δ), 8.5 mM aq. SDS (◊) and these aqueous SDS solutions with the addition of 0.1 M NaCl (□).

The cantilever equipment was not sufficiently sensitive to obtain pull-off forces in these solutions and thus the frictional forces could only be calculated directly for the case of dry and water lubricated contacts.

In the presence of salt it was assumed that electrical double layer repulsion is suppressed and the action of the SDS/NaCl solutions was simply to enhance wetting. In a parallel study with isopropanol/water mixtures [5] we found that there exists a linear relationship between $W_C^L$ and $\gamma_L$ where $W_C^L = 0$ when the liquid surface tension is equal to the critical wetting tension $\gamma_C$. The value of $\gamma_C$ in these alcohol/water mixtures was ca 22 mNm$^{-1}$ which is lower than $\gamma_S$ for PET because of the well known adsorption effect [20] and hence the complete inhibition of adhesion due to wetting occurs when $\gamma_C < \gamma_S$. Similar adsorption effects have been observed for SDS solutions [42] and this is reflected by the value of $\gamma_C$ of ca 10 mNm$^{-1}$ determined in this work from contact angle measurements on Mylar film. The measured $W_C^L$ (= 0.95 µN) in water gives the following expression for $W_C^L$ as a function of $\gamma_L$ assuming that a linear interpolation can be made between $\gamma_L = \gamma_C = 10$ mNm$^{-1}$ and $\gamma_L = \gamma_{water}$ 72.2 mNm$^{-1}$.

$$W_C^L/4\pi R = \gamma_{SL} = 0.42\, \gamma_L - 4.2 \tag{15}$$

The values of $\gamma_L$ for the SDS/salt solutions are virtually identical (see Table II) and substitution into Equation (15) gives $W_C^L = 0.37$ µN. Following the procedure used for water but with $W_C^L = 0.37$ µN, the frictional force as a function of applied normal load was calculated and the results plotted in Figure 9 (middle curve). The agreement is excellent considering the approximations made and supports the contention that only wetting forces operate in the presence of NaCl.

The corresponding values of $W_C^L$ in 3.4 and 8.5 mM SDS are 0.67 and 0.46 µN from Equation (15). The appropriate substitution in Equation (13) failed to give an adequate fit of the experimental data. If electrical double layer forces are effective then the total normal load W can be written as the sum of the applied normal load W*, the pull off force $W_C^L$ and an electrical term $W_{el}$ thus

$$W = W^* + W_C^L + W_{el} \tag{16}$$

To fit the experimental data it was necessary to set $W_{el}$ equal to -0.3 and -0.46 µN for the 3.4 and 8.5 mM SDS solutions respectively (middle and lower curves of Fig. 9). Bell & Peterson [43] have developed a comprehensive theory to account for the electrical double layer interaction forces between spheres. Peterson [44] has calculated for the asperity radii found in this work, that such values of $W_{el}$ can be generated using reasonable estimates for the surface potential at the contact. His calculations also show that the predominate repulsion occurs outside the contact zone.

## Lubrication in organic liquids

It is generally recognised that effective lubrication of polymers using liquids is very difficult to achieve except at high sliding velocities when fluid lubrication can operate. Bowers et al [45] suggested that the absorption sites are not highly energetic and have a low number density on polymer surfaces. The importance of these factors has been established by the improved lubrication shown by polymers after chemical treatments to induce charged sites on the surface [46,47]. Pascoe and Tabor [48] attributed the poor boundary lubrication of polymers to the similarity in properties of the boundary lubricants and polymers which are both generally long chain organic molecules. Rubenstein [49] considered that some lubricants lose their effectiveness because they plasticise the polymer and the resulting softening causes an increase in contact area. Further work by Cohen and Tabor [50] showed that for nylon in the presence of a film of water, there is an initial reduction in friction as the surface is softened and after prolonged periods in water the friction rises due to an increase in contact area. They also found a strong correlation between the extent of wetting by the lubricant and the reduction in polymer friction. For water in polymer-glass contacts it was concluded that a weakly bound film of water molecules was adsorbed to the hydrophilic glass surface and this promoted lubrication.

In the present work we have found that Equation (13) adequately describes the reduction in friction of PET fibres in water and SDS/salt solutions only on the basis of the wetting characteristics of the liquid. The model can be readily extended to determine the level of boundary lubrication since the equation contains interfacial shear strength terms. In the same way it should be possible to determine the effect of plasticising substances.

Figure 10 shows the friction as a function of applied normal load for PET fibres in n-decanol and fluorodecalin using the cantilever technique. The pull-off force for the dry fibres was 1.6 μN giving an asperity radius of 2.6 μm (Equation 4). The liquids completely wet Mylar film so that we conclude $W_C$ is zero. For K, $\alpha$ and $\tau_o$ equal to the values for those of the dry fibres the upper curve in Figure 10 is generated. This gives a good description of the data for n-decanol and we deduce that this liquid provides no effective boundary lubrication. The friction is lower than that in the dry state but this can be ascribed simply to the wetting action of the alcohol. The friction for perfluorodecalin is less than that for n-decanol and hence some boundary action by this liquid is detected. The bending modulus was found to be unchanged after immersion in the liquid and hence

K in Equation (13) will be the same as for the dry values. Taking R = 2.6 μm and $W_C^L$ = 0, the lower curve in Figure 10 was calculated from Equation (13) for $\tau_o$ = 3.4 x $10^6$ $Nm^{-2}$ and $\alpha$ = 0.2. These are about half of the values found for the clean fibres. Similar values have been found for other highly fluorinated surfaces such as PTFE ($\tau_o$ = 3.4 $\pm$ 1 x $10^6$ and $\alpha$ = 0.09 $\pm$ 0.01 [51]). Presumably fluorinated material has ingressed into the interface.

Figure 11 shows the results for n-heptane and n-octylamine using a different set of fibres. The lower curve was calculated using the same procedure as that which was described for the n-decanol data. In this case for the dry contact $W_C$ was found to be 0.16 μN and hence we have R = 0.25 μm. Taking $W_C^L$ = 0, gives a good fit of the data and, like n-decanol, no boundary action seems to be apparent. On wetting arguments alone the n-octylamine results should be similar to those for n-heptane since they both wet Mylar film. In n-octylamine, however, it was found that the bending modulus of the fibres was reduced e.g. the 10 sec values were 9.8 and 6.75 x $10^9$ $Nm^{-2}$ for the dry fibre and in n-octylamine respectively. Thus we would expect changes in K, $\alpha$ and $\tau_o$. We have not measured these parameters independently but for example the data can be fitted by taking $E_1$ = 0.13 x $10^9$ $Nm^{-2}$ and leaving $\alpha$ and $\tau_o$ unchanged. The results are plotted as the upper curve in Figure 11.

The results for n-octylamine are of interest because long chain amines have been studied in connection with chemically modified polymers to produce a negatively charged surface [46,47]. Their effectiveness has been ascribed to the binding of $NH_2$ groups to the anionic sites. We have found that the amines react rapidly with carbon dioxide in the atmosphere to form a solid carbonate. This would lead to a stronger interaction between $-NH_3^+$ and the negative sites. In the present work the measurements were carried out under nitrogen to avoid this problem but the effect seriously limited the scope of the present study.

## Solid boundary lubricants

In general solid lubricants are studied as monolayers [34,52,46,47] or deposited from solution leaving an unknown film thickness. Frewing [53] found that for large metal contacts at high loads and slow speeds excess lubricant is squeezed out leaving a unimolecular film which acts as the effective lubricant layer. For fibres where the contact dimensions are small and sometimes comparable with the film thickness the excess material is not expelled as efficiently. A large ploughing component can be unavoidable. The following data exemplify this effect. PET fibres were coated with stearic acid from a volatile solvent at concentrations of 0.05,

0.5 and 1% w/w. No independent measurement of film thickness was made but clearly the thickness is proportional to solution concentration. The friction as a function of applied normal load was measured using the hanging fibre equipment and the results are shown in Figure 12. The results for the clean fibres were fitted to Equation (13) using $W_C = 0.8$ µN and hence $R = 1.29$ µm with K, $\alpha$ and $\tau_0$ as before.

Also shown in Figure 12 is the theoretical curve (lower curve in the figure) using interface shear data for stearic acid ($\tau_0 = 0.6 \times 10^6$ and $\alpha = 0.038$ [34]) in Equation (13). The upper curve is that calculated for the clean fibres. The friction of the treated fibres is greater than the theoretical curve but the results show that as the film thickness decreases the difference becomes less. This suggests that there is indeed a substantial ploughing component of the friction which increases with film thickness. It did not prove possible by solvent deposition to generate accurate data which corresponded to the theoretical curve for stearic acid. Langmuir-Blodgett deposition was also found to be unsatisfactory because of poor monolayer transfer onto the fibres.

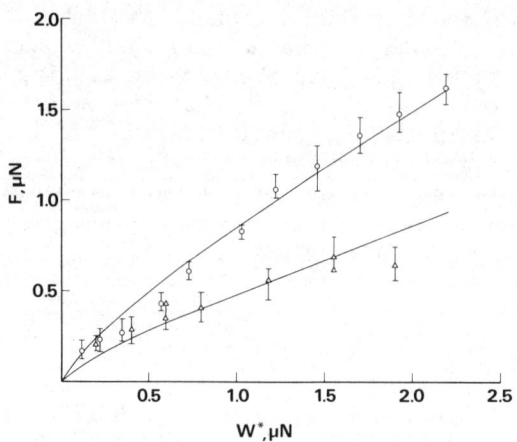

Figure 10. The frictional force (cantilever equipment) as a function of applied normal load for fibres immersed in n-decanol (O) and perfluorodecalin (Δ).

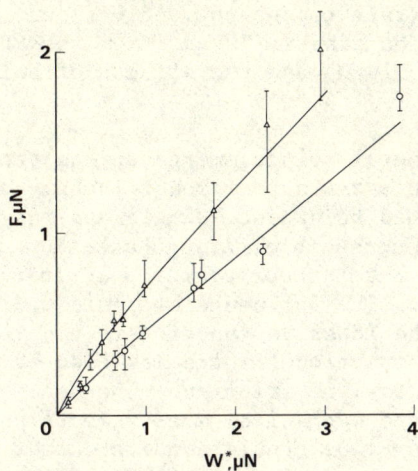

Figure 11. The frictional force (cantilever equipment) as a function of applied load for fibres immersed in n-octylamine (Δ) and n-heptane (O).

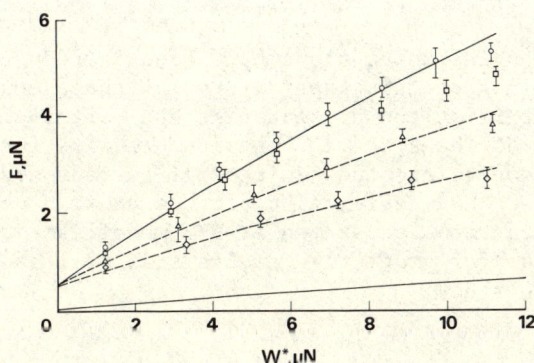

Figure 12. The frictional force (hanging fibre apparatus) as a function of applied normal load for fibres treated in 0% (O), 0.05% (◊) 0.5% (Δ) and 1% (□) w/w stearic acid solution in n-heptane.

## CONCLUSIONS

The paper has reviewed recent and current studies of the adhesion and friction of polyethylene-terephthalate monofilaments. The influence of surface topography and various liquids on these processes has been described. In addition, short descriptions of the dynamics of sliding and the effects of solid films have been given.

A number of general points emerge during the search for tools to predict the behaviour of monofilament contacts. These systems may be assumed to produce single asperity contacts whose contact geometry is accessible through adhesion measurements and the application of recent theories for describing the adhesion of elastic contacts. The influence of fluid media on the force required to break the adhesive junction is consistent with these models. The static frictional force may also be calculated using the adhesion model of friction. The necessary input data is taken from separate mechanical measurements and from the adhesion experiments. The studies indicate that in many cases the fluid simply attenuates the magnitude of the effective applied load or more accurately reduces the contact area. The fluid does not appear to actually ingress into the contact area for such fluids as water, saline solution, decanol, and octylamine. In other cases such as pure fluorodecalin the specific frictional work is actually reduced as well. For aqueous sodium dodecyl sulphate solutions a special effect is observed, electrical double layer forces seem to provide a load bearing capacity which counterbalances a part of the applied load.

Many substantial assumptions have been introduced in the analytical sections of this paper. Some of these could be justified in detail at the present time but certain features of the dynamics of the systems' behaviour require further study. Many of these aspects are the subjects of current work. The present paper has been designed to provide a broad account of a range of interactions generated at PET monofilament contacts and the potential for predictive analysis as well as identifying areas for future study.

## ACKNOWLEDGEMENTS

The authors are particularly grateful to Mr E O'Keefe (URPSL) for his assistance in some of the experiments described.

## REFERENCES

1. F.P. Bowden and D. Tabor, "Friction and Lubrication of Solids", Oxford UP, London 1954.
2. B.J. Briscoe and D. Tabor, in "Polymer Surfaces", D. T. Clark and W. J. Feast, Editors, pp. 1-23, John Wiley, New York, 1978
3. M. W. Pascoe and D. Tabor, Proc. R. Soc., $\underline{A235}$, 210 (1956).
4. B. J. Briscoe and S. L. Kremnitzer, J. Phys. D: Appl, Phys., $\underline{12}$, 505 (1979).
5. M. J. Adams, B. J. Briscoe and S. L. Kremnitzer, in "Microscopic Aspects of Adhesion and Lubrication, Proc. 34th Int. Conf. Societe de Chimie Physique", 14-18 Sept. 1981, Elsevier, Amsterdam, 1982.
6. M. J. Adams, B. J. Briscoe and A. Winkler, (1982) unpublished data.
7. M. J. Adams, B. J. Briscoe and A. Winkler, (1982), unpublished data.
8. M. J. Adams, B. J. Briscoe and E. O'Keefe (1982) unpublished data.
9. J. A. G. Taylor and J. Mingins, JCS Faraday Trans., $\underline{71}$, 1161, (1975).
10. B. McGhee, 'Ph.D Dissertation', Manchester, England, 1966.
11. W. C. Bigelow, D. L. Pickette and W. A. J. Zisman, Colloid Sci., $\underline{1}$, 513 (1946).
12. B. J. Briscoe, in "Polymer Surfaces", D. T. Clark and W. J. Feast, Editors, pp 25-46, John Wiley, New York, 1978.
13. H. G. Howell, K. W. Mieszkis and D. Tabor, in "Friction in Textiles", Butterworth's Scientific Publications, London 1959.
14. B. V. Derjaguin, V. M. Muller and P. Yu. Toporov. J. Colloid Interface Sci., $\underline{58}$, 314 (1975).
15. B. V. Derjaguin, V. M. Muller and P. Yu. Toporov, J. Colloid Interface Sci., $\underline{67}$, 378 (1978).
16. K. L. Johnson, K. Kendall and A. D. Roberts, Proc. R. Soc., $\underline{A324}$, 301, (1971).
17. J. N. Israelachvili, E. Perez and R. K. Tandon, J. Colloid Interface Sci., $\underline{78}$, 265 (1980).
18. D. Tabor, J. Colloid Interface Sci., $\underline{58}$, 2 (1977).
19. V. M. Muller, V. S. Yushchenko and B. V. Derjaguin, J. Colloid Interface Sci., 77, 91 (1980).
20. A. El-Shimi and E. D. Goddard, J. Colloid Interface Sci., $\underline{48}$, 242 (1974).
21. S. Wu, J. Polym Sci., $\underline{C34}$, 19 (1971).
22. S. Wu, J. Adhesion, $\underline{5}$, 39 (1973).
23. A. N. Gent and J. Schultz, J. Adhesion, $\underline{3}$, 281 (1972).
24. K. L. Mittal, in "Adhesion Science and Technology", L. H. Lee, Editor, Vol. 9A, pp. 129-168, Plenum Press, New York, 1975.
25. K. L. Mittal, Polym. Eng. Sci., $\underline{17}$, 467 (1977).

26. R. J. Good, in "Treatise on Adhesion and Adhesives".
    R. L. Patrick, Editor, Vol. I, pp. 9-24, Marcel Dekker,
    New York, 1967.
27. S. Wu, J. Macromol., Sci., C10, 1 (1974).
28. F. M. Fowkes, Org. Coat. Plast. Chem., 40, 13 (1979).
29. S. P. Timoshenko and J. N. Goodier, "Theory of Elasticity"
    McGraw-Hill, New York, 1970.
30. I. M. Ward, "Mechanical Properties of Solid Polymers",
    Wiley - Interscience, London, 1971.
31. A. R. Savkoor and G. A. D. Briggs, Proc. R. Soc., A356,
    103 (1977).
32. B. J. Briscoe and D. Tabor, Amer. Chem. Soc. Preprints, 21,
    10 (1976).
33. J. K. A. Amuzu, B. J. Briscoe and M. M. Chandri, J. Phys.
    D: Appl, Phys. 9, 133 (1976).
34. B. J. Briscoe, B. Scruton and R. F. Willis, Proc. Royal Soc.,
    A333, 99 (1973).
35. J. K. A. Amuzu, 'Ph.D Dissertation', University of
    Cambridge, England, 1976.
36. S. C. Scheier and W, J. Lyons, Text. Res. J., 35, 383 (1965).
37. F. L. Scardino and W. J. Lyons, Text, Res. J., 40, 559 (1970).
38. D. C. B. Evans, 'Ph.D Dissertation', University of Cambridge,
    England, 1975.
39. D. M. Rowson, Wear, 31, 213, (1975).
40. B. J. Briscoe, D. C. B. Evans and D. Tabor, J Colloid
    Interface Sci., 61, 9 (1977).
41. A. D. Roberts and D. Tabor, Proc. R. Soc., A325, 323 (1971).
42. C. A. Smolders, in "Physics and Physical Chemistry of Surface
    Active Substances" J. Th. G Overbeek, Editor, Chemistry
    Physics and Applications of Surface Active Substances, Vol II,
    pp. 343-348, Gordon and Breach Science Publishers, London 1965.
43. G. M. Bell and G. C. Peterson, J. Colloid Interface Sci., 41,
    542 (1972).
44. G. C. Peterson (1981), personal communication.
45. R. C. Bowers, W. C. Clinton and W. A. Zisman, Ind. Eng.
    Chem., 46, 2416 (1954).
46. T. Fort Jr, J. Phys. Chem., 66, 1136 (1962).
47. J. M. Senior and G. H. West, Wear 18, 311 (1971).
48. M. W. Pascoe and D. Tabor, Research S., 15, 8 (1955).
49. C. Rubenstein, J. Appl. Phys., 32, 1445 (1961).
50. S. C. Cohen and D. Tabor, Proc. Royal Soc. A291, 186 (1966).
51. B. J. Briscoe and D. C. B. Evans, Proc. Royal Soc., A380, 389
    (1982).
52. W. A. Zisman, in "Friction and Wear", R. Davies, Editor,
    pp. 110-148, Elsevier Publishing Co., Amsterdam, 1964.
53. J. J. Frewing, Proc. Royal Soc., A181, 23 (1942).

SURFACE ENERGETICS AND TRIBOLOGICAL PROPERTIES OF MINIATURE

POLYMER ELEMENTS

Z. Rymuza

Warsaw Technical University

Warszawa, Poland

The relationships between the surface energetics of polymer miniature elements and lubricants and the tribological properties of the journal steel/polymer microbearings are discussed. A method for the determination of the relative magnitudes of surface free energies of the real surfaces of polymer miniature bushes is described. This method is based on the analysis of the spreading dynamics of a drop of a lubricant on the bearing polymer surface.

## INTRODUCTION

A special application of polymers is their use as materials for elements of precision mechanisms. The rubbing elements are often very small and the specific loads can be very high. The surfaces of contacting elements are very smooth. The typical combinations of materials are steel/thermoplastics.

The tribological properties of the micropairs depend on the nature of the polymer material and the rubbing conditions. The friction coefficient and the wear rate are the most important tribological properties for the micropairs analyzed. The wear rate is a particularly significant property of the micropairs because of the correct work of the precision mechanism.

Some research works concerning the tribological properties of the rubbing miniature polymer elements are published but they do not explain the reasons for the observed friction and wear properties.[1,2,3]

The author has carried out these investigations to find out if some relationship between the tribological and surface properties of the rubbing elements (journal microbearings with steel shaft and polymer bush) exists. The effect of lubrication has been determined.

## SURFACE PROPERTIES

The hypothesis has been that the adhesion between the rubbing surfaces plays a dominant role in the friction and wear processes of the micropairs investigated. It was necessary therefore to estimate the surface energetics of the real polymer microelements, because the surface free energy of a polymer[4,5] depends not only on the chemical structure of macromolecules but also on the manufacturing conditions.[6] The miniature polymer bushings were manufactured in a mold at a temperature of 80°C with polished surfaces. The following plastics were used: PA 6 (Tarnamid T-27), 25% fibreglass-reinforced PA 6 (Itamid 25 G) and POMs (Tarnoform, Delrin 500 NC 10). The Batenfeld BSKM-10-HK extrusion machine was used.

An estimation of the surface free energy of the surface of the bearing hole of the real bush could not be carried out by classical methods (measurements of the wetting angle) because of the small radius of curvature of the surface analyzed. It was decided, therefore, to investigate the dynamics of spreading of a drop on the bearing surface of the polymer bush. It is known[7] that the spreading velocity of a drop depends mainly on the surface free energy of the solid, the surface tension of the liquid, viscosity and drop volume of the liquid and to a lesser extent on the surface roughness and environmental conditions.

Spreading studies of oil drops on the surface of the bearing hole of the polymer bushing were carried out for the same plastic-oil combination as used for the investigation of the tribological properties. The drop spreading was recorded with a movie camera.

The drop length measured along the bearing hole axis was assumed to be the spreading drop dimension. The mass of the drop was small (less than 2 mg). The drop spreading was nevertheless evaluated with the relative drop dimension increment given by

$$\lambda = \frac{l_i - l_o}{l_o} \qquad (1)$$

where $l_o$ is the initial drop dimension and $l_i$ the drop dimension at the i-th image. The drop length was measured along the bearing hole axis of the polymer bush.

A one-dimensional spreading of a liquid drop can be described by the relationship[8]

$$x = \left(\frac{3}{4} \frac{F}{\eta \rho} \frac{m}{a}\right)^{\frac{1}{3}} t^{\frac{1}{3}} \qquad (2)$$

where
- x — drop length,
- F — force of spreading,
- m — drop mass,
- η — viscosity of liquid,
- ρ — liquid density,
- a — breadth of drop,
- t — time of spreading.

The velocity of the change in the relative increment of the drop length, considering the relationship (2) can be expressed as

$$\frac{d\lambda}{dt} = \frac{1}{3}\left(1 + \frac{t}{t_o}\right)^{-\frac{2}{3}} \frac{1}{t_o} \qquad (3)$$

when $t = t_o \qquad x = \frac{l_o}{2}$

hence

$$t_o = \frac{2}{3} \frac{l_o^2}{b\,F}$$

where b is the drop height.

The ratio of the velocities of spreading of drops of the same liquid on two bushes made from different polymers is

$$\frac{\left(\frac{d\lambda}{dt}\right)_1}{\left(\frac{d\lambda}{dt}\right)_2} = \frac{F_1}{F_2} \left( \frac{1 + \frac{3}{2} \frac{b}{l_o^2 \eta} F_2 t}{1 + \frac{3}{2} \frac{b}{l_o^2 \eta} F_1 t} \right)^{\frac{2}{3}} \qquad (4)$$

Since for the oils applied

$$\frac{3}{2} \frac{b}{l_o^2 \eta} F t \gg 1$$

therefore

$$\frac{\left(\frac{d\lambda}{dt}\right)_1}{\left(\frac{d\lambda}{dt}\right)_2} = \left(\frac{F_1}{F_2}\right)^{\frac{1}{3}} \qquad (5)$$

Considering[9]

$$F = (\gamma_s - \gamma_{sl}) - \gamma_l$$

and

$$\gamma_{sl} = \gamma_s + \gamma_l - 2\phi(\gamma_s \gamma_l)^{\frac{1}{2}}$$

Since $\phi \approx 1$ for polymers and liquids used, therefore

$$F \approx 2\left(\gamma_s \gamma_l\right)^{\frac{1}{2}} - \gamma_l \approx 2\gamma_l^{\frac{1}{2}}\left(\gamma_s^{\frac{1}{2}} - \gamma_l^{\frac{1}{2}}\right) \qquad (6)$$

Taking into consideration Equation (5) we have

$$\frac{\left(-\frac{d\lambda}{dt}\right)_1}{\left(-\frac{d\lambda}{dt}\right)_2} = \left(\frac{\gamma_{s1}^{\frac{1}{2}} - \gamma_l^{\frac{1}{2}}}{\gamma_{s2}^{\frac{1}{2}} - \gamma_l^{\frac{1}{2}}}\right)^{\frac{1}{3}} \qquad (7)$$

The decrease in the $d\lambda/dt$ vs. time is presented in Figure 1.

One can see the changes in the spreading dynamics of the clock oil on POM polymers (Tarnoform and Delrin 500 NC 10), PA 6 (Tarnamid T-27) and PA 6 + 25% glass fibres (Itamid).

Taking into account the experimentally determined values of $d\lambda/dt$ (for the polymers used and three oil combinations) and relationship (7), the relative magnitudes of the surface free energies of the polymer materials were calculated.

The surface free energies of the applied POM materials are very close. The relative magnitudes of the surface free energies of the real surfaces of the bearing hole of the bushes manufactured using POM, PA 6 and PA 6 + 25% glass fibres have been found to be 1:1.1:5.7.

## TRIBOLOGICAL PROPERTIES

The investigation of the tribological properties (at $v \leqslant 0.067$ m/s and $p \leqslant 3$ MPa in dry conditions and $v \leqslant 0.182$ m/s and $p \leqslant 7$ MPa at lubrication) were carried out for the journal steel/polymer combination. It was found that the frictional properties are very good when the steel/POM combination is used; the friction coefficient is ca. 0.25. The friction coefficients of the steel/PA 6 and steel/PA 6 + 25% glass fibre were found to be ca. 0.4 and 0.55 respectively.

The wear rate of the bearings investigated was very high for the steel/PA 6 + 25% glass fibre bearings when high loads were applied ( >0.5 MPa at a sliding speed 0.067 m/s). The best wear properties were obtained for the steel/POM bearings.

The application of lubrication (one oil drop of ca. 0.04 ml) to the microbearings investigated is very important to make their tribological properties better. The friction coefficient of the lubricated bearings decreases to < 0.1 and in particular the wear rate decreases significantly. The wear rate decreases as the effect of lubrication is very high especially for the steel/PA 6 + 25% glass fibre bearings (Table I). The poorest efficiency of lubrications was found for the steel/POM bearings. The wear rate of the lubricated microbearings was highest for the steel/PA 6 + 25% glass fibre bearings and lowest for the steel/POM bearings.

The effect of lubrication was greatest when clock oils were applied as the lubricant. The poor efficiency of lubrication was found for mineral oil and glycerol.

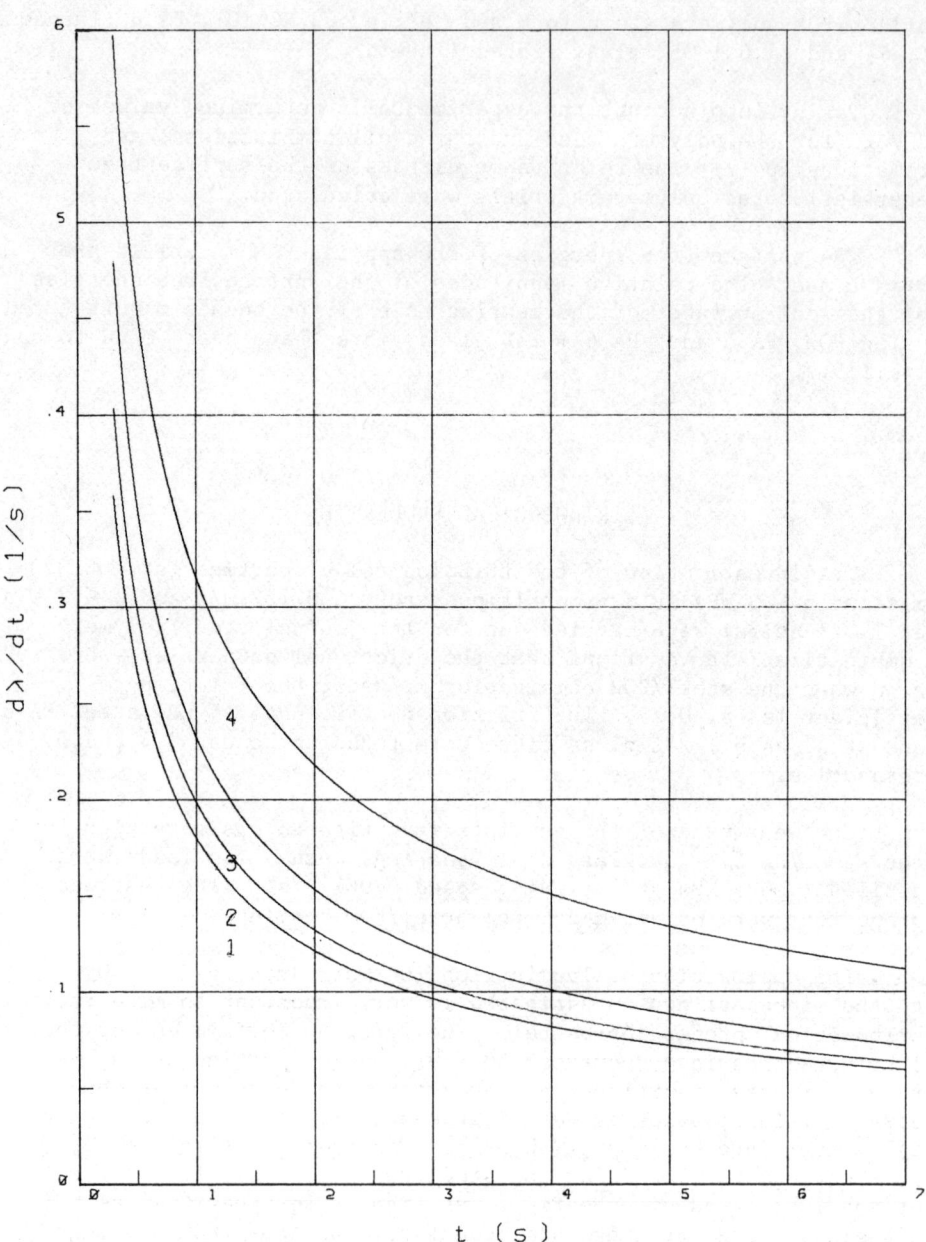

Figure 1. The computer plot of the velocity of spreading $d\lambda/dt$ vs time for a drop of clock oil Cuypers 4 on various polymer surfaces. 1 - Tarnoform, 2 - Delrin 500 NC 10, 3 - Tarnamid T-27, 4 - Itamid.

## TRIBOLOGICAL PROPERTIES AND SURFACE ENERGETICS

The relationship between the tribological properties and the surface properties of polymer materials and lubricants has been studied.

The relationship between the friction coefficient and the velocity of spreading of oil drops on the bearing hole of the bushes manufactured using POM, PA 6 and PA 6 + 25% glass fibre is presented in Figure 2. It is seen that the coefficient of friction increases with increasing velocity of spreading of the oil drop. The radial wear of the polymer bush as a function of velocity of spreading is presented in Figure 3.

Table I. The Average Efficiency $\xi$ of Lubrication (the Ratio of Radial Dry Wear to Lubricated Wear) of the Steel/Polymer Microbearings. Sliding Distance 10 km, Sliding Speed v = 0.067m/s, Specific Pressure p = 3 MPa.

| Material \ Lubricant | Mineral oil MWP | Silicon-mineral oil OKB-122-16 | Clock oil Cuypers 4 |
|---|---|---|---|
| PA 6 | 29 | 17 | 50 |
| PA 6 + 25% glass fibre | 47 | 33 | 70 |
| POM | 5.3 | 4.5 | 8.4 |

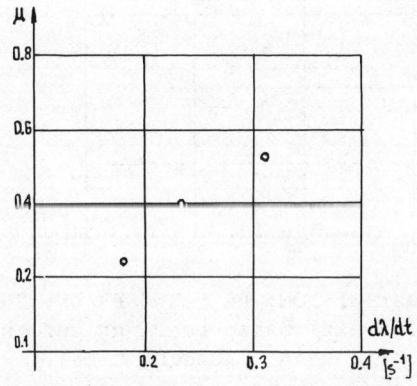

Figure 2. Coefficient of friction vs. velocity of spreading d$\lambda$/dt after 2 s of spreading time) for POM, PA 6 and PA 6 + 25% glass fibre materials (from left to right respectively).

Some interesting effects have been observed as a result of lubrication. For the materials and liquids used when $\phi \approx 1$ it can be easily demonstrated that the reduction in the work of adhesion $\Delta W$ (for the metal/polymer combination) as a result of the introduction of oil between the contacting steel/polymer surfaces can be expressed as

$$\Delta W = 2 \, (\gamma_s \gamma_l)^{\frac{1}{2}} \tag{8}$$

where $\gamma_s$ is the surface free energy of polymer and $\gamma_l$ the surface tension of the oil.

The relationship (8) has been determined[11] using the Young's equation (both for metal/polymer and metal/liquid/polymer combinations), relationship for $\gamma_{sl}$ (see above) and taking into consideration that the applied lubricants give contact angle $\theta \to 0$, on the steel surface.

The comparison between the decrease in the intensity of wear $\Delta(\frac{dz}{ds})$ (z - radial wear, s - sliding distance) and the reduction in the work of adhesion $\Delta W$ has been carried out. It has been found that with increasing $\Delta W$, the value $\Delta(\frac{ds}{dz})$ increases rapidly. The relationship for the POM - based bearings is presented in Figure 4.

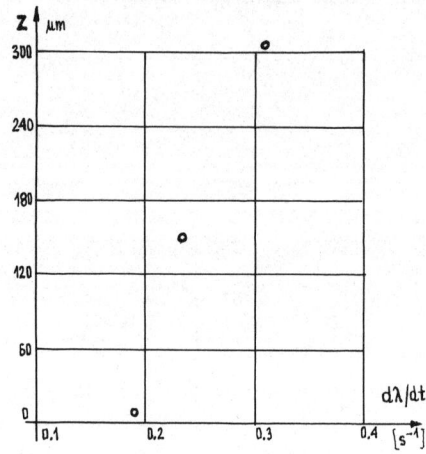

Figure 3. Radial wear of POM, PA 6 and PA 6 + 25% glass fibre (from left to right respectively) based bearings and the velocity of spreading $d\lambda/dt$ (after 2 s of spreading time).

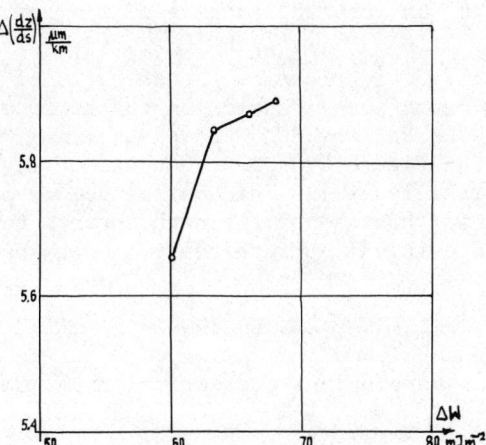

Figure 4. Decrease in wear intensity $\Delta(\frac{dz}{ds})$ by introducing oils between steel/POM surfaces rubbing against each other and the reduction in the work of adhesion $\Delta W$.

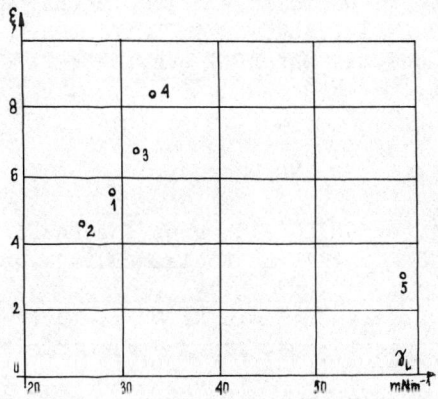

Figure 5. Relationship between the efficiency of lubrication $\xi$ and the surface tension of oils for steel/POM microbearings. 1 - mineral oil, 2 - silicon-mineral, 3-4 - clock oils of varying surface tensions.

When the oils with high surface tension, e.g., clock oils are used the plot of $\Delta(\frac{dz}{ds})$ as a function of $\Delta W$ is more inclined toward the $\Delta W$ axis. This effect is presumably related to the deteriorating wettability of the POM material because of its low surface free energy.

The efficiency of lubrication $\xi$ (the ratio of the dry radial wear to the lubricated wear) depends on the surface tension of the lubricants applied. For the POM - based bearings, the relationship is presented in the Figure 5. One can see the very poor efficiency of lubrication with glycerol. This may be due to poor wetting of the polymer material because of the high surface tension of the lubricant and low critical surface tension of the POM material.

## CONCLUSIONS

The following conclusions are derived from the studies presented here.

1. There exists a relationship between the tribological properties of the journal polymer miniature bearings and the surface energetics (expressed as the velocity of spreading $d\lambda/dt$) of the polymer materials and lubricants.
2. The friction coefficient and wear of microbearings are high when the velocity of spreading $d\lambda/dt$, on the bearing polymer surface is high.
3. The reduction in the intensity of wear as a result of lubrication correlates with the reducation in the work of adhesion.
4. The efficiency of lubrication is better when the surface free energy of the polymer material and the surface tension of the oil are higher.

## REFERENCES

1. F. Dürr, in "Actes du 10 e Congres International de Chronometrie", Vol. 3, p. 427, La Societé Suisse de Chronometrie, Genève, 1979.
2. G. Régnault, Dissertation ETH Nr 6345, Zürich, 1979.
3. M. Massin, Mecanique - Materiaux - Electricité, No. 330-331, 12 (1977).
4. K. L. Mittal, in "Adhesion Science and Technology", L. H. Lee, Editor, Vol. 9A, pp. 129-167, Plenum Press, New York, 1975.
5. K. L. Mittal, Polymer Eng. Sci., 17, 467 (1977).
6. H. Schonhorn, H. L. Frisch and G. L. Gaines, Polymer Eng. Sci., 17, 440 (1977).
7. B. D. Summ and V. V. Goryunov, "Physical-Chemical Principles of Wetting and Spreading", Khimya, Moscow, 1977.
8. E. A. Raud, B. D. Summ and E. D. Stschukin, Doklady Akademii Nauk SSSR, 205, 1134 (1972).

9. D. W. Krevelen, "Properties of Polymers: Their Estimation and Correlation with Chemical Structure", Elsevier, Amsterdam, 1976..
10. Z. Rymuza, Schmiertechnik + Tribologie, 25, 179 (1978).
11. Z. Rymuza, in "Proceedings of the I Symposium Intertribo 81", p. 187, VUZ, Bratislava, 1981.

CHARGE STATES IN POLYMERS: APPLICATION TO TRIBOELECTRICITY


C.B. Duke

Xerox Webster Research Laboratory
Xerox Square-114
Rochester, N.Y., 14644, USA


A mathematical model of localized molecular-ion, and molecular-exciton states in a dielectric medium is described. This model is shown to provide a quantitative description of the lineshapes and temperature dependence of photoemission and UV absorption from molecular glasses and pendant-group polymers. Its extension to describe electron transfer in these systems is indicated. Attention is focussed on polystyrene, poly(2-vinyl pyridine), and molecular glasses of 2-ethyl benzene, isopropylbenzene and 2-ethyl pyridine as model systems. Application of the model to describe contact charge exchange in copolymers of polystyrene, poly(methyl methacrylate) and poly(2-vinyl pyridine) correctly predicts the steady-state charge exchange among these materials.

## INTRODUCTION

The nature of the electronic states associated with injected charges and optical excitations in polymeric and molecular solids has been a subject of renewed interest in recent years both because of the increasing use of such materials in electronic devices and because a question has arisen concerning the appropriateness of traditional one-electron energy band theory for the description of these states.[1-3] In particular, it has been suggested that under many circumstances charges injected into polymeric and molecular materials are more analogous to localized molecular ions in solution than to the extended "Bloch" states which are thought to be associated with electronic motion in metals and common covalent semiconductors.[1,2] Consequently, a suspicion has been growing that semiconducting organic solids should not be regarded as "organic semiconductors" in the traditional sense, in spite of an extensive early literature in which energy-band models were applied to interpret measurements of their optical and transport properties.[4]

This paper is devoted to the presentation of a brief overview of a recently-developed "relaxation-localization" model of localized molecular-ion and exciton states in polymers and molecular glasses. This model was proposed initially to interpret photoemission measurements from two pendant-group polymers: polystyrene and poly(2-vinyl pyridine.)[2,5] It next was utilized in the prediction[1] and subsequent observation[6] of surface states of molecular solids as well as of the temperature dependence of photoemission and UV absorption linewidths of molecular films.[2,5,7] Having proven successful in describing the spectroscopic properties of typical pendant-group polymers and molecular glasses, the model most recently has been extended to provide a description of electron-transfer processes in both these materials and molecularly-doped polymers.[8] Therefore it affords a unified and experimentally-verified microscopic description of electron ionization, excitation and transfer processes in a variety of molecular and polymeric materials.

The basic concepts on which the relaxation-localization model is based are simple.[3,9] The molecular character of the solid state leads to weak interactions between the molecular entities and hence to a high degree of disorder. Contributions to the disorder are both static (e.g., local variations in composition and/or structure) and dynamic (e.g., thermally induced vibrations) in nature. The static disorder localizes both injected charges as molecular ions and injected excitations as molecular excitons. Once localized, these entities interact strongly with the (dynamic) charges which they induce in the surrounding dielectric medium. The induced charges in turn can be described in terms of the normal modes, called longitudinal polarization fluctuations, of the dielectric medium. Therefore the relaxation-localization model consists of molecular ions (excitons) in a disordered static potential interacting with longitudinal polarization fluctuations via electrostatic forces.[5,8,9] This model is a variant of familiar polaron hopping models,[10,11] differing from its predecessors primarily by vitrue of embodying a considerably more extensive description of the dielectric response of the molecular medium: A description which has been shown to be necessary to describe measured photoemission[2,5,7,9] and UV absorption[2,5] spectra.

We proceed by first considering the relaxation energy of an injected charge (exciton) in a molecular medium caused by its inducing other charge motion within the medium. Since this relaxation energy varies from one molecular-ion (exciton) site to another in a non-uniform medium, we next examine the consequences of these energy variations for the localization of the injected charge (exciton). Then, we indicate the nature of the manifestations of the interaction of these injected charges (excitons) with excitations of the medium on the observed properties associated with electronic charge motion in the medium, i.e., photoemission, UV absorption, and electron transfer from one site to another. We conclude with a description of the application of the relaxation-localization model to interpret contact charge exchange between various pendant-group copolymers.

## RELAXATION AND FLUCTUATIONS

When a charge (or optically-excited exciton) is injected into a molecular solid, it induces changes in the electronic charge density and atomic positions both on the molecular site which it occupies (intramolecular relaxation) and on neighboring molecular sites (intermolecular relaxation). This phenomenon is called "relaxation" and leads to a lowering of the energy of the composite system of added-charge-plus-molecular-solid by an amount called the "relaxation energy," $E_r$. The relaxation energy is formally defined for an injected charge (e.g., molecular ion) as the difference between the ground state energy of the ion and the Hartree-Fock molecular orbital eigenvalue (in the canonical basis) which corresponds to the free-ion state. Both the intramolecular and intermolecular contributions to the relaxation energy can be large for molecular solids: i.e., $E_r(\text{intra}) \simeq E_r(\text{inter}) \simeq 1\text{-}2$ eV (1,2,12). Therefore the energies of molecular anions (cations) in condensed molecular media are about 2-4 eV lower (higher) than those of the corresponding free-molecule orbitals.

An important feature of the intermolecular contributions to the relaxation energy is their dependence on the local atomic structure in the vicinity of the injected ion or exciton. For example, a surface ion exhibits only about 60% of the bulk relaxation energy because of the absence of polarizable neighboring species outside the condensed phase.[1,3] This variation in the value of the molecular-ion site energy at the surface relative to the bulk led to the prediction[1] and subsequent observation[6] of localized surface states for molecular films (specifically, anthracene). More generally, local variations in the composition and structure of polymers and molecular glasses cause spatial fluctuations in the site energies of molecular ions and excitons in these materials. For intrinsic molecular-ion states, the magnitude of these fluctuations may be inferred via analyses of the widths of valence-electron photoemission lines[2,5,7,9] or of contact charge exchange spectra.[2,13-15] Typically one finds $E_r \simeq 1$ eV for surfaces, and approximately 50% larger, i.e, 1-2 eV, in the bulk. The importance of these fluctuations is their creation[1-3] of localized molecular-ion states in polymers and glasses. In particular, intrinsic localized molecular-ion states are thought to be involved in the contact charge exchange behavior of pendant group polymers,[13-16] and extrinsic localized molecular-ion states dominate the transport properties of molecularly doped polymers.[17-19]

Another important feature of the intermolecular contributions to the relaxation energy is the dependence of their average value on the dielectric response of the medium. If $\varepsilon(\underline{q}, \omega)$ is the non-local dielectric function of the medium associated with wave vector $\underline{q}$ and frequency $\nu = \omega/2\pi$, then the longitudinal polarization fluctuations of the medium are defined by

$$\varepsilon[\underline{q}, \omega_\alpha(\underline{q})] = 0 \qquad (1)$$

in which $\hbar\omega_\alpha(\underline{q})$ is the energy of the fluctuation characterized by the wave vector $\underline{q}$ and branch index $\alpha$. In a molecular medium, there are typically many branches, $\alpha$, associated with various electronic and vibrational excitations. For example, in a pendant-group polymer like poly(2-vinyl pyridine) three types of branches occur:

$$\omega_t \sim 10^3 \text{ sec}^{-1} \text{ (torsional or backbone modes)}, \qquad (2a)$$

$$\omega_{IR} \sim 10^{13} \text{ sec}^{-1} \text{ (IR modes)}, \qquad (2b)$$

$$\omega_v \sim 10^{15} \text{ sec}^{-1} \text{ (valence electron modes)}. \qquad (2c)$$

The relaxation-localization model of charge motion in molecular media incorporates the interaction of each of these polarization-fluctuation normal modes with an injected charge (exciton) as well as the interaction of this charge with the intramolecular normal modes of the molecule on which it resides.[5,8] One must include <u>all</u> of these diverse types of modes in the model because each type exerts a distinct and characteristic influence on each of the three experimentally accessible measures of electronic motion in the material, i.e., photoemission, UV absorption, and electron transfer. We return to this topic in the section entitled TRANSPORT.

## LOCALIZATION

The local or extended nature of molecular-ion (or exciton) states in molecular solids is determined by a competition between fluctuations in the local site energies of these states (which tend to localize them) and the hopping integrals for inter-site excitation transfer (which tend to delocalize them). In order to define this fluctuation-induced localization concept more precisely, consider the model defined by the one-electron Hamiltonian

$$H = \sum_{n,\beta} \varepsilon_n(\beta) a^+_{n\beta} a_{n\beta} + \sum_{n \neq n'} V_{nn'} a^+_{n'\beta'} a_{n\beta}, \quad (3)$$

in which n is a site index; $\beta$ designates the molecular orbital; $V_{nn'}$ designate the intermolecular hopping integrals; and $a_{n\beta}$ is the annihilation operator for an electron occupying the orbital $\beta$ at the site n.

Both the molecular ion energies (i.e., site energies, $\varepsilon_n$) and the hopping integrals, ($V_{nn'}$) form distributions because of local variations in composition and structure. While a proper analysis of the resulting model is complicated,[20,21] the qualitative features of interest to us can be defined in terms of a mean site energy $\bar{\varepsilon}$, the rms deviation from this mean

$$\Delta \equiv \sqrt{\langle (\varepsilon_n - \bar{\varepsilon})^2 \rangle_{AV}}, \quad (4)$$

and a mean hopping integral, $\bar{V}$. Variations in the site energies from the mean (described by $\Delta$) are referred to as "diagonal disorder" whereas analogous variations of the hopping integrals are called "off-diagonal disorder." Similarily, if these variations are caused by local time-independent fluctuations in composition or structure one speaks of "static disorder" while if they are generated by thermal vibrations or longitudinal polarization fluctuations one employs the term "dynamic disorder." Whereas static diagonal and off-diagonal disorder can occur in polymers, only dynamic disorder should be present in molecular crystals. The important point for our present purpose, however, is the recognition that all sources of disorder must be considered in order to determine the localized or extended character of the electronic states associated with charges (or excitons) injected into polymers and molecular materials, and that these sources of disorder are described by choosing suitable distributions of the $\varepsilon_n$ and $V_{nn'}$ parameters in Equation (3) or an appropriate generalization thereof.

In terms of the simplification of Equation (3) defined by considering only nearest-neighbor hopping integrals (i.e., neglecting long-range hops), it can be shown[20,21] that injected charges are localized, i.e., they form molecular cations or anions within the solid, if

$$\Delta > cz\overline{V} \qquad (5)$$

in which z is the coordination number of the (presumably identical) molecular sites and c is a dimensionless number of the order of unity which depends both on the connectivity (i.e., dimensionality) of the molecular system and on the extent of off-diagonal disorder. Typically $c \simeq 2.5$ for (isotropic) three-dimensional systems, 1.5 for two-dimensional systems, and zero for one-dimensional systems.

Different classes of polymers and molecular solids exhibit systematically distinct ranges of values of $\Delta$ and $V$.[1] For typical pendant-group and molecularly doped polymers, one finds $0.1 \text{ eV} \lesssim \Delta \lesssim 1 \text{ eV}$, $\overline{V} \lesssim 0.1 \text{ eV}$, for motion along the polymeric backbone of pendant group polymers, and otherwise $\overline{V} << 0.1 \text{ eV}$.[1,3] Therefore inequality (5) is clearly satisfied for these materials, so injected charges form intrinsic (pendant-group polymers) or extrinsic (molecularly-doped polymers) local molecular-ion states.[1,3,13,19] A similar situation seems to prevail for molecular glasses, although the parameter values are not yet firmly established in this case. For bulk molecular crystals in the absence of defects the static disorder contribution to $\Delta$ vanishes so one expects $\Delta = c'T$ in which $c'$ is a constant and T is the temperature.[1,3] For the materials with which we are concerned here, therefore, inequality (5) is satisfied so that injected charges are more appropriately regarded as localized molecular anions and cations than as "electrons" and "holes", respectively, described by traditional one-electron energy band models.

## SPECTROSCOPY AND TRANSPORT

In this section the major results obtained by applying the relaxation-localization model to interpret photoemission, UV absorption, and transport experiments are indicated. The analysis of photoemission measurements is emphasized because the model has been tested most extensively in this case.

### Valence-Electron Photoemission

A useful starting point in the assessment of the nature of intrinsic valence-electron states in molecular solids is the comparison of the gas-phase and solid-state photoemission and UV absorption spectra of molecules whose radical cation geometries are expected to be the same in the two phases, e.g., $S_8$, $Se_8$, benzene, p-tetracyanoquinodimethane (TCNQ), and the polyacenes. One finds that in these cases the valence-electron photoemission spectra are shifted rigidly to lower binding energies by an intermolecular relaxation energies 1 eV $\lesssim$ $E_r$(inter) $\lesssim$ 2 eV and are broadened considerably.[2,5-7,22] This result indicates that the photogenerated hole states in both glassy and crystalline monomolecular solids are essentially those of molecular cations.[22] Indeed, suitable molecular-orbital models provide quite satisfactory descriptions both of the major peaks in the spectra[23] and of the intramolecular vibrational progressions observed in the gas phase.[24] In the case of less rigid molecules, e.g., triphenylamine and phenylenediamines, the geometry of the molecular cations can change between the gaseous and solid states, so that more complex differences occur between the valence electron photoelectron spectra characteristic of these two phases.[25] Even in these cases, however, the spectra of condensed molecular films are essentially those of localized molecular cations. Therefore the task of any model of molecular cation states in condensed media (both glassy and crystalline) is the prediction of the relaxation energy shifts and large increases in widths of the photoionization lines in the condensed relative to the gaseous phase.

It might be anticipated that the intermolecular relaxation energies would be simply the static polarization energies of a molecular cation in the condensed medium given, for example, by the Born model of solvation.[26] That such cannot be the case is immediately evident upon noting that nonpolar polystyrene and condensed ethyl-benzene exhibit the same relaxation-energy shifts, $E_r$(inter)=1.5eV, as polar poly(2-vinyl pyridine) and condensed 2-ethyl pyridine.[5] The static dielectric constants of the polar materials are over twice those of the non-polar ones, demonstrating that static models of the relaxation energy are clearly incompatible with the observations. The relaxation-localization model correctly predicts this result by revealing that only the high-frequency electronic branches of the longitudinal polarization spectrum, e.g., $\omega_v$ in Equation (2c), contribute to $E_r$(inter). The other two branches contribute to the widths of the photoemission lines in the condensed phases, not to their shifts.[2,5,8] The magnitudes of the $E_r$(inter) also are predicted correctly by the model.[2,5]

The analysis of the widths, $\sigma$, of the condensed phase photoionization lines is more complex because both static and dynamic disorder contribute to $\sigma$. Schematically, we can write

$$\sigma^2 = \sigma^2(\text{ins}) + \sigma^2(\text{intra}) + \sigma^2(\text{inhom}) + \sigma^2(\text{hom}). \qquad (6a)$$

The instrumental width, $\sigma$(ins), is typically less than about 0.02 eV for a well designed ultraviolet photoemission spectrometer.[5-7] The intramolecular contributions to the widths, $\sigma$(intra), vary from one molecular-ion state to another, lying in the range 0.2 eV $\lesssim \sigma$(intra) $\lesssim$ 0.4 eV.[25] Usually the predominent contribution to the width in the condensed state is the inhomogeneous width, $\sigma$(inhom), caused by spatial fluctuations of the electronic contributions to the relaxation energy.[1,5,6,8,13] Typically 0.4 eV $\lesssim \sigma$(inhom) $\lesssim$ 0.6 eV depending upon sample preparation conditions.[5-7] Finally, the homogeneous contributions may be written as

$$\sigma^2(\text{hom}) = \sigma^2(\text{IR}) + c^2 \kappa T \qquad (6b)$$

in which $\kappa$ is Boltzmann's constant and T is the temperature. The first term arises from infrared active vibrations on all molecules including the one on which the molecular ion is formed. The second is caused by thermally excited longitudinal polarization fluctuations. It was predicted by the relaxation-localization model a priori,[2,5,7] and subsequently observed for condensed isopropyl benzene.[7] The model predicts 0.1 eV $\lesssim \sigma$(IR) $\lesssim$ 0.3 eV, but the contributions to from $\sigma$(IR) and $\sigma$(inhom) have not yet been separated experimentally.

## UV Absorption

The analysis of intramolecular UV absorption using the relaxation-localization model is completely analogous to that of valence-electron photoemission. The major change is that charge densities appropriate for molecular excitons rather than molecular ions are used in the model expressions for the lineshapes.[2,5,27] This change leads to much smaller values of $E_r$(inter) $\simeq$ 0.1 eV and $\sigma$(inhom) $\simeq$ 0.1 eV in the case of UV absorption relative to photoemission. The expressions for the lineshapes also are somewhat more complicated.[5,28] Comparisons of gas-phase and solid-state spectra have been made for ethylbenzene relative to polystyrene and 2-ethyl pyridine relative to poly(2-vinyl pyridine).[2,5] They suggest localized molecular excitons in these two polymers. The data for UV absorption are far less extensive than those for photoemission, however, and certainly could be extended usefully.

A new feature in the solid state UV absorption relative to photoemission is the possibility of charge-transfer excitons associated with the transition of an electron from one molecular-ion site to an excited state on a neighboring site. In the special case that an excess electron (hole) on one site experiences photon-assisted hopping to an excited state on a neighboring (initially-neutral) site, the spectral density for photoexcitation is the two-site analogue of the single-site photoemission spectral density.[8,27] Although these processes obviously can provide detailed microscopic information on the wave functions and vibrational modes of the charge-transfer excited state, no systematic experimental studies suitable for detailed model analysis seem to have been performed.

## Transport

The relaxation-localization model recently has been applied to derive explicit expressions for electron (hole) transfer probabilities between two molecular sites in pendant-group and molecularly-doped polymers.[8] These probabilities are the input data used in multiple-hopping models of measured transport properties, e.g., drift mobilities and photoconductivity.[9,19,29] Consequently, they constitute the link between observed transport properties and the microscopic electronic structure of polymers. Indeed, the model has proven useful in quantifying the trade-offs between the electrical and mechanical properties of polymers used in electrophotographic copying machines and in suggesting improved materials for such applications.[8]

The model's important prediction is that of the influence of each of branches of the longitudinal polarization fluctuation spectrum on the electron-transfer probability.[8] The high-frequency electronic branches, Equation (2c), which are responsible for the photoemission relaxation energies, exert only a minor influence on electron-transfer by virtue of a temperature-independent prefactor. The low-frequency branches, Equation (2a), contribute to the activation energy for the electron transfer process as well as to the temperature-dependent widths of photoemission lines via Equation (6b). The IR branches, Equation (2b), cause the apparent activation energies to increase with increasing temperature and generate a rather more complicated expression for the transfer probability[19,29] than those commonly used in multiple-hopping or polaron[10,11] models. Thus, the relaxation-localization model has clarified the physical basis of electron-transfer processes in condensed media as well as provided a new, more accurate mathematical expression for the transfer probabilities themselves. Its predictions are consistent with available experimental data which, however, are sparse.[8,9] Since the model relates dielectric relaxation measurements (i.e., the longitudinal polarization fluctuation spectrum) to the electron-transfer rates, both quantities must be observed in a series of homologous materials in order to test the model's quantitative validity. Such a series of measurements has not yet been performed, and obviously would be of considerable interest.

## CONTACT CHARGE EXCHANGE

Although over a century of research has been devoted to contact charge exchange between insulators, prior to 1978 no model had been proposed which successfully predicted the quantitative magnitude of this charge exchange. In 1978 a phenomenological version of the relaxation-localization model was published which predicted quantitatively the polymer-polymer charge exchange among various copolymers of polystyrene and poly(methyl methacrylate) in terms of effective spatially-averaged densities of molecular ion-states per unit energy,[5,8] $\rho(E)$, for each polymer determined from metal-polymer contact charge exchange experiments.[16] This model also predicted correctly the difference in contact charge exchange between pigmented and unpigmented polymers,[16] and later was extended to encompass copolymers of poly(styrene)[15,30] and poly(methyl methacrylate) with poly(2-vinyl pyridine),[15,30] and copolymers of $(C_2F_3Cl)_m$ and $(C_2H_3F)_n$ (i.e., "Kel F-800").[31] In addition, new data for polystyrene, poly(methyl methacrylate) and polyethylene subsequently were reported which purportedly contradicted the relaxation-localization average density-of-states model interpretation.[32] The differences between the data sets were attributed to the presence of extrinsic states and to the absence of a sufficient number of contacts to achieve adequate spatial averaging,[33] however, so no contradiction of the original model has yet been established.

The relationship between the chemical properties of the polymers in question and the steady-state contact charge exchange is embedded in the parameters characteristic of the average density of states.[12,14,15] Typically $\rho(E)$ is the form:[5,8,16]

$$\rho(E) = \rho_A(E) + \rho_D(E), \tag{7a}$$

$$\rho_i(E) = (\pi \sigma_i^2)^{-1/2} \exp[-(E-\varepsilon_i)^2/\sigma_i^2], \qquad (7b)$$

in which D designates electron donors and A electron acceptors. Good donors exhibit small radical cation energies, $\varepsilon_D$, and good acceptors large radical anion energies, $\varepsilon_A$. Polar media lead to $\sigma_i^2$ which are large and which have a temperature dependence as described in Equation (6b) with large values of $c^8$. The structure of the polymer host enters by virtue of more rigid (e.g., crystalline) materials tending to reduce $\varepsilon_A$, to increase $\varepsilon_D$, and to reduce $\sigma_i^2$ for a given acceptor or donor species, <u>ceteris paribus</u>. The correlation of the resulting contact charge exchange properties with measures of chemical reactivity and molecular structure have been discussed elsewhere in the literature.[34,38] These correlations are consistent with triboelectric charge transfer occurring between relaxed molecular-ion states, typically intrinsic states localized on the pendant group moieties for polymers like polystyrene, poly(methyl methacrylate) and poly(2-vinyl pyridine). Thus, in polystyrene, for example, the donor states are π-electron radical cations on the phenyl moieties and the acceptor states are π-electron radical anions on these moieties.[12]

## SYNOPSIS

In this paper the development and major applications of a new relaxation-localization model of electronic states in molecular solids have been briefly outlined. This model affords a detailed description of valence-electron photoemission spectra which, moreover, have been utilized to verify explicitly several of its important predictions. It provides a similar description of UV absorption spectra and electron-transfer processes. Inadequate experimental data are available in these cases, however, to test the model critically. Finally a phenomenological version of the model has been shown to provide a quantitative description of steady-state contact charge exchange in polymers and to constitute a link between this charge exchange and the molecular architecture of pendant group polymers.

## REFERENCES

1. C.B. Duke, Surface Sci. <u>70</u>, 674 (1978).
2. C.B. Duke, Mol. Cryst. Liq. Cryst. <u>50</u>, 63 (1979).
3. C.B. Duke and L.B. Schein, Physics Today <u>33</u> (2), 42 (1980).
4. H. Meier, "Organic Semiconductors," Verlag Chemie, Weinheim, 1974.
5. C.B. Duke, W.R. Salaneck, T.J. Fabish, J.J. Ritsko, H.R. Thomas and A. Paton, Phys. Rev. B <u>18</u>, 5717 (1978).
6. W.R. Salaneck, Phys. Rev. Lett. <u>40</u>, 60 (1978).

7. W.R. Salaneck, C.B. Duke, W. Eberhardt, E.W. Plummer and H.J. Freund, Phys. Rev. Lett. 45, 280 (1980).
8. C.B. Duke and R.J. Meyer, Phys. Rev. B23, 2111 (1981).
9. C.B. Duke, in "Extended Linear Chain Conductors," J.S. Miller, Editor, Vol. 2, pp. 59-125, Plenum Press, New York, 1981.
10. H. Frohlich, Adv. Phys. 3, 325 (1954).
11. N.F. Mott and E.A. Davis, "Electronic Processes in Non-Crystalline Materials," University Press, Oxford, England, 1971, Chapt. 4.
12. C.B. Duke and T.J. Fabish, Phys. Rev. Lett. 37, 1075 (1976).
13. C.B. Duke and T.J. Fabish, Chem. Phys. Lett. 49,133 (1977).
14. T.J. Fabish and C.B. Duke, J. Appl. Phys. 48, 4256 (1977); in "Polymer Surfaces," D.T. Clark and W.J. Feast, Editors, pp. 109-119, Wiley-Interscience, New York, 1978.
15. C.B. Duke, J. Vac. Sci. Technol. 15, 157 (1978).
16. C.B. Duke, and T.J. Fabish, J. Appl. Phys. 49, 315 (1978).
17. H. Hoegl, J. Phys. Chem. 69, 755 (1965).
18. W.D. Gill, J. Appl. Phys. 43, 5033 (1972); in "Photoconductivity and Related Phenomena," J. Mort and D. Pai, Editors, pp. 303-334, Elsevier, Amsterdam, 1976.
19. G. Pfister, Phys. Rev. B 16, 3676 (1977).
20. P.W. Anderson, Rev. Mod. Phys., 50, 191 (1978).
21. D. Weaire and V. Srivastava, in "Amorphous and Liquid Semiconductors," W.E. Spear, Editor, pp. 286-290, G.G. Stevenson, Dundee, 1977.
22. C.B. Duke, W.R. Salaneck, A. Paton, K.S. Liang, N.O. Lipari and R. Zallen in "Structure and Excitations of Amorphous Solids," G. Lucovsky and F. Galeener, Editors, pp. 23-30, American Institute of Physics, New York, 1976.
23. C.B. Duke, Int. J. Quant. Chem: Quantum Chem. Symp. 13, 267 (1979).
24. C.B. Duke, Ann. N.Y. Acad. Sci. 313 166 (1978).
25. C.B. Duke, J. W-p Lin, A. Paton, W.R. Salaneck and K.L. Yip, Chem. Phys. Lett. 61, 402 (1979).
26. R.R. Dogonadze and A.A. Kornyshev, J. Chem. Soc. Farad. Trans. II 70, 1121 (1974).
27. C.B. Duke in "Tunneling in Biological Systems," B. Chance et al., Editors, pp. 31-66, Academic Press, New York, 1979.
28. C.B. Duke, N.O. Lipari and L. Pietronero, Chem. Phys. Lett. 30, 415 (1975).
29. H. Scher, in "Photoconductivity and Related Phenomena," J. Mort and D. Pai, Editors, pp. 71-115, Elsevier, Amsterdam, 1976,
30. T.J. Fabish and H.R. Thomas, Macromolecules 13, 1487 (1980).
31. T.J. Fabish, Crit. Rev. Solid State Mater. Sci. 8, 383 (1979).
32. G.A. Cottrell, J. Lowell and A.C. Rose Innes, J. Appl. Phys. 50, 374 (1979).

33. T.J. Fabish, C.B. Duke, M.L. Hair and H.M. Saltsburg, J. Appl. Phys. 51, 1247 (1980).
34. P.J. Cressman, G.C. Hartmann, J.E. Kuder, F.D. Saeva and D. Wychick, J. Chem. Phys. 61, 2740 (1974).
35. H.W. Gibson, J. Am. Chem. Soc. 97, 3832 (1975).
36. I. Shinohara, F. Yamomoto, H. Anzai and S. Endo, J. Electrostatics 2, 99 (1976).
37. H.W. Gibson and F.C. Bailey, Chem. Phys. Lett. 51, 352 (1977).
38. H.W. Gibson, F.C. Bailey, J.L. Mincer and W.H.H. Gunther, J. Poly Sci., Poly. Chem. Ed. 17, 2961 (1979).

# TRIBOELECTRIFICATION OF POLYMERS - A CHEMIST'S VIEWPOINT

Donald A. Seanor

Xerox Corporation
Joseph C. Wilson for Technology
800 Phillips Road
Webster, New York  14580

Contact electrification of polymers is a complex phenomenon. An understanding requires a knowledge of the driving force for charge transfer between the contacting (sliding) surfaces, the chemical nature of the surfaces before and after charge transfer, polymer friction and mechanics, as well as charge transport mechanisms. The driving force for charge transfer is related to differences in work function of the contacting surfaces; in turn this can be related to the ionization potential and Hammett σ function of the specific surfaces. Surface modification and mass transfer may also be involved. The role of surface states and molecular ion states in charge transfer and transport is discussed. Detailed understanding of the contacting event and frictional electrification also involves a knowledge of the polymer mechanical and dynamic properties. Such considerations lead to a time/temperature transform of the WLF type. Hazards and useful applications of contact electrification are briefly discussed.

## INTRODUCTION

Contact electrification has been known for over two thousand years. It is, however, only since the widespread use of synthetic polymers in the decade following World War Two that serious attempts have been made to understand and rationalize their contact electrification. Personal discomfort, explosions, dust collection, film and fiber manufacturing problems, etc. can all be attributed to contact electrification (Table I). Modern technologies such as electrospraying, electroprecipitation and xerography depend on contact electrification.

Table I. Hazards and Uses of Contact Electrification.

A. <u>Problem Areas</u>
   Hospital areas
      Shocks to patients in beds
      Spark-generated explosions in operating theaters under
         conditions were gas/oxygen ratios are within explosion
         limits
      Induction charging leading to sparking
   Powder, detonator, explosions caused by sparking; gas or liquid
      handling
   Fiber or film handling
      Accurate registration problems
      Fog marking, dust adhesion
      Film separation
      Increased friction and adhesion
   Sensitive electric instruments
      Meter sticking
      Cathode ray tubes
      Radiofrequency noise from sparks
   Printing industry - paper tracking, "ink flying"
   Light-sensitive areas - spark discharges
   Personal inconvenience

B. <u>Useful Areas</u>
   Xerography
   Electric separation
   High voltage generators
   Electrostatic precipitation
   Electrostatic microphones
   Electrostatic flocking

In reviews of contact electrification (1-11) there is repeated reference to the experimental problems associated with obtaining good reliable data on contact electrification, to the need for well defined surfaces, controlled atmospheres, and reproducible contacts. This point is reemphasized but not discussed further. It should, however, be noted that in industrial applications, surfaces are neither clean nor well defined. Contacts are seldom the intimate contacts of the type required for obtaining and understanding of the basic physics, and material is often transferred from one surface to the other (Figure 1). This is a major complication.

In the review which follows, no attempt is made to give complete coverage to the whole field of contact electrification. The subject matter is limited to selected topics considered to be of interest to the Symposium on Physicochemical Aspects of Polymer Surfaces. Subject matters to be discussed are: the driving force for contact electrification, the charge carriers and the resultant charged states, the possibility that adsorption is involved in charge transfer and frictional charging.

Before embarking on the main thesis, it is useful to consider the numbers involved in contact electrification. If it were possible to obtain a III plane of sodium chloride (all sodium ions or chlorine ions), there would be $6.25 \times 10^{14}$ ions per sq cm or a charge of $3 \times 10^5$ esu per sq. cm. The associated field would be $\sim 10^9$ volts $cm^{-1}$. Electrical breakdown would promptly occur, and the surface would be neutralized by ions of opposite sign. Charge densities in the 5-8 esu $cm^{-2}$ will, in general, be sufficient to create air breakdown (30 kVcm$^{-1}$) from a planar surface. Asperities such as dust particles, scratches, etc., with their associated non-uniform potential gradients would further lower breakdown fields resulting in even lower _measured_ charge densities. Usually, the charge measured in a contact electrification experiment approximates, or is less than, that causing air or dielectric breakdown and is at least four orders of magnitude less than the number of atoms $cm^{-2}$.

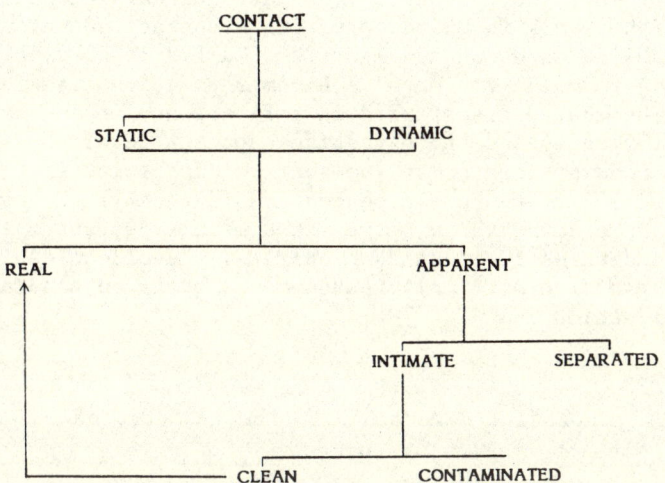

Figure 1. Nature of the contacting event

This observation invariably raises questions concerning the nature of contact electrification — Why is the observed charge density so low? Bearing in mind that charge transfer can only be measured after separation, commonly found explanations involve back transfer via a tunnelling mechanism[2] and air breakdown. Low values could also be due to non-equilibration[12], to the effective capacitance of the interface layer or perhaps to a real limitation in the number of available states within a reasonable energy range ($\sim \pm 1$ Ev) of each other.

## THE DRIVING FORCE

The generally accepted hypothesis of contact electrification is that it is driven by differences in contact potential. The analogy is made with the ideal metal-to-metal contacts shown in Figure 2. The key materials parameter is the Fermi level of the electrons in the metal. This is defined as the energy of the electrons at $0°K$. It, therefore, requires work $\phi$ to remove the electron from the metal and a further small amount of work, $qV_s$, to remove the electron to infinity from its image. $\phi$ is termed the work function of the metal and $V_s$ is its surface potential. While the electrochemical potential of the electrons, $\mu$, is strictly equal to $\phi + qV_s$, in contacting two metals the $qV_s$ terms will cancel. The driving force will then be equal to the difference in work function. However in the case of real surfaces the existence surface dipole layers or space charge regions will cause the work function to be different to that of the ideal metal. Consequently the electrochemical potential of the real surface will, in general, not be that of the ideal surface.

On contacting two ideal metals, A and B, electrons flow from the metal of lower work function (higher chemical potential) to the metal of higher work function, creating a contact potential $V_c = (\phi_B - \phi_A)/q$ as shown in Figure 2. In Figure 2, electrons flow from metal A to metal B. Metal A becomes positive with respect to metal B. The charge transferred is a function of the contact potential differences and the effective capacitance of the interface. The charge remains located at the surface of the metals. The contact potential difference can be measured directly. However, on separation, the capacitance between the metals decreases and the potential increases (for constant charge). Electrons can tunnel across the gap or electrical breakdown can occur on separation. A good approximation is:

$$Q = C_o V_c$$

where $C_o$ is a constant determined by the geometry of the contacting surfaces. Note the following points:

o The charge transferred is related directly to the contact potential difference.

o  The contact potential difference is related directly to the difference in <u>chemical potential</u> of the two contacting surfaces.

To a chemist, the latter point is extremely significant since chemical potential is a thermodynamic term, and can, in general, be related to all types of chemical reactions whether they be acid/base reactions, electrode reactions or phase transitions. Of particular interest will be the similarities to adsorption on semiconductors.

The next step in discussing the contact electrification of polymers is to consider the contact of a metal and a semiconductor in terms of band theory and band bending. The electrical properties of semiconductors are controlled by the incorporation of impurities. "n" type conduction is created by the addition of electron donors and "p" type conduction by addition of electron acceptors. In an intrinsic semiconductor the Fermi level is not a real state but lies in the middle of forbidden energy gap at an energy $(E_V + E_C)/2$. In a doped semiconductor the Fermi level lies closer to energy of the predominant dopant. On contacting with a metal an "n" type semiconductor loses electrons to the metal if the metal electrochemical potential lies below the donor level (Figure 3) and becomes positively charged. A "p" type material would become negatively charged if its electrochemical potential were below that of the metal. Charge flows until the surfaces are at the same potential. The difference in electrochemical potential is balanced by the change in surface potential (the contact potential). The charge transferred creates the contact potential, $V_C$. Analysis[3] suggests that the contact potential of the charge transferred, $\sigma$, per unit area, can be described in terms of the density of donor or acceptor states, N per cc and the penetration depth $\psi$ cms, over which the charge is distributed.

The charge transferred, $\sigma$ per sq cm is given by

$\sigma = qN\psi$

If Poisson's Equation

$d^2V/dx^2 = -Nq/\varepsilon$

is integrated under boundary conditions $x = 0$; $V = V_c$; $x = \psi$, $V = 0$

$V_c = Nq\psi^2/\varepsilon$

and  $\sigma = 2V_c\varepsilon/\psi$

The critical assumptions are that the density of states has a uniform distribution and that the penetration depth is finite. In early work

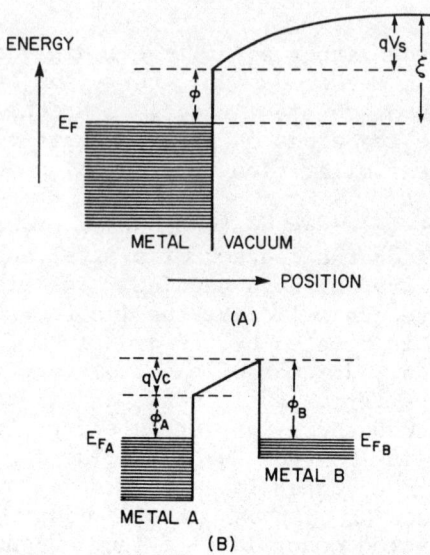

Figure 2. Contact of two ideal metals. A) $\phi$ is the work function. $\xi$ is the electrochemical potential = $\phi + q V_s$ where $V_s$ is the surface potential. B) On contacting metals A and B electrons flow from the metal of low work function to the metal of high work function until the two Fermi levels are the same. The contact potential Vc is $(\phi_B - \phi_A)/q$. Metal A is positive with respect to B. Real surfaces may have dipole layers, or space charge layers on the surface which will change the system energetics and cause the electrochemical potential to be different to the chemical potential of the electrons in the bulk metal.

with insulators[3], the density of states was related to the density of mobile charge carriers using the insulator conductivity. This resulted in penetration depths which could exceed the sample dimensions. The density of trapping states seems to be the more important parameter. The model can be extended to include surface states as well as bulk states or other nonuniform state distributions. Only if $V_c$ and $\sigma$ are known can the penetration depth and the density of states be calculated. If we consider the contact potential, $V_c$, acting across a distance equal to the penetration depth, $\psi$, to be equivalent to an effective surface capacitance, $C_s$, the injecting field $V_c/\psi$ is neutralized by the injected charge $\sigma/2\varepsilon$.

The concept for polymers will require modification since energy bands at the energies of interest are not likely to exist in polymers.

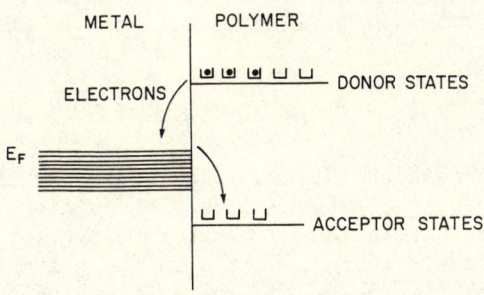

Figure 3. Charge transfer in a metal polymer contact. The electrochemical potential of the polymer, <E>, is defined as $(E_O + E_A)/2$. If this lies above that of the metal electrons move from donor states to the metal. If <E> is below that of the metal, electrons will flow from the metal to the acceptor states.

Polymers are disordered molecular materials.[13] The charge involved is likely to originate from, or end up in, <u>localized states</u>. The discussion will then be concerned with the nature of these localized states. Discussion of molecular materials[14] leads to the conclusion that the energy bands in molecular crystals will be narrow when compared to those of covalent semiconductors and of the order of 0.1 eV or less wide. Disordered materials contain many localized states.[15] Polymers are molecular materials with considerable disorder. If disorder is then introduced into the ordered array of molecules, the "band" becomes even narrower and the mean free path of the electron becomes less than the interatomic spacing. This is physically unacceptable. The concept that electrons spend most of their time immobile leads directly to the idea of localized states. If the residence time in a localized state exceeds the vibrational relaxation time, then these localized states are, strictly speaking, ionized states. If associated with specific molecules in molecular materials, the localized states are molecular ions and can be treated in quantum mechanical terms.[13] In polymers, these localized states may be associated with specific side groups or dipolar groups in the side chain or polymer backbone as discussed by Duke in this proceeding volume.

A note on terminology is required; in an "n" type semiconductor the donor state is neutral when the donor level is occupied and positive when empty. In the polymer situations to be described later, the sites are neutral to begin with, and after contacting, the groups losing an electron form radical cations, and those gaining an electron become radical anions. The electrochemical potential of such systems is defined as the centroid of the anion and cation energy levels.

i.e. $\langle E \rangle = (E_A + E_C)/2$.

Before discussing the states in more detail, their role in dictating the direction of charge transfer is discussed. Depending upon their energy in relation to the Fermi level of the metal electrons will transfer from donor states in the polymer if the Fermi level of the metal lies below the energy of the donor states (Figure 3). If the Fermi level of the metal lies above the electrochemical potential of the localized states, electrons will flow from the metal into the acceptor states. Charge continues to be transferred until either all the surface states are filled or until a sufficient surface potential is created so that no further charge transfer can occur. Note that when charge transfer stops, we have created a concentration gradient of charged species within the bulk (Figure 4).

The dynamics of charge transfer are complex. The possibility of surface states as well as bulk states must be considered. The dielectric is inhomogeneous at the microscopic level and the local density of states may be a function of bulk morphology and temporally fluctuating. In addition to dielectric relaxations there are also mechanical relaxations to consider. Thus while there is evidence that charge transfer rapidly reaches equilibrium such evidence is not definitive and the definition of equilibrium may require discussion. Dynamic effects appear to be related to the continuing change in real contact area. In other words, although the charge transferred can increase with time, concomitant changes in real area of contact also occur. The net result[11,16] is that the charge/

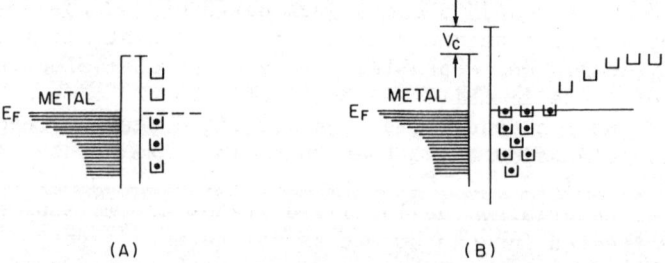

Figure 4. Limits to charge transfer. A) Number of surface states is the limiting factor. B) The surface potential limits further charge transfer.

real area of contact remains constant inferring rapid completion of charge transfer. Measurements against mercury[17] and with soft rubbers[18] suggest rapid transfer of charge. Repeated contacting often leads to increased charge being transferred, which can be explained in terms of slow diffusion from surface states into bulk states,[4,19] as from the state after contacting shown in Figure 4A or from surface potential limited charge transfer (Figure 4B). In each case (as illustrated in Figure 5), after separation the driving force is the charge concentration gradient. Vacancies are created in the surface region which can be refilled on recontacting.[2] Such a two-zone model,[4,19] incorporating variable times of contact and separation has recently been developed[20,21] although details have not yet been published. While back transfer may occur on separation, it is tempting to suggest that steady state filling of the surface states or of bulk states close to the surface is attained rapidly. Transfer to the bulk states further from the surface is the rate limiting step. The analogy in terms of chemical kinetics would be the sequence

$$A + B \xrightarrow{k_1} C$$

$$C \xrightarrow{k_2} D + B,$$

where A represents the infinite source or sink of electrons (the electrode), B represents an unfilled state near the surface, C, a filled B site, D represents a filled bulk state and the rate constant, $k_1$, is much greater than the rate constant $k_2$. Such a model results in rapid filling of the surface states and slow filling of bulk states. For insulators for which the dielectric relaxation times are long, it is unlikely that true equilibrium would be attained. The distribution of charge would not be uniform and would depend on the detailed charging history.

If the energy level of localized states in the polymer and the electrochemical potential of the polymer relative to the Fermi level in the metal is the driving force to contact electrification, the contact potential difference between a series of polymers and given metal should be a function of polymer ionization potential. This is because of the relationship[22] between the Fermi level, $E_F$, ionization potential, $I_p$ and polarizability, P.

$$E_F = \frac{I_p + P}{2}.$$

Additionally, for a given polymer with a uniform density of states there should be a linear relationship between the charge transferred from a series of metals and their work function. Experimentation confirms some of these predictions.

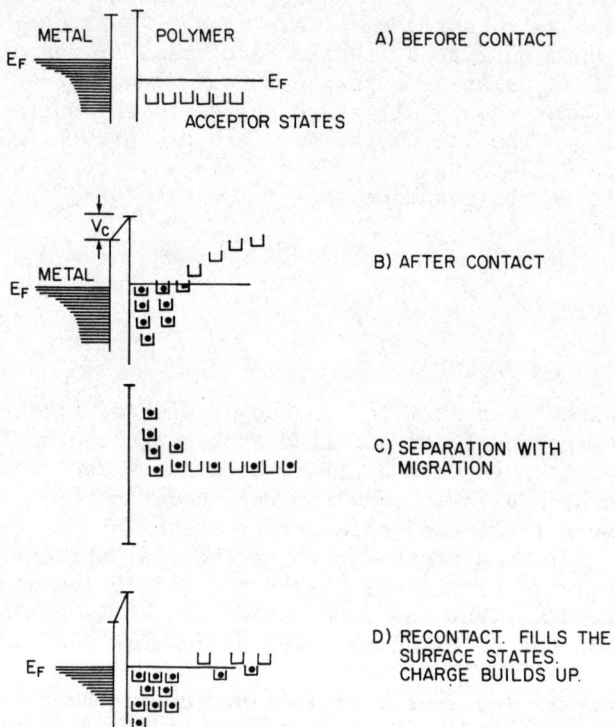

Figure 5. Repeated contacting leads to steady increase in the charge transferred.

Table II compares the work function of a series of polymers[24] with the ionization potential estimated from the ionization potential of similar monomers. The work function was obtained by direct measurement of contact potential of a polymer film using the vibrating capacitor technique[25] under vacuum. The reference electrode was gold, assumed to be 4.7 eV. Calculation of the ionization potential is based on the values of structurally similar model compounds given by Turner.[26] The relationship is reasonably linear but certainly not the 1 to 1 relationship which might be expected. Reasons for the small variation in work function are currently unexplained.

Table II. Ionization Potential Based Tribo Series.

|  | Ionization Potential (eV)A | Work Function (V)B | C |
|---|---|---|---|
| Polytetrafluoroethylene | 13.3 | 5.75 | 4.26 |
| Polyvinylchloride | 10.6 | 5.13 | 4.85 |
| Polyethylene | 10.0 | 4.90 | |
| Polycarbonate | 10.1 | 4.80 | 4.26 |
| Polymethycrylate | 10.0 | 4.40 | |
| Polymethylmethacrylate | 9.8 | 4.70 | |
| Polyethyleneoxide | 9.5 | 4.50 | |
| Polyamide (nylon 6,6) | 8.8-9.4 | 4.3-4.5 | 4.08 |

A) Based on values from Turner, Advances in Phys. Org. Chem, 4, 46, (1966)
B) Strella (unpublished) See D. Seanor in "Polymer Science," A.D. Jenkins, Editor, Vol. 2, p. 1267, North Holland, Amsterdam, 1972.
C) D.K. Davies, Advances in Static Electrifiction, 1, 10, 1970.

There are a number of papers in the literature dealing with the charge transferred to a fixed material from a series of metals.[27-30] Typical is the experimental data of Arridge[27] shown in Figure 6 which plots the charge transferred to nylon 6,6 as a function of the work function of a series of metals. While it is possible to set up a tribo series using contact potential differences as the ordering parameter, it is not possible without invoking a nonuniform or an energy-dependent density of states, to set up a consistent tribo series based solely on the quantity of charge transferred from a specific metal to different polymers. This is because the measured charge transferred can be defined in terms of an effective capacitance, $C_O$, of the interfacial region as well as the contact potential difference, $V_c$. Even in the event that back transfer does not occur on separation, the assumption that $C_O = Q/V_c$ is the same for all polymers is extremely tenuous. That it should fall into some regular order based in some way on polymer structure or morphology is very difficult to envisage. Considering Figure 4b the charge transferred is distributed across a penetration depth, $\chi$, in such a way that $q = \int_0^\chi \rho(x)dx$ as discussed previously. Considering that neither $\chi$ nor $\rho(x)$ (the charge density distribution) are known and correspondingly the relationship of $\chi$ and $\rho(x)$ to factors such as polymer structure or morphology are unknown, it is not surprising that attempts[2] to obtain self-consistent charge based tribo series have been unsuccessful. The work of Duke (this volume) is aimed at rationalizing nonlinear charge transfer-work/function relationships and insulator-insulator contact charge transfer in terms of energy dependent density of state curves.

## THE STATES INVOLVED

Clearly, a model which involves only the ionization potential of the likely molecular states must be oversimplified. It is perhaps easy to envision donor states in which an electron is removed from the highest occupied molecular orbital (HOMO) of the group involved. It is harder to envision acceptor states where the electron will end up in the lowest unoccupied molecular orbital. (LEMO) without invoking at least the electron affinity of the group. Certainly this will not be the same over a series of dissimilar polymers.

Gibson[29,30] has attempted to place the relationships between the charge transferred in contacting experiments and molecular structure onto a more rational basis. By analogy with electrochemistry, he suggests that the charge transferred, (q/m), will be governed in an exponential manner by the difference in energy of the donor and acceptor states. Using the well established relationships between the Hammett $\sigma$ function[31] and molecular energy levels, he obtains the relationships below where $E_{HOMO}$ and $E_{LUMO}$ refer to the energy of the highest occupied and lowest empty molecular orbitals, $\sigma_\chi$ is the Hammett constant. By defining the charge transferred, (q/m) in terms of the electrochemical potential, $E_F$ and the molecular energy levels the relationships are as shown ($m_1$, $m_2$ and $b_1$, $b_2$ are constants).

For positive charging:

$$\ln(q/m) = E_{HOMO} - E_F$$

$$E_{HOMO} = m_1\sigma_x + b_1$$

$$\therefore \ln(q/m) = E_F + m_2\sigma_x - b_2$$

For negative charging:

$$\ln(q/m) = E_F - E_{LUMO}$$

$$E_{LUMO} = -m_2\sigma_x + b_2$$

$$\therefore \ln(q/m) = E_F + m_2\sigma_x - b_2,$$

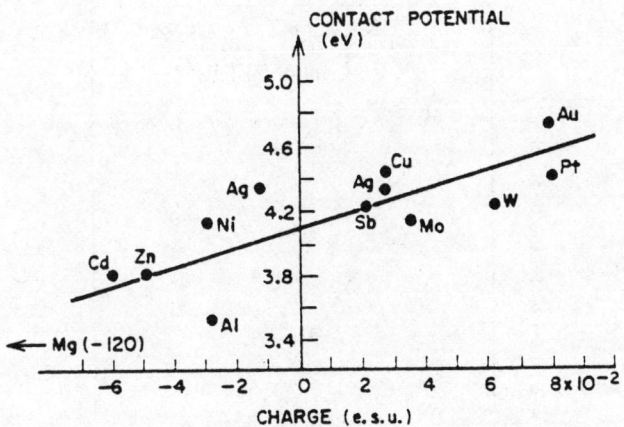

Figure 6. Charging of nylon 6.6 against metals of different work function. Arridge[27]

i.e., for positive charging the charge transferred will decrease with increasing $\sigma_x$. The charge will increase with increasing $\sigma_x$ for negative charging. His results for substituted anils,

$$\underset{OH}{\bigcirc} - C = N - \bigcirc - X$$

which charge positively against steel and for substituted polystyrenes which charge negatively against zinc powder are shown in Figure 7. The same prediction holds true for polymers of the form

$$-(\underset{Y}{\overset{X}{\underset{|}{C}}}-CH_2)-_n$$

if an effective $\sigma_i = \sigma_{ix} + \sigma_{iy}$ is used, (Figure 8) (where X and Y are summarized in Table III).

It has also been shown[32] that there is a linear relationship between ionization potential and $\sigma_i$. Thus, the data of Figure 8 can be transformed to give a linear relationship between $ln(q/m)$ and ionization potential. These relationships are open to the criticism that they do not allow the sign of the charge transferred to change with metal work function with a particular polymer. This is because the electrochemical potential is left undefined in their model. In other cases, the relationship observed between (q/m) and $\sigma_x$ is linear rather than logarithmic.[33]

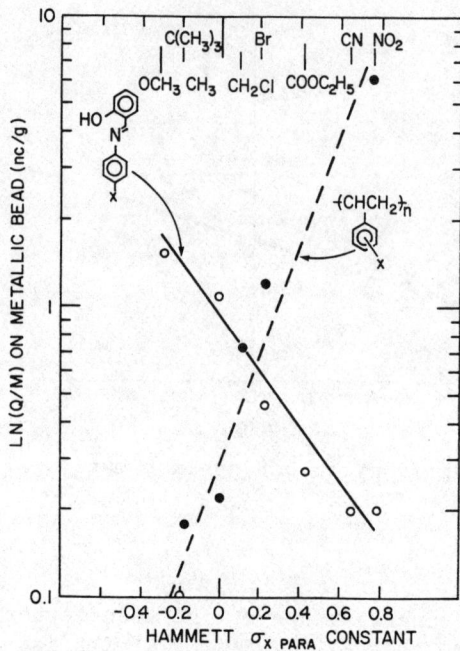

Figure 7. Linear free energy relations and contact charging.[29] The anils charge positively against steel, O. The substituted polystyrenes charge negatively against zinc powder, O.

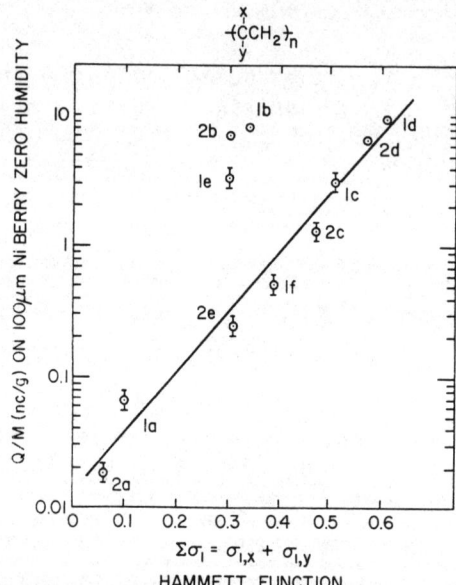

Figure 8. Charging of substituted polyolefins and the Hammett function. The polymers charge negatively against nickel berry. The sample numbers correspond to those in Table III.[30]

Table III. Hammett σ Functions $[\sigma_I = \sigma_{Ix} + \sigma_{Iy}]$ for Polymers of the Structure.

$$\left[ CH_2 - \underset{Y}{\overset{X}{\underset{|}{\overset{|}{C}}}} - \right]_n \quad \text{(from Ref. 30)}$$

| 1, X = H | $\sigma_I$ | 2, X = $CH_3$ | $\sigma_I$ |
|---|---|---|---|
| (a) Y = $C_6H_5$, | + 0.1 | (a) Y = $C_6H_5$, | + 0.06 |
| (b) Y = COOH, | + 0.34 | (b) Y = COOH, | + 0.03 |
| (c) Y = CI, | + 0.51 | (c) Y = CI, | + 0.47 |
| (d) Y = CN, | + 0.01 | (d) Y = CN, | + 0.57 |
| (e) Y = OH, | + 0.30 | (e) Y = $COOCH_3$ | + 0.31 |
| (f) Y = $OOCCH_3$ | + 0.39 | | |

The most detailed discussion of the states involved has been in a series of papers by Duke, Fabish, et al.[34-38] Basically, the argument is that the states involved are, indeed, molecular and molecular ion states associated with the particular groups in the polymer are created. In particular, since localized states are associated with polarization fluctuations, polarizable side groups in non-polar polymers are important since in general they have energies different to the predominantly carbon/hydrogen backbone. These groups would, in the gas phase, have well-defined energy states. However, in the solid polymer, the energy levels are broadened by the local temporally fluctuating environment. The presence of local disorder, dipoles, time-dependent polarization and polarizability, and local density fluctuations, causes a broadening in such molecular group energies of several electron volts. In this model, the distribution of energy states is represented by a Gaussian distribution for each of the molecular ion states. The center of each Gaussian distribution is located at the solid molecular anion and cation energy level for the acceptor and donor states, respectively. The width of the distribution is determined by the local environment which, in the most general terms, includes both intermolecular and intramolecular contributions. Fitted distributions based on experimental measurements are shown in Figure 9.

According to the Duke-Fabish model,[37] the charge transferred is controlled by the way in which the distribution is sampled. Thus, the first criterion is the relationship between the center of the energy state distributions of the anion and cation and the Fermi level of the metal. The centroid of the anion and cation energies effectively defines the polymer electrochemical potential[37] and determines the sign of the charge transferred. The actual distribution of energy levels within $\sim$ 0.4 eV ($\sim$ 15kT) of the metal Fermi level determines the quantity of charge transferred. This is because the probability of transfer requires both electrons and vacant states to be available with approximately the same energy, since the

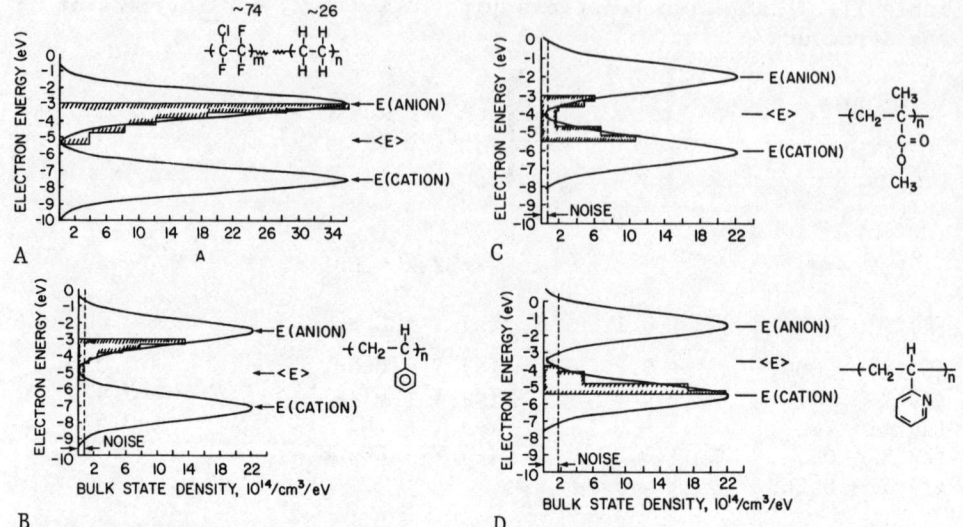

Figure 9. Charge state distributions in polymers.[37] A) Poly(chlorotrifluoro ethylene coethylene). B) Polystyrene. C) Polymethylmethacrylate. D) Poly(vinyl pyridine.)

probability of transfer decays exponentially with the difference in energy between initial and final states. It should also be noted that the probability of electron transfer, $P_{trans}$ is also a function of distance between sites, $R_{ij}$ as well as the energy difference, $\Delta E_{ij}$.

$$P_{trans} = \gamma_o \exp{-2\alpha R_{ij}} \exp{-\Delta E_{ij}/kT}$$

where $\gamma_o$ is of the order of the phonon frequency, and $\alpha$ is the rate at which the electron wave function decreases with distance from its parent site.[39]

Key to this model is the assumption that the states do not communicate with each other. Once a set of states is filled, the charge remains (relatively) localized; it does not actively seek out the site of lowest energy. Thus, it is possible to fill one set of sites by contacting with metal 1 and a second set of sites by contacting with metal 2. Provided the Fermi energy difference between the metals is sufficient not to populate the same states, the experiment appears to give the same charge transfer regardless of the order of contacting.[35,36] In respect to the so-called tunnelling window,[37] the Duke-Fabish model differs greatly from the more widely acknowledged model. It is based on tunnelling into localized non-communicating states within a narrow

energy distribution about the metal Fermi level in the case of
metal/polymer contacts. The model more typically used considers
a random distribution of communicating sites so that all sites
would be filled until the Fermi levels match.

Non communication between sites seems to be implied in a series
of papers dealing with photoemission from polymer surfaces.[40-43]
First, it is shown (Figure 10) that there is no sharp onset of photoemission; significant photoemission begins to occur at about 4 eV
photon energy. If the polymer is previously exposed to low energy
electrons, the quantum yield for photoemission is enhanced at low
photon energies, suggesting that the electrons populate low energy
surface states. Prior exposure to high energy light causes polyethylene (only "dirty" polyethylene charges; other behavior suggests
the presence of extrinsic states) which previously charged negatively
to charge positively suggesting that the high energy irradiation
causes electrons to be promoted to higher energy states, from which
they transfer to the metal. The lack of evidence for enhanced negative charging, which might be expected if the empty levels were filled,
appears to support the Duke-Fabish sampling/non-communicating state
model. Certainly, the electrons promoted by the UV irradiation have
not recombined with the positive sites which must have been created.
Nor does the particular metal appear to sample the positive sites.

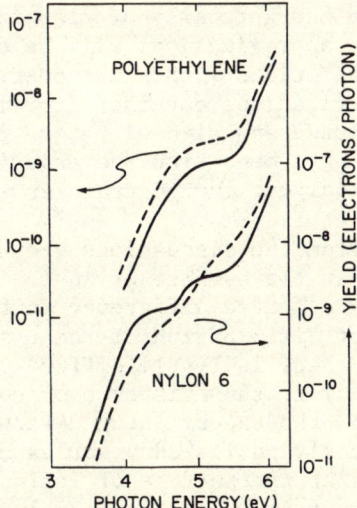

Figure 10. The impact of contact charging on photoemission from
polymers;[43] — before contact, --- after contact; nylon/mercury, poly-
ethylene/aluminum.

Figure 10 shows the effect on photoemission quantum yield after contacting nylon with mercury and polyethylene with aluminum.[43] Assuming these results are not caused by material transfer, the enhanced emission from polyethylene and extension to emission at lower incident photon energy is consistent with the negative charging observed. The nylon case is more complex and perhaps more significant. The quantum yield for low energy photons decreased, while at high energies the quantum yield increased. The net charge transferred was positive. The implication is that both positive and negative states coexist after contact. While there are more positive states than negative states, the communication between them is slow. This is strong support of the localized, non-communicating, state hypothesis. This hypothesis when extended to insulator/insulator contacts may help explain some of the anomalies discussed by Harper[2] who showed that predictions from a charge transfer based tribo series were not consistent in magnitude or sign.

While there does not seem to be a rigorous means of predicting the charge transferred in a metal/polymer contacting event, the experimentally determined energy distributions of Fabish, et al. appear to allow a prediction of polymer-to-polymer contacts which agrees to within a factor of two with experiment.[38] The basis for the prediction is illustrated in Figure 11. It is shown that the acceptor states in polymer 1 are aligned in energy with the donor states in polymer 2 over only a limited energy range. Charge transfer can take place only between the energetically overlapping donor and acceptor states and, to a first approximation, will be controlled by the smaller of the two state distributions within the overlapping energy distributions. Using the distributions obtained from metal/polymer contacts to calculate the magnitude and sign of the charge transferred between two polymers the good agreement with the experimentally determined values of the polymer/polymer charge transfer is shown in Table IV.

It is not clear from the discussions whether or not the states involved are the same at the surface or in the bulk. From a mathematical viewpoint, there is little difference in the formulation of the charging and subsequent distribution of the transferred charge. Practically speaking, there is little difference. However, from a more fundamental viewpoint, the existence of surface and bulk states is of interest although the experimental determination is difficult. Von Seggern[44] has recently published a series of papers demonstrating the existence of traps at the surface of Teflon ™ FEP, which are shallower than the bulk traps, whereas in polyethylene he shows the reverse to be true. There are apparently no surface traps associated with Mylar ™ polyethylene terephthalate.

The nature of surface traps in polymers is of interest and since not widely studied, worthy of speculation. Unlike covalently bonded semiconductors, molecular materials do not have the possibility of

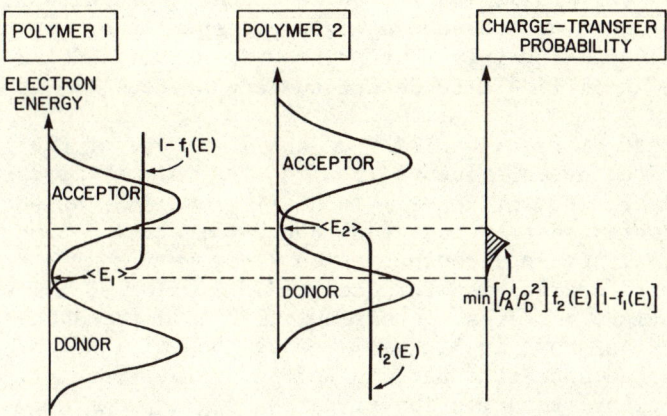

Figure 11. Contacting of two polymers.[38] Charge transfer is limited by the lower density of states in the overlapping donor state/acceptor state energy level region.

Table IV. Polymer/Polymer Charge Transfer ($\mu$ Coulombs/gm)[38].

| System | Calculated | Measured |
|---|---|---|
| Polystyrene/Poly(styrene co methlmethacrylate) (15/85) | 9 | 11 |
| 65/35 Copolymer/15/85 Copolymer | 5 | 1.3 |
| Pigmented 65/35 Copolymer/15/85 Copolymer | 24 | 14 |

free valences (dangling bonds) at the surface. Therefore, in the absence of surface free radicals intrinsic surface states seem unlikely. Disregarding extrinsic surface states for the moment, "quasi intrinsic" surface states could arise from the differences in local environment at the surface. Loss of half the dielectric sphere at the surface will change the energetics of the local environment as well as allowing for easier relaxation processes at the surface. It is possible for surface morphology to be very different to bulk morphology. Thus the probability of electron transfer at the surface could be quite different to transfer to and within the bulk. Extrinsic differences related to chemical impurities and polymer surface reactions can also be envisaged. Small molecules tend to diffuse to the surface, the products chemical reactions, particularly oxidation and even the products of chain termination reactions could be greater at the surface. For example the experiments of Von Seggern[44] could speculatively be explained in terms of Teflon™ FEP minimizing its surface energy or relaxing more readily at the surface, the presence of oxidized species at the surface of polyethylene and the fact that Mylar™ is stabilized and very polar would make it difficult to detect surface states.

The lack of predictability of the magnitude of charge transfer is illustrated by experiments in which the polymer is chemically doped with acceptor or donor molecules. Lowell[45] has shown that the charge transferred to octadecanol doped polyethylene increases with increasing octadecanol concentration. However, only about 1/40 of the predicted charge transfer occurred. Similarly, while oxidation of polyethylene surfaces increases both the contact potential and charge transferred as shown in Figure 12, only about 1 in 30 of the carbonyl groups created appear to be involved in charge transfer.[46] However, it is not clear that creation of carbonyl groups is necessarily the only factor involved. Changes in charge carrier trapping in oxidized polyethylene have been attributed to loss of terminal vinyl groups on oxidation[47,48] rather than to increase in the number of carbonyl groups.

## ADSORBED GASES AND ION TRANSFER

The hypotheses for charge transfer discussed in the earlier sections are based on good experiments conducted in controlled atmospheres with clean surfaces. The electrochemical potential difference driving force and the relevance of molecular ion states seems very plausible. However, from time-to-time, the suggestion is made that ion transfer and adsorbed layers might be involved in contact electrification.[49] Few good experiments have been made to determine these effects in polymers, which certainly cannot be ignored for surfaces in non-vacuum environments.

The similarity of the contact potential hypothesis and the theory of boundary layer chemisorption on semiconductors[50-53] cannot easily be dismissed. The concepts also apply to organic materials[54] and the interaction of oxygen with poly-N-vinylcarbazole, for example, has been well established.

In the boundary layer theory of chemisorption on semiconductors it is assumed that when a molecule of gas is adsorbed on a solid, charge transfer occurs resulting in the formation of a potential barrier at the surface. As adsorption proceeds, the potential barrier builds up, the chemical potential of the surface changes until at equilibrium the chemical potential at the surface of the solid is the same as the chemical potential of the adsorbate gas.

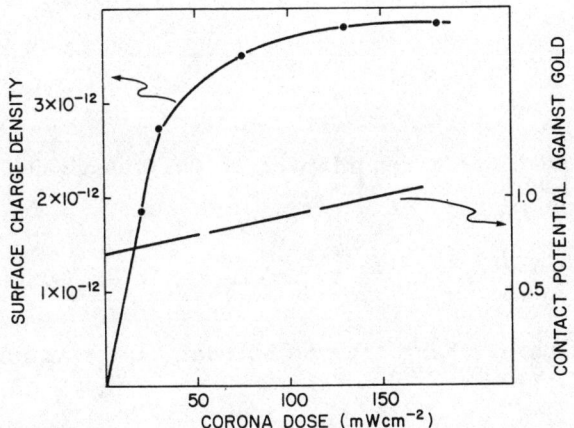

Figure 12. Effect of corona oxidation on the contact electrification of polyethylene against gold.[46]

At the point when adsorption ceases, the difference between the bulk chemical potential of the solid and the surface is balanced by a potential difference between the bulk and the surface. The surface is effectively at the chemical potential of the adsorbate. The amount of adsorption depends intimately on the electronic properties of the solid. For example the term "depletive" chemisorption is used to describe the adsorption oxygen (an electron acceptor) on "n" type zinc oxide (an electron donor). Equilibrium is reached when no further electrons are available at the surface[53] and the electrical conductivity has dropped. The similarity to contact charging is obvious.

The mathematical treatment is also similar to the simple model of contact between a metal and a semiconductor discussed earlier. At equilibrium, let the number of adsorbed ions per unit area equal $N_f$, the thickness of the boundary layer be $\ell$, the density of acceptor or donor states be $n_o$ and the charge density $\rho$.

Applying Poisson's Equation:

$$\frac{d^2V}{dx^2} = \frac{4\pi\rho}{\varepsilon\varepsilon_o}$$

where x is the distance into the solid. Integrating between limits

$$x = 0, V = V_f; \quad x = \ell, V = 0$$

we have

$$V_f = \frac{2\pi\rho\ell^2}{\varepsilon\varepsilon_o}$$

Assuming that each state contributes to the charge density

$$\rho = q\, n_o$$

$$V_f = 2\pi q\, n_o \ell^2/\varepsilon\varepsilon_o$$

Since the total charge in the boundary layer equals that in the adsorbed layer,

$$N_f = n_o \ell$$

$$V_f = \frac{2\pi q\, N_f^2}{n_o \varepsilon\varepsilon_o}$$

Measurement of $V_f$ and $N_f$ can then yield $n_o$, the density of states. For a typical semiconductor, $V_f = 1$ volt, $n_o = 10^{18}$ per cc, $\rho = 2.5 \times 10^{-6}$ cms, $N_f \sim 2.5 \times 10^{12}$ per square cm and $\sigma = 4 \times 10^{-7}$ coulombs cm$^{-2}$ i.e. $\sim 1\%$ of the surface is covered with adsorbed gas in depletive chemisorption. For comparision a polymer might acquire a charge of $10^9$ electronic charges per sq cm ($1.6 \times 10^{-10}$ coulombs cm$^{-2}$) with a contact potential difference of 1 volt. This corresponds to a penetration depth of $3 \times 10^{-4}$ cms and a density of states = $5 \times 10^{12}$ per cm$^3$.

In both cases, we can legitimately ask what the penetration depth is, and whether or not the site distribution is uniform. For example, introducing $\rho = \rho(x)$ will completely change the form of the relationships, particularly if we allow different densities at the surface and in the bulk.

If ion transfer or adsorbed layers were involved, what might be expected?

First, as shown in Figure 13, if two different surfaces covered with the same adsorbate are contacted, although the chemical potentials are the same on contact, a concentration gradient will be set up. After contact, the concentrations on each will adjust until at the new equilibrium:

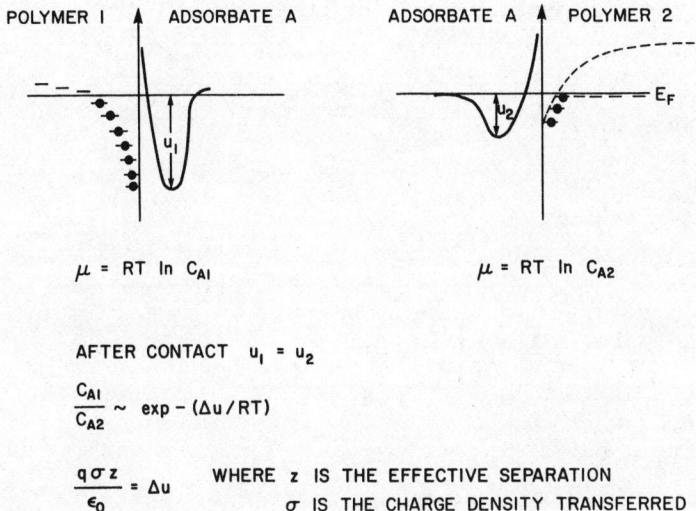

Figure 13. Model for the involvement of ion transfer in the contact electrification of two polymers in a gaseous environment.

$$\frac{n_1}{n_2} \sim \exp - \frac{(U_2 - U_1)}{kT}$$

and the charge transferred $\sigma$ is given by:

$$\frac{q\sigma z}{\varepsilon_o \varepsilon} = \Delta U,$$

where z is the effective distance of separation. For concentrations differing by a factor of 5, $\Delta U \sim 0.04$ volts after reequilibration and the charge transferred for an effective separation of 100Å is calculated to be $3.5 \times 10^{-10}$ coulombs/sq cm ($2.2 \times 10^9$ charges/sq cm).

For identical substrates the chemical potential before and after contact remains the same. There should be no charge transfer However, if there are temperature differences, which would be the case in sliding contact, there would be a readjustment in concentration and charge transfer.

Assuming thermodynamic equilibrium $\mu_1 = \mu_2$.

Then,

$$n_i = \exp - U_1/T_1$$

$$n_2 = \exp - U_1/T_2$$

and

$$\frac{q\sigma z}{\varepsilon\varepsilon_o} = U_1 \left\{ \frac{1}{T_1} - \frac{1}{T_1 + \Delta T} \right\}^2 = \frac{\Delta T}{T_1^2} U_i$$

where $\Delta T = T_2 - T_i$

From which

$$\sigma = \frac{\varepsilon\varepsilon_o}{zq} \frac{U_1 \Delta T}{T_1^2}$$

For

$$U = 1 \text{ eV}, \Delta T = 1°C \; T_1 \sim 300°K \text{ and } z = 30 \text{ Å}$$

$$\sigma = \sim 10^{-9} \text{ Coulombs/cm}^2$$

which is by no means insignificant, and has profound implications for frictional charging, particularly in moist environments. The implied reduction in charge transferred as temperature increases has been noted by Ruckdeschel[55] using glass beads on metal substrates.

Water vapor has been shown to have a marked influenced on contact electrification, acting to increase conductivity as well as to decrease the breakdown strength of air. As a result of adsorbed water on both surfaces, contact potential differences will be lower leading to less transfer of charge. Water vapor also increases polymer conductivity, $\sigma$, according to a relationship[56]

$$\log \sigma = \log \sigma_{dry} + \alpha w_{ADS}$$

where $w_{ADS}$ is the weight of water adsorbed per unit weight of polymer. Such a relationship implies that water acts by a complex combination of increased local dielectric constant, increased number of charge carriers, increased local molecular motion (plasticization) and increased charge carrier mobility.[13] Thus in moist atmospheres not only is the amount of charge transferred lower but the potential achieved on separation is lower and the charge dissipation rates are higher.

## CONTACTS, SLIDING AND FRICTION

Of fundamental interest is the relationship between the total charge transferred in a contacting event and the real contact area. It is, unfortunately, rather difficult to measure the actual contact area; if the real contact area can be measured, then some anamolies can often be rationalized, particularly when it is recognized that polymers are viscoelastic materials.

The relationship between load and real contact area depends on the rheological properties of the material. For elastic materials, the true contact area should vary as the load to the 2/3 power. For a plastic material, the area of contact should vary linearly with load. Once these considerations are taken into account, the charge transferred per unit area of real contact appears to be independent of load. For example, Davies showed the charging of polyethylene terephthalate to vary as the 2/3 power of the load.[57] Since the area of contact should also vary in like manner, the real charge density is independent of load. Similarly, the observation that the charge density transferred to a plastic insulator by a sliding metal contact varies as the half power of the load[20] is rationalized by noting that the width of the track varies as the half power of the load.

In view of the industrial importance of fiber spinning and film extrusion and handling, it is not surprising that there have been many studies of the relationship between rubbing and electrification.[3]

Empirical observations each as those shown in Figure 14[58,59] can be analytically described by Equations:

$$q = q_o (1 - \exp - BW)$$

where B is a constant, W is the work against friction and $q_o$ is the limiting charge density which is a function of load.

By relating W to friction coefficient, $\mu$, load, L, and slip velocity, $\Delta V$, a relationship between charge and friction can be derived[8]

$$q = \frac{q_e(L)}{B^2 \Delta V} (1 - \exp - B\mu L \Delta V^2 t)$$

which quite adequately describes the experimental curves. Given that the friction coefficient is a function of real contact area and that the real contact area should be related to the contact time and molecular relaxation processes, rate effects described in terms of polymer rheology should be observed. Lowell[60] has used the term "molecular stirring" to describe processes by which filled traps are carried into the bulk of the polymer and unfilled traps are brought to surface. While the thermodynamics were not defined and the broader consequences not pursued, molecular relaxation processes are clearly implicit in this mechanism.

It is, therefore, not surprising that rate and temperature effect are noted in the contact charging of polymers.[61-63] Figure 15 shows the frictional electrification of a polycarbonate film when run in contact with a chromium tip at different speeds and temperatures.[63]

As anticipated, rate and temperature effects are noted with close correlation between the electrostatic charge and friction force being obtained.[64] Applying the time-temperature superposition principle[65] i.e. the Williams-Landel-Ferry transform, the master curve shown in Figure 16 was obtained. From these results, it is clear that intimate contact of the surfaces is involved in contact electrification and that maxima in friction and contact electrification are to be expected when the molecular relaxation processes, duration of contact and frequency of contact are, so to speak, in resonance. Current thinking also suggests such processes may also be involved in the transport of charge into and through the bulk of the polymer.[13]

## SUMMARY

On the basis of many varied experiments involving contact electrification, a number of conclusions may be drawn.

o  There is a reasonable understanding of the driving forces, role of contacts, atmospheric effects, etc. when polymers are contacted by metals.

o  Charge transfer involves movement of electrons, resulting in molecular ion states, which may or may not be different at surface or within the bulk.

o  The energy states of the molecular ions extend over a wide range, which is associated with the local structure (geometry, dipolar/polar, crystalline/amorphous, local density fluctuations).

o  Many observations can be rationalized in terms of polymer molecular structure and relaxation behavior.

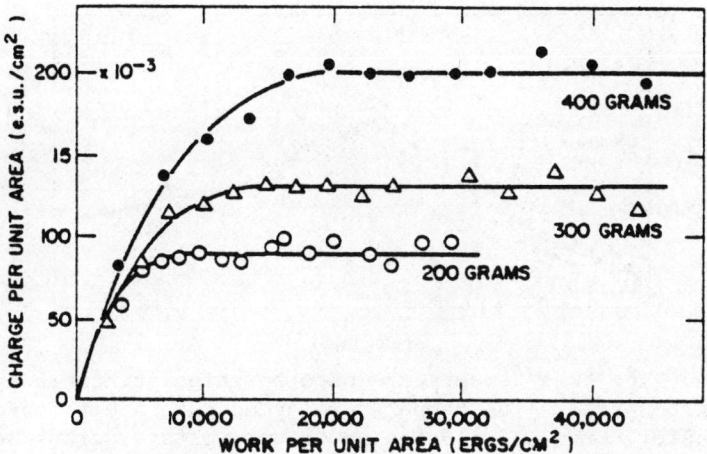

Figure 14. Charge generated as a function of work against friction. Cellulose triacetate on chrome plated steel.[58]

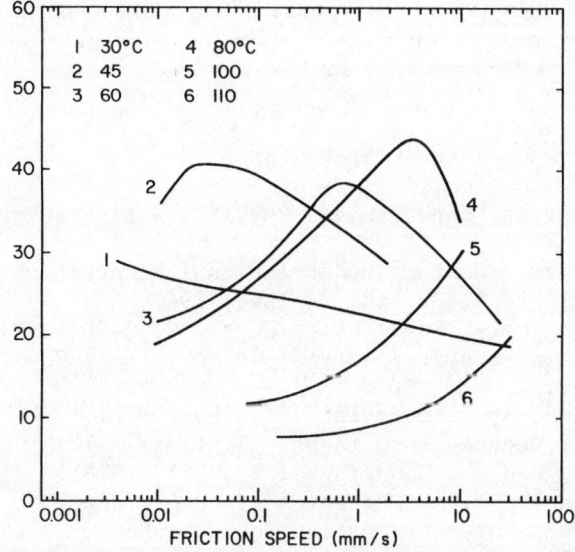

Figure 15. The effect of friction speed and temperature on the charging of polycarbonate by chrome plated steel.[63] The output voltage is plotted.

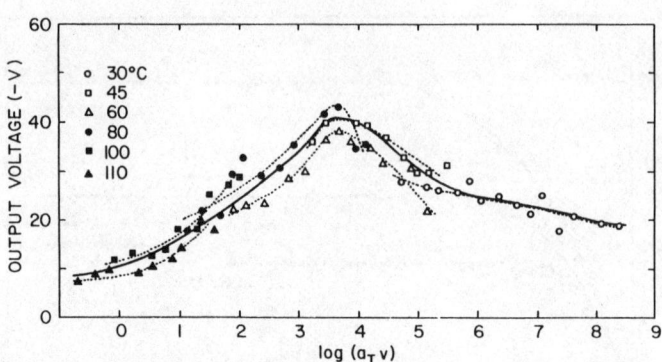

Figure 16. The time/temperature superposition principle applied to polycarbonate.[63] The temperature/speed data are transformed to a single master curve and plotted as output voltage versus log $a_T v$ where $a_T$ is equal to A exp - E/RT.

On the other hand, it is by no means clear what determines or limits the amount of charge transferred. There is little knowledge of the spatial distribution of the transferred charge (or the molecular sites involved). The role of impurities is incompletely understood and neither metal/polymer nor polymer/polymer charging is predictable, from first principles.

Understanding of the contact electrification phenomenon has come a long way to the point that industrial processes can be based on controlled charging of polymers. Application of modern analytic tools and instrumentation should allow an even greater depth of understanding to be achieved in the future.

## REFERENCES

1. L. B. Loeb, in "Handbuch der Physik", Springer-Verlag, Berlin, 1958.
2. W. R. Harper, "Contact and Frictional Electrification", Oxford University Press, Oxford, England, 1967.
3. D. J. Montgomery, Solid State Phys., 9, 139 (1959).
4. H. Krupp, in "Static Electrification", Inst. Phys. Conf. Ser., 11, 1 (1971).
5. J. Fuhrmann, J. Electrostat., 4, 109 (1977).
6. H. Bauser, Dechema Monographs, 72, 11 (1974).
7. W. Ruppel, Dechema Monographs, 72, 321 (1974).
8. D. A. Seanor, Polymer-Plastics Technol. Eng., 3, 69 (1974).
9. S. P. Hersh, Polymer-Plastics Technol. Eng., 3, 29 (1974).
10. M. W. Williams, J. Macromol. Sci.-Rev. Macromol. Chem., C14, 251 (1976).
11. J. Lowell and A.C. Ross-Innes, Advances in Physics, 29, 947 (1980).

12. F. R. Ruckdeschel and L. P. Hunter, J. App. Phys., 48, 4899 (1977).
13. D. A. Seanor, in "Polymer Science", A.D. Jenkins, Editor, Vol. 2, p. 1187, North Holland, Amsterdam, 1972.
14. F. Gutmann and L. E. Lyons, "Organic Semiconductors", Wiley and Sons, New York, 1967.
15. N. F. Mott, Advances in Physics, 16, 49 (1967).
16. K. P. Homewood and A. C. Ross-Innes, Inst. Phys. Conf. Ser., 48, 233 (1979).
17. J. A. Medley, Brit. J. App. Phys., 4, 528 (1953).
18. G.A. Cottrell J. Phys. D, 11, 681 (1975).
19. H. Bauser, W. Klöpfer and H. Rabenhorst, Advances in Static Electricity, Vol. 1, p. 2, Auxilia, Brussels, 1970.
20. A. Wählin and G. Bäckstrom, J. App. Phys., 45, 2058 (1974).
21. M. Henneke, R. Hofmann and J. Fuhrmann, J. Electrostatics, 6, 15 (1979).
22. J. Fuhrmann and J. Kürschner, J. Electrostatics, 10, 115 (1981).
23. L. E. Lyons, J. Chem. Soc., 5001 (1957).
24. S. Strella - unpublished. See Ref. 12, p. 1267.
25. W. A. Zisman, Rev. Sci. Instrum., 3, 367 (1932).
26. D. W. Turner, Adv. in Physical Organic Chem., 4, 46 (1966).
27. R.G.C. Arridge, Brit. J. App. Phys., 18, 1311 (1967).
28. D. K. Davies, Inst. Phys. Conf. Ser., 4, 29 (1967).
29. H. Gibson, J. Amer. Chem. Soc., 97, 3832 (1975).
30. H. Gibson and F.C. Bailey, Chem. Phys. Letters, 51, 352 (1977).
31. L.D. Hammett, "Physical Organic Chemistry", 2nd Ed., McGraw Hill, New York, 1970.
32. H. Gibson, Can. J. Chem., 55, 2637 (1977).
33. P. J. Cressman, G. C. Hartmann, J. E. Kuder, F.D. Saeva and D. Wychick, J. Chem. Phys., 61, 2740 (1974).
34. C. B. Duke, This proceedings volume and references therein.
35. T. J. Fabish, H. M. Saltsburg and M. L. Hair, J. App. Phys., 47, 930 (1976).
36. T. J. Fabish, H. M. Saltsburg and M. L. Hair, J. App. Phys., 47, 940 (1976).
37. T. J. Fabish and C. B. Duke, J. App. Phys., 48, 4256 (1977).
38. C. B. Duke and T. J. Fabish, J. App. Phys., 49, 315 (1978).
39. J. Mort, G. Pfister and S. Grammatica, Solid State Comm., 18, 693 (1976).
40. Y. Murata, Jap. J. App. Phys., 18, 1 (1979).
41. S. Kittaka and Y. Murata, Jap. J. App. Phys., 18, 515 (1979).
42. Y. Murata and S. Kittaka, Jap. J. App. Phys., 18, 421 (1979).
43. Y. Murata, T. Hodoshima and S. Kittaka, Jap. J. App. Phys., 18, 2215 (1979).
44. H. VonSeggern, J. App. Phys., 52, 4081, 4086 (1981).
45. J. Lowell, J. Phys. D., 12, 2217 (1979).
46. M. Selders, F. K. Dolezalek, D. Frenzel and H. Rabenhorst, J. Electrostatics, 10, 315 (1981).
47. P. J. Locke, Dechema. Monograph, 72, 87 (1974).
48. D. K. Davies, J. Phys. D., 14, L61 (1981).

49. P.S. H. Henry, Brit. J. App. Phys., 4, 531 (1951).
50. P. Aigrain and C. Dugas, Z. Electrochem., 56, 363 (1952).
51. K. Hauffe and H. J. Engel, Z. Electrochem., 56, 366 (1952).
52. P. B. Weisz, J. Chem. Phys., 20, 1483 (1952); 21, 1531 (1953).
53. F. S. Stone, in "Chemistry of the Solid State", W. E. Garner, Editor, p. 367, Butterworths, London, 1955.
54. J. Mitzuguchi, Jap. J. App. Phys., 20, 713 (1981).
55. F. R. Ruckdeschel and L. P. Hunter, J. App. Phys., 46, 4416 (1975).
56. B. Rosenberg, Nature, 193, 364 (1962).
57. D. K. Davies, J. Phys. D., 6, 117 (1973).
58. R. G. Cunningham, J. App. Phys., 37, 1734 (1966).
59. R. G. Cunningham and T. J. Coburn, J.App. Phys., 37, 2931 (1966).
60. J. Lowell, J. Phys. D., 9, 1571 (1976).
61. K. Ohara, Inst. Phys. Conf. Ser., 48, 257 (1979).
62. K. Ohara, J. Electrostatics, 4, 233 (1978).
63. K. Ohara, J. Electrostatics, 9, 107 (1980).
64. K. Ohara, J. App. Pol. Sci., 21, 1409 (1977).
65. M. L. Williams, R. E. Landel and J. D. Ferry, J. Amer. Chem. Soc., 77, 3701 (1955).

ELECTRIC CONTACT PERFORMANCES AND ELECTRICAL CONDUCTION MECHANISMS

OF AN ELASTOMERIC CONDUCTIVE POLYMER

T. Tamai

Sophia University

Tokyo, 102, Japan

The conductive polymer is usually composed of dispersed electrically conductive particles such as carbon black or metal in a durable polymer matrix such as silicone rubber. As it is possible to obtain conductivity and elasticity in simple materials, this is useful for high density micro-sized electrical contacts as connector or as key-board switch. However, in spite of the excellent properties, it is difficult to obtain low electrical resistivity. In order to approach this problem, performances of the contacts between the polymer surface and metal were studied from the view point of surface contamination. The electrical conduction mechanism and a method to get low resistivity were also studied. The results show that the surface of the polymers containing carbon black shows contamination resistance properties because of their chemical stability. The conduction mechanism changes with particle content. For large content, electric current flows through particles contacting each other; and for small content, the current passes through the gap between particles by Schottky conduction. Thus, as the resistance consists of particle resistance, constriction resistance and gap resistance, the resistance can be controlled by particle material, their content and size.

## INTRODUCTION

Elastomeric conductive polymers have been used for various electrical applications. In general, conductive polymers consist of conductive particles dispersed into a durable polymer matrix such as silicone rubber. Elastomeric conductive polymers are very suitable for high density micro-sized electrical connector or keyboard switches, because conductivity and elasticity can be obtained in one simple material.[1-3]

However, since this type of conductive polymer has high resistivity compared with a metal conductor, the application is limited to low level current or voltage transmission. It is difficult but important to develope conductive polymer with low electrical resistivity.[4-6]

On the other hand, on the surface of the contacts which are exposed to the air, the reaction with gases in the air produces various kinds of contaminant films. This is noticeable in base metals, and is also observed in noble metals. Even on Au surface, adsorbent films are produced, and their influence cannot be ignored.[7,8] Within these contaminant films produced at the contact surface and present at the interface of surfaces in contact with each other, the resistivity is much higher than that of bulk material. Thus, the contact resistance increases, the contact reliability is greatly degraded, and the contact performance and the lifetime are impaired.[9-13]

Accordingly the present study deals with the corrosion property of the contact interface between the polymer and a Au plated circuit board. Good corrosion resistance resulting from elasticity of the polymer was found. Namely, the elastomeric conductive polymer maintained an air-tight contact for the Au plated circuit board, and such a contact prevents the ingress of corrosive gases into the contact interface.

In the present paper, moreover, in order to clarify also electrical properties of elastomeric conductive polymer, electrical conduction was determined for different content of dispersed conductive particle based on the previous investigation.[14] It was shown that for large particle content, current flows through particles which contact each other; and for small particle content, the current passes through the gap between particles by Schottky effect. Therefore, in order to obtain low resistivity, conductive particles should contact each other.

## EXPERIMENTAL

### Specimen

As conductive particles, carbon black particles prepared from acetylene black were used, its size was measured approximately 420 Å; and as the elastomeric polymer, in view of its durability and chemical stability, silicone rubber was applied. The silicone rubber had a density of 1.20 g/cm$^3$, a specific heat of 0.35 cal/g °c, and a resistivity of $9 \times 10^{16}$ Ω-cm. For the elastomeric conductive specimen, the carbon black particles were dispersed into silicone rubber by the rolling process. The particle contents were selected to be 5 wt%, 12 wt%, 20 wt%, and 40 wt%. The specimens were cured under heating, and its size was measured 10 mm (width) × 1 mm (thickness) × 20 mm (length).

## Measurement procedure

The corrosion property of contact interface between the conductive polymer with 40 wt% carbon particle and Au plated printed circuit board was examined under a corrosive environment of $H_2S$ 3 ppm for 1000 hours at 40 °c and 80 % relative humidity. The specimen arrangement is shown in Figure 1; two printed circuit boards hold the conductive polymer in a sandwich configuration at a deformation rate of 20 % in compression. After 1000 hours test, to evaluate the air-tight property of the polymer on the Au plated circuit board, discoloration and contact resistance for the surface of Au plate were examined. The contact resistance was measured by using Au probe with 1 mm$\phi$ as the other contact member under flowing a current of 0.15 mA and a contact load of 10 g. Corroded parts due to defects in the Au plate show a high level contact resistance because of contaminant films. On the contrary, for the uncorroded part, contact resistance is maintained at a low level. Therefore, contact resistance measurements can be used to evaluate the corrosion of the contact surface.

To examine the effect of particle content on the resistivity, the relationship between resistivity and carbon particle content was measured by the experimental arrangement shown in Figure 2. Au plated on copper substrate was used as the other contact mem-

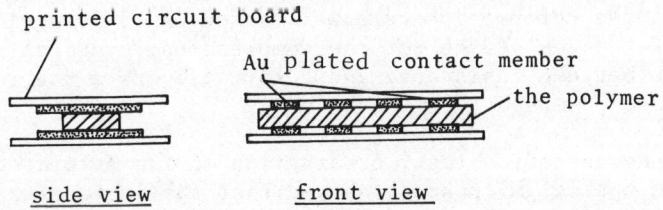

Figure 1. Experimental arrangement for studying corrosion behavior.

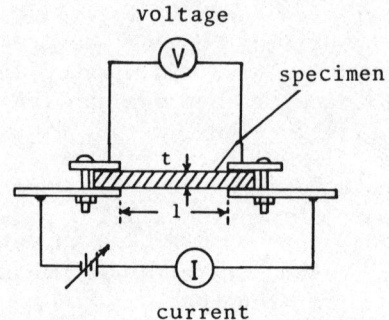

1 mm (t) × 10 mm (1) × 10 mm (width)
Figure 2. Circuit for measuring resistivity.

ber. The Au plate had a thickness of few μm which is commonly used on electric contacts. The particle distribution in the polymer matrix was observed with the scanning electron microscope. Moreover, relationship between current and voltage was measured for each carbon content. In this case average temperature on the surface of polymer was also measured by a thermocouple. To clarify the electrical conduction, the specimen was cooled from 25 °c to -200 °c by the cryostat system.[15]

## CORROSION PROPERTY OF CONTACT INTERFACE BETWEEN THE POLYMER AND PRINTED CIRCUIT BOARD

Prior to the corrosion test, to obtain the optimum compressive load, the relationship between compressive deformation of the conductive polymer in the sandwich configuration, deformation rate, and electrical resistance were measured as shown in Figure 3. From the resistance data as a function of deformation rate in this figure, the deformation rate was chosen as 20 % in view of stability of the resistance for the compression. Hysteresis in the relationship between resistance and deformation rate for repeated stress was not observed.

After the corrosion test, observation of the Au plated surface with an optical microscope showed that the surface exposed to the corrosive gas changed color to brown, and the surface which was the boundary of the area exposed to corrosive gas and covered by the polymer changed color to blue, and the surface covered by

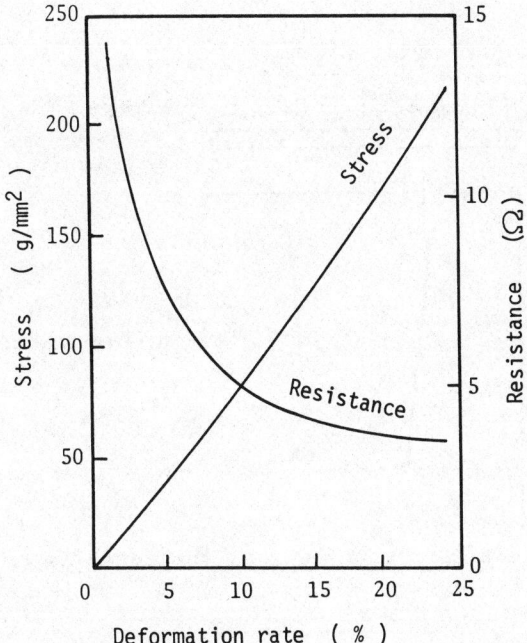

Figure 3. Change in resistance and stress versus compressive deformation, (carbon black: 40 wt%).

the polymer was not discolored. This discoloration indicates the growth of corrosive contaminant films on the Au plated surface due to defects and porosities in the plate. These contaminant films may be considered as $Cu_2S$ caused by underlying metal of copper. Adsorbed vapor was observed also along the boundary for high humidity of 80 % R.H..

Next, different contact resistance values were measured for the three parts of the Au plated contact circuit board as shown in Figure 4. As expected the lowest contact resistance was shown for the surface which had been contacted by the conductive polymer. The middle level of contact resistance was measured on the surface which was the boundary of the area exposed to corrosive gas and covered by the polymer. The highest value was measured at the surface exposed to the corrosive environment.

From the above results, it is concluded that the contact resistance 10 m$\Omega$ measured at the surface contacted by the polymer is very similar to that for the clean Au surface. Therefore, the contact between the polymer and Au plated surface has a corrosion resistant property. The silicone polymer has a little air flow in the matrix, however, the elasticity of the polymer prevents ingress of the corrosive gas into the contact interface; and as a result, the Au plated surface is not corroded. This means that the

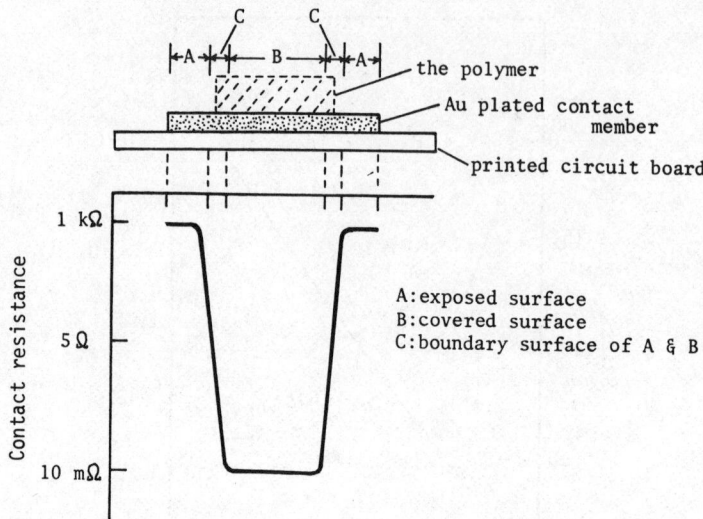

Figure 4. Effect of the polymer on the corrosion of Au plated printed circuit board.

conductive polymer can prevent corrosion of the metal contact surface. Even if thinner Au plated contacts which contain many porosities and defects, or base metals, are brought in contact with the polymer, a none corrosive contact surface may be obtained. And therefore such combination of the polymer and metal contact can realize high reliability connectors and switches.

## DEPENDENCE OF RESISTIVITY ON CONDUCTIVE PARTICLE CONTENT

Resistivity data as a function of the particle content are shown in Figure 5. The curve in this figure can be divided into three regions; the lowest resistivity region 1, sharp change rate region 2, and higher resistivity region 3. These three regions may be caused by different gap distance between particles.

Therefore, we tried to evaluate the gap distance. By assuming a smaller carbon particles enclosed in a larger rubber sphere surface, and the density of the carbon particle and the rubber to be 2 g/cm$^3$ and 1 g/cm$^3$ respectively. the following equation has been deduced.[5]

$$s = [((104.7/L) + 0.52)^{1/3} - 1] d_a \tag{1}$$

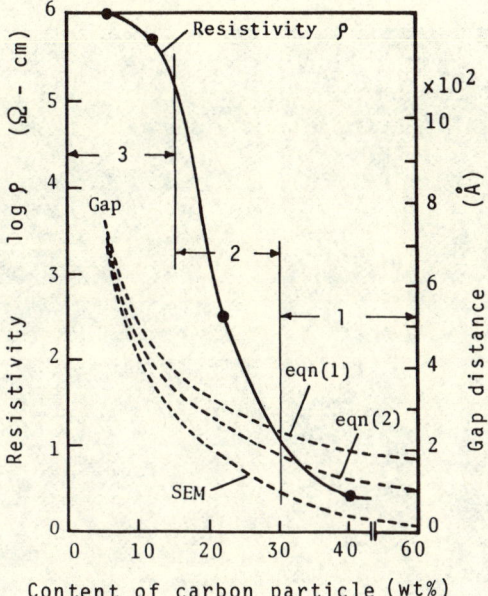

Figure 5. Changes in resistivity and gap distance as a function of carbon particle content. (Adapted from T. Tamai, Fig. 2, Ref.14.)

Where s is the gap distance, L is the particle content in percent, and $d_a$ is the radius of the carbon particle. Moreover, if an arrangement in which the center of particle is considered to be located at a corner of a cube in the rubber matrix, the gap distance is given by following equation.

$$s = [0.806(L)^{-3/2} - 1] d_a \qquad (2)$$

According to these equations, the gap distance is given by only $d_a$ and L. Particle gap distances between carbon particles for various contents as derived from Equations (1) and (2) are also shown in Figure 5 along with the change in resistivity.

According to Equation (2), all particles contact each other completely at a particle content of 87 wt%. Moreover, at particle content of 40 wt%, each particle may approach to within a distance of less than 100 Å. However, according to the scanning electron micrographs, as shown in Figure 6, distribution of cohered massive particle was recognized. The gap distance measured from the scanning micrographs are shown in Figure 5. From the data of gap distance vs. particle content shown in Figure 5, Equation (2) is more practical than Equation (1). Therefore, in the lowest resistivity region 1, electric current paths may be formed by contact of particles.

carbon black content
(1):5wt%, (2):12wt%
(3):20wt%, (4):40wt%

1.9 m

Figure 6. Distribution state of the particles in the polymer.
(Adapted from T. Tamai, Fig. 3, Ref. 14.)

## ELECTRICAL PROPERTY OF THE CONDUCTIVE POLYMER

Electrical properties such as change in resistivity vs. voltage, change in current vs. voltage, and the change in resistivity vs. temperature corresponding to the three regions of resistivity ( see Figure 5 ) considered above were examined, and electrical conductions were clarified as follows.

### Resistivity Region 1

The relationship between voltage and current for the specimen which contained 40 wt% carbon particle is shown in Figure 7. The changes in resistivity and surface temperature as a function of voltage are also shown in Figure 7. The resistivity was deduced from the Ohm's law and the geometrical dimensions for the specimen. It is apparent from Figure 7 that resistivity increases at 1.5 V and ohmic linear relationship disappears. This characteristic feature is caused by the rise in temperature due to Joule

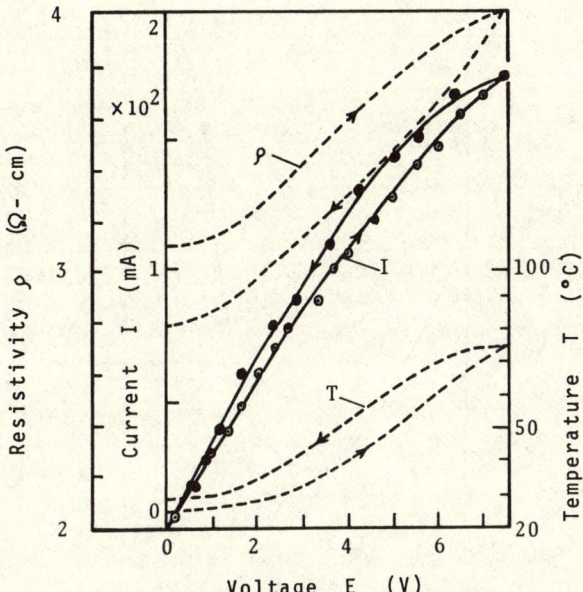

Figure 7. Electrical properties for the polymer containing 40 wt% carbon black. (Adapted from T. Tamai, Fig. 5, Ref. 14.)

heat. The rise in surface temperature corresponds well to the change in resistivity. This characteristic feature may be affected by two factors; namely, thermal expansion of the silicone rubber and the thermal change of resistivity of the particle. Thermal expansion of silicone rubber is larger than that of the carbon particles. Therefore, since particles in the rubber matrix should be separated from each other by the thermal expansion of the rubber, the resistivity increases. On the contrary, since the resistivity of carbon black has negative coefficient of temperature, resistivity decreases with rise in temperature. However, change in the thermal expansion of the rubber is far larger than the change in resistivity of the carbon black.

To clarify the above discussion, the specimen was subjected to a wide temperature change ranging from 25 °c to -200 °c. The resistivity was measured at each temperature. Change in resistivity as a function of temperature is shown in Figure 8. From Figure 8 it is found that resistivity has a minimum level of 0.48 $\Omega$-cm at -70 °c. This agrees with the hardening temperature of silicone rubber of -70 °c. Therefore, thermal contraction of the rubber is remarkable until -70 °c. Accordingly, the gap between particles is reduced, and at -70 °c particles contact each other completely in the close pack situation. Thus decrease in resistivity may be reached at the minimum level. For temperatures lower than -70 °c,

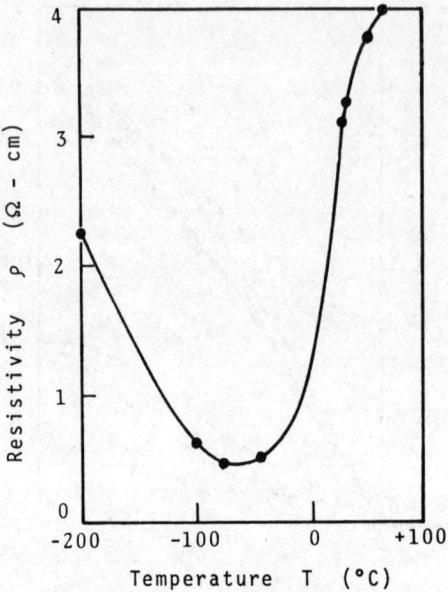

Figure 8. Change in resistivity versus temperature, (carbon black : 40 wt%). (Adapted from T. Tamai, Fig. 6, Ref. 14.)

the resistivity characteristic of carbon black appears as increase in resistivity due to the negative coefficient of temperature.

As a result, for carbon black particle content of 40 wt% or more, namely Region 1 in Figure 5, many complex current paths should be established by contact of particles in the rubber matrix.

## Resistivity Region 2

Relationships between voltage and current and the change in resistivity as a function of voltage in the Region 2 of Figure 5 are shown in Figure 9. These relationships are very different from the relationships in Region 1, e.g. resistivity decreases with increase in voltage. In this region, since the gap between particles is about 200 - 1000 Å, the current should pass through the gap by some electrical conduction mechanism. From the relationship in Figure 9 and gap distance, the Schottky effect may be the dominant conduction mechanism. According to the Schottky effect,[15-17] the relationship between Schottky current, $I_S$, and voltage, $E$, is expressed by $\log I_S \propto E^{1/2}$. If the Schottky effect exists, the relationship between current and voltage in Figure 9 should be linear in coodinates consisting of ordinate $E^{1/2}$ and abscissa $\log I_S$.

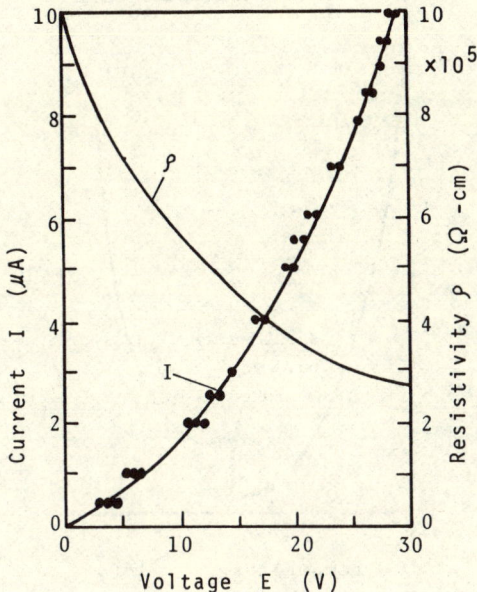

Figure 9. Electrical properties for the polymer containing 12 wt% carbon black. (Adapted from T. Tamai, Fig. 7, Ref. 14.)

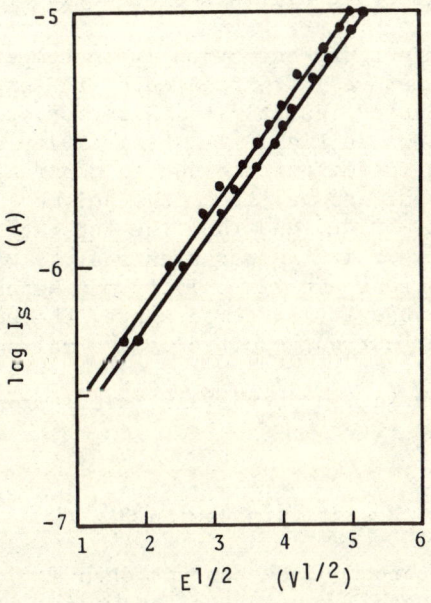

Figure 10. Log $I_s$ - $E^{1/2}$ plot, (carbon black: 12 wt%). (Adapted from T. Tamai, Fig. 8, Ref. 14.)

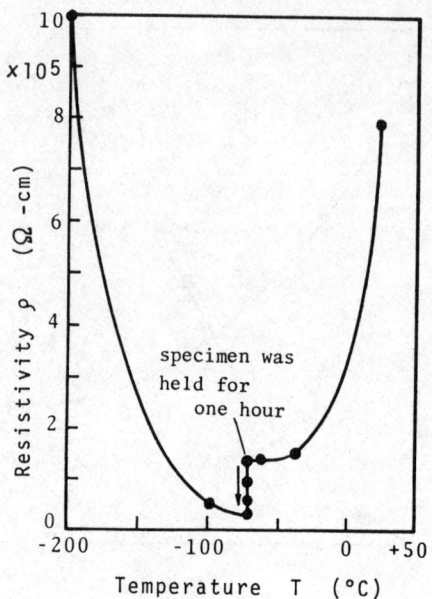

Figure 11. Change in resistivity versus temperature, (carbon black: 12 wt%). (Adapted from T. Tamai, Fig. 9, Ref. 14.)

The linear relationship shown in Figure 10 which is recalculated from the data in Figure 9 confirms the Schottky effect.

The Schottky effect depends strongly on temperature. Schottky current $I_s$ and temperature T are related by the expression $\log(I_s/T^2) \propto 1/T$. Measured resistivity characteristic as a function of temperature is shown in Figure 11. Resistivity decreases until -70 °c with decrease in temperature due to contraction of the polymer matrix as mentioned above. After the polymer hardened, it was held for one hour at -70 °c, and thus the gap between the particles may be fixed. Resistivity increases sharply with decreasing temperature. This increase in resistivity may be caused by the Schottky effect, because at low temperature, the conduction electrons which serve for current excited by thermal energy may vanish.

As a result, in Region 2, the electrical property is dominated by the Schottky effect.

### Resistivity Region 3

In case of low carbon black content such as 5 wt%, namely, for Region 3, relationship between voltage and current is linear with constant resistivity of $10^6 \Omega$-cm as shown in Figure 12. For

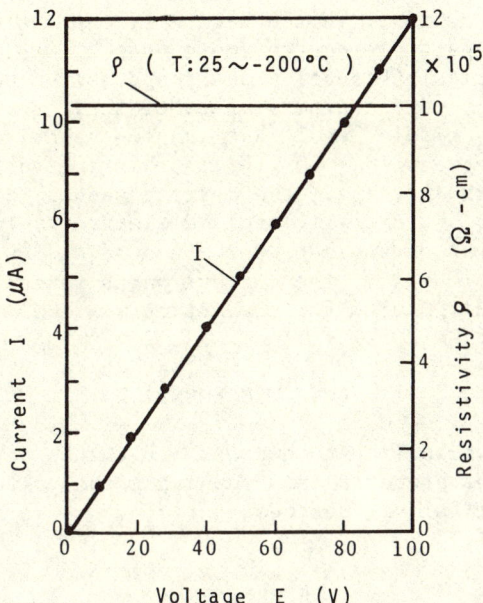

Figure 12. Electrical properties for the polymer containing 5 wt% carbon black. (Adapted from T. Tamai, Fig. 10, Ref. 14.)

the wide temperature change shown in Figure 12 the resistivity holds a constant level. In order to explain the behavior of polymer with very low carbon particle content, the gap distance between particles must become too long for electrons to pass. Relationships in Figure 12 show a kind of insulating property of the polymer itself.

## CONCLUSIONS

The electrical properties of the elastomeric conductive polymer consisting of carbon black particles in silicone rubber for application as an elastomeric connector and switch is described in this paper. Moreover, the chemical corrosion properties of the contact between the polymer and metal surface were examined.

The polymer prevents corrosion of the surface of metal brought into contact. This property was examined by using $H_2S$ corrosive gas. Elasticity of the polymer prevents ingress of the corrosive gas into the contact interface, and as result, air-tight contact interface can be achieved by the polymer.

The resistivity of the polymer changes sharply with particle content. The resistivity is divided into three regions as a function of particle content. In the Region 1, the resistivity is at a minimum level at particle content more than 30 wt%. For this resistivity, electrical conduction was found as the current paths are established by many contacts of particles. In Region 2, the resistivity shows middle level, but changes senstively with particle content, between 15 - 30 wt%. In this region, the Schottky conduction is dominant, i.e., the current passes through the gap between particles by thermally excited electrons. In Region 3, the resistivity is that of an insulator, because at low particle content ( less than 15 wt% ), the current paths vanish and the silicone matrix is only medium for conduction.

## ACKNOWLEGEMENTS

The author would like to express his thanks to SHIN-ETSU POLYMER CO., LTD. for preparing and supplying the polymer specimen, and for very fruitful discussions.

## REFERENCES

1. S. Mori, J. Inst. Electronics Communication Engrs. Japan, 63, 415 (1980).
2. D. I. Amey and R. P. Moore, in "Proc. Holm Seminar on Electrical Contacts, Illinois Inst. of Tech., Chicago, Ill.", 51 (1977).
3. T. P. Piccirillo, C. A. Dalamangas, L. S. Buchoff and R. Hasan, in "Proc. Holm Seminar on Electrical Contacts, Illinois Inst. of Tech., Chicago, Ill.", 71 (1976).
4. L. K. H. van Beek and B. I. C. F. van Pul, J. Appl. Polymer Sci. 6, 651 (1962).
5. M. H. Polley and B. B. S. T. Boonstra, Rubber Chem. Technol., 36, (1957).
6. B. B. S. T. Boonstra and E. M. Dannenberg, Ind. Eng. Chem., 46, 218 (1954).
7. C. A. Haque, IEEE Trans. P. H. P., PHP-9, 58 (1973).
8. S. Yamazaki, T. Nagata, Y. Kishimoto and N. Kanno, in "Proc. 8th Int. Conf. Electrical Contact Phenomena, Tokyo", 282 (1976).
9. T. Tamai, in "Surface Contamination:Genesis, Detection and Control", K.L. Mittal, Editor, Vol.2, p.967, Plenum Press, New York, 1979.
10. R. Holm, "Electric Contacts", 4th Edition, Springer, Berlin, 1976.
11. W. H. Abbott, IEEE Trans. P. H. P., PHP-5, 195 (1969).
12. M. Antler, IEEE Trans. P. M. P., PMP-2, 59 (1966).
13. T. Tamai and K. Tsuchiya, Trans. Inst. Elec. Engrs. Japan, 93-A, 237 (1973).
14. T. Tamai, IEEE Trans. C. H. M. T., CHMT-5, 56 (1982).
15. T. Tamai and K. Tsuchiya, Trans. C. H. M. T., CHMT-1, 54 (1978)
16. J. G. Simmons, J. Appl. Phys., 35, 2472 (1964).
17. E. Holm, in "Proc. 4th Int. Conf. Electrical Contact Phenomena, England", 12 (1968).

**Part V
Crazing, Fracture
and Morphology**

SURFACE FREE ENERGIES AND FRACTURE SURFACE ENERGIES OF GLASSY POLYMERS

Lieng-Huang Lee

Wilson Center for Technology
Xerox Corporation
Webster, New York 14580

The importance of energetics in both the bond-forming and the bond-breaking processes is stressed. Surface free energy and other energetic terms related to the bond-breaking process are discussed in detail. Basic fracture mechanics concepts are introduced as a framework in defining several important energetic terms. Effective fracture surface energy, $\Gamma$, fracture energy, $G_c$, and fracture toughness, $K_c$, are clearly defined. Three parameters affecting the energetic terms discussed in this paper are molecular weight, crack speed and molecular orientation of polymers. Glassy polymers are compared with metals and non-metals by the fracture energies.

The results indicate that glassy polymers can be toughened by increasing the molecular weight, or by applying the biaxial orientation, or by reinforcing with elastomers.

## I. INTRODUCTION

The bond-forming process between two surfaces is considered to be at a reversible thermodynamic equilibrium, and the maximum work of adhesion, $W_A$, can be expressed as[1]

$$W_A = \gamma_1 + \gamma_2 - \gamma_{12}, \tag{1}$$

where $\gamma_1$, $\gamma_2$ are the surface free energies of the respective surfaces,[2] and $\gamma_{12}$ is the interfacial free energy per unit area. Similarly, the work of cohesion, $W_c$, is defined as

$$W_c = 2\gamma_1. \tag{2}$$

when one of the two surfaces is liquid, Equation (1) becomes the Dupre's equation.[2] Dupre's original conception was that for liquid-solid adhesion, $W_A$ is also equal to the work of separation on the ground that the separated interface is clear-cut and reproducible. Therefore, if a liquid polymer or an adhesive wets and spreads fully on a solid substrate, $W_A$ should be the ideal adhesive strength.[3] Then, with the values of surface free energy of solids, we can estimate both the work of adhesion and the work of cohesion. Several reviews[4-6] list both surface free energies and interfacial free energies of various polymers.

Contrary to the bond-forming process, the bond-breaking process is, in reality, irreversible, and the energy required to break an adhesive joint is several magnitudes greater than the work of adhesion. The work of deformation, $W_d$, plays a major role in the fracture energy, while the surface free energy of a solid $\gamma_s$ exerts only a limited influence to the fracture energy.

One of the purposes of this paper is to differentiate between surface free energy and fracture surface energy. Thus, several basic principles of fracture mechanics are discussed with reference to the Griffith's energy-balance concept[7] and the Irwin-Orowan's plastic-zone concept.[8-9] We also define several frequently misused terms, e.g., effective fracture surface energy, fracture energy and fracture toughness. Actual values of related parameters are presented to illustrate the applicability of fracture mechanics to the selection of polymers for structural materials.

In the following sections, we shall discuss three modes of fracture and several fracture criteria. Later, we shall enumerate several major variables affecting effective fracture surface energy, fracture energy and fracture toughness of polymers.

## II. BASIC CONCEPTS OF FRACTURE MECHANICS

### 1. Modes of Separation

There are three modes of separation[10-12] at the crack tip as shown in Figure 1:

<u>Mode I (Opening mode)</u>. The tensile component of stress is applied in the y direction, normal to the faces of the crack, either under plane-strain (thick plate) or plane-stress (thin plate) conditions (Figure 2).

<u>Mode II (Sliding mode)</u>. The shear component of stress is applied normal to the leading edge of the crack either under plane-strain or plane-stress condition.

<u>Mode III (Tearing mode)</u>. The shear component of stress is applied parallel to the leading edge of the crack (antiplane strain). This mode corresponds to mutual shearing parallel to the crack front.

Of the three modes, the opening mode is by far the most common to crack propagation in brittle solids. This would appear to be consistent with the simple model of crack extension by progressive stretching and rupture of cohesive bond across the plane. Genuine shear fractures do occur in some cases, for instance, in the rupture of highly plastic metals and polymers (or adhesives), where ductile tearing is involved.[10]

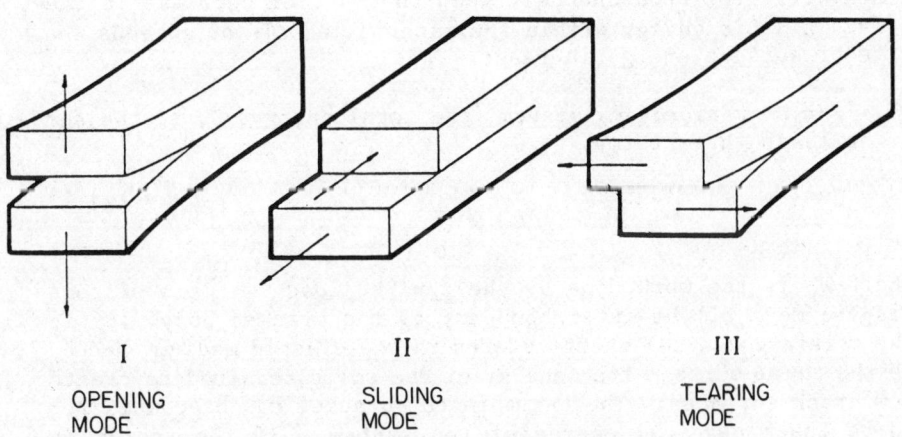

I  
OPENING MODE

II  
SLIDING MODE

III  
TEARING MODE

Figure 1. Modes of separation during fracture.

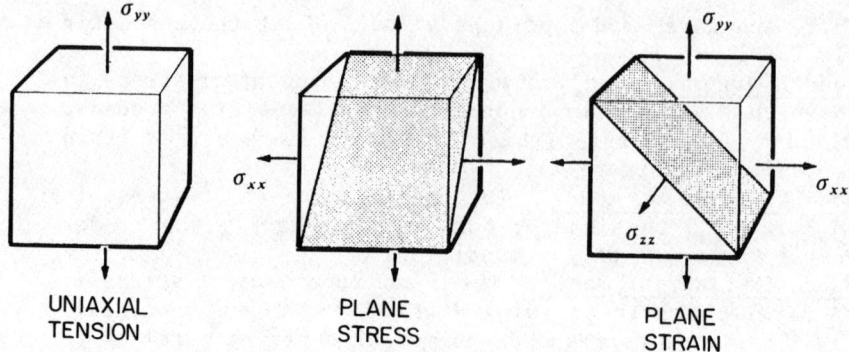

Figure 2. Stress states.

## 2. Griffith's Energy-balance Concept

Griffith[7] considered a crack system in terms of a reversible thermodynamic process. The crack is assumed to be an ellipse of vanishing minor axis (Figure 3). Hooke's law presumably applies up to the corners of the crack. The crack plane is perpendicular to the applied load. Furthermore, the crack will grow spontaneously when the rate of decrease in the stored elastic energy within the material equals or exceeds the rate of surface energy increase.

For a static crack system, the total energy, U, is the sum of at least three terms:

$$U = (-W_L + U_E) + U_S, \qquad (3)$$

where $W_L$ is the work done by the applied load in terms of some displacement of the outer boundary of the cracked body. $U_E$ is the strain potential energy stored in the elastic medium, and $U_S$ is the total surface free energy of the solid required to create new crack surfaces. The terms in the bracket can be considered to be the mechanical energy of the system. If the system is dynamic, a kinetic energy term should be added to Equation (3).

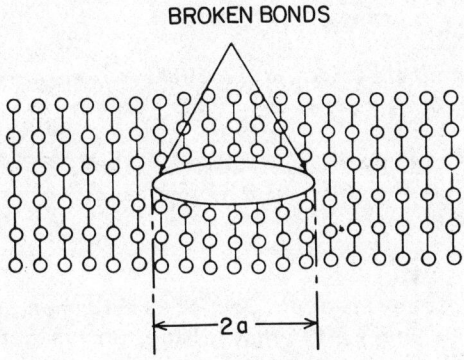

Figure 3. Elliptical model of crack used by Griffith.

Thermodynamic equilibrium of the crack propagation can be attained by balancing the mechanical energy and surface energy terms over a virtual crack extension, $\delta a$. As the crack propagates, the mechanical energy must decrease, and the surface energy must increase. Thus, the first term in Equation (3) favors crack propagation, while the second opposes it. This is the Griffith's energy-balance concept. It can be stated that at equilibrium

$$dU/da = 0. \qquad (4)$$

Griffith applied the Inglis analysis[13] to treat a narrow elliptical 'crack' ($b \rightarrow 0$) in a remote, uniform tensile stress field. He also utilized a result from linear elastic theory; namely, that for any body under constant applied load during crack propagation,

$$W_L = 2U_E \text{ (constant load)}. \qquad (5)$$

For the two-dimensional case of a thin plate in plane stress, Griffith obtained:

$$U_E = \frac{\pi a^2 \sigma^2}{E}, \qquad (6)$$

and for the three-dimensional case of a thick plate, in plane strain, he derived:

$$U_E = \frac{\pi a^2 \sigma^2 (1-\nu^2)}{E}, \qquad (7)$$

where $\nu$ is the Poisson's ratio; $\sigma$ is the applied stress; 2a is the crack length and E is the modulus of elasticity.

For the surface energy of the crack system, Griffith gave the following relation for unit width of crack:

$$U_S = 4a\gamma_s, \qquad (8)$$

where $\gamma_s$ is the surface free energy of solid per unit area. Then, the total system energy for the plane stress case becomes

$$U = -\frac{\pi a^2 \sigma^2}{E} + 4a\gamma_s. \qquad (9)$$

For the Griffith equilibrium condition,

$$dU/da = 0,$$

the fracture stress at the critical condition becomes

$$\sigma_c = \left[\frac{2E\gamma_s}{\pi a}\right]^{1/2}. \qquad (10)$$

for the constant load, plane stress condition. The Griffith equation relates the material fracture strength, $\sigma_c$, to a rheological property, E, a surface property, $\gamma_s$, and a microgeometry, a. According to Griffith, the fracture criterion is that if the applied stress exceeds the critical level, the crack is free to propagate spontaneously without limit, i.e.,

$$\sigma \geq \sigma_c. \qquad \text{(crack propagation)}$$

Thus, Equation (10) is a necessary condition for elastic crack propagation. Drucker[14] and Craggs[15] have pointed out that the Griffith criterion is a necessary but not sufficient condition for crack propagation. For a crack which becomes unstable and propagates spontaneously, it is true that the energy-balance relationship is satisfied, but the converse is not necessarily true. A second condition appears to be necessary. Drucker[14] suggests that the true local maximum stress must reach a critical value, or Mylonas[16] argues that the local strain must

# SURFACE FREE ENERGIES OF GLASSY POLYMERS

reach a critical value. However, Craggs[15] pointed out that from a design point of view, it is not unreasonable, by assuming the worst, to take the Griffith criterion of fracture as a sufficient one.

## 3. Irwin-Orowan's Modification of Griffith's Concept

For materials which display slight ductility so that there is a plastic flow at the crack tip during crack propagation, Griffith's equation no longer applies, and other approximations are required to define the stress field at the crack tip. After 1950, Irwin[9] and Orowan,[8] working independently, proposed two hypothetical zones for the crack system (Figure 4): the outer, surrounding zone, consisting of exclusively linear elastic material, which transmits the applied forces to the inner crack-tip zone, where the nonlinear separation processes dominate.

For a stationary crack of length, c, and width, unity, the tip is surrounded by a nonlinear zone of dimension $\ll$ a and the energetics is extended by an increment of $\delta$ a. The system energetics may be treated as two separate (but not independent) parts. First, if we exclude the inner zone as being negligibly

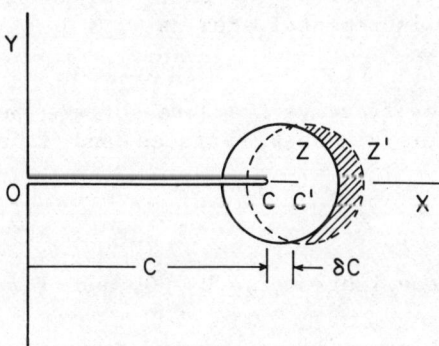

Figure 4. Irwin-Orowan Basic Zone Model.

small, the mechanical energy variation term, $\delta(-W_L + U_E)$, for the system is assumed to be for the ideally linearized system. Second, we can postulate that the nonlinear forces which do work ($\delta U_S$) in creating the new fracture surfaces should include a dissipative component in the process. Similar to Equation (3), we obtain

$$\delta U \simeq \delta(-W_L + U_E) + \delta U_S$$
$$\simeq (d(-W_L + U_E)/da)\delta a + (dU_S/da)\delta a. \quad (11)$$

Let

$$G = -d(-W_L + U_E)/da. \quad (12)$$

Then

$$\delta U = -G\delta a + (dU_S/dc)\delta a, \quad (13)$$

Here G is basically an energy, or the crack extension force, and G does not depend on the loading type. We may be able to consider the fixed-grips case as one of the general situations. Then, Equation (12) becomes

$$G = -(\partial U_E/\partial a)_U. \quad (14)$$

Thus, generally, G is the strain-energy-release rate per unit width of the crack front. G is defined as the derivative of the mechanical energy released with respect to the crack area rather than the length, but since we shall be dealing mostly with plane cracks in plates of unit thickness, we may conveniently take the incremental area as $\delta a \times 1$.

We now define the effective fracture surface energy, $\Gamma$, as the energy in creating two new surfaces and it's quantity becomes

$$-2\Gamma = -dU_S/da. \quad (15)$$

At the thermodynamic equilibrium, $\delta U = 0$, and $G = G_c$, and

$$G_c = 2\Gamma. \quad (16)$$

$G_c$ is the fracture energy. It has also been called the critical strain-energy release rate (per unit width of the crack front) for unstable crack extension, or the "toughness", or the crack resistance force of the material. The effective fracture surface energy, $\Gamma$, has also been called "fracture toughness", or "work of fracture:, or "fracture resistance". These misnomers have added much confusion to the real meanings of G and $\Gamma$. On the basis of the Irwin-Orowan's concept, we can derive a new criterion for fracture; that is, the crack will only extend when G is equal to or exceeds the fracture energy. This value of G is $G_c$. The criterion for fracture in any system must then be

$$G \geq G_c = 2\Gamma.$$

Since the effective fracture surface energy includes chiefly the dissipative component, or the work of plastic deformation, $W_d$, and a minor component, surface free energy of a solid, $\gamma_s$, we may write

$$G_c = 2(\gamma_s + W_d). \tag{17}$$

Similarly (to Griffith's equation), Orowan derived the following equation:

$$\sigma_c = \left(\frac{EG_c}{\pi a}\right)^{1/2} = \left[\frac{2E(\gamma_s + W_d)}{\pi a}\right]^{1/2}. \tag{18}$$

Thus, the fracture criterion remains the same as that for the Griffith's criterion:

$$\sigma \geq \sigma_c.$$

## 4. Linear Elastic Crack-tip Stress Field

The important feature of most stress analyses of cracks is the nature of the stress field near the crack tip. Irwin[17] in 1958 used Westergaard analytical method[18] to derive the following equations for the stresses at the tip of the crack under Mode I condition (see Figure 1):

$$\sigma_x = \frac{K_I}{(2\pi r)^{1/2}} \cos(\theta/2) \left[1-\sin(\theta/2)\sin(3\theta/2)\right], \tag{19a}$$

$$\sigma_y = \frac{K_I}{(2\pi r)^{1/2}} \cos(\theta/2)\left[1+\sin(\theta/2)\sin(3\theta/2)\right], \quad (19b)$$

$$\tau_{xy} = \frac{K_I}{(2\pi r)^{1/2}} \sin(\theta/2)\cos(\theta/2)\cos(3\theta/2), \quad (19c)$$

$$\sigma_z = \nu(\sigma_x + \sigma_y) \text{ (plain strain)}, \quad (19d)$$

$$\sigma_{xz} = 0 \text{ (plain strain)}, \quad (19e)$$

$$\tau_{xz} = \tau_{yz} = 0. \quad (19f)$$

In the above equations, r and $\theta$ are the usual polar coordinates measured from the crack tip. K is a parameter influenced by the configuration, loading, and elastic properties of the specimen. K has been called the stress-intensity factor with a unit of stress X(length)$^{1/2}$. $K_I$ refers to K under Mode I condition.

K can be determined by a limiting process with $\theta = 0$ from Equation (19a):

$$K_I = \lim_{r \to 0} (2\pi r)^{1/2} \sigma_x, \quad (20)$$

where r is the distance from the tip of the crack. For an infinitely sharp elastic crack, with an internal crack length of 2a in an infinitely wide plate, the stress-intensity factor, K, is defined as

$$K = \sigma(\alpha \pi a)^{1/2}, \quad (21)$$

where $\alpha$ is the geometric parameter depending on the specimen and crack geometry; for a central crack in a thin plate, $\alpha = 1.0$. At $K_c$, a critical value of K, the existing crack may indeed become unstable and start to propagate. Thus, another fracture criterion is

$$K \geqq K_c \quad (22)$$

Here, $K_c$ is defined as the fracture toughness of the material. $K_c$ has also been called the critical stress-intensity factor. Therefore, stress is to strength as the stress-intensity factor is to fracture toughness.

The parameter, $K_c$, is a measure of the strength required for crack propagation in the material. Thus, it is expected that, in the rate-dependent materials, such as polymers, $K_c$ will vary with the loading rate, environment and temperature. We shall discuss some of these relationships in a later section.

Frequently, $K_c$,[19] not $G_c$, is used by designers of structural materials. As one of the fracture criteria, the fracture condition for an infinitely large cracked plate would be

$$K \geq K_c = \sigma (\pi a)^{1/2}. \qquad (23)$$

- $K$: Material selection
- $K_c$: (Material selection)
- $\sigma$: Design stress
- $a$: Allowable flaw size

The designer can either fix $K_c$, the material, or the design stress, $\sigma$, or the flaw size, $a$, and then estimate the other two parameters.

## 5. Relationships between G and K

Irwin[9] demonstrated that a simple relationship exists between the stress state near the crack tip and the rate of release of elastic strain energy, G, with respect to the crack areas of an isotropic material. For a central crack in a thin plate with a crack length of 2a,

$$K = \sigma (\pi a)^{1/2} \qquad (24)$$

and

$$\sigma = \left(\frac{EG}{\pi a}\right)^{1/2}. \qquad (25)$$

Thus,

$$K = (EG)^{1/2}. \qquad (26)$$

At the critical condition for Mode I (see Figure 1) and thin plate (plane stress),

$$K_{Ic} = (EG_{IC})^{1/2}. \qquad (27)$$

Similarly, for a central crack in a thick plate (plain strain),

$$K = \left(\frac{EG}{1-\nu^2}\right)^{1/2} \tag{28a}$$

and

$$K_{Ic} = \left(\frac{EG_{Ic}}{1-\nu^2}\right)^{1/2}. \tag{28b}$$

These relationships between G and K are not coincidental. On the basis of a detailed analysis by Irwin,[9] these relationships have been shown to be valid.

6. Additivity of G and K

Fracture energies for different modes of fracture have been shown to be additive.[12] For thin plates (plane stress):

$$G = K_I^2/E + K_{II}^2/E + K_{III}^2(1+\nu)/E \tag{29}$$

and for thick plates (plane strain):

$$G = K_I^2(1-\nu^2)/E + K_{II}^2(1-\nu^2)/E$$

$$+ K_{III}^2(1+\nu)/E. \tag{30}$$

Since the opening mode (Mode I) is the most common type of fracture of brittle materials, e.g., glassy polymers, we shall limit our discussion to $G_{Ic}$ and $K_{Ic}$ in the following sections.

III. EXPERIMENTAL DETERMINATION OF FRACTURE ENERGY

Based on linear elastic fracture mechanics,[10] the displacement, $\ell$, in the direction of a wedge-load, P, is

$$\ell = CP, \tag{31}$$

where C is the compliance of the specimen. Under a fixed load condition, any change in $\ell$ may be related to a change in the crack length, a (in the case of an edge crack), by

$$\delta\ell = P\frac{dC}{da} \delta a. \tag{32}$$

The strain energy, $\epsilon = 1/2\, P\, \delta\ell$, is related to G:

$$\epsilon = 1/2 \, P \, \delta \ell = G \, b \, \delta a, \tag{33}$$

where b is the specimen thickness. From Equations (32) and (33), we obtain

$$G = \frac{P^2}{2b} \frac{dC}{da}. \tag{34}$$

Once dC/da is determined for a specific specimen geometry, G can be obtained to relate it directly to P and other dimensional parameters. Based on the above relationships, the double-cantilever beam has been the one most successfully used for measuring the fracture energy of material.

## IV. PARAMETERS AFFECTING EFFECTIVE FRACTURE SURFACE ENERGY, AND FRACTURE TOUGHNESS OF POLYMETHYL METHACRYLATE

In the following sections, we shall discuss the effects of molecular weight, crack speed and molecular orientation on effective fracture surface energy, fracture energy and fracture toughness of glassy polymers. Polymethyl methacrylate will be used for illustration. Other variables, e.g., temperature, sample thickness, etc., have been well discussed in the literature,[21] we do not include them in this section.

1. Effect of Molecular Weight on Effective Surface Fracture Energy, Fracture Energy and Fracture Toughness of Polymethyl Methacrylate

The effect of molecular weight on effective fracture surface energies of glassy polymers has been discussed by Kusy and Katz.[20] According to them, the total effective fracture surface energy measurement depends on three molecular events: the amount of sliding of one macromolecule past another (i.e., glide), the number of fractures (or breaks) of covalent bonds and the extent of physical interpenetration of networks (i.e., entanglements). Kusy and Katz[20] propose that the fracture surface energy, $\Gamma$, consists of three terms:

$$\Gamma = \Gamma_o + \Gamma_1 + \Gamma_2, \tag{35}$$

where $\Gamma_o$ is the cohesive surface free energy (as $\gamma_s$ in the Griffith's equation); $\Gamma_1$ is the fracture surface energy of breaking and $\Gamma_2$ is the fracture surface energy of deformation (as $W_d$ in the Irwin-Orowan equation). Since $\Gamma_1$ and $\Gamma_2$ are

Table I. Effect of Molecular Weight on Effective Fracture Surface Energy and Fracture Toughness of Polymethyl Methacrylate.

| $M_v \times 10^{-5}$ | $\Gamma^*$ ($10^5 \text{erg/cm}^2$) | $E^*$ ($10^{10} \text{dyne/cm}^2$) | $G_C$ (calc) ($10^5 \text{erg/cm}^2$) | $K_{IC}$ (calc) ($10^7 \text{dyne/cm}^{3/2}$) |
|---|---|---|---|---|
| 0.98 | 1.14 | 2.29 | 2.28 | 7.2 |
| 1.1 | 1.24 | 2.45 | 2.48 | 7.8 |
| 1.8 | 1.33 | 2.47 | 2.66 | 8.1 |
| 4.2 | 1.45 | 2.53 | 2.90 | 8.6 |
| 13 | 1.50 | 2.93 | 3.00 | 9.4 |
| 30 | 1.35 | 3.06 | 2.70 | 9.1 |
| 60 | 1.56 | 2.92 | 3.12 | 9.5 |

*Data of fracture surface energy and modulus were adopted from Table IIB-II by J. P. Berry in "Fracture Processes in Polymeric Solids," B. Rosen, Editor, p. 213, Wiley, New York, 1964.

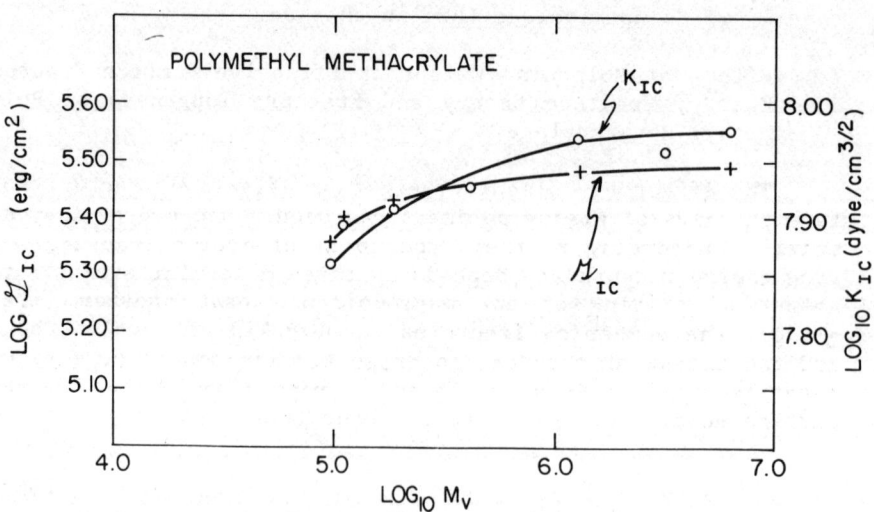

Figure 5. Relationship between fracture energy and viscosity-average molecular weight.

influenced by the polymer structure, the molecular weight can affect the overall fracture surface energy.

The effect of molecular weight on fracture energy and fracture toughness of polymethyl methacrylate (PMMA) is illustrated with the data calculated from the effective fracture surface energy obtained by Berry[21] and shown in Table I. Both $G_{IC}$ and $K_{IC}$ are plotted versus the logarithmic viscosity-average molecular weight ($\ln M_v$) of PMMA. Both curves rise to a plateau (Figure 5).

2. Effect of Crack Speed on Effective Fracture surface Energy, Fracture Energy and Fracture Toughness of Polymethyl Methacrylate

The effect of crack speed on fracture toughness of PMMA has been tabulated by Marshall and Williams.[22] From their data, we calculated the effective fracture surface energy and the fracture energy as shown in Table II. In Fig. 6, fracture toughness[22] is shown to increase with the increase in crack speed, $\dot{a}$. Williams[23] used the analytic method to relate $K_{IC}$ and the strain rate at the tip of a propagating crack by the following equation:

$$\dot{\epsilon} = \pi \epsilon_y^3 \left[ E(t)/K_{Ic} \right]^2 \dot{a}, \qquad (36)$$

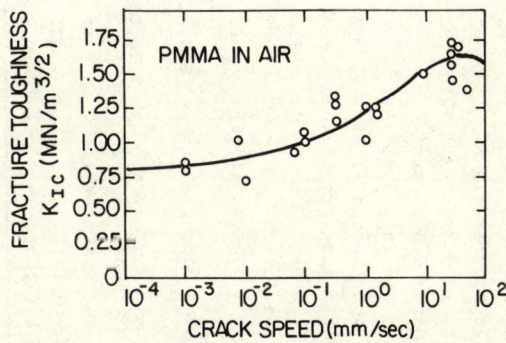

Figure 6. Dependence of fracture toughness on crack speed (Data of G.P. Marshall and J.G. Williams)[22].

Table II. Effective Fracture Surface Energies, Fracture Energies and Fracture Toughness of Polymethylmethacrylate*.

| Author(s) | Test Method | $\dot{a}$ (mm/sec) | $\Gamma$ x $10^2$ or x $10^5$ (Joule/cm$^2$) (erg/cm$^2$) | $G_c$ (calc) x $10^5$ erg/cm$^2$ | $K_{IC}$ (calc) (MN/M$^{3/2}$ or x $10^8$ dyne/cm$^{3/2}$) |
|---|---|---|---|---|---|
| Benbow and Roesler (1956) | PC(c) | $\sim 10^{-2}$ | 4.9 | 9.8 | [1.69]† |
| Benbow (1961) | CNC(c) | $10^{-2}$ | 4.2 | 8.4 | [1.56] |
| Berry (1961) | SEN | Instability | 3.0 | 6.0 | [1.12] |
| Berry (1963) | PC | $10^{-3}$ | 1.4 | 2.8 | (0.76)† |
| van den Boogaart (1966) | PC | $\sim 10^{-3}$ | 1.65 | 3.3 | (0.83) |
| Broutman and McGarry (1965) | PC | 1.25 | 1.25 | 2.5 | (0.99) |
| Broutman and Kobayashi* | TC | 1.25 | 2.0 | 4.0 | (1.26) |
| Davidge and Tappin (1968) | B | Instability | (3.65) | (7.3) | 1.94 |
| Olear and Erdogan (1968) | CN | Instability | (1.39) | (2.78) | 1.19 |
| Key, Katz and Parker (1968) | SEN | Instability | (1.15–2.7) | (2.3–5.4) | 1.09–1.66 |
| Kies (1953) | SEN | Instability | 6.15 | (12.3) | [1.19] |
| Svensson (1961) | CNC(c) | $\sim 10^{-2}$ | 4.5 | 9.0 | [1.16] |
| Vincent and Gotham (1966) | SEN PC, I | $6.8 \times 10^{-2}$ $4.3 \times 10^1$ | 1.5–3.4 | 3.0–6.8 | [0.93–1.6] |
| Williams, Radon, and Turner (1968) | SEN, DEN, B | $\sim 2.5 \times 10^{-3**}$ | (2.2–3.5) | (4.4–7.0) | 1.13–1.43 |
| Williams, Randon, and Turner (1968) | SEN, DEN, B | – | (2.11–3.0) | (4.22–6.0) | 1.48–1.75 |
| Irwin and Kies (1954) | CN | Instability | 4.4 | 8.8 | [1.6] |
| Fujishiro (1971) | PC | $\sim 10^{-2}$ | 0.923 | 1.85 | (0.63) |
| Higuchi (1965) | DEN | Instability | 3.6 | 7.2 | [1.7] |

†() = Converted value-using "derived" E (via Williams [1972]).  * Private communication.
[ ] = Converted value-using quoted E.  ** "Apparent" speed.

CNC = centre notch cleavage  SEN = single edge notch  I = impact
PC = parallel cleavage  (c) = with compression  B = bending
SEN = tapered cleavage  CN = centre notched  DEN = double edge notch

* Ref: G.P. Marshall and J.G. Williams, J. Mat. Sci. <u>8</u>, 138 (1973).

where $\dot{\varepsilon}$ is the effective strain rate; $\varepsilon_y$ is the yield strain and $E(t)$ is the time-dependent modulus.

3. Effect of Molecular Orientation on Effective Fracture Surface Energy

For the study of the effect of orientation on fracture surface energy, PMMA sheets were biaxially oriented in their planes.[21] Effective fracture surface energies for three mutually perpendicular planes (Figure 7) were measured by Berry.[21] Their results were used to calculate fracture energies (Table III). Three directions of the oriented sheet have much higher effective fracture surface energies than that of the d-direction.

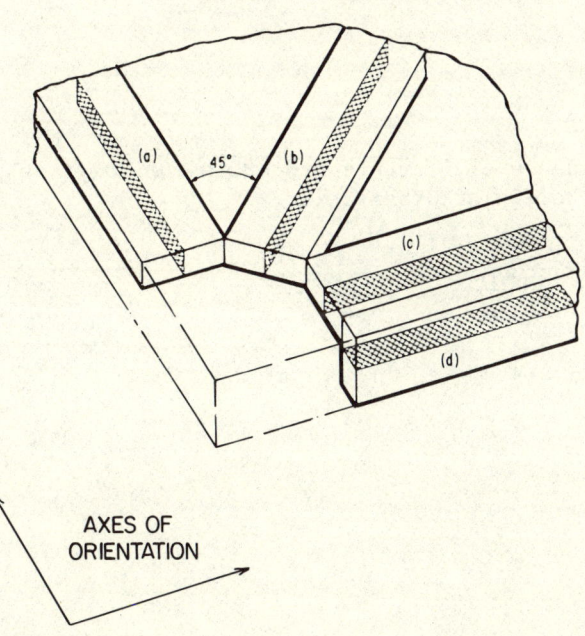

Figure 7. Dependence of fracture energy on molecular orientation (Ref. J.P. Berry in "Fracture Processes in Polymeric Solids," B. Rosen, Editor p. 213, Wiley, New York (1964).

All but the d-direction have fracture surface energies three to four times higher than that of the unoriented film. The same trend applies to the fracture energy. Thus, biaxial orientation, in general, increases the crack resistance of polymeric films.

## V. RELATIVE RANKING OF FRACTURE ENERGIES OF GLASSY POLYMERS

Since fracture energy is affected by the molecular weight, crack speed, molecular orientation, etc., ideally we should rank fracture energies of glassy polymers under the same conditions. Since this is not possible when literature values[24,25] are compared, we attempt to use the range of fracture energies instead of a single value for each polymer.

Table III. Dependence of Fracture Properties of Poly(methyl methacrylate) on Molecular Orientation.

| Orientation | $E \times 10^{-5}$ (psi) | $\Gamma \times 10^{-5}$ (erg/cm$^2$) | $G_c$ (calc) $\times 10^{-5}$ (erg/cm$^2$) |
|---|---|---|---|
| a | 4.65 | 6.5 | 13.0 |
| b | 4.30 | 5.8 | 11.6 |
| c | 4.70 | 7.8 | 15.6 |
| d | — | 0.17 | 0.34 |
| Unoriented | 4.45 | 1.4 | 2.8 |

Ref: J.P. Berry in "Fracture Processes in Polymeric Solids", B. Rosen, Editor, p 213, Wiley, New York (1964).

# SURFACE FREE ENERGIES OF GLASSY POLYMERS

In Figure 8, we can readily rank the following polymers with a decreasing fracture energy:

polyvinylchloride[26] > polycarbonate [27,28] > cellulose acetate[17] > polystyrene[24] > polymethyl methacrylate > unmodified epoxy resin.[25]

In the general ranking including metals and composites, polyvinyl chloride appears to be as tough as some metals.

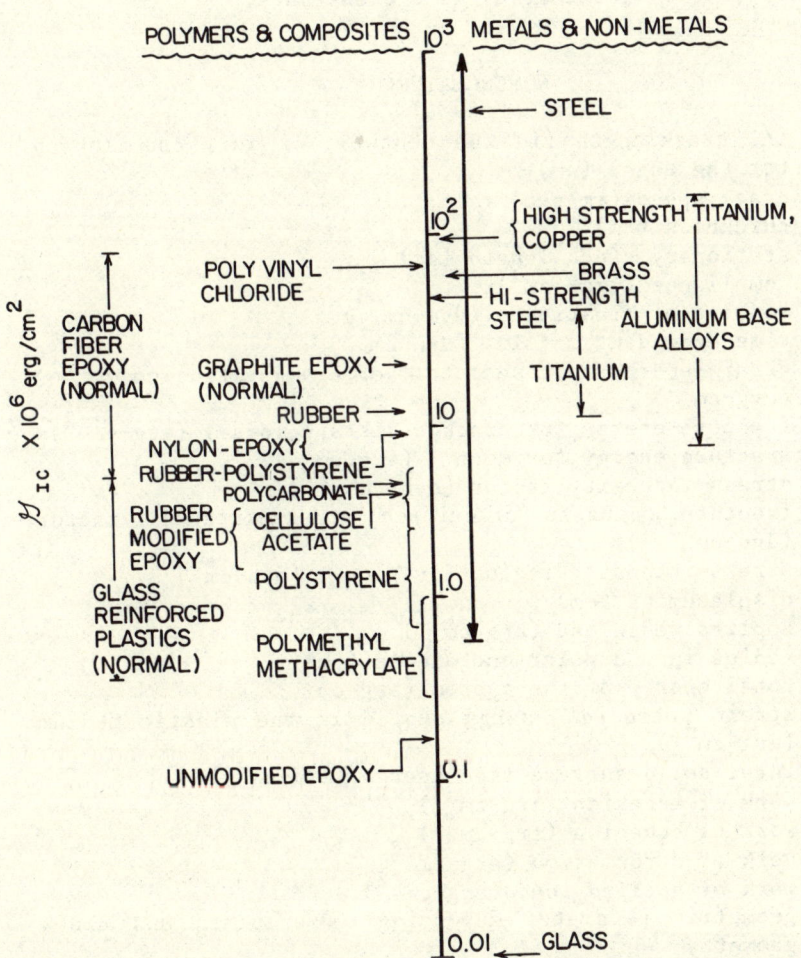

Figure 8. Relative ranking of fracture energies of glassy polymers versus metals and nonmetals.

## VI. CONCLUSIONS

In this brief survey, we have introduced several basic concepts of fracture mechanics. We have defined several important parameters, e.g., effective fracture surface energy, fracture energy and fracture toughness. For design purposes, fracture toughness of polymers should be the most useful parameter. Three variables, e.g., molecular weight, crack speed and orientation were discussed with polymethyl methacrylate as an example. Fracture energies of glassy polymers were compared with those of metals and non-metals.

The results indicate that glassy polymers can be toughened by increasing the molecular weight, or by applying the biaxial orientation, or by reinforcing with elastomers.

## NOMENCLATURE

$a$ = 1/2 cracklength (in the center), or full cracklength (at the edge) (cm)
$\dot{a}$ = crack speed (mm/sec)
$b$ = thickness (cm)
$c$ = stationary crack length (cm)
$C$ = compliance (erg-cm$^3$)
$E$ = modulus of elasticity (dyn-cm$^{-2}$)
$E(t)$ = time-dependent modulus (dyn-cm$^{-2}$)
$G$ = strain-energy release rate, (or crack extension force) (erg-cm$^{-2}$)
$G_c$ = fracture energy (or crack resistant force) (erg-cm$^{-2}$)
$G_I^c$ = fracture energy for Mode I (erg-cm$^{-2}$)
$K$ = stress-intensity factor (dyn-cm$^{-3/2}$)
$K_c$ = fracture toughness (or critical stress-intensity factor (dyn-cm$^{-3/2}$)
$K_I$ = stress-intensity factor for Mode I (dyn-cm$^{-3/2}$)
$\ell$ = displacement (cm)
$P$ = applied wedgeload (erg cm$^{-2}$)
$r$ = radius in the polar coordinate (cm)
$U$ = total energy of the system (erg-cm$^{-2}$)
$U_E$ = strain potential energy stored in the elastic medium (erg-cm$^{-2}$)
$U_s$ = total solid surface free energy (erg-cm$^{-2}$)
$W_A$ = work of adhesion (erg-cm$^{-2}$)
$W_c$ = work of cohesion (erg-cm$^{-2}$)
$W_d$ = work of deformation (erg-cm$^{-2}$)
$W_L$ = work of applied load (erg-cm$^{-2}$),
$\alpha$ = geometric parameter depending on specimen and crack geometry
$\gamma$ = surface free energy per unit area (dyn-cm$^{-1}$)
$\gamma_s$ = surface free energy of solid per unit area (dyn-cm$^{-1}$)

$\Gamma$ = effective fracture surface energy (erg-cm$^{-2}$)
$\Gamma_0$ = cohesive surface free energy (erg-cm$^{-2}$)
$\Gamma_1$ = fracture surface energy of breaking (erg-cm$^{-2}$)
$\Gamma_2$ = fracture surface energy of deformation (erg-cm$^{-2}$)
$\epsilon$ = strain energy (erg-cm$^{-2}$)
$\dot{\epsilon}$ = effective strain rate (cm$^{-1}$)
$\epsilon_y$ = yield strain
$\theta$ = angle of the polar coordinate (deg)
$\nu$ = Poisson's ratio
$\sigma$ = applied tensile stress (dyn-cm$^{-2}$)
$\sigma_c$ = fracture stress, or critical tensil stress at fracture (dyn-cm$^{-2}$)
$\tau$ = shear stress (dyn-cm$^{-2}$)

## REFERENCES

1. L.H. Lee, Editor, "Recent Advances in Adhesion," p. 35, Gordon and Breach, 1973.
2. A. Dupre, "Theorie Mechanique de la Chaleur", p. 393, Gauthier-Villars, Paris 1869.
3. J.E. McNutt, Adhesives Age, 7, 24 Oct. 1964.
4. S. Wu, in "Recent Advances in Adhesion", L. H. Lee, Editor, p. 45, Gordon and Breach, 1973.
5. S. Wu, J. Macromol. Sci.-Revs. Macromol. Chem., C10, 1 (1974).
6. T. Hata and T. Kasemura, in "Adhesion and Adsorption of Polymers," L.H. Lee, Editor, Plenum Press, New York 1979.
7. A.A. Griffith, Phil. Trans., Roy. Soc. 221, 163(1920).
8. D. Orowan, Reports Prog. Phys. XII, 185 (1948).
9. G.R. Irwin, J. Appl. Mech., 24, 361 (1957).
10. B.R. Lawn and T.R. Wilshaw, "Fracture of Brittle Solids," Cambridge University Press, London, 1975.
11. A.S. Tetelman and A.J. McEvily, Jr. "Fracture of Structural Materials," Wiley, New York 1967.
12. G.P. Anderson, S.J. Bennett and K.L. DeVries, "Analysis and Testing of Adhesive Bonds," Academic Press, New York, 1977.
13. C. Inglis, Trans. Inst. Naval Archit., London, 55, 219 (1913).
14. D.C. Drucker, Report No. SSC-69, Ship Structure Committee, Dept. of Navy, Washington, D.C.
15. J.W. Craggs, "Fracture of Solids," pp. 51-63, Wiley Interscience, New York, 1963.
16. C. Mylonas, in "Proceedings of 11th International Congress of Applied Mechanics", Munich, 1964.
17. G.R. Irwin, in "Handbuch der Physik", Vol. VI, pp. 551-590, Springer-Verlag, Berlin 1958.
18. H.M. Westergaard, Trans. ASME, J. Appl. Mech., 61, 49 (1939).

19. R.W. Hertzberg, "Deformation and Fracture Mechanics of Engineering Materials," Wiley, New York 1976.
20. R.P. Kusy and M.J. Katz, Polymer, 19, 1345 (1978).
21. J.P. Berry, in "Fracture Processes in Polymeric Solids," B. Rosen, Editor, p 213, Wiley, New York, 1964.
22. G.P. Marshall and J.G. Williams, J. Materials Sci. 8, 138 (1973).
23. J.G. Williams, Int. J. Fract. Mech., 8, 393 (1972).
24. R.S. Seth and D.H. Page, J. Materials Sci., 9, 1745 (1974).
25. W.D. Bascom, R.L. Cottington and C.O. Timmons, Appl. Polymer Sym. 32, 165 (1977).
26. R.J. Ferguson, G.P. Marshall and J.G. Williams, Polymer 14, 451 (1973).
27. D.G. Legrand, J. Appl. Polymer Sci., 13, 2129 (1969).
28. S. Arad, J.C. Randon and L.E. Culver, J. Appl. Polymer Sci., 17, 1467 (1973).

# THE WORK OF ADHESION AND THE FRACTURE SURFACE ENERGY OF EPOXY-POLYCARBONATE ADHESIVE JOINTS

B.W. Cherry

Department of Materials Engineering
Monash University
Victoria, 3168. Australia

When an adhesive cures, shrinkage stresses are set up at the adhesive/adherend interface, and the relief of these shrinkage stresses by the propagation of a crack adds an additional component to the strain energy release rate which drives the crack. Since the effect of the "shrinkage stress strain energy release rate" is to lower the apparent fracture surface energy for the joint, a technique has been devised to measure its magnitude for an epoxy/polycarbonate joint system. The value obtained has been confirmed by comparison with a system in which the relief of shrinkage stresses can not contribute to crack extension.

Since shrinkage stresses play a dominant role in the environmental fracture of joints with a negative work of adhesion, it was of interest to see whether they could cause failure in a system for which this was positive. Although contact angle measurements showed that the work of adhesion of epoxy/polycarbonate bonds in water was only $+29\,\mathrm{mJ\,m^{-2}}$, the shrinkage stresses did not bring about disbonding on prolonged exposure to water.

## INTRODUCTION

The apparent contradiction involved in invoking the work of adhesion, defined by the methods of reversible thermodynamics, to explain the strength of adhesive joints which fail by a highly irreversible fracture process has been largely resolved by such workers as Andrews and Kinloch[1,2], Gent and Schulz[3], Gledhill and Kinloch[4] and Kaelble[5]. They have suggested that contrary to the earlier views of Orowan[6], the adhesive fracture surface energy ($R_A$) is not represented by an expression of the form

$$R_A = w_a + P$$

where $w_a$ is the reversible work of adhesion and P the energy dissipated irreversibly in bringing about the plastic deformation of the crack tip. They have suggested that the fracture surface energy is (following the generalised fracture mechanics of Andrews[7] better represented by an expression of the form

$$R_a = w_a \, f(\dot{a}, T, \sigma_o \ldots) \qquad (1)$$

where the term $f(\dot{a}, T, \sigma_o \ldots)$ represents some as yet undetermined function of the crack propagation rate, the temperature, the remotely applied stress and possibly other factors. Equation (1) thus demonstrates the dependence of the fracture surface energy on the work of adhesion and provides a link between the thermodynamic and fracture mechanic approaches to joint rupture.

Although Equation (1) may represent a theoretical relation between the thermodynamics of joint formation and the actual strength of an adhesive joint, its utility in practice is hindered by a number of major difficulties, and it is the object of the present paper to discuss two of these.

$R_a$ in Equation (1) is the fracture surface energy for a crack which is propagating stably at a constant velocity. $R_a$ can only be related to the strength of the joint by the standard methods of fracture mechanics if all the sources of energy which are available to assist propagation are well characterised. In particular, if the propagation of a crack allows the relief of the stresses set up by the differential shrinkage of adhesive and adherend during curing, then this additional component of crack extension force can reduce the stress which must be applied to the system by external agencies to cause the crack to propagate. It has been suggested[8] that whereas the classical condition for crack propagation is

$$G_c = R_a$$

where $G_c$ is the critical fixed grip strain energy release rate resulting from an applied force; in a system in which shrinkage stresses are being relieved by the action of the propagation of a crack, the criterion should be written

$$G_c' = R_a - G_s \tag{2}$$

where $G_s$ is the shrinkage strain energy release rate and $G_c'$ is the actual fixed grip strain energy release rate at which the crack propagates. Consequently before Equation (1) can be used for the prediction of joint strength, it is necessary to be able to determine $G_s$. One object of this paper is to describe a technique for the determination of the shrinkage strain energy release rate and to demonstrate is validity.

A second problem which arises as a result of the presence of shrinkage stresses in an adhesive joint, is the role which these stresses play in assisting the disbonding of joints which are thermodynamically unstable.

Gledhill and Kinloch[4] have calculated the work of adhesion for an epoxy/aluminium joint in air and in water and have suggested that since

$$(w_a)_{air} = \gamma_a + \gamma_s - \gamma_{as} = 232 \text{ mJ m}^{-2}$$

$$\text{and } (w_a)_{water} = \gamma_{aw} + \gamma_{sw} - \gamma_{as} = -137 \text{ mJ m}^{-2} \tag{3}$$

a joint which is formed in air should spontaneously disbond in water. Cherry and Thomson[9] have however pointed out that the results of the work of Mostovoy, Ripling and Bersch[10] suggest that there is a critical strain energy release rate below which crack propagation will not take place. For epoxy/aluminium joints in water, Mostovoy, Ripling and Bersch have suggested that this minimum strain energy release rate is of the order of 1 J m$^{-2}$. It is suggested that the discrepancy between the results of Mostovoy, Ripling and Bersch and the predictions of Gledhill and Kinloch lies in the fact that Equation (3) is only concerned with the free energy difference between the initial (bonded) state and the final (disbonded) state of the system and neglects the energy which must be expended to drive the system at a given rate over the activation energy barrier which may lie between these states. On this reasoning the 1 J m$^{-2}$ figure for the minimum strain energy release rate to achieve disbonding, represents the necessary energy dissipation to bring about the disbonding with observable kinetics. If this is the case, then Cherry and Thomson[9] have queried how a free energy decrease of 137 mJ m$^{-2}$ can bring about a process requiring 1 J m$^{-2}$ and have suggested that the additional source of energy for the propagation of an interfacial crack may derive from the relief of shrinkage stresses. The results of Cherry and Thomson[9,11] suggest therefore that in the case of a joint such as epoxy/aluminium which is unstable thermodynamically when immersed in water, the joint may still remain intact unless an applied stress (which may well derive from shrinkage stresses) activates the bond scission process so that it can occur within a reasonable time scale. In particular Thomson[12], has shown that an epoxy/aluminium joint which has been formed in such a way as to minimise the shrinkage stresses (the substrate is stretched prior to joint formation and then allowed to contract after joint formation so that the shrinkage of the substrate matches that of the adhesive)will remain intact although immersed in water for very long periods.

What is not revealed by the earlier work of Cherry and Thomson is whether a joint which is thermodyanmically stable in water can be disrupted by the operation of shrinkage stresses acting alone. It is to this problem that the present paper is also addressed.

## EXPERIMENTAL

### Materials

Epoxy/polycarbonate joints were used throughout this study. They were formed using the technique of Cherry and Hang[12]. The casting resin Araldite D with a diamine hardener HY 951, both manufactured by Ciba-Geigy was chosen as the adhesive component. This resin has a low viscosity and can be cast easily at room temperature. The problem of trapped air bubbles formed during pouring is minimised by the low viscosity. The resin is fairly transparent allowing easy location of the crack tip and has a high stress-optical activity making it suitable for photo-elastic stress analysis. A relatively low curing temperature of 30°C was used and thus the adhesive should exhibit comparatively small shrinkage stresses.

The polycarbonate was Lexan 144 manufactured by General Electric. Before any measurements were made the polycarbonate surface was degreased with methyl alcohol, polished with fine grit sand-paper (No. 600) and cleaned again with methyl alcohol.

### The Work of Adhesion of the Epoxy/Polycarbonate Joint

The work of adhesion for an immersed joint is given by $(w_a)_w$ where

$$(w_a)_w = \gamma_{aw} + \gamma_{sw} - \gamma_{as} \qquad (3)$$

In Equation (3), $\gamma_{aw}$ and $\gamma_{sw}$ are the interfacial energies for the adhesive and the substrate respectively in contact with the immersing liquid (in this experiment water).

The Young-Dupré equation gives

$$\gamma_{aw} = \gamma_a - \gamma_w \cos \theta_{aw}$$

and  $\gamma_{sw} = \gamma_s - \gamma_w \cos \theta_{sw}$

where $\theta_{aw}$ and $\theta_{sw}$ are the contact angles made by the immersing liquid on the solidifed adhesive and the substrate. Since Owens and Wendt[13] have written the interfacial energy of a system as

$$\gamma_{as} = \gamma_a + \gamma_s - 2(\gamma_a^d \gamma_s^d)^{\frac{1}{2}} - 2(\gamma_a^p \gamma_s^p)^{\frac{1}{2}} \qquad (4)$$

where the superscripts d and p refer to the dispersion and polar components of the surface energy, Equation (2) may be rewritten as

$$(w_a)_w = 2(\gamma_a^d \gamma_s^d)^{\frac{1}{2}} + 2(\gamma_a^p \gamma_s^p)^{\frac{1}{2}} - \gamma_w(\cos \theta_{aw} + \cos \theta_{sw}) \qquad (5)$$

Hence from Equation (5) it can be seen that $(w_a)_w$ can be deduced from a knowledge of the polar and dispersion force components of the adhesive and substrate and the contact angles made by the immersing liquid on them.

The polar and dispersion force components of the surface energy of a solid can be determined using Owens and Wendt's Equation (4) written in the form

$$w_a = 2(\gamma_s^d \gamma_{\ell_1}^d)^{\frac{1}{2}} + 2(\gamma_s^p \gamma_{\ell_1}^p)^{\frac{1}{2}} = \gamma_{\ell_1}(1 + \cos \theta_{\ell_1}) \qquad (6)$$

where $\gamma_s^p$ and $\gamma_s^d$ are the polar and dispersion force components of the surface energy of the solid and $\theta_{s\ell_1}$ is the contact angle made by liquid $\ell_1$ on the surface of the solid. Hence if the contact angles made by two liquids (whose polar and disperson force contributions to the surface energy are known) are measured it is possible to determine the components of the surface energy of the solid.

Glycerol and water were used as the test liquids and contact angles were measured from photographs of drops of the liquid on each of the surfaces being examined. The results are shown in Table I.

Table I.

| Liquid | $\gamma_\ell$ | $\gamma_\ell^p$ | $\gamma_\ell^d$ | Contact Angle on epoxy | Contact Angle on Polycarbonate |
|---|---|---|---|---|---|
| Water | 72.2 | 50.2 | 22.0 | 52° | 81° |
| Glycerol | 64.0 | 30.0 | 34.0 | 49° | 66° |

By the simultaneous solution of Equation (6) for the two liquids, the components of the surface energies of the polycarbonate and the epoxy were found to be as given in Table II.

Table II.

| Solid | $\gamma^p$ | $\gamma^d$ | $\gamma$ |
|---|---|---|---|
| Epoxy | 33.5 | 13.3 | 46.8 |
| Poly-carbonate | 5.8 | 33.5 | 39.3 |

By substitution in Equation (3) or Equation (5) the work of adhesion for an epoxy/polycarbonate joint is determined as $w_a$ = 70 mJ m$^{-2}$ $(w_a)_w$ = 29 mJ m$^{-2}$.

## The Fracture Surface Energy

If a crack spreads from the centre of an interface then, since at points remote from the spreading crack, the adhesive is still constrained to the dimensions of the adhesive, little relief of shrinkage strain is possible, at least during the early stages of crack propagation. The disadvantage of determining the adhesive fracture surface energy by means of a crack spreading from the centre of an interface is that, in general, crack propagation in such circumstances is unstable and consequently only one determination of the fracture surface energy is possible for each specimen destroyed.

Wang, Kwei and Zupko[14] have described experiments carried out on butt-jointed epoxy-aluminium plates which contained single cracks at the centre of the bond surface. They used the results of Sih and Rice[15] and Erdogan[16] to show that the stress intensity factors for crack propagation are given by

$$K_1 = \sigma^\infty (a)^{\frac{1}{2}} / \cosh \pi \varepsilon \tag{7}$$

$$\text{and } K_2 = 2 \varepsilon \sigma^\infty (a)^{\frac{1}{2}} / \cosh \pi \varepsilon = 2 \varepsilon K_1$$

where $K_1$ and $K_2$ are the stress intensity factors for mode 1 and mode 2 crack propagation respectively, $\sigma^\infty$, is the remotely applied stress perpendicular to the interface, 2a is the crack length and $\varepsilon$ is the "bimaterial constant" defined by

$$\varepsilon = \frac{1}{2\pi} \ln \frac{G_2 \eta_1 + G_1}{G_1 \eta_2 + G_2}$$

$G_1$ and $G_2$ are the shear moduli of the adhesive and adherend respectively and $\eta_1$ and $\eta_2$ are given by $\eta = (3-\nu) / (1+\nu)$ where $\nu$ is Poisson's ratio. Hence it can be seen from Equation (7) that $K_1$ and hence $K_2$ can be determined from a plot of fracture strength against crack length for such a specimen, if the mechanical constants of the materials are known.

Malyshev and Salganik[17] have shown that crack propagation will take place at an adhesive interface under the conditions

$$G = \frac{(\pi/8)[G_1 + G_2(3-4\nu_1)][G_2 + (3-4\nu_2) G_1]}{G_1 G_2 [G_1(1-\nu_2) + G_2(1-\nu_1)]} \left[ K_1^2 + K_2^2 \right] \tag{8}$$

Hence if $K_1$ and $K_2$ ($= 2\varepsilon K_1$) have been determined as described above then simple substitution of their values into Equation (8) allows an immediate calculation of the critical strain energy release rate and therefore fracture surface energy for the adhesive/adherend system under investigation.

An experimental arrangement, very similar to that developed by Wang, Kwei and Zupko[14] was used for the formation of an adhesive bond between plates of polycarbonate and epoxy resin. The mould is shown in Figure 1. In order to form the joint, the polycarbonate sheet is clamped at the bottom of the mould and the liquid adhesive poured into the mould and allowed to cure. Before bond formation, the polycarbonate surface was treated in exactly the same way as for the previous experiments.

The cracks at the interface were created by placing small strips of Teflon (PTFE) tape at the centre of the substrate before the adhesive was poured into the mould. The Teflon strip was prepared with a chamfer in order to locate a sharp crack tip at the adhesive-adherend interface, and this preparation was carried out by cutting the strip from a bevelled block of Teflon using a sledge microtome. The strip was not removed before the joint was tested.

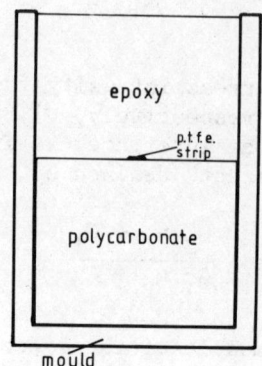

Figure 1. Epoxy/polycarbonate butt joint with centre flaw.

The mechanical properties of adhesive and substrate were determined by applying strain gauge rosettes to flat plates of both adhesive and substrate and determining the strain response to small applied stresses in the elastic range. Poisson's ratios for the materials were calculated from the slopes of the graphs of axial to lateral strain. The results of the determination of the mechanical constants are shown in Table III.

Table III.

| Property | Epoxy Resin | Polycarbonate |
|---|---|---|
| E | 2.27 GPa | 2.51 GPa |
| $\nu$ | 0.37 | 0.33 |
| G (calc.) | 828 kPa | 944 kPa |
| $\eta$ (calc.) | 1.92 | 2.01 |

$$\varepsilon = 1.9 \times 10^{-3}$$

From equation (7) a graph of log $\sigma^{\infty}$ against log a should have a slope of -0.5 and intersect the a = 1 ordinate at a value of $\sigma^{\infty}$ = $K_1 \cosh \pi \varepsilon$. A graph of the stress at which the crack propagated, plotted against crack length on logarithmic scales is shown in Figure 2. The line of best fit is shown in the Figure. The slope of the graph is -0.46 with a correlation coefficient of 0.94. The value of the stress intensity factor is then found to be 260 ± 38 kPa m$^{\frac{1}{2}}$. Substitution in Equation (8) ($K_2$ can be regarded as negligible) gives a value for $R_A$ of 156 Jm$^{-2}$.

# ADHESION AND FRACTURE SURFACE ENERGY

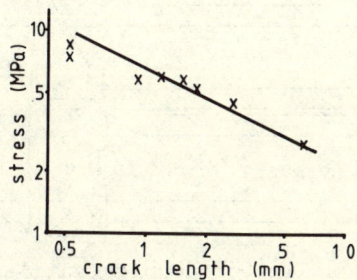

Figure 2. Epoxy/polycarbonate butt joint. Variation of failure stress with interfacial crack length.

## The Shrinkage Strain Energy Release Rate

When a crack propagates inwards from the end of a joint, then the propagation of the crack permits the adhesive to contract and hence shrinkage strains are relieved. The application of Equation (2) to such a system would therefore suggest that the measured critical strain energy release rate for crack propagation is lowered by the relief of such strains. Consequently one method for the determination of R is to apply a compressive force to the substrate so that its dimensions decrease and allow the adhesive to shrink to a state in which there are no shrinkage stresses. Therefore if such a compressive force is applied to the adherend, it is anticipated that $G_c'$ will increase as the shrinkage stresses are progressively reduced and that if the shrinkage stresses are completely eliminated $G_c'$ will reach a maximum value at which it equals R.

A double cantilever beam, consisting of a layer of epoxy resin bonded to another layer of polycarbonate (Figure 3) was used. In order to prevent buckling when the compressive force is applied a stainless steel stiffener was inserted in the adherend. The crack at the interface was initiated by placing a piece of Teflon tape (0.1 mm thick) on the polycarbonate surface near the loading edge, before the bonding process.

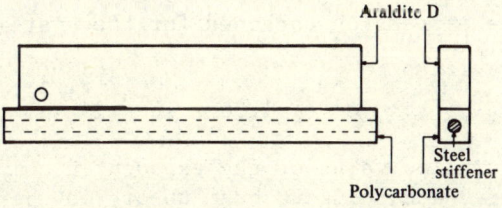

Figure 3. Epoxy/polycarbonate double cantilever beam adhesive joint system.

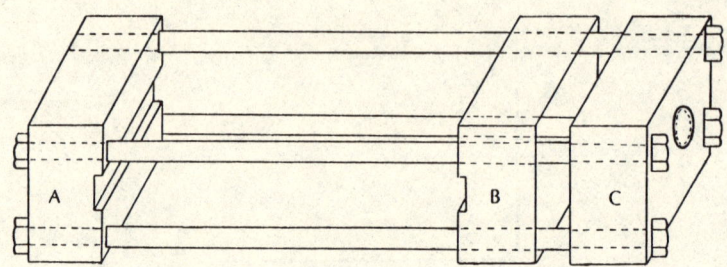

Figure 4. Compression Rig for double cantilever beam adhesive joint specimen adherend.

The compression rig was made of steel and is shown in Figure 4. Blocks A and C were fixed on the base while block B was free to slide along the supporting bars. A screw was used to displace the moveable block which in turn applies the desired amount of compressive force on the adherend. The applied force was measured by means of a set of four strain gauges fixed on the supporting bars.

The specimen mounted in the compression rig was loaded in an Instron testing machine and the fracture toughness, $G_c'$ was determined for varying compressive forces, using the compliance technique developed by Berry[19]. Berry suggested that the compliance C, is related to the crack length, $\ell$, by an expression of the form:

$$C = A\ell^n$$

where A and n can be determined from experimental data. The fracture toughness was then calculated from the following equation :

$$G_c' = \frac{1}{2} P_c^2 \frac{dC}{dA}$$

The stability of crack growth in this work allowed several measurements to be made on each specimen. A crosshead speed of 0.2 mm/min was used giving an average crack front velocity of 1 mm/min.

A compliance expression obtained for the system used in this work is

$$C = 1.36 \times 10^{-2} \ell^{2.21}$$

The relatively low value of the exponent n compared with an expected value of 3 according to beam theory may be explained by the fact that the system is not ideally elastic and the thickness of the specimen is quite considerable. Hoagland[20] has observed that n decreases with an increase in the beam depth and reported a value of

2.38 for a beam depth of 25 mm in his investigation of the plane strain fracture toughness of metals.

The variation of fracture toughness with the applied compressive force is represented in Figure 5. Due to the contribution of the strain energy released by the relief of shrinkage stresses when the crack propagates, the observed fracture toughness of the joint was lower than its true value. As the compressive force F on the adherend is increased, the effect of shrinkage stresses is reduced until finally a maximum value of $G_c'$ was obtained at $F_c = 800$ N. This maximum value of $G_c'$, where $G_c'$ is the measured fracture toughness, is assumed to correspond to a state of zero shrinkage stress. It reflects the true fracture surface energy for opening mode crack extension and may be equated to $R_I$. It may be seen that $(G_c')_{max}$ is approximately twice the value of $(G_c')_{initial}$, the value it assumes in the absence of an applied compressive stress. After the peak value, any further compression on the adherend would only cause an additional shear stress and therefore $G_c'$ again decreases.

The results from Figure 5 suggest that the true fracture surface energy is approximately $120\ J\,m^{-2}$. This compares well, considering the experimental scatter involved with the value of $156\ J\,m^{-2}$ obtained by the previous method.

The results also suggest that the strain energy release rate deriving from the relief of shrinkage strains is of the order of $50\ J\,m^{-2}$, which reduces the applied strain energy release rate to a value of approximately $70\ J\,m^{-2}$ for the propagation of a crack in the absence of any compressive stress being applied to the substrate.

## The Stability of Epoxy/Polycarbonate Joints in Water

Although the work of adhesion for an epoxy/polycarbonate joint immersed in water is $29\ mJ\,m^{-2}$, the fracture surface energy for such a system is not known because the form of $f(\dot{a}, T, _o\ldots)$ in Equation (1) is unknown. It is therefore of interest to examine whether the strain-energy release rate of $50\ J\,m^{-2}$ which derives from the relief of shrinkage strains will cause the propagation of a crack in such a system.

A series of epoxy/polycarbonate joints was prepared using precisely the same preparation technique for the polycarbonate that had been used in all the other experiments. These joints were immersed in water. Six months have so far elapsed without any signs of disbonding. It will be recollected that in the case of an epoxy aluminium joint[11] nucleation of crack propagation occurred within 100 days of immersion for joints which were subjected to shrinkage stresses. It was also observed[19] that when the shrinkage stresses were eliminated there was no crack propagation.

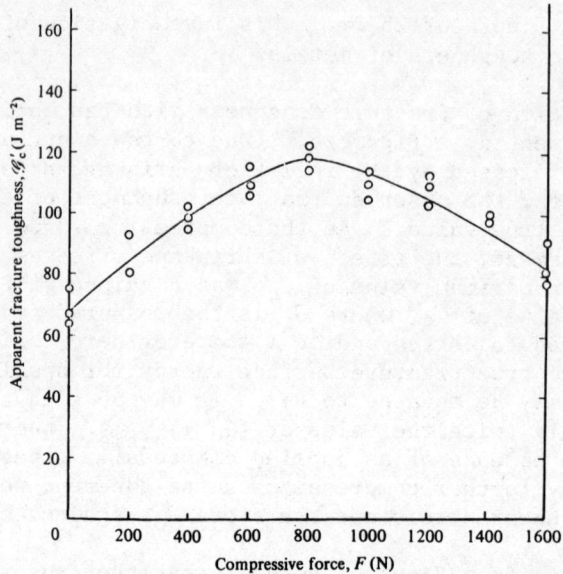

Figure 5. Epoxy/polycarbonate double cantilever beam adhesive joint. Variation of apparent fracture toughness with compressive force applied to adherend.

## CONCLUSIONS

The system which was chosen for this study was one in which the shrinkage of the adhesive was very small. Even so it was found that the effect of the shrinkage stresses was to reduce the crack extension force by over 40%. It may readily be seen that in many commercial adhesive systems the reduction in strength below that which is theoretically possible by the operation of shrinkage stresses may be very much greater. It is suggested that shrinkage stresses may often be the major cause of weakness in an adhesive joint.

It is not known whether the epoxy/polycarbonate joints in water will eventually disbond. The actual fracture surface energy of the system is currently unknown, although the positive value of the work of adhesion may indicate that the fracture surface energy could be higher than the shrinkage strain energy release rate. What this study has done is to emphasise the different roles of thermodynamics and kinetics in the failure process. Although a joint may be thermodynamically unstable (as in the case of the epoxy/aluminium joint), the operation of a shrinkage stress is necessary to cause the crack to propagate at a rate which is discernible on the time-scale of the observer. In the case of a joint which is thermodynamically stable, the action of a shrinkage stress was not great enough to bring about failure in the same time-scale.

## ACKNOWLEDGEMENTS

The author acknowledges the experimental assistance provided by Mr. S. Tay in the work described in Sections 2.2 and 2.3. The work described in Section 2.4 is partially based on work carried out by Mrs. P.T.T. Hang for her M.Eng.Sc. thesis[21].

## REFERENCES

1. E.H. Andrews and A.J. Kinloch, Proc.Roy.Soc. A332, 385 (1973).
2. E.H. Andrews and A.J. Kinloch, Proc.Roy.Soc. A332, 401 (1973).
3. A.N. Gent and J. Schultz, J.Adhesion, 3, 281 (1972).
4. R.A. Gledhill and A.J. Kinloch, J.Adhesion, 6, 315 (1974).
5. D.H. Kaelble, J.Appl.Polym.Sci., 18, 1869 (1974).
6. E. Orowan, Reports on Progress in Physics, 12, 185 (1948/49).
7. E.H. Andrews, J.Mater.Sci., 9, 887 (1974).
8. B.W. Cherry and P.T.T. Hang, in "Proceedings 1974 Conference of the Australian Fracture Group", R.H. Leicester, Editor, pp.41-50, Australian Fracture Group, Melbourne, Australia.
9. B.W. Cherry and K.W. Thomson, in "Fracture Mechanics and Technology", Vol.2, G.C. Sih, Editor, pp.723-737, Sitjhoff and Noordhoff, Netherlands, 1977.
10. S. Mostovoy, E.J. Ripling, and C.F. Bersch, J. Adhesion, 3, 125(1971).
11. B.W. Cherry and K.W. Thomson, in "Adhesion I", K.W. Allen, Editor, pp.251-268, Applied Science Publishers, London 1977.
12. K.W. Thomson, "The conjoint action of stress and water on epoxy aluminium joints", M.Eng.Sc.thesis, Monash University, Australia, 1977.
13. D.K. Owens and R.C. Wendt., J.Appl.Poly.Sci., 13, 1741 (1969).
14. T.T. Wang, T.K. Kwei and H.M. Zupko, Int. J. Fracture Mechanics, 6, 127 (1970).
15. G.C. Sih and J.R. Rice, J.Appl.Mechanics, 31, 477 (1964).
16. F. Erdogan, J.Appl.Mechanics, 30, 232 (1963).
17. B.M. Malyshev and R.L. Salganik, Int.J.Fract.Mech., 1, 114 (1965).
18. B.W. Cherry, "Polymer Surfaces",p.92, Cambridge University Press, Cambridge, England, 1981.
19. J.P. Berry, J.Appl.Phys., 34, 62 (1963).
20. R.G. Hoagland, J. Basic Engineering, 89, 525 (1967).
21. P.T.T. Hang., "Shrinkage stresses in adhesive joints", M.Eng.Sc. thesis, Monash University, Australia 1975.

# THE VARIATION OF POLYMER MORPHOLOGY AND STRUCTURE THROUGH SURFACE INTERACTIONS

Jerome B. Lando

Department of Macromolecular Science
Case Western Reserve University
Cleveland, Ohio  44106

We have been exploring several methods of preparing unique polymer structures through variation of surface interactions between a polymerizing monomer and a suitable substrate. We have termed the first of these techniques "epitaxial polymerization." It involves the epitaxial crystallization of a monomer on a crystalline substrate such as an alkali halide, and subsequent solid state polymerization. A number of new monomer and polymer crystalline phases have been produced by this technique. Monomers being investigated include cyclic sulfur-nitrogen compounds, cyclic posphazines, and diacetylenes. Monomers that do not ordinarily polymerize in the solid state can be polymerized by "epitaxial polymerization."

A second general class of polymerization reactions that are being investigated are soap-like monomers that can form condensable monolayers at the gas-water interface, such as vinyl stearate, α-octadecylacrylic acid, and a variety of diacetylene monomers. In the latter case monomer spreading conditions can control the morphology of the polymer formed by polymerization at the gas-water interface. Multilayers formed by the Langmuir-Blodgett technique have also been investigated. The alternation of pure layers of different monomers in a number of cases has been found to alter the structure within individual polymerized layers. In addition we have developed a unique method of varying the side chain packing in multilayers which leads to different polymer structures that should have unusual properties.

## INTRODUCTION

A topochemical effect in a solid state reaction is any effect on the structure and properties of the product or the kinetics of the reaction that can be directly attributed to the geometric arrangement of the reacting groups or the distance between those groups. The degree of topochemical control in a solid state reaction can vary greatly depending upon the particular system investigated.[1] Reactions to be discussed here, in which there is a crystallographic correlation between the reactant and the resulting product, can occur in solid solution or with the nucleation and growth of a product phase.

Systematic investigation of solid state polymerization reactions began with the discovery that crystalline acrylamide polymerizes when exposed to ionizing radiation.[2,3] Since that time investigators have been intrigued by the possibility of producing polymers with unusual structures, morphologies and properties through solid state polymeriztion. Although these goals have been fullfilled to some extent in the past twenty-five years, a pervasive problem has been a lack of control over monomer morphology and structure. In simple terms we are "stuck" with the structures that nature gives us. In the following paper several methods will be discussed that allow variation of monomer structure and morphology. Both methods involve the control of surface interactions during monomer crystallization and polymerization.

The first involves a technique we have termed epitaxial polymerization.[4] Monomers are crystallized on crystalline substrates. Lattice matching allows the variation of monomer structure and morphology, yielding different polymer structures and morphologies upon polymerization.

The second method involves use of the Langmuir Blodgett[5] technique with modification. Soap-like monomers are spread in a volitile solvent at the gas-water interface, a lateral pressure is applied, and the monolayer is either polymerized by ultraviolet radiation[6] or picked up as multilayers and polymerized by ultraviolet or ionizing radiation.[7] Control of spreading conditions and/or polymerization conditions can then yield polymers with unusual properties.

## EXPERIMENTAL

<u>Epitaxial Polymerization</u>: Tetrasulfur tetramide ($S_4N_4$) was sublimed at 100°C and $10^{-5}$ Torr. The vapor was passed through $Ag_2S$ glass wool at 210°C forming $S_2N_2$. The hot $S_2N_2$ vapor condensed and crystallized on $KC\ell$, $KI$, $KBr$ $NaC\ell$ $KF$ and $RbI$ single crystals at -78°C.[4,8] Polymerization occured during heating to room temperature. After suitable preparation[4,8] Jeol Jem 100B

transmission electron microscope was used to obtain micrographs and electron diffraction photographs.

Monolayer and Multilayer Polymerization: Monolayers at the gas-water interface of the diacetylene monomer $CH_3(CH_2)_{15}C\equiv C-C\equiv C-(CH_2)_8COOH$ and its lithium salt were formed on a 8 x 10 cm Teflon trough. Polymerization was obtained in a nitrogen atmosphere by exposing the monolayer to a short-wave length mercury lamp. Monolayers were deposited onto glass plates for observation in the optical microscope by dipping the slides with their surfaces parallel to the water surface down through the polymerized film. Bilayers were formed on electron microscope grids for optical and electron microscopy.

The apparatus used for the preparation of multilayers of vinyl monomers has been described elsewhere[9]. It consists of a Teflon coated double trough connected by a canal with sliding doors. Multilayer formation was accomplished by dipping the substrate through the monolayer at the air-water interface. In order to get multilayer films with unusual structure special techniques were followed as described below.

The X form (head-tail side group arrangement) of the cadmium salt of poly(octadecylacrylic acid) was prepared by spreading the monomeric acid at the air-water interface, the water being $10^{-3}$ N in cadmium ions and at a pH of 9. A backing material (gold shadowed aluminum foil, germanium) is dipped into the water through the interface. The slide is moved into the second trough through the canal and polymerized by UV radiation while beneath the water surface. The sample is brought up through the clean water surface and the procedure is repeated to give as many polymerized X layers as desired. If the unpolymerized monomer is brought up through the interface Y structures (head-head, tail-tail side chain packing) result.

Alternating layers of octadecylacrylic acid and octadecylacrylamide as well as mixed multilayers of the two monomers were prepared and polymerized. Both alternating bilayers and alternating monolayers were prepared using the double trough. No attempt was made to inhibit the turn around. Thus packing of these polar materials was head-head, tail-tail.

Characterization of these unusual multilayer samples was accomplished using small angle X ray diffraction, electron diffraction and Fourier transform infrared spectroscopy.[10,11]

RESULTS and DISCUSSION

Epitaxial Polymerization: $S_2N_2$, deposited on KCl, KI, NaCl and KBr all produced polymer oriented along the ⟨110⟩ direction of the substrate.[8] However, no orientation was observed using the highly hydroscopic KF and RbI substrates. Fourier transform infrared studies of $(SN)_x$ produced on these substrates showed a high degree of degradation by water.[12] The $(SN)_x$ crystals produced on KCl, KI, NaF, NaCl and KBr are rectangular platelets. The sharpness of the edges on KCl indicate rapid reaction while warming, whereas on some of the other substrates rounded edges were observed, indicating sublimation of $S_2N_2$ before complete polymerization. This type of difference can be attributed to a differing catalytic effect on polymerization of these substrates.

The most striking changes are differences in structure of $(SN)_x$ crystals on these substrates as well as a difference in morphology from conventionally obtained $(SN)_x$. The different crystal structures are shown in Table I. Examination of the (110) spacings of the salts indicated that lattice matching was responsible for the appearance of three new phases of $(SN)_x$, the $\gamma$, $\delta$ and $\varepsilon$ phases.

It should be noted that none of the $(SN)_x$ produced by direct epitaxial polymerization was fibrous in nature as is the conventional material in the normal $\alpha$ form.[13-16] Not only is the $\alpha$ form produced on KCl not fibrous it contains none of the extensive crystal twinning observed in the conventional polymerization. Thus greater crystal perfection is obtained by epitaxial polymerization. The conventional fibrous $\alpha(SN)_x$ can be produced if a second deposition of $S_2N_2$ is made on $(SN)_x$ crystals produced by epitaxial polymerization. This is shown in Figure 1. Note that an epitaxial effect is retained - the fibers are oriented.

It should be noted that epitaxial polymerization can be a general method of obtaining new polymer structures and morphologies. Other systems are currently under investigation.

<u>Monolayer and Multilayer Polymerization</u>: A reaction scheme for the polymerization of a diacetylene monolayer is shown in Figure 2. In the following work the lithium salt was studied and the surface pressure during polymerization was 10 dynes/cm.

Three different morphologies of polymerized monolayers were obtainable depending on the spreading solvent and spreading conditions.[6] In dilute chloroform solutions monolayers appeared as spherulites in the optical microscope under crossed polymers, having blue, yellow and black sectors. From dilute benzene-chloroform solution small blue and yellow mosaic block structures were obtained. When the monomer was spread from chloroform solution after almost saturating the nitrogen above the substrate with chloroform we obtained large blue or yellow crystals millimeters in lateral dimensions but only 35Å thick. These optical micrographs are shown in Figures 3, 4 and 5.

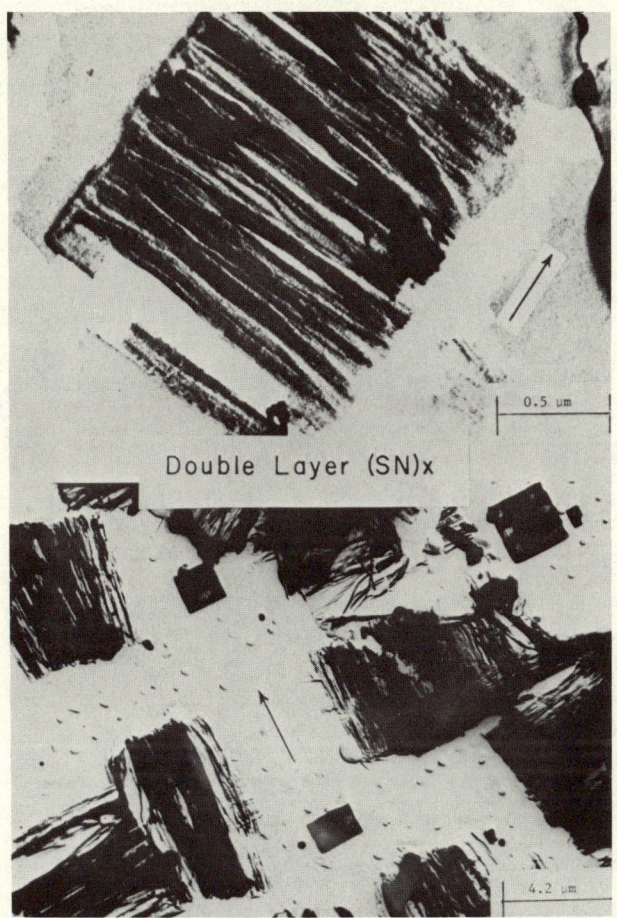

Figure 1. (a) Double-layer morphology of polythiazyl deposited on KI. The second layer preferentially grows on top of the first layer, and clearly shows a fibrous nature. (b) Electron micrograph as above, indicating the orientation of the chain is along both (110) directions of the substrate, but with only one direction per crystal. Some of the second deposit grows on previously unoccupied regions on the substrate as normal rectangular platelets (arrow points to one of three in micrograph).

Table I. Crystal Forms of $(SN)_x$ Produced by Epitaxial Polymerization[4].[a]

| Form | Crystal Class | a | b | c | γ | Space Group | Substrate |
|---|---|---|---|---|---|---|---|
| α | Monoclinic | 0.415 | 0.764 | 0.444 | 110 | $P2_1/a$ | KCl |
| γ | Orthorhombic | 0.920 | 1.072 | 0.493 | | $P2_1 2_1 2_1$ | KI, NaF |
| δ | ... | 0.680 | ... | 0.524 | | $P2_1$ [b] | NaCl |
| ε | ... | 1.392 | ... | 0.581 | | $P2_1$ [b] | KBr |

[a] Units are in nanometers and degrees, and c is the chain axis.
[b] Full space group unknown.

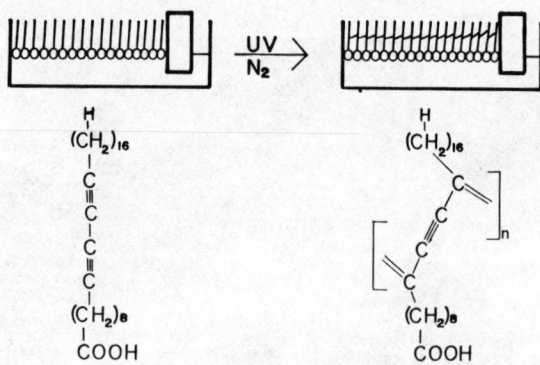

Figure 2. Diacetylene monolayer reaction scheme.

Figure 3. Optical micrograph of spherulitic monolayer.

Figure 4. Optical micrograph of mosaic block monolayer.

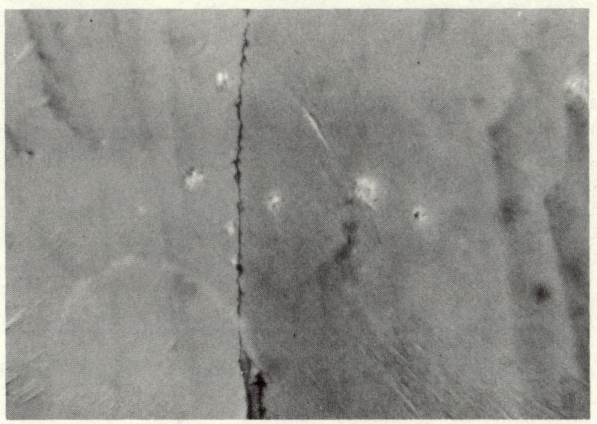

Figure 5. Optical micrograph of two large polymerized crystals.

The difference in color in these optical microscope has been shown to be caused by differences in the chain direction. The fact that birefringence is observed in these monolayers demonstrates the great optical anisotropy of these materials. The locus of the chain axis in the crystallites that make up the spherulites was shown to be left handed or right handed spirals eminating from the sperulite nucleus.[6] It should be emphasized that, although morphology of the polymer can be varied by varying the monomer spreading conditions,[17] the crystal structure of the monolayers remain the same.

Polymerized head-to-head bilayers, (70Å thick prepared as previously described,[6] could be spread across macroscopic holes (0.5 by 0.5 mm holes). This stability is related to the fact that no crystallographic register between chains is observed. Thus crystal or spherulite boundries do not superimpose and the two layers are held together and reinforced by polar interactions between the head groups.

Unusual Multilayers: In previous work we demonstrated that multilayers of poly(vinyl stearate) which normally occur as X layers (head to tail) can be prepared with the side chains packed in Y layers (head-head-tail-tail).[18] By the technique described in the experimental section we have obtained an X structure for multilayers of the polymerized cadmium salt of octadecylacrylic acid, a material that normally forms Y structures. As with vinyl stearate[18] the turn around of the cadmium salt of octadecylacrylic acrylic acid monomer layers during the up trip through the gas-water interface is precluded by polymerizing each layer underneath the water surface. This has been verified by changes in the infrared spectrum[10,11] and by x-ray diffraction measurements, which show that the repeat corresponds to 28Å not 56Å as would be expected for head-head-tail-tail layers.

We have also prepared alternating layers of octadecylacylic[10] acid and octadecylacrylamide monomer and polymer layers. Table II lists the basic small angle X-ray spacings for pure octadecylacrylic acid (ODA acid) and pure octadecylacrylamide (ODA amide) monomer and polymer multilayers and Table III gives the same information for alternating and mixed layers of these two materials.

TABLE II

| Compound | Side Chain Repeat (Å) | | Multilayer Structure |
|---|---|---|---|
| | Monomer | Polymer | |
| ODA Amide | 59 | 62 | Y |
| ODA Acid | 50 | 59 | Y |

TABLE III

| Compound | Alternation Type | Side Chain Repeat(Å) | | Multilayer Structure |
|---|---|---|---|---|
| | | Monomer | Polymer | |
| ODA Amide and ODA Acid | bilayers | 108 | 120 | $Y_1 + Y_2$ |
| | monolayer | 56 | 58 | Y |
| | none(mixed monolayers) | 61 | 63 | Y |

It can be seen from these data that the observed spacings of the alternating and mixed layers are what one would expect from consideration of the pure materials. It should be noted that electron diffraction from the pure materials and the alternating bilayers, which indicate the packing of the molecules perpendicular to the side chains, yield somewhat different spacings. For example, in the polymers side chain packing is in all cases order-disorder hexagonal, but the basic d-spacing is 3.84Å for the amide, 4.04Å for the acid, and 4.25Å for the alternating bilayers.

## CONCLUSION

Techniques for the controlled variation of the structure and morphology of solid state polymerized materials have been demonstrated. Such unusual properties may have a variety of uses.

## ACKNOWLEDGEMENT

The partial financial support of this work by the National Science Foundation under DMR 77-13001 and DMR 81-11441 and by the Army Research Office under contract number DAAG 29-80C-0099 is gratefully acknowledged.

## REFERENCES

1. J. B. Lando, Contemporary Topics in Polymer Science, 2, 189 (1977)
2. R. B. Mesrobrian, P. Ander, D. S. Ballantine and G. L. Dienes, J. Chem. Phys., 22, 565 (1954).
3. R. Schulz, A. Henglein, H. E. von Steinwehr and H. Bambauer, Agnew. Chem., 67, 232 (1955).
4. S. E. Rickert, J. B. Lando, A. J. Hopfinger and E. Baer, Macromolecules, 12, 1053 (1979).
5. K. B. Blodgett and I. Langmuier, Phys. Rev., 51, 964 (1937).
6. D. R. Day and J. B. Lando, Macromolecules, 13, 1478,(1980).
7. A. Cemel, T. Fort, Jr. and J. B. Lando, J. Polymer Sci., A-1, 10, 2061 (1972).
8. S. E. Rickert, H. Ishida, J. B. Lando, J. L. Koenig and E. Baer, J. Appl. Physics, 51, 5194 (1980).
9. M. Puterman, T. Fort, Jr. and J. B. Lando, J. Colloid and Interface Sci., 47, 705 (1974).
10. A. Banerjie, Ph. D. Thesis, Case Western Reserve University, 1977
11. T. Hanschen, MS Thesis, Case Western Reserve University, 1981 .
12. H. Ishida, S. E. Rickert, J. Hopfinger, J. B. Lando, E. Baer and J. L. Koenig, J. Appl. Phys. 51 (10), 5188 (1980).
13. M. Boudelle, Ph.D. Thesis, University of Lyon, France,1974 .
14. R. H. Baughman, R. R. Chance, and M. J. Cohen, J. Chem. Phys. 64, 1869,(1976.).
15. R. H. Baughman, P. A. Apgar, R. R. Chance, A. G. MacDiarmid, and A. F. Garito, J. Chem. Phys. 66, 401(1977).
16. R. J. Young and R. H. Baughman, J. Mater. Sci. 13, 55 (1978).
17. D. R. Day and J. B. Lando, Macromolecules, 13, 1483 (1980).
18. V. Enkelmann and J.B. Lando, J. Polymer Sci.,Chem. Ed. 15,1843(1977).

ABOUT THE CONTRIBUTORS

Here are included biodata of only those authors who have contributed to this volume. Biodata of contributors to Volume 2 are included in that volume.

*M. J. Adams* is affiliated with the Unilever Research as a Scientist working on industrial applications of materials science. Joined Unilever subsequent to receiving Ph.D. degree in 1975. Current research interests are concerned with the tribology of powders and fibers.

*David L. Allara* is a Technical Staff Member at Bell Laboratories, Murray Hill, NJ which he joined in 1969. He received his Ph.D. degree in Physical Organic Chemistry in 1964 from UCLA and had research and faculty appointments before coming to Bell Labs. He has over 50 publications and is an Editorial Board Member for Advances in Chemistry and Symposium Series (American Chemical Society), and Surface and Interface Analysis. His research interests are chemical kinetics and thermochemistry, surface chemistry, surface spectroscopy, and polymer interfaces.

*S. Bodoff* is with Bell Laboratories, Murray Hill, NJ.

*Jean Pol Boutique* is a graduate student at the Facultes Universitaires Notre Dame de la Paix, Nemur, Belgium. His work is related to the electronic structure of conducting organic solids.

*Brian J. Briscoe* has been a Lecturer in Interface Science, Department of Chemical Engineering and Chemical Technology, Imperial College, London, since 1978. Prior to his present position, he was Oppenheimer Fellow in Surface Science and Assistant Director of Research at the Cavendish Laboratory, Cambridge University. His research interests include the surface and mechanical properties of organic polymers with particular emphasis on adhesion, friction, lubrication and wear.

*Brian Cherry* is currently Associate Professor of Materials Engineering, Monash University, Victoria, Australia. He had his academic training at Trinity College, Cambridge, England and spent some time as Goldsmiths Research Fellow at Churchill College, Cambridge. His main research interests are in the fields of surfaces of both polymers and metals. In addition to completing recently a book Polymer Surfaces, he is currently in charge of a feasibility study into the establishment of an Australian National Corrosion Centre.

*Emil Chibowski* is an adjunct in the Department of Physical Chemistry, Maria Curie-Sklodowska University, Lublin, Poland. Holds a Doctor of Chemistry degree and the field of research is physical chemistry of surfaces especially the electrochemical aspects.

*David Clark* is Professor and Chairman, Department of Chemistry, University of Durham, Durham, England which he joined in 1964. He received his Ph.D. degree in 1964 from the University of Sheffield and was recipient of the Turner Prize for Research. He was a Fullbright Scholar at Caltech, 1964-1965. In 1969 he established the first ESCA research group in the U.K. with particular emphasis on study of organic systems. He has been interested in a number of research areas but currently his research is predominantly on the experimental and theoretical investigation of core hole state phenomena particularly on polymeric systems. He has about 200 research papers to his credit and has given numerous invited lectures at international symposia, Gordon Conferences and NATO Summer Schools.

*Jean-Pierre Dauchot* is a permanent researcher at the University of Mons, Belgium and his research interest is in the catalysis of thin films and electronic devices like gas sensors. He received his Ph.D. in 1975.

*Jean-Pol Delrue* is with the IRIS (Institute for Research in Interface Science) group in Namur, Belgium. He received his Ph.D. degree in 1978 followed by postdoctoral fellowship at Stanford University. His research concerns mainly catalysis on thin films, surface and interface characterization of solid state devices and has published several papers on thin films.

*Mumtaz A. Dinno* is Associate Professor of Physics and Astronomy at the University of Mississippi. Prior to coming to Mississippi he was Assistant Professor of Physics at VPI&SU, Blacksburg, VA. He received his B.S. degree from the University of Baghdad, Iraq and Ph.D. from the University of Louisville. He has published a number of papers and has edited the proceedings of two conferences. He is the Founder of the Conference Series on Membrane Biophysics: Structure and Function in Epithelia. His research interests include thin film solid state physics, membrane biophysics and interfacial phenomena.

# ABOUT THE CONTRIBUTORS

*Charles B. Duke* is a Senior Research Fellow and Manager, Molecular and Organic Materials Research at the Xerox Research Laboratories in Webster, NY, and an Adjunct Professor of Physics at the University of Rochester. He is very active in the American Vacuum Society and served as its president in 1979. He is Chairman of the Board of Editors of the J. Vac. Sci. Technol., and a member of the editorial boards of Surface Science, Critical Reviews of Solid State and Materials Science, Advances in the Mechanics and Physics of Surfaces, and Surface Science Reports. He is a Fellow of the American Physical Sociiety and has served on the Governing Board of the American Institute of Physics since 1976. In 1979 he received the M. W. Welch Award in Vacuum Science and Technology administered by AVS. Recently, his main research interests have been in the areas of semiconductor physics, electron-solid scattering, the electronic structure of organic materials, and applications of surface chemistry to microelectronics. He has written over 200 technical papers on a variety of topics, as well as a monograph on electron tunneling in solids.

*M. A. Fortes* is in the faculty, Departmento de Metalurgia, Instituto Superior Técnico, and CEMUL (Centro de Mecânica e Materials da Universidade Técnica de Lisboa), Lisboa, Portugal. He has been active in a number of research areas including wetting of surfaces and has many publications to his credit.

*Donald F. Gerson* is currently at the Genex Corp. in Gaithersburg, MD. Prior to coming to Genex, he was with the Basel Institute for Immunology, Basel, Switzerland. He is a biophysicist and his areas of current research include the application of surface physics to biological problems, and the measurement of intracellular pH and its relation to cellular physiology. He has also been interested in the development of new instrumentation for biological research.

*A. Gupta* is Technical Group Supervisor of the Materials Group at Jet Propulsion Laboratory, Pasadena, CA. He received his Ph.D. degree in 1974 from California Institute of Technology. His research interests are polymer photochemistry and energy migration phenomena in aromatic polymers.

*Moshe Gur-Aryeh* is a Chemist in the Research Center of Nir-Lat in Negev, Israel. He holds a B.Sc. degree.

*J. R. Hallman* is Director of the Engineering Division at Lakeland Community College in Mentor, OH where he has been since Jan. 1981. Prior to his present position, he was Professor and Head, Chemical Engineering Technology, Nashville State Technical Institute. He received his Ph.D. degree in Chemical Engineering in 1971 from Oklahoma University and has had a number of years of industrial experience. He has presented or published some 30 papers dealing with engineering technology education and his research in polymer ignition and polymer absorption to radiation.

*Andre Hecq* is currently doing his Ph.D. on the mechanism of sputtering deposition techniques and has already published several papers.

*Frank J. Holly* is Professor of Ophthalmology and Biochemistry in charge of resident research at the Texas Tech University Health Sciences Center. He received his Ph.D. degree in Physical Chemistry in 1962 from Cornell University followed by industrial appointments and ten years at the Eye Research Institute of the Retina Foundation in Boston. He has written over 70 papers on the results of his pioneering studies in basic dental research, biomaterial research including the biocompatibility of contact lenses and electrets, lacrimal and corneal physiology, surface chemistry of biointerfaces, bioadhesion, and cellular interaction. He has served as consultant to the National Institutes of Health and is a Council Member of the International Society for Contact Lens Research. He is the inventor of several drugs, a therapeutic dentrifice, and has developed new diagnostic techniques for the detection of early dry eye states.

*Jack E. Houston* is a staff member in the Surface Physics Division at Sandia National Laboratories, Albuquerque, NM. He received his Ph.D. in Physics in 1965 from Oklahoma State University. His interests revolve around obtaining a broad understanding of the nature of solid surfaces, and his present research deals with determining the relationship between line shapes in various electron-spectroscopic techniques and the detailed chemical environment of surfaces and adsorbed species.

*A. L. Huston* is a graduate student in the Department of Chemistry, University of California, Riverside, CA.

*Hatsuo Ishida* is presently Assistant Professor of Macromolecular Science at Case Western Reserve University and is also Director of C. Richard Newpher Polymer Composite Processing Laboratory which has recently been established at this university. He received his Ph.D. in Macromolecular Science from the same university where he applied FT-IR and laser Raman spectroscopy to his research on composite materials. He has been active in the molecular studies of glass fiber/matrix interface of composites and his latest interest involves the correlation of molecular structure and mechanical property of composites made by various processing conditions.

*Kazuhiko Itagaki* is with the USA Cold Regions Research and Engineering Laboratory, Hanover, NH, which he joined in 1964. He received his Ph.D. in Physics in 1960 from Hokkaido University, Sapporo, Japan. His interests are in surface and interfacial properties of single crystals and polycrystalline aggregates of ice and snow.

# ABOUT THE CONTRIBUTORS

*Bronislaw Janczuk* is an Adjunct in the Department of Physical Chemistry, UMCS, Lublin, Poland. Holds a Doctor of Chemistry degree and his research interests include wettability of surfaces and film properties.

*H. H. G. Jellinek* is Professor Emeritus (effective June 1982), Clarkson College of Technology, Potsdam, NY. He received his Ph.D. degree in Inorganic and Physical Chemistry from London, England, 1942 and another Ph.D. in Colloid and Physical Chemistry from Cambridge, England, 1945. For his original scientific contributions, he was awarded Sc.D. (Cambridge) in 1964. He has about 200 publications (including books and patents) to his credit and is the sole editor of yearly <u>Advances</u> <u>in</u> <u>Degradation</u> <u>and</u> <u>Stabilization</u> <u>of</u> <u>Polymers</u>. In 1982, he received the Letter of Appreciation for teaching and research from Ronald Reagan, President of the United States.

*D. R. Jennison* is with the Sandia National Laboratories, Albuquerque, NM.

*Yoonok Kang* is a Research Chemist at Eastman Kodak Co. She received her Ph.D. degree in 1981 in Chemistry from VPI&SU, Blacksburg, VA with a dissertation on the characterization of polymeric membranes.

*J. Kelber* is with the Sandia National Laboratories, Albuquerque, NM. He received his Ph.D. degree in Inorganic Chemistry in 1979 from the University of Illinois (Urbana). His research interests include the physics of the Auger process, the application of photoelectron spectroscopy to the study of polymer surfaces, and the adhesion of metals to polymer surfaces.

*Jack L. Koenig* is Professor of Macromolecular Science, Director of the NSF sponsored Materials Research Laboratory program, and Director of the Molecular Spectroscopy Laboratory at Case Western Reserve University where he has been since 1963. He received his Ph.D. degree in 1960 from the University of Nebraska for his research in theoretcical spectroscopy, followed by employment at Du Pont in Wilmington. In 1973, he joined the NSF in the Division of Materials Research as Program Director of the Solid State Chemistry and Polymer Science Section. His interests include high resolution solid state NMR, Fourier transform infrared spectroscopy and Raman spectroscopy and is well known for his basic work in spectroscopic characterization of polymeric materials. He has over 200 publications to his credit.

*S. L. Kremnitzer* is currently Plant Engineer at BP Chemicals, Carrington Lancs, England. Studied for a Ph.D. degree in Surface Physics of Fiber Systems at the Cavendish Laboratory, Cambridge University,England and received the degree in 1978.

*Jerome B. Lando* is Professor and Chairman, Department of Macromolecular Science, Case Western Reserve University which he joined in 1965. He received his Ph.D. degree in Polymer Science in 1963 from the Polytechnic Institute of New York followed by two years at Camile Dreyfus Laboratory, Research Triangle Institute. From Jan. 1, 1974 to June 30, 1974 he was Visiting Professor, University of Mainz, West Germany, Alexander Von Humbold Senior American Scientist Awardee. He is the coauthor (with S. Maron) of the book <u>Fundamentals of Physical Chemistry</u> published in 1974 and has 97 publications to his credit. He is a Fellow of the American Physical Society and serves on the Editorial Advisory Board of Macromolecules. His research interests are organic reactions in the solid state, solid state polymerization, polymer crystal structure, synthesis of stereoregular polymers, and pyroelectric and piezoelectric polymers.

*L. H. Lee* is a Senior Scientist at Xerox Corp., Webster, NY. He received his Ph.D. degree in Chemistry from Case Institute of Technology. He is the editor of six books, author of over 50 papers, and holds 25 U.S. patents. He has been a member of the Advisory Board of the Journal of Adhesion, and Chairman, Division of Organic Coatings and Plastics Chemistry of the American Chemical Society.

*R. H. Liang* is a Technical Group Leader in the Materials Group at Jet Propulsion Laboratory, Pasadena, CA. His major research interests are laser Raman spectroscopy, polymer photochemistry, and polymer degradation processes.

*Douglas R. Lloyd* is Assistant Professor of Chemical Engineering at the University of Texas at Austin which he joined in 1981. Prior to his current position, he was (since 1978) in the faculty in the Chemical Engineering Department at VPI&SU. He received his Ph. D. degree in Chemical Engineering from the University of Waterloo, and his research interests include the preparation, characterization and application of synthetic polymer membranes.

*Julia Lucki* is affiliated with the Department of Polymer Technology, The Royal Institute of Technology, Stockholm, Sweden which she joined in 1969. She is a specialist in the field of polymer photodegradation and photostabilization, and has authored 20 original publications.

*James E. McGrath* is Professor of Chemistry and Co-Director of the Polymer Materials and Interfaces Laboratory at VPI&SU, Blacksburg, VA which he joined in 1975. He has coauthored what is considered a definitive book on Block Copolymers (1977) and has over 70 contributions in the literature, including 20 U.S. patents. He is a consultant to industry and government and has been very active in the Polymer Division (Secretary 1978-1980) of the American Chemical Society. He also initiated three experimentally oriented ACS short courses in polymer chemistry which are taught on the VPI&SU campus.

## ABOUT THE CONTRIBUTORS

*Bernard Miller* has been Associate Director of Research at Textile Research Institute, Princeton, NJ, since 1970. He received his Ph.D. degree in Chemistry from McGill University followed by a number of industrial and academic appointments. He was a Fiber Society National Lecturer, 1973-1974 and has served on the Executive Committees of the Information Council on Fabric Flammability and the North American Thermal Analysis Society. In 1977, he received the Harold DeWitt Smith Medal of the American Society for Testing and Materials for his work in fiber and textile measurements. His major fields of interest are the thermal and combustion behavior of polymers, fabric flammability and the surface properties of fibrous materials.

*Kashmiri Lal Mittal,** is presently employed at the IBM Corporation in Hopewell Junction, NY. He received his M.Sc. (First Class First) in 1966 from Indian Institute of Technology, New Delhi, and Ph.D. in Colloid Chemistry in 1970 from the University of Southern California. In the last eight years, he has organized and chaired a number of very successful international symposia and in addition to this two-volume set, he has edited eleven more volumes as follows: Adsorption at Interfaces, and Colloidal Dispersions and Micellar Behavior (1975); Micellization, Solubilization, and Microemulsions, Volumes 1 & 2 (1977); Adhesion Measurement of Thin Films, Thick Films, and Bulk Coatings (1978); Surface Contamination: Genesis, Detection, and Control, Volumes 1 & 2 (1979); Solution Chemistry of Surfactants, Volumes 1 & 2 (1979); Solution Behavior of Surfactants - Theoretical and Applied Aspects, Volumes 1 & 2 (1982). In addition to these volumes he has published about 50 papers in the areas of surface and colloid chemistry, adhesion, polymers, etc. He has given many invited talks on the multifarious facets of surface science, particularly adhesion, on the invitation of various societies and organizations in many countries all over the world, and is always a sought-after speaker. He is a Fellow of the American Institute of Chemists and Indian Chemical Society, is listed in American Men and Women of Science, Who's Who in the East and other reference works. He is or has been a member of the Editorial Boards of a number of scientific and technical journals.

*D. F. Moore* is a mechanical engineer, consultant, distinguished academic, author and specialist in Tribology. He received his Ph.D. degree in Mechanical Engineering from Pennsylvania State University and a D.Sc. degree for published work from the National University of Ireland. He is a world-wide authority on tribology and has published about 80 papers and technical reports in lubrication technology, friction, wear, impact and energy degrading, and is the author of five textbooks on Tribology (including two Russian translations) and one on Thermodynamics. He has served as Chairman of an ASTM task group on skid-resistance standards and a corresponding member of C.I.R.P. He has been a NATO Senior Fellow, a Royal Irish

---

*As the editor of this two-volume set.

Academy Fellow, a Gast professor at the Technische Universitat Munchen, and a postdoctoral Fellow at the Technische Hogeschool, Delft.

*C. A. Murray* is with Bell Laboratories, Murray Hill, NJ.

*Somasak Naviroj* is currently a Ph.D. candidate in the Department of Macromolecular Science, Case Western Reserve University, Cleveland. He is presently studying the molecular structure of composite interfaces using FT-IR and structure of silanes in aqueous solution by laser Raman spectroscopy.

*A. W. Neumann* is Professor in the Department of Mechanical Engineering, University of Toronto, Canada. His primary area of interest is surface thermodynamics and its application to technological and biomedical problems.

*S. N. Omenyi* is currently working for Marshal Space Flight Center in Huntsville, AL, and received Ph.D. from the University of Toronto.

*Jean-Jacques Pireaux* has since 1979 been a Research Associate National Fund for Scientific Research (Belgium) in Namur which he joined in 1972. He is a physicist and received his Ph.D. in 1976. He has been working with the XPS technique since 1972 and has published about 40 papers including ten in the field of XPS study of valence bands of polymers.

*C. I. Poser* is presently with 3M Co., St. Paul, MN.

*W. Pradellok* is a Visiting Scholar at the University of Massachusetts, Amherst, where he did postdoctoral work with Professor Vogl. His research interests are polymer synthesis and scale-up of polymer processing techniques.

*Jan F. Rabek* is with the Department of Polymer Technology, The Royal Institute of Technology, Stockholm, Sweden which he joined in 1971. He is a specialist in the field of polymer photochemistry and has been working in this field since 1960. He is the author or coauthor of six books on photochemistry, ESR spectroscopy and experimental methods in polymer chemistry, and has authored 80 original publications and review papers. He was a Visiting Professor in Japan and China in 1976 and 1981 respectively.

*Bengt Rånby* is Professor of Polymer Technology, The Royal Institute of Technology, Stockholm, Sweden which he joined in 1961. He received his doctor's degree in Physical Chemistry in 1952 from the University of Uppsala, Sweden and was in the United States (1956-1961). He has specialized in free radical reactions of polymers studied by ESR and photochemical methods, polymer degradation and oxidation, singlet oxygen reactions with polymers, and modification of polysaccharides by graft copolymerization. He has published about 250 original and review papers and is the author or coauthor of five books. He has been Visiting Professor in Japan and U.S.A. and has been awarded honorary doctors degrees in Poland and Finland.

*Joseph Riga* is currently a senior staff member at Namur working mainly on molecular solids and polymers using XPS. He obtained his Ph.D. degree from the Facultes Universitaires Notre Dame de la Paix, Namur, Belgium with a thesis on the electronic structure of organic disulfides.

*Robert Rye* has since 1974 been at the Sandia National Laboratories, Albuquerque, and has been actively involved in the surface chemical aspects of the Fusion Power Program, and most recently in the development of the ability to obtain detailed chemical information from Auger electron spectroscopy. Before joining Sandia, he was a member of the Chemistry Department (1968-1974) at Cornell University. He received his Ph.D. degree in 1968 from Iowa State University. He has held several positions in both the Surface Science Division of the AVS and the Colloid and Surface Chemistry Division of the ACS.

*Zygmunt Rymuza* is Assistant Professor in the Department of Precision Engineering at the Warsaw Technical University, Warsaw, Poland. He is specially interested in the friction, wear and lubrication of polymer micropairs and has over 50 publications in the general area of tribology.

*Isaac C. Sanchez* is a Group Leader of the Polymer Characterization and Standards Group at the National Bureau of Standards, Washington, DC which he joined in 1977. Prior to his current position, he was since 1972 in the faculty of the Polymer Science and Engineering Department at the University of Massachusetts, where he still retains academic ties through an adjunct professorship. He received his Ph. D. degree in Physical Chemistry in 1969 from the University of Delaware. He has published about 30 scientific papers and has given many invited lectures in a broad range of topics including theories of polymer cyrstallization and vitrification and theory of interfacial tensions. In 1980 he won the Department of Commerce Bronze Medal for his scientific contributions in the area of polymer solution theory. He has been very active in the American Chemical Society and American Physical Society.

*G. W. Scott* is Associate Professor of Chemistry, University of California, Riverside. His major research interests are picosecond spectroscopy in condensed media and study of intramolecular energy transfer processes.

*D. A. Seanor* is currently Technical Specialist/Project Manager in the Advanced Product and Technology Department of the Xerox Corp., Webster, NY. He received his Ph.D. degree from the University of Bristol followed by a postdoctoral fellowship at the National Research Council of Canada. His publications and patents are in the areas of surface chemistry, electrical properties of polymers, and the xerographic process. He is the coeditor of the book Photoconductivity in Polymers - An Interdisciplinary Approach and has edited the book Electrical Properties of Polymers to be published in 1982.

*Michael Shepelev* is Chief Chemist at Nir-Lat in Negev, Israel. Prior to his current position, he was Head of Laboratory in the Scientific Research Institute of Rubber and Latex Goods, Moscow, USSR. He is the author of 140 published articles, reports and inventions, and holds a Ph.D. degree.

*Olga Shepelev* is Chief Technologist at Nir-Lat in Negev, Israel. Prior to her current position, she was Senior Scientific Worker in the Scientific Research Institute of Rubber and Latex Goods, Moscow, USSR. She holds a Ph.D. degree and has authored 40 articles and reports.

*C. M. Sliepcevich* has since 1963 held the rank of George Lynn Cross Research Professor of Engineering, University of Oklahoma which he joined in 1955. He received his Ph.D. degree in Chemical Engineering in 1948 from the University of Michigan. He has been active in a variety of fields including energy scattering, high pressure reaction kinetics, flame dynamics, cryogenics, thermodynamics, heat and mass transfer, extractive metallurgy and desalination. He has published more than 150 technical papers, three volumes of mathematical tables and a text book on Thermodynamics in three parts. He has been the recipient of a number of awards and honors including the Curtis McGraw Research Award; the Ipatieff Prize; the George Westinghouse Award; Engineer of the Year; University of Michigan's Sesquicentennial Award for Distinguished Alumni. In 1972, he was elected to membership in the National Academy of Engineering and in 1974 was inducted into the Oklahoma Hall of Fame.

*R. P. Smith* is a Ph.D. student at the University of Toronto, Canada.

*Terutaka Tamai* is in the Faculty of Science and Technology, Sophia University, Tokyo which he joined in 1965. He received his Ph.D. degree in Electrical Engineering from the University of Tokyo. His research interests are in electrical contact phenomena and related surface subjects.

*Jacques Verbist* is Co-Director of the Laboratoire de Spectroscopie Electrotronique, Namur, Belgium which has a broad research program of electronic structure of solids and interfaces. He obtained his Ph.D degree in 1969 from the Louvain University, Belgium. He learned the XPS technique at Berkeley in 1971 and was in charge of the method at Louvain until he was appointed Professor of Chemistry at Namur in 1972.

*O. Vogl* is Professor in the Department of Polymer Science and Engineering, University of Massachusetts, Amherst, MA. He is a member of the National Academy of Sciences and is recognized internationally in the field of polymers. His research interests are synthesis of polymerizable absorbers and antioxidants and development of synthetic polymeric drugs. He is a prolific author and has many research publications to his credit.

*Hans-Dietrich Weigmann* is currently Associate Director of Research at Textile Research Institute in Princeton, NJ which he joined in 1963. He received his Ph.D. in Organic Chemistry at Technische Hochschule in Aachen in 1960 and carried out postdoctoral work at the German Wood Research Institute in Aachen and at the Textile Research Institute in Princeton. His interest is in the chemistry and physics of natural fibers and solvent interactions with textile fibers.

*J. Reed Welker* is President of Applied Technology Corp., Norman, OK. Before his current position, he has held a number of academic, industrial and consultative positions. He received his Ph.D. degree in Chemical Engineering in 1965 from the University of Oklahoma and has published over 40 papers on flame behavior and fire research. He completed graduate work in fire research and is best known for his work on radiation heat transfer to and from flames, fire behavior, and atmospheric dispersion.

*James P. Wightman* is Professor of Chemistry at VPI&SU, Blacksburg, VA. He received his Ph.D. degree in Chemistry from Lehigh University and was a Research Associate at Penn State before joining the faculty at Virginia Tech in 1962. He was a Visiting Professor (1975-1976) at the University of Bristol, Bristol, England. He has coauthored 54 publications and is a member of the Advisory Board of the J. Colloid Interface Sci. His research interests include the adsorption of gases and liquids on solids and the characterization of solid surfaces.

*Wieslaw Wojcik* is an Adjunct in the Department of Physical Chemistry, UMCS, Lublin, Poland. He holds a Doctor of Chemistry degree and is interested in the physical chemistry of mineral processing.

*Aleksander M. Wrobel* is Head of the Plasma Laboratory at the Centre of Molecular and Macromolecular Studies of the Polish Academy of Sciences in Lodz. He received his Ph.D. degree in Physical Chemistry of Polymers from the Technical University of Lodz. He spent a year (1971/72) at the University of Liverpool with Prof. C. H. Bamford working in free radical polymerization and a year (1978) at the Ecole Polytechnique, Montreal where he was involved in plasma polymerization and plasma treatment. He has been the recipient of the Prize of the Research Secretary of the Polish Academy of Sciences.

SUBJECT INDEX

Pages 1-580 appear in Volume 1
Pages 581-1234 appear in Volume 2

Abrasion Resistance of Nylon 6
  Fibers, 843
Abrasive Wear, 418
Acid-Base
  modern theory of, 585-587
Acid-Base Interactions
  and filler-polymer systems,
    739
  and polymer adhesion, 583-601
Acidity-Basicity of Polymers,
  599-600
Acrylamide Graft Copolymerization
  of polyethylene, 924
Acrylate Polymers
  adhesion of metal deposits
    on, 1125-1137
Acrylic Copolymer Surfaces
  photoxidation degradation
    of, 293-302
  transmission IR spectra
    of, 298
Adhesion
  and surface topography,
    672-676
  at low roughness, 676-680
  elastic and viscoelastic,
    669-686
  fundamental, 1037
  mechanism(s) of, 1037-1040
  of dry fibers, 431-434
  of ice to polymers and other
    surfaces, 241-251
  of metallized acrylates, ABS
    copolymers, modified poly-
    olefins and polyvinyl-

Adhesion, contd.
  chloride, (table), 1136
  of platelets, 996, 1004
  of polymeric particles,
    609-612
  of polymers (see also Polymer
    Adhesion)
    electrostatic forces in,
      605-612
    molecular forces in, 605-612
    role of acid-base inter-
      actions in, 583-601
  of wet fibers, 434-435
  practical, 1037
  role of interface in,
    1036-1040
Adhesive Joining
  of Al to polyester fiberglass
    composite, 1139-1159
    bond durability in various
      environments, 1147-1158
  of Al to styrene 1169-1179
    bond durability in various
      environments, 1167-1178
Adsorbed Gases and Ion Transfer
  in Polymers, 496-501
Adsorption
  and contact angle studies,
    613-622
  measurement by ellipsometry,
    615
  of alcohols on polypropylene
    and polycarbonate, 613-622
  of BSA on polymeric sub-
    strates, 926, 934

Adsorption, contd.
  of BSA on silicone materials, 631
  of native plasma proteins on polymer surfaces, 1011-1029
  of polymers on inorganic surfaces, 590-599
    acid-base nature of, 590-593
    enthalpies of, 593-599
  of surfactants on fibers, 654-656
  of water on polypropylene and polycarbonate, 613-622
  on modified silicone surfaces, 625-635
Advancing Contact Angle, 115-119, 144, 189-192, 629, 913-916
Agglutination, 945
n-Alkanes, Surface Tension of, 176
Aluminum
  adhesive joining to polyester fiberglass composite, 1139-1159
  adhesive joining to polystyrenes, 1161-1179
Anils, Charging of, 490
ATR-IR Spectroscopy of Polymer Films, 204, 803, 808, 932
Auger Spectra
  of methane, neopentane and n-pentane, 87
  of pigmented polyurethane paint surface, 1058
Auger States (Localized) in Polyethylene, 83-89
Autoadhesion of Etched Polyethylene Films, 689-697
  effect of relative humidity, 692
Autohesion of Latex, 377-383
  effect of surfactant concentration, 380

Benzophenone
  solidification front velocity of, 166
  surface tension of, 161, 168

Benzoyl Chloride
  and improvement in adhesion to biological tissue, 947
Bibenzyl
  solidification front velocity of, 166
  surface tension of, 161-170
Biological Adhesion
  and surface energetics, 895-908
Biological Environment
  polymer surface interactions in, 943-952
Biological Tissue
  adhesion of commercial adhesives to, 947
  polymer adhesion to, 946
Biosurface
  adhesion of polymer adhesives to, 950
Bisphenol A Polycarbonate
  photodegradation of, 17-31
Blood
  compatibility, 933
  interactions of, with multiphase polymers, 985
  platelets, 953
Blood at Interfaces
  behavior of, 1003-1008
Boundary Lubrication of Polymers, 403-406
Breaking Twist Angle Properties of Nylon 6 filaments, 849

Calcium Carbonate
  plasma treatment, 740
  silane treatment, 740
Cantilever Apparatus for Adhesion and Friction Measurements, 430
Cationic Surfactants
  sorption of, on fibers, 647-651
Cell
  medium, interfacial free energies of, table, 238
  surface, interfacial free energies of, 237-239

# INDEX

Cellular Responses
  surface energy dependence of, 906-908
Chemical Wear Mechanisms, 402
Clot Formation, 933, 1004
Cellulose Fibers
  water adsorption of, 645
Cellulosetriacetate
  charging of, 503
Charge States in Polymers
  application to tribology, 463-473
Charge Transfer
  in metal-polymer contact, 483
  in polymer-polymer contact, 495
Chemical Bonding Mechanism of Adhesion, 1039-1040
Chemical Derivatization and ESCA Analysis, 795, 870, 1048, 1201
Competitive Ablation and Polymerization, 10
Conductive Polymers
  and corrosion, 509
  electrical conduction mechanism of, 507-520
  electrical contact performance of, 507-520
  electrical property of, 514-519
  resistivity of, 512-514
Contact Angle (see also Wettability)
  advancing, 115-119, 144, 189-192, 629, 913-916
  as an indicator of surface polarization, 150-151
  dynamic, 134-136
  hysteresis, 115-134, 145-149, 155-170, 914-916
  measurement, 912
  on rough surfaces, 119-126
  receding, 115-119, 189-192, 629, 913-916
Contact Angles on
  epoxy, 550
  high density polyethylene, 927-928, 930-931
  hydrogels, 141-153

Contact Angles on, contd.
  low density polyethylene, 617, 927-928, 930-931
  membranes, 825-831
  methylmethacrylate-hydroxyethylmethacrylate copolymers, 916
  polycarbonate, 550, 617
  poly(dimethylsiloxane), 918
  polymethylmethacrylate, 915
  polypropylene, 617
  silicone elastomers, 629, 633-634
  silicone elastomers and other materials, (table), 629
  surface grafted PVA films, 813
Contact Charge Exchange, 472-473
Contact Electrification, 477-504
  driving forces for, 480-488
  effect of corona oxidation on, 497
  effect of friction on, 503
  hazards and uses of, 478
  impact of, on photoemission from polymers, 493
  states involved in, 479
Contact Event
  nature of, 479
Copper-Polymer Systems
  interfacial phenomena in, 1069-1073
Copper (Electroless) Adhesion (Peel Strength) to PVC, 1115-1122
Corrosion
  of contact interface between conductive polymer and metal, 509-512
Cowan Plot (Limiting Current), 832
Critical Strain Energy Release Rate, 530
Critical Surface Tension of Wetting (see also Surface Tension of)
  as a function of aging time, 299
  of hydrogels, 144

Critical Surface Tension of
Wetting, contd.
of plasma polymerized isomeric
fluorobenzenes, 15
of polyethyleneterephthalate,
665
of polymeric solids, (table),
144
of a variety of materials,
(table), 235

Degradation of Polymers
studied by ESCA, 3
Degradation between Ethylene-
Acrylic Acid Copolymers and
Lead-Tin Alloys
studied by ESCA, 1181-1196
Derivatization Methods to Enhance ESCA Capabilities, 795,
870, 1048, 1201
Diacetylene Monolayer Polymerization, 562
Diffusion Theory of Adhesion,
1039
Double Cantilever Beam, 553-554
Durability
factors influencing, 1035
of aluminum to polyester
fiberglass composite joints,
1147-1158
of aluminum to polystyrene
joints, 1167-1178
of epoxy-polycarbonate joints,
555-556
of plastic welded joints, 725-
735
Durability (Joint) Factor, 1144
Dynamic Contact Angles, 134-136
Dynamic Friction, 438-441

Effective Fracture Surface
Energy, 530
Efficiency of Lubrication,
457-460
and surface tension of oils,
459
Elastohydrodynamic Lubrication,
421-424

Elastomers
friction and wear mechanisms
in, 420-421
tribological properties of,
413-424
Electrical Conduction Mechanisms,
of an elastomeric conductive
polymer, 507-520
Electrical Contact Performances,
507
Electrical Properties of Conductive Polymers, 514-520
Electrodialysis Cell, Schematic
of, 835
Electrodialytic Membranes, 818
Electrokinetic Measurements on
Fibers, 651-654
Electroless Metallization, 1060,
1213-1223
Electron Mean Free Paths, 4-7
Electrostatic
forces in polymer adhesion,
605-612
theory of adhesion, 1039
Ellipsometry and Adsorption
Studies, 615
Epitaxial Polymerization,
559-562
Epoxy
contact angle on, 550
mechanical constants of, 552
surface energies of, 550
Epoxy-Polycarbonate Adhesive
Joints
fracture surface energy of,
550-553
stability in water, 555-556
work of adhesion of, 548-550
ESCA (XPS) Analysis or Spectra
of
AAm grafted low density
polyethylene, 933
AAm grafted high density
polyethylene, 933
benzyl sulfide, 47
cyclooctasulfur, 47
ethylene-acrylic acid
copolymers, 1184, 1187,
1193, 1195

# INDEX

ESCA (XPS) Analysis or Spectra of, contd.
  extracted polyurethanes, 969
  fluorinated polymers, 777-790
  fluoropolymers, 59-79
  fractured surfaces of epoxide paints on Al, 1046
  gold substrate with thin polyparaxylylene layer, 5
  interfacial region of crosslinked polystyrene paint on galvanized steel, 1069
  Kevlar®, 869-877
  lead tin alloys, 1184, 1187, 1188, 1193, 1195
  metals on polymers, 1083
  Pellethane tubing, 976
  pentafluorobenzene plasma polymer, 12
  phenyl sulfide, 47
  polycarbonate samples, 23-29
  polyethers, 960-961
  poly-p-phenylene, 48
  poly-p-phenylene sulfide, 47-49
  poly(p-xylylene) and its derivatives, 753-770
  polyethyleneterephthalate, 794
  polystyrene and polystyrene with Ni, 1097
  polyvinyl alcohol with Ni, 1098
  polyurethane extracts, 969
  segmented polyether polyurethane ureas, 953, 963-964
  sulfonated polysulfone membranes, 350-353
  unfluorinated polymer films, 789
  vapor deposited Cu on polystyrene, 1071
  vapor deposited Cu on polyvinyl alcohol, 1071
  vinyl copolymer surfaces, 1200

ESCA, Applications of
  in detecting locus of failure, 1045

ESCA, Applications of, contd.
  in studying modification, degradation and synthesis of polymer surfaces, 3-31
  in studying degradation between EAA copolymer and lead tin alloys, 1181-1196
  in surface analysis (see ESCA Analysis of)
  in understanding interfacial interactions, 1047-1049, 1069-1071, 1093-1099

ESR Spectroscopy of Polymer Films, 804, 806, 808, 809, 811

Ethylene-Acrylic Acid Copolymers
  joints with lead tin alloys, 1181-1196

Extracted Polyurethanes
  XPS studies of, 969

Extrusion Welding, 711

Fatigue Wear, 419

Fibers
  determination of surface forces of, 638-645
  dry adhesion of, 431-434
  dynamic friction of, 438-441
  electrokinetic properties of, 651-654
  friction in aqueous sodium dodecyl sulfate, 441-444
  interaction with surface active agents, 654-656
  soiling of, 642-644
  sorption of cationic surfactants on, 647-651
  static friction of, 435-438
  water vapor adsorption on, 645-647
  wet adhesion of, 434-435

Fiber Forming Polymers
  physicochemical surface properties of, 637-656

Filled Polymer Systems
  and acid-base considerations, 739

Filler
  effect of, on the adhesion of copper to PVC, 1115-1122

Filler, contd.
  surface modification of, using microwave plasma treatment, 739
Flexing Device, 846
Flexural Fatigue
  of Nylon 6 fibers, 843
  test, 848
Flow Microcalorimetric Measurements, 598
Fluorination Apparatus, Schematic of, 776
Fluorination (Mild) of Polymers, 773-790
Fluorobenzenes
  plasma polymerization of, 14-16
Fluorocarbon Films
  analysis of, using XPS, 53-79
  produced by sputtering of a PTFE bulk cathode, 53
Fluoropolymers
  XPS data of, 59-79
Fouling Properties of Membranes, 817-820
Fracture
  modes of separation during, 525
  resistance, 531
  toughness, 531
  work of, 531
Fracture Energy
  definition of, 530
  experimental determination of, 534-535
  of glass polymers, relative ranking, 540-541
Fracture Mechanics
  basic concepts of, 525
Fracture Surface Energy
  of epoxy-polycarbonate adhesive joints, 545, 550-553
  of glassy polymers, 523
Friction
  and contact charging, 501-504
  dynamic, 438-441
  of plasma modified fibers, 654
  welding, 709

Friction of Polymers (see also Tribology of Polymers), 389-395, 417-421
FT-IR
  spectrum of methanol precipitate of Pellethane, 979-980
  study of glass/silane interface and silane interphase, 91-104
Functional Groups on Polymer
  determination of, using ESCA sensitive elements, 794-800
Fundamental or Basic Adhesion, 1037

Glassy Polymers
  fracture energies of, relative ranking, 540-541
  fracture surface energies of, 523
  surface free energies of, 523
Glow Discharge Cleaning, 899
Graft Copolymerization, 804, 812, 881-891, 925
Griffith Energy Balance Concept, 526-529

Hamaker Constant of Polymers, (table), 641
Hammett Function of Polymers, 491
Hanging Fiber Friction Apparatus, 431
Heated Tool Butt Welding, 701
Heat Sources
  monochromatic emissive power of, 307
Hexadecylpyridinium Chloride
  adsorption on fibers, 647-648
Human Plasma Proteins
  diffusion coefficients of, 1019
Hydrogels
  wettability of, 141-153
Hydrophobic Polymer
  surface modification of, 923

Ice Adhesion
  to polymers and other surfaces, 241-251

# INDEX

Icephobic Coatings, 248
Implant Surface Properties
  time-related changes in, 905-906
Industrial Atmosphere
  effect on bond durability, 1155-1173
Inelastic Electron Tunneling Spectroscopy (IETS)
  and its role in adhesion, 1055-1056
Interfaces
  blood behavior at, 1003-1008
Interfacial
  catalysis and surface (interface) degradation, 268-272
  degradation of lead-tin alloy - EAA joints, 1181
  interactions in metallized plastics, 1065-1073, 1079-1084, 1093-1099
  interactions studies, 1035-1085, 1093-1099
  shear strength, 393-395
  wear processes, 398-399
Interfacial Phenomena, Tools for Investigation of, 1040-1059
  optical and electron microscopy methods, 1040-1043
  other methods, 1054-1059
    autoradiography
    IETS
    mass spectrometry
    metal-insulator-semiconductor
    picosecond Raman gain spectroscopy
  surface analytical methods, 1043-1045
  vibrational methods, 1050-1054
Internal Reflection Infrared Spectroscopy, 34-36
  of tissue faces, 903
Inverse Gas Chromatography
  surface characterization of polymers by, 659-665
Ionization Potentials of Polymers, (table), 487
Irwin-Orowan Basic Zone Model, 529-531

Isotactic Polypropylene
  IR reflection spectra of, 274
Isotactic Polystyrene
  thermooxidative degradation of, 263

Kevlar®
  tensile strength of single fibers of, 874, 876
Kevlar® Surfaces
  adhesion to epoxy resin matrices, 861
  characterization of, 861
  chemical modification of, 861, 864-869
Kinetics of Wetting, 134-136

Langmuir-Blodgett Multilayers, 820-822
Latex Autohesion, 377-383
Latex Films Vulcanization
  surface phenomena in, 367-375
Lattice Field Model, 177
Lead-Tin Alloys
  interaction with ethylene-acrylic acid copolymers, 1181-1196
Lewis Acid-Base Interaction Parameter
  and critical mixing times, 745
  and dispersion quality index, 746
Liquid Polymers
  surface thermodynamics of, 173-380
Localized States in Polyethylene, 83-89
Lubrication
  elastohydrodynamic, 421-424
  in organic liquids, 444-445
  of polymers, 402-406, 421-424

Mechanical Interlocking Mechanism of Adhesion, 1037-1038
Mechanisms of Adhesion, 1037-1040
Membrane Modification
  using oriented monolayer assemblies, 817-840

Membranes
  contact angle decay of, 827–830
  fouling propensity tests of, 833–838
  limiting current evaluations of, 830–833
  SEM of, 824

Metallized Plastics (Polymers), 1059–1084, 1093–1099, 1115–1122, 1125–1137, 1213–1223
  interfacial interactions in, 1065–1073, 1079–1084, 1093–1099

Metal-Polymer Interactions and bond durability, 1035–1085

Metal-Polymer Boundaries
  analysis of, using ultrasonic interface waves, 1101–1114

Metal-Insulator-Semiconductor Technique, 1056–1057

$\gamma$-Methacryloxypropyltrimethoxysilane
  interphase, studies by FT-IR, 91–104
  surface characteristics of, 91–104

Microtransmission Spectra of Tissue Faces, 903

Miniature Polymer Elements
  surface energetics and tribological properties of, 451–460

Modification of Polymer Surfaces, 739, 749–771, 773–791, 793–800, 801–815, 817, 843–859, 861–878, 1062–1065
  as studied by ESCA, 3
  by chemical means, 801–815, 861–878
  by fluorination, 773
  by monolayers, 817
  by microwave plasma treatment, 739
  methods for, 1062–1065

Molecular Forces
  in adhesion of polymers, 605–612

Monolayer
  and multilayer polymerization, 561–567
  assemblies for modification of membranes, 817

Multiphase Polymers
  interaction with blood, 985

Native Plasma Proteins
  adsorption on polymer surfaces, 1011–1029

Nylon 6 Fibers
  sunlight induced modification of, 843

Nylon 6,6
  charging of, against metals, 489

Optical and Electron Microscopy Methods, 1040–1043

Oriented Monolayer Assemblies to modify membranes, 817–840

Particle Engulfment and Interfacial Tensions, 155

Peeling, Viscoelastic, 680–685

Peel Strength of a Welded Joint, 723

Pentafluorobenzene
  plasma polymerization of, 12–14

Photooxidative Degradation of Acrylic Copolymer Surfaces, 293–302

Pigments
  acid-base sites in, 593

Plasma
  activation, 793
  and conventional polymers, 212–214
    surface energy properties and densities of, (table), 213
  induced polymerization, 9–10
  oxidation of poly(p-xylylene) and its derivatives, 749
  polymerization vs sputtering, 77–78
  polymers, general features of, 14–16

# INDEX

Plasma, contd.
  reactive species in a typical, 10-12
  reactor, schematic of, 751
Plasma Polymerization
  of fluorobenzene, 14-16
  of hydrocarbon monomers, 205-212
  of organosilicon monomers, 200-212
  of pentafluorobenzene, 12-14
  of thin polymer films, 197
  of ultrathin polymer films, 7-16
  schematic of reactive sequences in, 8
Plastic Film Welded Joints, 717-735
  durability of, 725-735
  welding technology of, 719-725
Plastics Metallization, 1059-1085, 1093-1099, 1115-1122, 1125-1137, 1213-1223
  general aspects of, 1059-1060
  interfacial interactions of, 1065-1073
  surface modification of polymers for, 1060-1065
Platable Toner Technology, 1215
Platelet
  adhesion, 996, 1004
  reactivity, 985-997
    measurement of, 989
  retention test, 957
Polycarbonate
  adsorption of alcohol on, 613
  adsorption of water on, 613
  charging of, 503
  contact angle on, 550, 617
  mechanical constants of, 552
  photodegradation of, 17-31
  surface energy of, 550
Poly(chloro-trifluoroethylene coethylene)
  charge state distribution in, 492
Polydienes
  photooxidation of, 284-286
Polydimethylsiloxane
  surface tension of, 179

Polyester Filaments
  characterization of, 183-195
  water adsorption of 645
  wettability of, 183-195
Polyester Fiberglass Composite
  adhesive joining to Al, 1139-1159
Polyethylene
  adhesion to metals, 1073-1079
  contact angle on, 617
  contact electrification of, 497
Polyethylene, contd.
  films and autoadhesion, 689-697
  localized Auger states in, 83-89
  radiation-induced graft copolymerization on, 925
  self-adherence, 690
  tensile shear strength of metals deposited on, 1081
  welded joints of, 728, 732-733
Polyethylene terephthalate
  adhesion, friction, lubrication of (filaments), 425-428
  advancing contact angle of SDS solution on, 442
  determination of functional groups on, 795-800
  plasma activation of, 794-795
  surface free energy values of, (table), 665
Polyimide
  Rutherford back scattering spectrum of, 1216
  surface study of, 1065
Polyisobutylene
  surface tension of, 180
Polymer(s)
  adsorption of, on inorganic surfaces, 593-599
  charge state distributions in, 492
  charge state in, 463-473
  chemical modification of, 861-878
  filament wetting, 183-195
  filler interactions, 739
  film forming, 637

Polymer(s) contd.
  films
    chemical characterization of, using ESCA, 793-800
    fluorination of, 773-790
  fracture surface energies of, 523
  friction, 389-395
    two term model of, 389
    interfacial, 393
  Hamaker constant of, (table), 641
  Hammett functions of, (table), 491
  interactions with other materials, 1035-1085
  ionization potentials of, (table), 487
  lubrication, 402-406
  metal interfacial interactions, 1065-1084, 1093-1099
  metallized, 1060-1084, 1093-1099, 1115-1122, 1125-1137, 1213-1223
  morphology and structure through surface interactions, 559-567
  particle surface tension, 164, 168, 170
  photoemission from, 493
  plasma activation of, 793
  polymer charge transfer, 495
  reflectance characteristics of, 305
  surface acidity-basicity of, 587-590
  surface characterization by inverse gas chromatography, 659-665
  surface free energies of, 523
  surface tensions, (see also Surface Tension of), 110-113
  transmittance characteristics of, 305
  wear, 396-402
    classification of, processes, 396
  welding of, 699-713, 717-735
  work function of, (table), 487

Polymer Adhesion (see also Adhesion)
  to biological tissue, 946
  and acid-base interactions, 583-601
Polymethylmethacrylate
  charge state distribution in, 492
  fracture surface energy, fracture energy and fracture toughness of, 535-540
Polymeric Particles
  adhesion of, 609-612
Polymerization
  epitaxial, 559-562
  monolayer and multilayer, 561-567
  plasma (see Plasma Polymerization)
Polymerized Monolayers, 561-567
Polymer Powders
  electroless metallization of, 1213-1223
Polymer Surfaces
  degradation and stability, 255-279, 283-291, 293-302
  interaction with several radiation sources, 305-345
  interactions in the biological environment, 943-952
  introduction of reactive groups on, 801-814
  modification methods, (table), (see also Surface Modification), 1060-1065
  to attain blood compatibility, 923-939
  oxidation reaction of unsaturated polymers, 283-291
  oxidative degradation, 260-268
  photochemical degradation at, 16-31
  photodegradation, 258-260
  plasma protein adsorption on, 1011-1029
  reactions at, 16-31
  reflectance-transmittance characteristics of, 305-345
  spectroscopy of, using surface enhanced Raman effect, 33-42

# INDEX

Polymer Surfaces, contd.
  thermal degradation, 258
Polymer Surface Modification, 739, 749-771, 773-791, 793-800, 801-815, 817, 843-859, 861-878, 1062-1065
  as studied by ESCA, 3
  by chemical means, 861-878, 801-815
  by fluorination, 773
  by microwave plasma treatment, 739
  by monolayers, 817
  methods for, 1062-1065
Polymer-Water Interfaces
  dynamics of, 911-921
  effect of adsorbed solute at, 148-149
Polynorborne
  photooxidation of, 286-287
Polyolefins
  adhesion of metallized, 1136
  charging of, 490
Poly-p-phenylene Sulfide
  XPS study of, 45-51
Polypropylene
  adsorption of alcohol on, 613
  adsorption of water on, 613
  contact angle on, 617
  films, welded joints of, 729-730, 732
  oxidation, 267
Polystyrene
  adhesive joining to Al, 1161-1179
  charge state distribution in, 492
  charging of, 490
Polysulfone (Sulfonated) Membranes
  characterization of, 347-365
Polytetrafluoroethylene
  tensile strength of metals deposited on, 1080
  wettability of, 217-227
Polyurethane Extracts
  XPS studies of, 969
Polyvinyl Chloride
  adhesion of electroless copper to, 1115-1122

Polyvinyl Chloride, contd.
  adhesion of metallized, 1136
  film, welded joints of, 728, 732, 735
Poly(vinyl pyridine)
  charge state distribution in, 492
Poly(p-xylylene) and its Chlorinated Derivatives
  plasma oxidation of, 749-771
Practical Adhesion, 1037
Printed Circuit Board
  corrosion of, 512
Protein
  adsorption and clot formation, 933-937
  adsorption on polymer surfaces, 1011-1029
  binding, energetics and mechanism of, 1025-1028
  interfacial free energies of, 237

Radiation Sources
  and reflectivity of polymers, (table), 315, 321
  and transmittivity of polymers, (table), 315, 322
  interaction with polymer surfaces, 305-345
Reagents Containing ESCA Sensitive Elements, 795
Receding Contact Angle, 115-119, 189-192, 629, 913-916
Reflectance Characteristics of Polymers, 305
Reflectivity of Polymers
  and radiation sources, (table), 315, 321
Relaxation - Localization Model of Molecular Ion and Exciton States in Polymers, 464-473
Resistivity of Conductive Polymers, 512-514
Roller Band Welding, 706
Rubberlike Materials
  tribological properties of, 413-423

Rutherford Back Scattering
  application to polymer surfaces, 1215

Salt Fog Cycle
  effect on bond durability, 1153-1155, 1170-1172
Sea Coast Atmosphere
  effect on bond durability, 1156-1157, 1175-1176
Segmented Polyether Polyurethane Ureas, 953-967, 971
  surface activity towards blood platelets, 953
  XPS analysis of, 953, 963-964
Self-adherence of Polyethylene, 690
SEM of
  blend of polyisoprene and poly (n-propyl p-styrene sulfonate), 995
  fractured EAA and lead tin alloys, 1192
  metal implants, 908
  Nylon 6 fiber, 854-858
  HDPE before and after graft copolymerization, 935
  platelets, 998-1000
Separating Seam Welding, 707
Shrinkage Strin Energy Release Rate, 553
Silane Layers
  influence of substrate on, 91
  surface characteristics of, 91
Silicone (Modified) Surfaces
  adsorption of BSA on, 631
  adsorption on, 625-635
  contact angle on, 629, 633-634
$(SN)_x$, Crystal Forms of
  produced by epitaxial polymerization, 564
Sodium Dodecyl Sulfate
  its effect on friction of polymers, 441-443
Soiling of Fibers, 642-644
Solid Boundary Lubrication, 445-447
Solidification Front Technique
  its use in determining interfacial tensions, 155-170

Solubility Behavior of
  ungrafted and grafted wool, 889
Specific Protein Interactions, 1003
Spreading (see also Wettability), 453-460
Spreading of Liquids, Edge Effects, 130-134
Sputtering of PTFE Bulk Cathode, 55, 74
Sputtering vs Plasma Polymerization, 77-78
Steel/Polymer Microbearing, 451-460
Stereochemical Changes
  at water-gel and vapor-gel interfaces, 146
Strain Energy Release Rate, 530
Stress Intensity Factor, 532, 551
Stress States, 526
Sulfonated Polysulfone Membranes, 347-365
  electrical properties of, 350, 354-365
  ESCA of, 350-353
  IR absorption due to bound water in, 354
  SEM microphotograph of, 348
Surface Acidity and Basicity of Polymers, 587-590
Surface Active Agents
  influence of, on film autohesion, 380, 383
  interaction with fibers, 654-656
Surface Analysis Methods, 273, 1043-1045
  comparison of, (table), 1044
Surface Energetics
  and biological adhesion, 895-908
  and cellular responses, 906-908
  and tribological properties, 451-460
  theory of adhesion, 1038-1039

# INDEX

Surface Enhanced Raman Effect
  and spectroscopy of polymer
  surfaces, 33-42
Surface Induced Homopolymerization, 91
Surface Induced Modification of
  Nylon 6 Fibers, 843-859
Surface Modification
  by chemical means, 801-815,
    861-878
  by fluorination, 773
  by microwave plasma treatment,
    739
  by monolayers, 817
  methods for, 1062-1065
Surface Tension of
  a variety of materials,
    (table), 235
  benzophenone, 161, 168
  bibenzyl, 161, 168
  epoxy, 550
  lubricating oils, 459
  n-alkanes, 176
  plasma polymer films deposited
    from hydrocarbon monomers,
    (table), 207
  plasma polymer films deposited
    from organosilicon monomers,
    (table), 202, 211
  plasma polymers and their conventional counterparts,
    (table), 213
  polycarbonate, 550
  polyethyleneterephthalate,
    435, 665
  polymer particles, 164, 168,
    170
  proteins, (table), 237
  thymol, 160-161
Surface Tension of Polymers,
  (see also Surface Tension of,
  and Critical Surface Tension
  of Wetting), 110-113, 176-
  181, 197-214, 229-239, 434-
  435
Surface Thermodynamics of Liquid
  Polymers, 173-180
Surface Topography
  and adhesion, 672-676

Synthesis of Polymers
  as studied by ESCA, 3

Teflon Powder
  zeta potential of, 218
Thermal Radiation
  and polymer surface activity,
    308
Thermoplastics
  weldability rating of, 704
Thymol, Surface Tension of,
  160-161
Transfer Wear, 399-401
Transmittance Characteristics
  of polymers, 305
Transmittivity of Polymers and
  Radiation Sources, (table),
  315, 322
Triboelectrification of Polymers, 463-473, 477-504
  a chemist's view, 477-504
Tribology of Polymers, 387-407,
  413-423, 425-448, 451-460
  an overview, 388-407
  meaning of, 388, 414

Ultimate Tensile Properties of
  Nylon 6 Filament, 849
Ultrasonic Interface Waves
  use in analyzing metal-
    polymer boundaries, 1101-
    1114
Ultrasonic Welding, 709-711
Ultrathin Polymer Films
  plasma synthesis of, 7-17
Unsaturated Polyesters and Their
  Composites
  photooxidation of, 287-291
Unsaturated Polymers
  surface oxidation reactions
    of, 283-291
Unusual Multilayers, 566

Vacuum Metallization of Polymers, 1080
Vibrational Spectroscopic
  Methods in Adhesion Science,
  1050-1054
Vinyl Copolymer Surfaces
  analysis by XPS, 1199

Viscoelastic Peeling, 680-685
Vulcanization of Latex Films, 367-375, 381-382

Water Adsorption on
  cellulose and polyester fibers, 645-647
  polypropylene and polycarbonate, 613-622
Water Stability of
  epoxy-polycarbonate joints, 555-556
Wear of Polymers (see also Tribology), 396-402
Weak Boundary Layers, 1040
Weld Testing, 712
Welded Joints of Plastic Films, 717-735
Welding of Polymers, 699-713, 717-735
  methods of, 700
Wettability of Polymer Surfaces (see also Contact Angles on, and Surface Tension of), 107-136, 141-153, 183-195, 197-214, 217-227, 229-239
  kinetics of, 134-136

Wool
  graft copolymerization of vinyl monomers onto, 881-891
  solubility behavior of ungrafted and grafted, 889
Work Function of Polymers, (table), 487
Work of Adhesion
  acid-base contribution to, 589
  of epoxy-polycarbonate adhesive joints, 545, 548-550

Young's Equation, 108
  modification of, 113-115

Zeta Potential
  and friction of plasma modified fibers, 654
  of fibers, 649-654
  of grafted polyvinyl alcohol, 814
  of Teflon powder, 218
Zisman Plot, 901